Digital Speech

Digital Speech

Coding for Low Bit Rate Communication Systems

Second Edition

A. M. Kondoz
University of Surrey, UK.

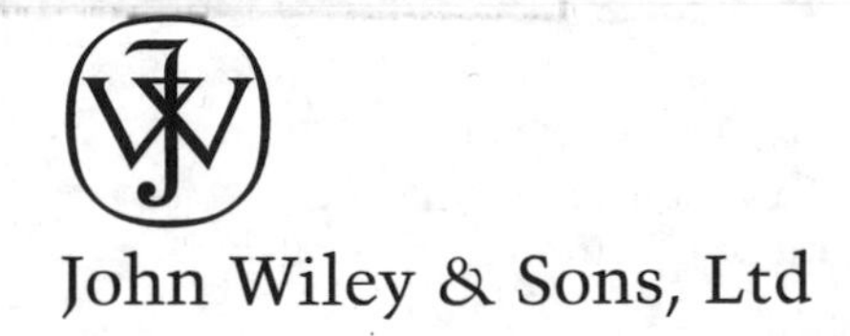

John Wiley & Sons, Ltd

Telephone (+44) 1243 779777

Email (for orders and customer service enquiries): cs-books@wiley.co.uk
Visit our Home Page on www.wileyeurope.com or www.wiley.com

Other Wiley Editorial Offices

John Wiley & Sons Inc., 111 River Street, Hoboken, NJ 07030, USA

Jossey-Bass, 989 Market Street, San Francisco, CA 94103-1741, USA

Wiley-VCH Verlag GmbH, Boschstr. 12, D-69469 Weinheim, Germany

John Wiley & Sons Australia Ltd, 33 Park Road, Milton, Queensland 4064, Australia

John Wiley & Sons (Asia) Pte Ltd, 2 Clementi Loop #02-01, Jin Xing Distripark, Singapore 129809

John Wiley & Sons Canada Ltd, 22 Worcester Road, Etobicoke, Ontario, Canada M9W 1L1

Wiley also publishes its books in a variety of electronic formats. Some content that appears in print may not be available in electronic books.

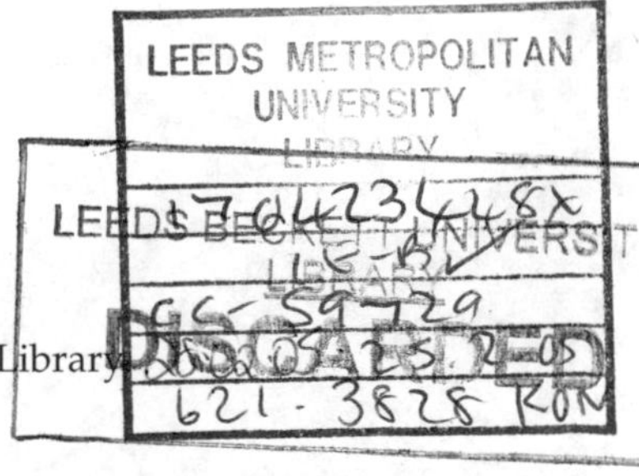

British Library Cataloguing in Publication Data

A catalogue record for this book is available from the British Library

ISBN 0-470-87008-7 (PB)

Typeset in 11/13pt Palatino by Laserwords Private Limited, Chennai, India
Printed and bound in Great Britain by Antony Rowe Ltd, Chippenham, Wiltshire
This book is printed on acid-free paper responsibly manufactured from sustainable forestry in which at least two trees are planted for each one used for paper production.

To my mother Fatma,
my wife Münise, and our children Mustafa and Fatma

Contents

Preface

Speech has remained the most desirable medium of communication between humans. Nevertheless, analogue telecommunication of speech is a cumbersome and inflexible process when transmission power and spectral utilization, the foremost resources in any communication system, are considered. Digital transmission of speech is more versatile, providing the opportunity of achieving lower costs, consistent quality, security and spectral efficiency in the systems that exploit it. The first stage in the digitization of speech involves sampling and quantizations. While the minimum sampling frequency is limited by the Nyquist criterion, the number of quantifier levels is generally determined by the degree of faithful reconstruction (quality) of the signal required at the receiver. For speech transmission systems, these two limitations lead to an initial bit rate of 64 kb/s – the PCM system. Such a high bit rate restricts the much desired spectral efficiency.

The last decade has witnessed the emergence of new fixed and mobile telecommunication systems for which spectral efficiency is a prime mover. This has fuelled the need to reduce the PCM bit rate of speech signals. Digital coding of speech and the bit rate reduction process has thus emerged as an important area of research. This research largely addresses the following problems:

- Although it is very attractive to reduce the PCM bit rate as much as possible, it becomes increasingly difficult to maintain acceptable speech quality as the bit rate falls.
- As the bit rate falls, acceptable speech quality can only be maintained by employing very complex algorithms, which are difficult to implement in real-time even with new fast processors with their associated high cost and power consumption, or by incurring excessive delay, which may create echo control problems elsewhere in the system.
- In order to achieve low bit rates, parameters of a speech production and/or perception model are encoded and transmitted. These parameters are however extremely sensitive to channel corruption. On the other hand, the systems in which these speech coders are needed typically operate

on highly degraded channels, raising the acute problem of maintaining acceptable speech quality from sensitive speech parameters even in bad channel conditions. Moreover, when estimating these parameters from the input, speech contaminated by the environmental noise typical of mobile/wireless communication systems can cause significant degradation of speech quality.

These problems are by no means insurmountable. The advent of faster and more reliable Digital Signal Processor (DSP) chips has made possible the easy real-time implementation of highly complex algorithms. Their sophistication is also exploited in the implementation of more effective echo control, background noise suppression, equalization and forward error control systems. The design of an optimum system is thus mainly a trading-off process of many factors which affect the overall quality of service provided at a reasonable cost.

This book presents some existing chapters from the first edition, as well as chapters on new speech processing and coding techniques. In order to lay the foundation of speech coding technology, it reviews sampling, quantizations and then the basic nature of speech signals, and the theory and tools applied in speech coding. The rest of the material presented has been drawn from recent postgraduate research and graduate teaching activities within the Multimedia Communications Research Group of the Centre for Communication Systems Research (CCSR), a teaching and research centre at the University of Surrey. Most of the material thus represents state-of-the-art thinking in this technology. It is suitable for both graduate and postgraduate teaching. For lecturing purposes, electronic versions of the figures are available at ftp://ftp.wiley.co.uk/pub/books/kondoz. It is hoped that the book will also be useful to research and development engineers for whom the hands-on approach to the base band design of low bit-rate fixed and mobile communication systems will prove attractive.

Ahmet Kondoz

Acknowledgements

I would like to thank Doctors Y. D. Cho, S. Villette, N. Katugampala and K. Al-Naimi for making available work in their PhDs during the preparation of this manuscript.

1

Introduction

Although data links are increasing in bandwidth and are becoming faster, speech communication is still the most dominant and common service in telecommunication networks. The fact that commercial and private usage of telephony in its various forms (especially wireless) continues to grow even a century after its first inception is obvious proof of its popularity as a form of communication. This popularity is expected to remain steady for the foreseeable future. The traditional plain analogue system has served telephony systems remarkably well considering its technological simplicity. However, modern information technology requirements have introduced the need for a more robust and flexible alternative to the analogue systems. Although the encoding of speech other than straight conversion to an analogue signal has been studied and employed for decades, it is only in the last 20 to 30 years that it has really taken on significant prominence. This is a direct result of many factors, including the introduction of many new application areas.

The attractions of digitally-encoded speech are obvious. As speech is condensed to a binary sequence, all of the advantages offered by digital systems are available for exploitation. These include the ease of regeneration and signalling, flexibility, security, and integration into the evolving new wireless systems. Although digitally-encoded speech possesses many advantages over its analogue counterpart, it nevertheless requires extra bandwidth for transmission if it is directly applied (without compression). The 64 kb/s Log-PCM and 32 kb/s ADPCM systems which have served the many early generations of digital systems well over the years have therefore been found to be inadequate in terms of spectrum efficiency when applied to the new, bandwidth limited, communication systems, e.g. satellite communications, digital mobile radio systems, and private networks. In these and other systems, the bandwidth and power available is severely restricted, hence signal compression is vital. For digitized speech, the signal compression is achieved via elaborate digital signal processing techniques that are facilitated by the

Digital Speech. A. Kondoz

ISBN 0-470-87007-9 (HB)

rapid improvement in digital hardware which has enabled the use of sophisticated digital signal processing techniques that were not feasible before. In response to the requirement for speech compression, feverish research activity has been pursued in all of the main research centres and, as a result, many different strategies have been developed for suitably compressing speech for bandwidth-restricted applications. During the last two decades, these efforts have begun to bear fruit. The use of low bit-rate speech coders has been standardized in many international, continental and national communication systems. In addition, there are a number of private network operators who use low bit-rate speech coders for specific applications.

The speech coding technology has gone through a number of phases starting with the development and deployment of PCM and ADPCM systems. This was followed by the development of good quality medium to low bit-rate coders covering the range from 16 kb/s to 8 kb/s. At the same time, very low bit-rate coders operating at around 2.4 kb/s produced better quality synthetic speech at the expense of higher complexity. The latest trend in speech coding is targeting the range from about 6 kb/s down to 2 kb/s by using speech-specific coders, which rely heavily on the extraction of speech-specific information from the input source. However, as the main applications of the low to very low bit-rate coders are in the area of mobile communication systems, where there may be significant levels of background noise, the accurate determination of the speech parameters becomes more difficult. Therefore the use of active noise suppression as a preprocessor to low bit-rate speech coding is becoming popular.

In addition to the required low bit-rate for spectral efficiency, the cost and power requirements of speech encoder/decoder hardware are very important. In wireless personal communication systems, where hand-held telephones are used, the battery consumption, cost and size of the portable equipment have to be reasonable in order to make the product widely acceptable.

In this book an attempt is made to cover many important aspects of low bit-rate speech coding. In Chapter 2, the background to speech coding, including the existing standards, is discussed. In Chapter 3, after briefly reviewing the sampling theorem, scalar and vector quantization schemes are discussed and formulated. In addition, various quantization types which are used in the remainder of this book are described.

In Chapter 4, speech analysis and modelling tools are described. After discussing the effects of windowing on the short-time Fourier transform of speech, extensive treatment of short-term linear prediction of speech is given. This is then followed by long-term prediction of speech. Finally, pitch detection methods, which are very important in speech vocoders, are discussed.

It is very important that the quantization of the linear prediction coefficients (LPC) of low bit-rate speech coders is performed efficiently both in terms of bit rate and sensitivity to channel errors. Hence, in Chapter 5, efficient quantization schemes of LPC parameters in the form of Line Spectral Frequencies are formulated, tested and compared.

In Chapter 6, more detailed modelling/classification of speech is studied. Various pitch estimation and voiced – unvoiced classification techniques are discussed.

In Chapter 7, after a general discussion of analysis by synthesis LPC coding schemes, code-excited linear prediction (CELP) is discussed in detail.

In Chapter 8, a brief review harmonic coding techniques is given.

In Chapter 9, a novel hybrid coding method, the integration of CELP and harmonic coding to form a multi-modal coder, is described.

Chapters 10 and 11 cover the topics of voice activity detection and speech enhancements methods, respectively.

2

Coding Strategies and Standards

2.1 Introduction

The invention of Pulse Code Modulation (PCM) in 1938 by Alec H. Reeves was the beginning of digital speech communications. Unlike the analogue systems, PCM systems allow perfect signal reconstruction at the repeaters of the communication systems, which compensate for the attenuation provided that the channel noise level is insufficient to corrupt the transmitted bit stream. In the early 1960s, as digital system components became widely available, PCM was implemented in private and public switched telephone networks. Today, nearly all of the public switched telephone networks (PSTN) are based upon PCM, much of it using fibre optic technology which is particularly suited to the transmission of digital data. The additional advantages of PCM over analogue transmission include the availability of sophisticated digital hardware for various other processing, error correction, encryption, multiplexing, switching, and compression.

The main disadvantage of PCM is that the transmission bandwidth is greater than that required by the original analogue signal. This is not desirable when using expensive and bandwidth-restricted channels such as satellite and cellular mobile radio systems. This has prompted extensive research into the area of speech coding during the last two decades and as a result of this intense activity many strategies and approaches have been developed for speech coding. As these strategies and techniques matured, standardization followed with specific application targets. This chapter presents a brief review of speech coding techniques. Also, the requirements of the current generation of speech coding standards are discussed. The motivation behind the review is to highlight the advantages and disadvantages of various techniques. The success of the different coding techniques is revealed in the description of the

Digital Speech. A. Kondoz
© 2004 John Wiley & Sons, Ltd ISBN 0-470-87007-9 (HB)

many coding standards currently in active operation, ranging from 64 kb/s down to 2.4 kb/s.

2.2 Speech Coding Techniques

Major speech coders have been separated into two classes: waveform approximating coders and parametric coders. Kleijn [1] defines them as follows:

- **Waveform approximating coders:** Speech coders producing a reconstructed signal which converges towards the original signal with decreasing quantization error.
- **Parametric coders:** Speech coders producing a reconstructed signal which does not converge to the original signal with decreasing quantization error.

Typical performance curves for waveform approximating and parametric speech coders are shown in Figure 2.1. It is worth noting that, in the past, speech coders were grouped into three classes: waveform coders, vocoders and hybrid coders. Waveform coders included speech coders, such as PCM and ADPCM, and vocoders included very low bit-rate synthetic speech coders. Finally hybrid coders were those speech coders which used both of these methods, such as CELP, MBE etc. However currently all speech coders use some form of speech modelling whether their output converges to the

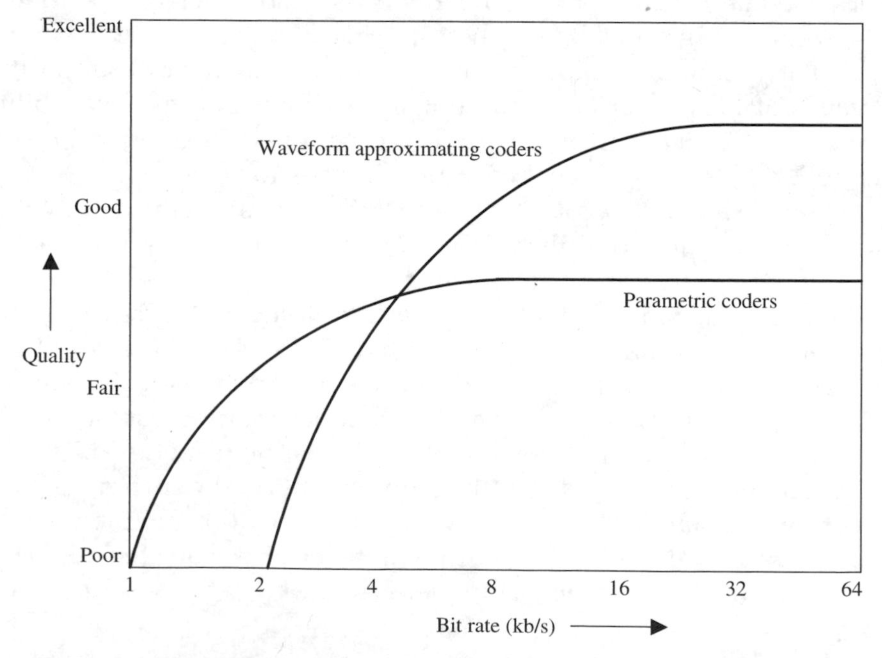

Figure 2.1 Quality vs bit rate for different speech coding techniques

original (with increasing bit rate) or not. It is therefore more appropriate to group speech coders into the above two groups as the old waveform coding terminology is no longer applicable. If required we can associate the name hybrid coding with coding types that may use more than one speech coding principle, which is switched in and out according to the input speech signal characteristics. For example, a waveform approximating coder, such as CELP, may combine in an advantageous way with a harmonic coder, which uses a parametric coding method, to form such a hybrid coder.

2.2.1 Parametric Coders

Parametric coders model the speech signal using a set of model parameters. The extracted parameters at the encoder are quantized and transmitted to the decoder. The decoder synthesizes speech according to the specified model. The speech production model does not account for the quantization noise or try to preserve the waveform similarity between the synthesized and the original speech signals. The model parameter estimation may be an open loop process with no feedback from the quantization or the speech synthesis. These coders only preserve the features included in the speech production model, e.g. spectral envelope, pitch and energy contour, etc. The speech quality of parametric coders do not converge towards the transparent quality of the original speech with better quantization of model parameters, see Figure 2.1. This is due to limitations of the speech production model used. Furthermore, they do not preserve the waveform similarity and the measurement of signal to noise ratio (SNR) is meaningless, as often the SNR becomes negative when expressed in dB (as the input and output waveforms may not have phase alignment). The SNR has no correlation with the synthesized speech quality and the quality should be assessed subjectively (or perceptually).

Linear Prediction Based Vocoders

Linear Prediction (LP) based vocoders are designed to emulate the human speech production mechanism [2]. The vocal tract is modelled by a linear prediction filter. The glottal pulses and turbulent air flow at the glottis are modelled by periodic pulses and Gaussian noise respectively, which form the excitation signal of the linear prediction filter. The LP filter coefficients, signal power, binary voicing decision (i.e. periodic pulses or noise excitation), and pitch period of the voiced segments are estimated for transmission to the decoder. The main weakness of LP based vocoders is the binary voicing decision of the excitation, which fails to model mixed signal types with both periodic and noisy components. By employing frequency domain voicing decision techniques, the performance of LP based vocoders can be improved [3].

Harmonic Coders

Harmonic or sinusoidal coding represents the speech signal as a sum of sinusoidal components. The model parameters, i.e. the amplitudes, frequencies and phases of sinusoids, are estimated at regular intervals from the speech spectrum. The frequency tracks are extracted from the peaks of the speech spectra, and the amplitudes and frequencies are interpolated in the synthesis process for smooth evolution [4]. The general sinusoidal model does not restrict the frequency tracks to be harmonics of the fundamental frequency. Increasing the parameter extraction rate converges the synthesized speech waveform towards the original, if the parameters are unquantized. However at low bit rates the phases are not transmitted and estimated at the decoder, and the frequency tracks are confined to be harmonics. Therefore point to point waveform similarity is not preserved.

2.2.2 Waveform-approximating Coders

Waveform coders minimize the error between the synthesized and the original speech waveforms. The early waveform coders such as companded Pulse Code Modulation (PCM) [5] and Adaptive Differential Pulse Code Modulation (ADPCM) [6] transmit a quantized value for each speech sample. However ADPCM employs an adaptive pole zero predictor and quantizes the error signal, with an adaptive quantizer step size. ADPCM predictor coefficients and the quantizer step size are backward adaptive and updated at the sampling rate.

The recent waveform-approximating coders based on time domain analysis by synthesis such as Code Excited Linear Prediction (CELP) [7], explicitly make use of the vocal tract model and the long term prediction to model the correlations present in the speech signal. CELP coders buffer the speech signal and perform block based analysis and transmit the prediction filter coefficients along with an index for the excitation vector. They also employ perceptual weighting so that the quantization noise spectrum is masked by the signal level.

2.2.3 Hybrid Coding of Speech

Almost all of the existing speech coders apply the same coding principle, regardless of the widely varying character of the speech signal, i.e. voiced, unvoiced, mixed, transitions etc. Examples include Adaptive Differential Pulse Code Modulation (ADPCM) [6], Code Excited Linear Prediction (CELP) [7, 8], and Improved Multi Band Excitation (IMBE) [9, 10]. When the bit rate is reduced, the perceived quality of these coders tends to degrade more for some speech segments while remaining adequate for others. This shows that the assumed coding principle is not adequate for all speech types. In order to circumvent this problem, hybrid coders that combine different coding principles to encode different types of speech segments have been introduced [11, 12, 13].

A hybrid coder can switch between a set of predefined coding modes. Hence they are also referred to as multimode coders. A hybrid coder is an adaptive coder, which can change the coding technique or mode according to the source, selecting the best mode for the local character of the speech signal. Network or channel dependent mode decision [14] allows a coder to adapt to the network load or the channel error performance, by varying the modes and the bit rate, and changing the relative bit allocation of the source and channel coding [15].

In source dependent mode decision, the speech classification can be based on fixed or variable length frames. The number of bits allocated for frames of different modes can be the same or different. The overall bit rate of a hybrid coder can be fixed or variable. In fact variable rate coding can be seen as an extension of hybrid coding.

2.3 Algorithm Objectives and Requirements

The design of a particular algorithm is often dictated by the target application. Therefore, during the design of an algorithm the relative weighting of the influencing factors requires careful consideration in order to obtain a balanced compromise between the often conflicting objectives. Some of the factors which influence the choice of algorithm for the foreseeable network applications are listed below.

2.3.1 Quality and Capacity

Speech quality and bit rate are two factors that directly conflict with each other. Lowering the bit rate of the speech coder, i.e. using higher signal compression, causes degradation of quality to a certain extent (simple parametric vocoders). For systems that connect to the Public Switched Telephone Network (PSTN) and associated systems, the quality requirements are strict and must conform to constraints and guidelines imposed by the relevant regulatory bodies, e.g. ITU (previously CCITT). Such systems demand high quality (toll quality) coding. However, closed systems such as private commercial networks and military systems may compromise the quality to lower the capacity requirements. Although absolute quality is often specified, it is often compromised if other factors are allocated a higher overall rating. For instance, in a mobile radio system it is the overall average quality that is often the deciding factor. This average quality takes into account both good and bad transmission conditions.

2.3.2 Coding Delay

The coding delay of a speech transmission system is a factor closely related to the quality requirements. Coding delay may be algorithmic (the buffering of speech for analysis), computational (the time taken to process the stored speech samples) or due to transmission. Only the first two concern the speech coding subsystem, although very often the coding scheme is tailored such that transmission can be initiated even before the algorithm has completed processing all of the information in the analysis frame, e.g. in the pan-European digital mobile radio system (better known as GSM) [16] the encoder starts transmission of the spectral parameters as soon as they are available. Again, for PSTN applications, low delay is essential if the major problem of echo is to be minimized. For mobile system applications and satellite communication systems, echo cancellation is employed as substantial propagation delays already exist. However, in the case of the PSTN where there is very little delay, extra echo cancellers will be required if coders with long delays are introduced. The other problem of encoder/decoder delay is the purely subjective annoyance factor. Most low-rate algorithms introduce a substantial coding delay compared with the standard 64 kb/s PCM system. For instance, the GSM system's initial upper limit was 65 ms for a back-to-back configuration, whereas for the 16 kb/s G.728 specification [17], it was a maximum of 5 ms with an objective of 2 ms.

2.3.3 Channel and Background Noise Robustness

For many applications, the speech source coding rate typically occupies only a fraction of the total channel capacity, the rest being used for forward error correction (FEC) and signalling. For mobile connections, which suffer greatly from both random and burst errors, a coding scheme's built-in tolerance to channel errors is vital for an acceptable average overall performance, i.e. communication quality. By employing built-in robustness, less FEC can be used and higher source coding capacity is available to give better speech quality. This trade-off between speech quality and robustness is often a very difficult balance to obtain and is a requirement that necessitates consideration from the beginning of the speech coding algorithm design. For other applications employing less severe channels, e.g. fibre-optic links, the problems due to channel errors are reduced significantly and robustness can be ignored for higher clean channel speech quality. This is a major difference between the wireless mobile systems and those of the fixed link systems.

In addition to the channel noise, coders may need to operate in noisy background environments. As background noise can degrade the performance of speech parameter extraction, it is crucial that the coder is designed in such a way that it can maintain good performance at all times. As well as maintaining good speech quality under noisy conditions, good quality background noise regeneration by the coder is also an important requirement (unless adaptive noise cancellation is used before speech coding).

2.3.4 Complexity and Cost

As ever more sophisticated algorithms are devised, the computational complexity is increased. The advent of Digital Signal Processor (DSP) chips [18] and custom Application Specific Integrated Circuit (ASIC) chips has enabled the cost of processing power to be considerably lowered. However, complexity/power consumption, and hence cost, is still a major problem especially in applications where hardware portability is a prime factor. One technique for overcoming power consumption whilst also improving channel efficiency is digital speech interpolation (DSI) [16]. DSI exploits the fact that only around half of speech conversation is actually active speech thus, during inactive periods, the channel can be used for other purposes, including limiting the transmitter activity, hence saving power. An important subsystem of DSI is the voice activity detector (VAD) which must operate efficiently and reliably to ensure that real speech is not mistaken for silence and vice versa. Obviously, a voice for silence mistake is tolerable, but the opposite can be very annoying.

2.3.5 Tandem Connection and Transcoding

As it is the end to end speech quality which is important to the end user, the ability of an algorithm to cope with tandeming with itself or with another coding system is important. Degradations introduced by tandeming are usually cumulative, and if an algorithm is heavily dependent on certain characteristics then severe degradations may result. This is a particularly urgent unresolved problem with current schemes which employ post-filtering in the output speech signal [17]. Transcoding into another format, usually PCM, also degrades the quality slightly and may introduce extra cost.

2.3.6 Voiceband Data Handling

As voice connections are regularly used for transmission of digital data, e.g. modem, facsimile, and other machine data, an important requirement is an algorithm's ability to transmit voiceband data. The waveform statistics and frequency spectrum of voiceband data signals are quite different from those of speech, therefore the algorithm must be capable of handling both types. The consideration of voiceband data handling is often left until the final stages of the algorithm development, which may be a mistake as end users expect nonvoice information to be adequately transported if the system is employed in the public network. Most of the latest low bit-rate speech coders are unable to pass voiceband data due to the fact they are too speech specific.

Other solutions are often used. A very common one is to detect the voiceband data and use an interface which bypasses the speech encoder/decoder.

2.4 Standard Speech Coders

Standardization is essential in removing the compatibility and conformability problems of implementations by various manufacturers. It allows for one manufacturer's speech coding equipment to work with that of others. In the following, standard speech coders, mostly developed for specific communication systems, are listed and briefly reviewed.

2.4.1 ITU-T Speech Coding Standard

Traditionally the International Telecommunication Union Telecommunication Standardization Sector (ITU-T, formerly CCITT) has standardized speech coding methods mainly for PSTN telephony with 3.4 kHz input speech bandwidth and 8 kHz sampling frequency, aiming to improve telecommunication network capacity by means of digital circuit multiplexing. Additionally, ITU-T has been conducting standardization for wideband speech coders to support 7 kHz input speech bandwidth with 16 kHz sampling frequency, mainly for ISDN applications.

In 1972, ITU-T released G.711 [19], an A/μ-Law PCM standard for 64 kb/s speech coding, which is designed on the basis of logarithmic scaling of each sampled pulse amplitude before digitization into eight bits. As the first digital telephony system, G.711 has been deployed in various PSTNs throughout the world. Since then, ITU-T has been actively involved in standardizing more complex speech coders, referenced as the G.72x series. ITU-T released G.721, the 32 kb/s adaptive differential pulse code modulation (ADPCM) coder, followed by the extended version (40/32/24/16 kb/s), G.726 [20]. The latest ADPCM version, G.726, superseded the former one. Each ITU-T speech coder except G.723.1 [21] was developed with a view to halving the bit rate of its predecessor. For example, the G.728 [22] and G.729 [23] speech coders, finalized in 1992 and 1996, were recommended at the rates of 16 kb/s and 8 kb/s, respectively. Additionally, ITU-T released G.723.1 [21], the 5.3/6.3 kb/s dual-rate speech coder, for video telephony systems. G.728, G.729, and G.723.1 principles are based on code excited linear prediction (CELP) technologies. For discontinuous transmission (DTX), ITU-T released the extended versions of G.729 and G.723.1, called G.729B [24] and G.723.1A [25], respectively. They are widely used in packet-based voice communications [26] due to their silence compression schemes. In the past few years there has been standardization activities at 4 kb/s. Currently there two coders competing for this standard but the process has been put on hold at the moment. One coder is based on the CELP model and the other

Table 2.1 ITU-T narrowband speech coding standards

Speech coder	Bit rate (kb/s)	VAD	Noise reduction	Delay (ms)	Quality	Year
G.711 (A/μ-Law PCM)	64	No	No	0	Toll	1972
G.726 (ADPCM)	40/32/24/16	No	No	0.25	Toll	1990
G.728 (LD-CELP)	16	No	No	1.25	Toll	1992
G.729 (CSA-CELP)	8	Yes	No	25	Toll	1996
G.723.1 (MP-MLQ/ACELP)	6.3/5.3	Yes	No	67.5	Toll/ Near-toll	1995
G.4k (to be determined)	4	–	Yes	~55	Toll	2001

is a hybrid model of CELP and sinusoidal speech coding principles [27, 28]. A summary of the narrowband speech coding standards recommended by ITU-T is given in Table 2.1.

In addition to the narrowband standards, ITU-T has released two wideband speech coders, G.722 [29] and G.722.1 [30], targeting mainly multimedia communications with higher voice quality. G.722 [29] supports three bit rates, 64, 56, and 48 kb/s based on subband ADPCM (SB-ADPCM). It decomposes the input signals into low and high subbands using the quadrature mirror filters, and then quantizes the band-pass filtered signals using ADPCM with variable step sizes depending on the subband. G.722.1 [30] operates at the rates of 32 and 24 kb/s and is based on the transform coding technique. Currently, a new wideband speech coder operating at 13/16/20/24 kb/s is undergoing standardization.

2.4.2 European Digital Cellular Telephony Standards

With the advent of digital cellular telephony there have been many speech coding standardization activities by the European Telecommunications Standards Institute (ETSI). The first release by ETSI was the GSM full rate (FR) speech coder operating at 13 kb/s [31]. Since then, ETSI has standardized 5.6 kb/s GSM half rate (HR) and 12.2 kb/s GSM enhanced full rate (EFR) speech coders [32, 33]. Following these, another ETSI standardization activity resulted in a new speech coder, called the adaptive multi-rate (AMR) coder [34], operating at eight bit rates from 12.2 to 4.75 kb/s (four rates for the full-rate and four for the half-rate channels). The AMR coder aims to provide enhanced speech quality based on optimal selection between the source and channel coding schemes (and rates). Under high radio interference, AMR is capable of allocating more bits for channel coding at the expense of reduced source coding rate and vice versa.

The ETSI speech coder standards are also capable of silence compression by way of voice activity detection [35–38], which facilitates channel

Table 2.2 ETSI speech coding standards for GSM mobile communications

Speech coder	Bit rate (kb/s)	VAD	Noise reduction	Delay (ms)	Quality	Year
FR (RPE-LTP)	13	Yes	No	40	Near-toll	1987
HR (VSELP)	5.6	Yes	No	45	Near-toll	1994
EFR (ACELP)	12.2	Yes	No	40	Toll	1998
AMR (ACELP)	12.2/10.2/7.95/ 7.4/6.7/5.9/ 5.15/4.75	Yes	No	40/45	Toll ~ Communi-cation	1999

interference reduction as well as battery life time extension for mobile communications. Standard speech coders for European mobile communications are summarized in Table 2.2.

2.4.3 North American Digital Cellular Telephony Standards

In North America, the Telecommunication Industries Association (TIA) of the Electronic Industries Association (EIA) has been standardizing mobile communication based on Code Division Multiple Access (CDMA) and Time Division Multiple Access (TDMA) technologies used in the USA. TIA/EIA adopted Qualcomm CELP (QCELP) [39] for Interim Standard-96-A (IS-96-A), operating at variable bit rates between 8 kb/s and 0.8 kb/s controlled by a rate determination algorithm. Subsequently, TIA/EIA released IS-127 [40], the enhanced variable rate coder, which features a novel function for noise reduction as a preprocessor to the speech compression module. Under noisy background conditions, noise reduction provides a more comfortable speech quality by enhancing noisy speech signals. For personal communication systems, TIA/EIA released IS-733 [41], which operates at variable bit rates between 14.4 and 1.8 kb/s. For North American TDMA standards, TIA/EIA released IS-54 and IS-641-A for full rate and enhanced full rate speech coding, respectively [42, 43]. Standard speech coders for North American mobile communications are summarized in Table 2.3.

2.4.4 Secure Communication Telephony

Speech coding is a crucial part of a secure communication system, where voice intelligibility is a major concern in order to deliver the exact voice commands in an emergency.

Standardization has mainly been organized by the Department of Defense (DoD) in the USA. The DoD released Federal Standard-1015 (FS-1015) and FS-1016, called 2.4 kb/s LPC-10e and 4.8 kb/s CELP coders, respectively [44–46]. The DoD also standardized a more recent 2.4 kb/s speech coder [47], based

Table 2.3 TIA/EIA speech coding standards for North American CDMA/TDMA mobile communications

Speech coder	Bit rate (kb/s)	VAD	Noise reduction	Delay (ms)	Quality	Year
IS-96-A (QCELP)	8.5/4/2/0.8	Yes	No	45	Near-toll	1993
IS-127 (EVRC)	8.5/4/2/0.8	Yes	Yes	45	Toll	1995
IS-733 (QCELP)	14.4/7.2/3.6/1.8	Yes	No	45	Toll	1998
IS-54 (VSELP)	7.95	Yes	No	45	Near-toll	1989
IS-641-A (ACELP)	7.4	Yes	No	45	Toll	1996

Table 2.4 DoD speech coding standards

Speech coder	Bit rate (kb/s)	VAD	Noise reduction	Delay (ms)	Quality	Year
FS-1015 (LPC-10e)	2.4	No	No	115	Intelligible	1984
FS-1016 (CELP)	4.8	No	No	67.5	Communication	1991
DoD 2.4 (MELP)	2.4	No	No	67.5	Communication	1996
STANAG (NATO) 2.4/1.2 (MELP)	2.4/1.2	No	Yes	>67.5	Communication	2001

on the mixed excitation linear prediction (MELP) vocoder [48] which is based on the sinusoidal speech coding model. The 2.4 kb/s DoD MELP speech coder gives better speech quality than the 4.8 kb/s FS-1016 coder at half the capacity. A modified and improved version of this coder, operating at dual rates of 2.4/1.2 kb/s and employing a noise preprocessor, has been selected as the new NATO standard. Parametric coders, such as MELP, have been widely used in secure communications due to their intelligible speech quality at very low bit rates. The DoD standard speech coders are summarized in Table 2.4.

2.4.5 Satellite Telephony

The international maritime satellite corporation (INMARSAT) has adopted two speech coders for satellite communications. INMARSAT has selected 4.15 kb/s improved multiband excitation (IMBE) [9] for INMARSAT M systems and 3.6 kb/s advanced multiband excitation (AMBE) vocoders for INMARSAT Mini-M systems (see Table 2.5).

2.4.6 Selection of a Speech Coder

Selecting the best speech coder for a given application may involve extensive testing under conditions representative of the target application. In general, lowering the bit rate results in a reduction in the quality of coded speech.

Table 2.5 INMARSAT speech coding standards

Speech coder	Bit rate (kb/s)	VAD	Noise reduction	Delay (ms)	Quality	Year
IMBE	4.15	No	No	120	Communication	1990
AMBE	3.6	No	No	–	–	–

Quality measurements based on SNR can be used to evaluate coders that preserve the waveform similarity, usually coders operating at bit rates above 16 kb/s. Low bit-rate parametric coders do not preserve the waveform similarity and SNR-based quality measures become meaningless. For parametric coders, perception-based subjective measures are more reliable. The Mean Opinion Score (MOS) [49] scale shown in Table 2.6 is a widely-used subjective quality measure.

Table 2.7 compares some of the most well-known speech coding standards in terms of their bit rate, algorithmic delay and Mean Opinion Scores and Figure 2.2 illustrates the performance of those standards in terms of speech quality against bit rate [50, 51].

Linear PCM at 128 kb/s offers transparent speech quality and its A-law companded 8 bits/sample (64 kb/s) version (which provides the standard for the best (narrowband) quality) has a MOS score higher than 4, which is described as Toll quality. In order to find the MOS score for a given

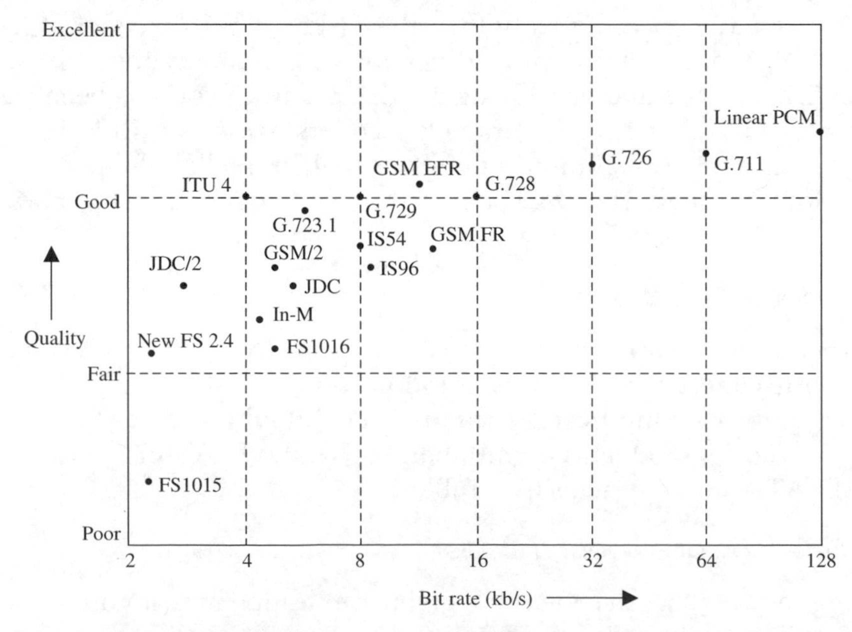

Figure 2.2 Performance of telephone band speech coding standards (only the top four points of the MOS scale have been used)

Table 2.6 Mean Opinion Score (MOS) scale

Grade (MOS)	Subjective opinion	Quality
5 Excellent	Imperceptible	Transparent
4 Good	Perceptible, but not annoying	Toll
3 Fair	Slightly annoying	Communication
2 Poor	Annoying	Synthetic
1 Bad	Very annoying	Bad

Table 2.7 Comparison of telephone band speech coding standards

Standard	Year	Algorithm	Bit rate (kb/s)	MOS*	Delay+
G.711	1972	Companded PCM	64	4.3	0.125
G.726	1991	VBR-ADPCM	16/24/32/40	toll	0.125
G.728	1994	LD-CELP	16	4	0.625
G.729	1995	CS-ACELP	8	4	15
G.723.1	1995	A/MP-MLQ CELP	5.3/6.3	toll	37.5
ITU 4	–	–	4	toll	25
GSM FR	1989	RPE-LTP	13	3.7	20
GSM EFR	1995	ACELP	12.2	4	20
GSM/2	1994	VSELP	5.6	3.5	24.375
IS54	1989	VSELP	7.95	3.6	20
IS96	1993	Q-CELP	0.8/2/4/8.5	3.5	20
JDC	1990	VSELP	6.7	commun.	20
JDC/2	1993	PSI-CELP	3.45	commun.	40
Inmarsat-M	1990	IMBE	4.15	3.4	78.75
FS1015	1984	LPC-10	2.4	synthetic	112.5
FS1016	1991	CELP	4.8	3	37.5
New FS 2.4	1997	MELP	2.4	3	45.5

* The MOS figures are obtained from formal subjective tests using varied test material (from the literature). These figures are therefore useful as a guide, but should not be taken as a definitive indication of codec performance.
+ Delay is the total algorithmic delay, i.e. the frame length and look ahead, and is given in milliseconds.

coder, extensive listening tests must be conducted. In these tests, as well as the 64 kb/s PCM reference, other representative coders are also used for calibration purposes. The cost of extensive listening tests is high and efforts have been made to produce simpler, less time-consuming, and hence cheaper, alternatives. These alternatives are based on objective measures with some subjective meanings. Objective measurements usually involve point to point comparison of systems under test. In some cases weighting may be used to give priority to some system parameters over others. In early speech coders, which aimed at reproducing the input speech waveform as output, objective measurement in the form of signal to quantization noise ratio was used. Since the bit rate of early speech coders was 16 kb/s or greater (i.e. they incurred only a small amount of quantization noise) and they did not involve complicated signal processing algorithms which could change the shape of the speech waveform, the SNR measures were reasonably accurate. However at lower bit rates where the noise (the objective difference between the original input and the synthetic output) increases, the use of signal to quantization noise ratio may be misleading. Hence there is a need for a better objective measurement which has a good correlation with the perceptual quality of the synthetic speech. The ITU standardized a number of these methods, the most recent of which is P.862 (or Perceptual Evaluation of Speech Quality). In this standard, various alignments and perceptual measures are used to match the objective results to fairly accurate subjective MOS scores.

2.5 Summary

Existing speech coders can be divided into three groups: parametric coders, waveform approximating coders, and hybrid coders. Parametric coders are not expected to reproduce the original waveform; they reproduce the perception of the original. Waveform approximating coders, on the other hand, are expected to replicate the input speech waveform as the bit rate increases. Hybrid coding is a combination of two or more coders of any type for the best subjective (and perhaps objective) performance at a given bit rate.

The design process of a speech coder involves several trade-offs between conflicting requirements. These requirements include the target bit rate, quality, delay, complexity, channel error sensitivity, and sending of nonspeech signals. Various standardization bodies have been involved in speech coder standardization activities and as a result there have been many standard speech coders in the last decade. The bit rate of these coders ranges from 16 kb/s down to around 4 kb/s with target applications mainly in cellular mobile radio. The selection of a speech coder involves expensive testing under the expected typical operating conditions. The most popular testing method is subjective listening tests. However, as this is expensive and time-consuming, there has been some effort to produce simpler yet reliable objective measures. ITU P.862 is the latest effort in this direction.

Bibliography

[1] W. B. Kleijn and K. K. Paliwal (1995) 'An introduction to speech coding', in *Speech coding and synthesis* by W. B. Kleijn and K. K. Paliwal (Eds), pp. 1–47. Amsterdam: Elsevier Science

[2] D. O'Shaughnessy (1987) *Speech communication: human and machine*. Addison Wesley

[3] I. Atkinson, S. Yeldener, and A. Kondoz (1997) 'High quality split-band LPC vocoder operating at low bit rates', in *Proc. of Int. Conf. on Acoust., Speech and Signal Processing*, pp. 1559–62. May 1997. Munich

[4] R. J. McAulay and T. F. Quatieri (1986) 'Speech analysis/synthesis based on a sinusoidal representation', in *IEEE Trans. on Acoust., Speech and Signal Processing*, 34(4):744–54.

[5] ITU-T (1972) *CCITT Recommendation G.711: Pulse Code Modulation (PCM) of Voice Frequencies*. International Telecommunication Union.

[6] N. S. Jayant and P. Noll (1984) *Digital Coding of Waveforms: Principles and applications to speech and video*. New Jersey: Prentice-Hall

[7] B. S. Atal and M. R. Schroeder (1984) 'Stochastic coding of speech at very low bit rates', in *Proc. Int. Conf. Comm*, pp. 1610–13. Amsterdam

[8] M. Schroeder and B. Atal (1985) 'Code excited linear prediction (CELP): high quality speech at very low bit rates', in *Proc. of Int. Conf. on Acoust., Speech and Signal Processing*, pp. 937–40. Tampa, FL

[9] DVSI (1991) *INMARSAT-M Voice Codec*, Version 1.7. September 1991. Digital Voice Systems Inc.

[10] J. C. Hardwick and J. S. Lim (1991) 'The application of the IMBE speech coder to mobile communications', in *Proc. of Int. Conf. on Acoust., Speech and Signal Processing*, pp. 249–52.

[11] W. B. Kleijn (1993) 'Encoding speech using prototype waveforms', in *IEEE Trans. Speech and Audio Processing*, 1:386–99.

[12] E. Shlomot, V. Cuperman, and A. Gersho (1998) 'Combined harmonic and waveform coding of speech at low bit rates', in *Proc. of Int. Conf. on Acoust., Speech and Signal Processing*.

[13] J. Stachurski and A. McCree (2000) 'Combining parametric and waveform-matching coders for low bit-rate speech coding', in *X European Signal Processing Conf.*

[14] T. Kawashima, V. Sharama, and A. Gersho (1994) 'Network control of speech bit rate for enhanced cellular CDMA performance', in *Proc. IEE Int. Conf. on Commun.*, 3:1276.

[15] P. Ho, E. Yuen, and V. Cuperman (1994) 'Variable rate speech and channel coding for mobile communications', in *Proc. of Vehicular Technology Conf.*

[16] J. E. Natvig, S. Hansen, and J. de Brito (1989) 'Speech processing in the pan-European digital mobile radio system (GSM): System overview', in *Proc. of Globecom*, Section 29B.

[17] J. H. Chen (1990) 'High quality 16 kbit/s speech coding with a one-way delay less than 2 ms', in *Proc. of Int. Conf. on Acoust., Speech and Signal Processing*, pp. 453–6.

[18] E. Lee (1988) 'Programmable DSP architectures', in *IEEE ASSP Magazine*, October 1988 and January 1989.

[19] ITU-T (1988) *Pulse code modulation (PCM) of voice frequencies*, ITU-T Rec. G.711.

[20] ITU-T (1990) *40, 32, 24, 16 kbit/s adaptive differential pulse code modulation (ADPCM)*, ITU-T Rec. G.726.

[21] ITU-T (1996) *Dual rate speech coder for multimedia communications transmitting at 5.3 and 6.3 kbit/s*, ITU-T Rec. G.723.1.

[22] ITU-T (1992) *Coding of speech at 16 kbit/s using low-delay code excited linear prediction*, ITU-T Rec. G.728.

[23] ITU-T (1996) *Coding of speech at 8 kbit/s using conjugate-structure algebraic-code-excited linear prediction (CS-ACELP)*, ITU-T Rec. G.729.

[24] ITU-T (1996) *A silence compression scheme for G.729 optimised for terminals conforming to ITU-T V.70*, ITU-T Rec. G.729 Annex B.

[25] ITU-T (1996) *Dual rate speech coder for multimedia communications transmitting at 5.3 and 6.3 kbit/s. Annex A: Silence compression scheme*, ITU-T Rec. G.723.1 Annex A.

[26] O. Hersent, D. Gurle, and J. Petit (2000) *IP Telephony: Packet-based multimedia communications systems*. Addison Wesley

[27] J. Thyssen, Y. Gao, A. Benyassine, E. Shylomot, H. -Y. Su, K. Mano, Y. Hiwasaki, H. Ehara, K. Yasunaga, C. Lamblin, B. Kovest, J. Stegmann, and H. -G. Kang (2001) 'A candidate for the ITU-T 4 kbit/s speech coding standard', in *Proc. of Int. Conf. on Acoust., Speech and Signal Processing*. May 2001. Salt Lake City, UT

[28] J. Stachurski and A. McCree (2000) 'A 4 kb/s hybrid MELP/CELP coder with alignment phase encoding and zero phase equalization', in *Proc. of Int. Conf. on Acoust., Speech and Signal Processing*, pp. 1379–82. May 2000. Istanbul

[29] ITU-T (1988) *7 khz audio-coding within 64 kbit/s*, ITU-T Rec. G.722.

[30] ITU-T (1999) *Coding at 24 and 32 kbit/s for hands-free operation in systems with low frame loss*, ITU-T Rec. G.722.1

[31] ETSI (1994) *Digital cellular telecommunications system (phase 2+); Full rate speech transcoding*, GSM 06.10 (ETS 300 580-2).

[32] ETSI (1997) *Digital cellular telecommunications system (phase 2+); Half rate speech; Half rate speech transcoding*, GSM 06.20 v5.1.0 (draft ETSI ETS 300 969).

[33] ETSI (1998) *Digital cellular telecommunications system (phase 2); Enhanced full rate (EFR) speech transcoding*, GSM 06.60 v4.1.0 (ETS 301 245), June.

[34] ETSI (1998) *Digital cellular telecommunications system (phase 2+); Adaptive multi-rate (AMR) speech transcoding*, GSM 06.90 v7.2.0 (draft ETSI EN 301 704).

[35] ETSI (1998) *Digital cellular telecommunications system (phase 2+); Voice activity detector (VAD) for full rate speech traffic channels*, GSM 06.32 (ETSI EN 300 965 v7.0.1).

[36] ETSI (1999) *Digital cellular telecommunications system (phase 2+); Voice activity detector (VAD) for full rate speech traffic channels*, GSM 06.42 (draft ETSI EN 300 973 v8.0.0).

[37] ETSI (1997) *Digital cellular telecommunications system; Voice activity detector (VAD) for enhanced full rate (EFR) speech traffic channels*, GSM 06.82 (ETS 300 730), March.

[38] ETSI (1998) *Digital cellular telecommunications system (phase 2+); Voice activity detector (VAD) for adaptive multi-rate (AMR) speech traffic channels*, GSM 06.94 v7.1.1 (ETSI EN 301 708).

[39] P. DeJaco, W. Gardner, and C. Lee (1993) 'QCELP: The North American CDMA digital cellular variable rate speech coding standard', in *IEEE Workshop on Speech Coding for Telecom*, pp. 5–6.

[40] TIA/EIA (1997) *Enhanced variable rate codec, speech service option 3 for wideband spread spectrum digital systems*, IS-127.

[41] TIA/EIA (1998) *High rate speech service option 17 for wideband spread spectrum communication systems*, IS-733.

[42] I. A. Gerson and M. A. Jasiuk (1990) 'Vector sum excited linear prediction (VSELP) speech coding at 8 kb/s', in *Proc. of Int. Conf. on Acoust., Speech and Signal Processing*, pp. 461–4. April 1990. Albuquerque, NM, USA

[43] T. Honkanen, J. Vainio, K. Jarvinen, and P. Haavisto (1997) 'Enhanced full rate speech coder for IS-136 digital cellular system', in *Proc. of Int. Conf. on Acoust., Speech and Signal Processing*, pp. 731–4. May 1997. Munich

[44] T. E. Tremain (1982) 'The government standard linear predictive coding algorithm: LPC-10', in *Speech Technology*, 1:40–9.

[45] J. P. Campbell Jr and T. E. Tremain (1986) 'Voiced/unvoiced classification of speech with applications to the US government LPC-10e algorithm', in *Proc. of Int. Conf. on Acoust., Speech and Signal Processing*, pp. 473–6.

[46] J. P. Campbell, V. C. Welch, and T. E. Tremain (1991) 'The DoD 4.8 kbps standard (proposed Federal Standard 1016)', in *Advances in Speech Coding* by B. Atal, V. Cuperman, and A. Gersho (Eds), pp. 121–33. Dordrecht, Holland: Kluwer Academic

[47] FIPS (1997) *Analog to digital conversion of voice by 2,400 bit/second mixed excitation linear prediction (MELP)*, Draft. Federal Information Processing Standards

[48] A. V. McCree and T. P. Barnwell (1995) 'A mixed excitation LPC vocoder model for low bit rate speech coding', in *IEEE Trans. Speech and Audio Processing*, 3(4):242–50.

[49] W. Daumer (1982) 'Subjective evaluation of several efficient speech coders', in *IEEE Trans. on Communications*, 30(4):655–62.

[50] R. V. Cox (1995) 'Speech coding standards', in *Speech coding and synthesis* by W. B. Kleijn and K. K. Paliwal (Eds), pp. 49–78. Amsterdam: Elsevier Science

[51] W. Wong, R. Mack, B. Cheetham, and X. Sun (1996) 'Low rate speech coding for telecommunications', in *BT Technol. J.*, 14(1):28–43.

3

Sampling and Quantization

3.1 Introduction

In digital communication systems, signal processing tools require the input source to be digitized before being processed through various stages of the network. The digitization process consists of two main stages: sampling the signal and converting the sampled amplitudes into binary (digital) codewords. The difference between the original analogue amplitudes and the digitized ones depend on the number of bits used in the conversion. A 16 bit analogue to digital converter is usually used to sample and digitize the input analogue speech signal. Having digitized the input speech, the speech coding algorithms are used to compress the resultant bit rate where various quantizers are used. In this chapter, after a brief review of the sampling process, quantizers which are used in speech coders are discussed.

3.2 Sampling

As stated above, the digital conversion process can be split into sampling, which discretizes the continuous time, and quantization, which reduces the infinite range of the sampled amplitudes to a finite set of possibilities. The sampled waveform can be represented by,

$$s(n) = s_a(nT) \quad -\infty < n < \infty \tag{3.1}$$

where s_a is the analogue waveform, n is the integer sample number and T is the sampling time (the time difference between any two adjacent samples, which is determined by the bandwidth or the highest frequency in the input signal).

Digital Speech. A. Kondoz

ISBN 0-470-87007-9 (HB)

The sampling theorem states that if a signal $s_a(t)$ has a band-limited Fourier transform $S_a(j\omega)$ given by,

$$S_a(j\omega) = \int_{-\infty}^{\infty} s_a(t)e^{-j\omega t}dt \tag{3.2}$$

such that $S_a(j\omega) = 0$ for $|\omega| \geq 2\pi W$ then the analogue signal can be reconstructed from its sampled version if $T \leq 1/2W$. W is called the *Nyquist frequency*.

The effect of sampling is shown in Figure 3.1. As can be seen from Figures 3.1b and 3.1c, the band-limited Fourier transform of the analogue signal which is shown in Figure 3.1a is duplicated at every multiple of the sampling frequency.

This is because the Fourier transform of the sampled signal is evaluated at multiples of the sampling frequency which forms the relationship,

$$S(e^{j\omega T}) = \frac{1}{T}\sum_{n=-\infty}^{\infty} S_a(j\omega + j2\pi n/T) \tag{3.3}$$

This can also be interpreted by looking into the time domain sampling process where the input signal is regularly (at every sampling interval) multiplied

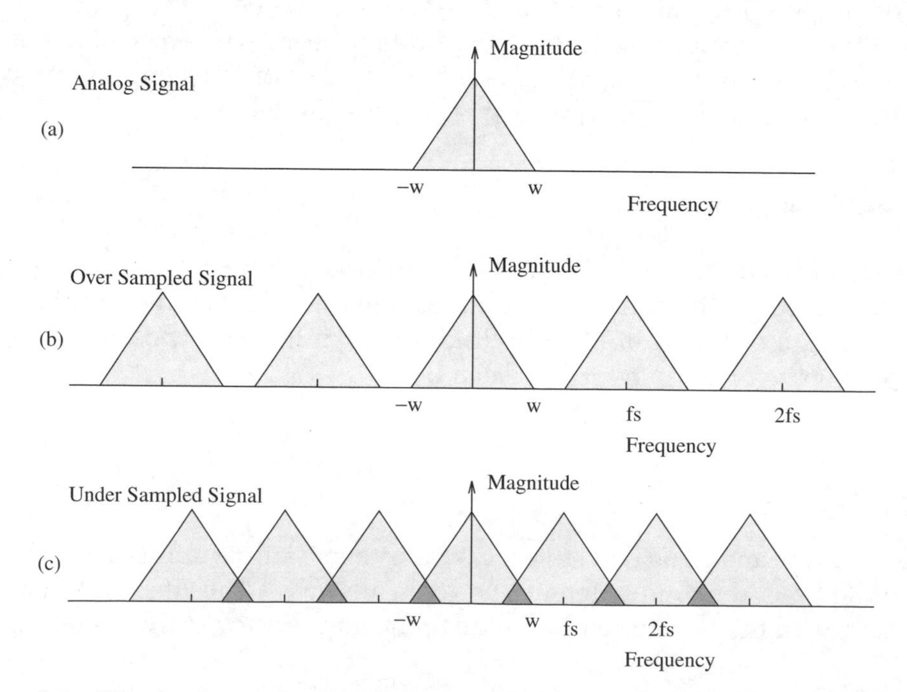

Figure 3.1 Effects of sampling: (a) original signal spectrum, (b) over sampled signal spectrum and (c) under sampled signal spectrum

with a delta function. When converted to the frequency domain, the multiplication becomes convolution and the message spectrum is reproduced at multiples of the sampling frequency.

We can clearly see that if the sampling frequency is less than twice the Nyquist frequency, the spectra of two adjacent multiples of the sampling frequencies will overlap. For example, if $\frac{1}{T} = f_s < 2W$ the analogue signal image centred at $2\pi/T$ overlaps into the base band image. The distortion caused by high frequencies overlapping low frequencies is called *aliasing*. In order to avoid aliasing distortion, either the input analogue signal has to be band-limited to a maximum of half the sampling frequency or the sampling frequency has to be increased to at least twice the highest frequency in the analogue signal.

Given the condition $1/T > 2W$, the Fourier transform of the sampled sequence is proportional to the Fourier transform of the analogue signal in the base band as follows:

$$S(e^{j\omega T}) = \frac{1}{T} S_a(j\omega) \quad |\omega| < \frac{\pi}{T} \tag{3.4}$$

Using the above relationship, the original analogue signal can be obtained from the sampled sequence using interpolation given by [1],

$$s_a(t) = \sum_{n=-\infty}^{\infty} s_a(nT) \frac{sin[\pi(t - nT)/T]}{\pi(t - nT)/T} \tag{3.5}$$

which can be written as,

$$s_a(t) = \sum_{n=-\infty}^{\infty} s_a(nT) sinc(\phi) \tag{3.6}$$

where $\phi = \pi(t - nT)/T$.

Therefore, if the sampling frequency is at least twice the Nyquist frequency, the analogue signal can be recovered completely from its sampled version by adding together *sinc* functions centred on each sampling point and scaled by the sampled value of the analogue signal. The $sinc(\phi)$ function in the above equation represents an ideal low pass filter. In practice, the front end band limitation before sampling is usually achieved by a low pass filter which is less than ideal and may cause aliasing distortion due to its roll-off characteristics. In order to avoid aliasing distortion, the sampling frequency is usually chosen to be higher than twice the Nyquist frequency. In telecommunication networks the analogue speech signal is band-limited to 300 to 3400 Hz and sampled at 8000 Hz. This same band limitation and sampling is used throughout this book unless otherwise specified.

3.3 Scalar Quantization

Quantization converts a continuous-amplitude signal (usually 16 bit, represented by the digitization process) to a discrete-amplitude signal that is different from the continuous-amplitude signal by the quantization error or noise. When each of a set of discrete values is quantized separately the process is known as scalar quantization. The input–output characteristics of a uniform scalar quantizer are shown in Figure 3.2.

Each sampled value of the input analogue signal, which has an infinite range (16 bit digitized), is compared against a finite set of amplitude values and the closest value from the finite set is chosen to represent the amplitude. The distance between the finite set of amplitude levels is called the quantizer step size and is usually represented by Δ. Each discrete amplitude level x_i is represented by a codeword $c(n)$ for transmission purposes. The codeword $c(n)$ indicates to the de-quantizer, which is usually at the receiver, which discrete amplitude is to be used.

Assuming all of the discrete amplitude values in the quantizer are represented by the same number of bits B and the sampling frequency is f_s, the

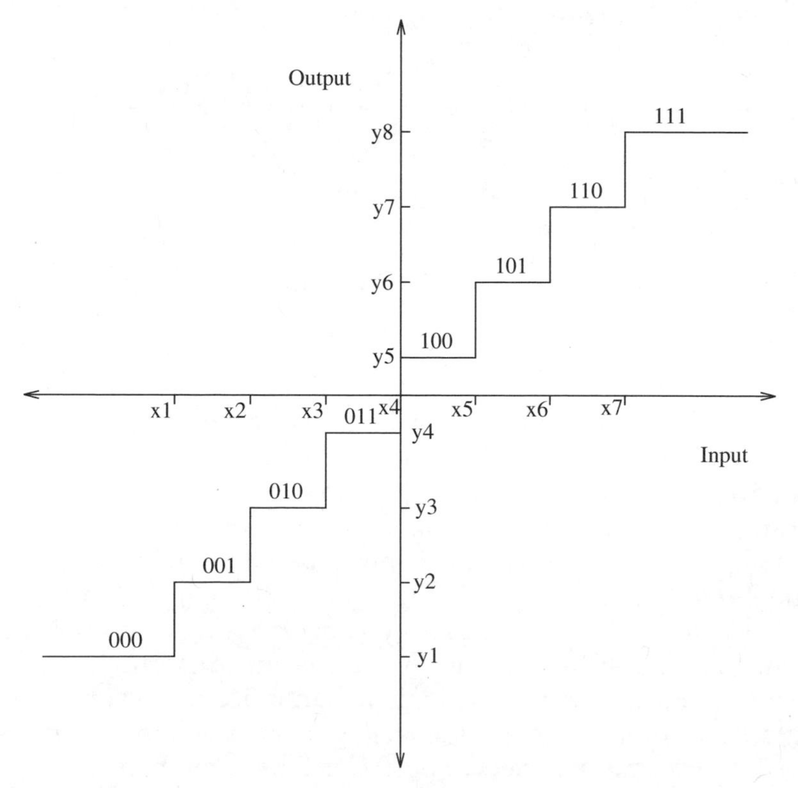

Figure 3.2 The input–output characteristics of a uniform quantizer

channel transmission bit rate is given by,

$$T_c = Bf_s \quad bits/second \tag{3.7}$$

Given a fixed sampling frequency, the only way to reduce the channel bit rate T_c is by reducing the length of the codeword $c(n)$. However, a reduced length $c(n)$ means a smaller set of discrete amplitudes separated by larger Δ and, hence, larger differences between the analogue and discrete amplitudes after quantization, which reduces the quality of reconstructed signal. In order to reduce the bit rate while maintaining good speech quality, various types of scalar quantizer have been designed and used in practice. The main aim of a specific quantizer is to match the input signal characteristics both in terms of its dynamic range and probability density function.

3.3.1 Quantization Error

When estimating the quantization error, we cannot assume that $\Delta_i = \Delta_{i+n}$ if the quantizer is not uniform [2]. Therefore, the signal lying in the ith interval,

$$x_i - \frac{\Delta_i}{2} \leq s(n) < x_i + \frac{\Delta_i}{2} \tag{3.8}$$

is represented by the quantized amplitude x_i and the difference between the input and quantized values is a function of Δ_i. The instantaneous squared error, for the signal lying in the ith interval is $(s(n) - x_i)^2$. The mean squared error of the signal can then be written by including the likelihood of the signal being in the ith interval as,

$$E_i^2 = \int_{x_i-\frac{\Delta_i}{2}}^{x_i+\frac{\Delta_i}{2}} (x - x_i)^2 p(x)dx \tag{3.9}$$

where $s(n)$ has been replaced by x for ease of notation and $p(x)$ represents the probability density function of x. Assuming the step size Δ_i is small, enabling very fine quantization, we can assume that $p(x)$ is flat within the interval $x_i - \frac{\Delta}{2}$ to $x_i + \frac{\Delta}{2}$. Representing the flat region of $p(x)$ by its value at the centre, $p(x_i)$, the above equation can be written as,

$$E_i^2 = p(x_i) \int_{-\frac{\Delta_i}{2}}^{\frac{\Delta_i}{2}} y^2 dy = \frac{\Delta_i^3}{12} p(x_i) \tag{3.10}$$

The probability of the signal falling in the ith interval is,

$$\Gamma_i = \int_{x_i-\frac{\Delta_i}{2}}^{x_i+\frac{\Delta_i}{2}} p(x)dx = p(x_i)\Delta_i \tag{3.11}$$

The above is true only if the quantization levels are very small and, hence, $p(x)$ in each interval can be assumed to be uniform. Substituting (3.11) into (3.10) for $p(x_i)$ we get,

$$E_i^2 = \frac{\Delta_i^2}{12}\Gamma_i \tag{3.12}$$

The total mean squared error is therefore given by,

$$E^2 = \frac{1}{12}\sum_{i=1}^{N}\Gamma_i\Delta_i^2 \tag{3.13}$$

where N is the total number of levels in the quantizer. In the case of a uniform quantizer where each step size is the same, Δ, the total mean squared error becomes,

$$E^2 = \frac{\Delta^2}{12}\sum_{i=1}^{N}\Gamma_i = \frac{\Delta^2}{12} \tag{3.14}$$

where we assume that the signal amplitude is always in the quantizer range and, hence, $\sum_{i=1}^{N}\Gamma_i = 1$.

3.3.2 Uniform Quantizer

The input–output characteristics of a uniform quantizer are shown in Figure 3.2. As can be seen from its input–output characteristics, all of the quantizer intervals (steps) are the same width. A uniform quantizer can be defined by two parameters: the number of quantizer levels and the quantizer step size Δ. The number of levels is generally chosen to be of the form 2^B, to make the most efficient use of B bit binary codewords. Δ and B must be chosen together to cover the range of input samples. Assuming $|x| \leq X_{max}$ and that the probability density function of x is symmetrical, then,

$$2X_{max} = \Delta 2^B \tag{3.15}$$

From the above equation it is easily seen that once the number of bits to be used, B, is known, then the step size, Δ, can be calculated by,

$$\Delta = \frac{2X_{max}}{2^B} \tag{3.16}$$

The quantization error $e_q(n)$ is bounded by,

$$-\frac{\Delta}{2} \leq e_q(n) \leq \frac{\Delta}{2} \tag{3.17}$$

In a uniform quantizer, the only way to reduce the quantization error is by increasing the number of bits. When a uniform quantizer is used, it is assumed that the input signal has a uniform probability density function varying between $\pm X_{max}$ with a constant height of $\frac{1}{2X_{max}}$. From this, the power of the input signal can be written as,

$$P_x = \int_{-X_{max}}^{X_{max}} x^2 p(x) dx = \frac{X_{max}^2}{3} \tag{3.18}$$

Using the result of (3.14), the signal to noise ratio can be written as,

$$SNR = \frac{P_x}{P_n} = \frac{X_{max}^2/3}{\Delta^2/12} \tag{3.19}$$

Substituting (3.16) for Δ we get,

$$SNR = \frac{P_x}{P_n} = 2^{2B} \tag{3.20}$$

Taking the log,

$$SNR(dB) = 10 \log_{10}(2^{2B}) = 20B \log_{10}(2) = 6.02B\ dB \tag{3.21}$$

The above result is useful both in determining the number of bits needed in the quantizer for certain signal to quantization noise ratio and in estimating the performance of a uniform quantizer for a given bit rate.

3.3.3 Optimum Quantizer

When choosing the levels of a quantizer, positioning of these levels has to be selected so that the quantization error is minimized. In order to maximize the ratio of signal to quantization noise for a given number of bits per sample, levels of the quantizer must be selected to match the probability density function of the signal to be quantized. This is because speech-like signals do not have a uniform probability density function, and the probability of smaller amplitudes occurring is much higher than that of large amplitudes. Consequently, to cover the signal dynamic range as accurately as possible, the optimum quantizer should have quantization levels with nonuniform spacing. The input–output characteristics of a typical nonuniform quantizer where the step size of the quantizer intervals is increasing for higher input signal values is shown in Figure 3.3. The noise contribution of each interval depends on the probability of the signal falling into a certain quantization interval. The nonuniform spacing of the quantization levels is equivalent to a nonlinear compressor $C(x)$ followed by a uniform quantizer. The nonlinear

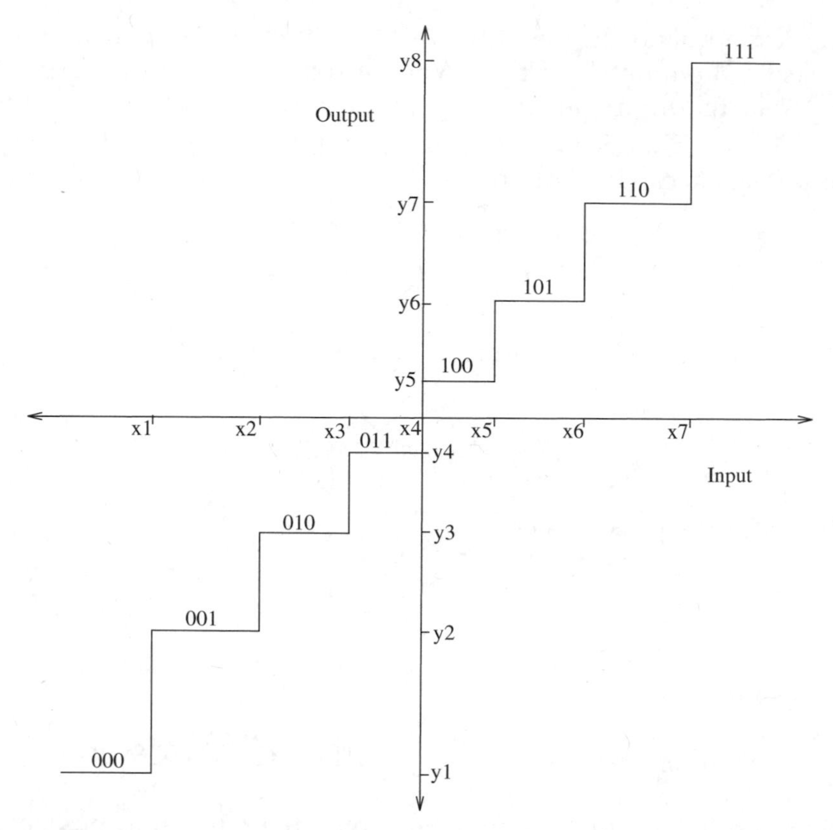

Figure 3.3 The input–output characteristics of a nonuniform quantizer

compressor, $C(x)$, compresses the input samples depending on their statistical properties. In other words, the less likely higher sample values are compressed more than the more likely low amplitude samples. The compressed samples are then quantized using a uniform quantizer. The effect of compression is reversed at the receiver by applying the inverse $C^{-1}(x)$ expansion to the de-quantized samples. The compression and expansion processes do not introduce any signal distortions.

It is quite important to select the best compression–expansion combination for a given input signal probability density function. Panter and Dite [3] used analysis based on the assumption that the quantization is sufficiently fine and that the amplitude probability density function of the input samples is constant within the quantization intervals. Their results show significant improvement in the signal to noise ratio over uniform quantization if the input samples have a *peak* to root mean squared (*rms*) ratio greater than 4.

In designing an optimum quantizer, Max [4] discovered how to optimally choose the output levels for nonuniform input quantizer levels. His analysis required prior knowledge of the probability density function together with the

Table 3.1 Max quantizer input and output levels for 1, 2, 3, 4, and 5 bit quantizers

Max quantizer thresholds									
1 bit		2 bit		3 bit		4 bit		5 bit	
i/p	*o/p*	*i/p*	*o/p*	*i/p*	*o/p*	*i/p*	*o/p*	*i/p*	*o/p*
0.0000	0.7980	0.0000	0.4528	0.0000	0.2451	0.0000	0.1284	0.0000	0.0659
		0.9816	1.5100	0.5006	0.7560	0.2582	0.3881	0.1320	0.1981
				1.0500	1.3440	0.5224	0.6568	0.2648	0.3314
				1.7480	2.1520	0.7996	0.9424	0.3991	0.4668
						1.0990	1.2560	0.5359	0.6050
						1.4370	1.6180	0.6761	0.7473
						1.8440	2.0690	0.8210	0.8947
						2.4010	2.7330	0.9718	1.0490
								1.1300	1.2120
								1.2990	1.3870
								1.4820	1.5770
								1.6820	1.7880
								1.9080	2.0290
								2.1740	2.3190
								2.5050	2.6920
								2.9770	3.2630

variance, σ_x^2, of the input signal but made no assumption of fine quantization. The quantizer input–output threshold values for 1–5 bit Max quantizers are tabulated in Table 3.1 [4]. The quantizers in Table 3.1 are for a unit variance signal with a normal probability density function. Each quantizer has the same threshold values in the corresponding negative side of the quantizer.

Nonuniform quantization is advantageous in speech coding, both in coarse and fine quantization cases, for two reasons. Firstly, a nonuniform quantizer matches the speech probability density function better and hence produces higher signal to noise ratio than a uniform quantizer. Secondly, lower amplitudes, which contribute more to the intelligibility of speech, are quantized more accurately in a nonuniform quantizer.

In speech coding, Max's quantizer [4] is widely used to normalize the input samples to unit variance, which guarantees the input dynamic range. In many other cases, specific nonuniform quantizers are designed by optimizing the quantizer intervals using a large number of samples of the signal to be quantized. Although, these specific quantizers are not generally applicable, they give the best performance for a given signal with a given probability density function and variance. In cases where the variance of the signal has a large dynamic range, the variance of the signal is transmitted separately at known time intervals enabling a unit variance nonuniform quantizer to be used. These quantizers are called forward adaptive nonuniform quantizers.

3.3.4 Logarithmic Quantizer

As was discussed above, an optimum quantizer is advantageous if the dynamic range (or variance) of the input signal is fixed to a small known range. However, the performance of such a quantizer deteriorates rapidly as the power of the signal moves away from the value that the quantizer is designed for. Although, this can be controlled by normalizing the input signal to unit variance, this process requires the transmission of the signal variance at known time intervals for correct scaling of the de-quantized signal amplitudes.

In order to cater for the wide dynamic range of the input speech signal, Cattermole [2] suggested two companding laws called A-Law and μ-Law Pulse Code Modulation (PCM). In both schemes, the signal to quantization noise performance can be very close to that of a uniform quantizer, but their performances do not change significantly with changing signal variance and remain relatively constant over a wide range of input speech levels. When compared with uniform quantizers, companded quantizers require fewer bits per input sample for a specified signal dynamic range and signal to quantization noise ratio. In a companding quantizer, quantizer levels are closely spaced for small amplitudes which progressively increase as the input signal range increases. This ensures that, when quantizing speech signals where the probability density function is zero mean and maximum at the origin, the frequently occurring small amplitudes are more accurately quantized than the less frequent large amplitudes, achieving a significantly better performance than a uniform quantizer.

The A-Law compression is defined by:

$$A_{Law}(x) = \frac{Ax}{1+\log_{10}(A)} \quad for\ 0 \le x \le \frac{1}{A} \tag{3.22}$$

$$A_{Law}(x) = \frac{1+\log_{10}(Ax)}{1+\log_{10}(A)} \quad for\ \frac{1}{A} \le x \le 1 \tag{3.23}$$

where A is the compression parameter with typical values of 86 for 7 bit (North American) PCM and 87.56 for 8 bit (European) PCM speech quantizers.

The μ-Law compression on the other hand is defined by:

$$\mu_{Law}(x) = sign(x) \frac{V_o \log_{10}\left[1 + \frac{\mu|x|}{V_o}\right]}{\log_{10}[1+\mu]} \tag{3.24}$$

where V_o is given by $V_o = L\sigma_x$ in which L is the loading factor and σ_x is the *rms* value of the input speech signal.

A typical value of the compression factor μ is 255. The above expressions show that the A-Law is a combination of a logarithmic curve for large amplitudes and a linear curve for small amplitudes. The μ-Law on the other hand is not exactly linear or logarithmic in any range but it is approximately linear for small amplitudes and logarithmic for large amplitudes. A comparison made in [5] between a μ-Law quantizer and an optimum quantizer showed that the optimum quantizer can be as much as 4 dB better. However, an optimum quantizer may have more background noise when the channel is idle and its dynamic range is limited to a smaller input signal range. For these two reasons, logarithmic quantizers are usually preferred.

3.3.5 Adaptive Quantizer

As we have seen from the already discussed quantization schemes, the dynamic range of the input signal plays a crucial role in determining the performance of a quantizer. Although, the probability density function of speech can easily be estimated and used in a quantizer design process, the variations in its dynamic range, which can be as much as 30 dB, reduces the performance of any quantizer. This can be overcome by controlling the dynamic range of the input signal. As was briefly mentioned earlier, one way of achieving this is by estimating the variance of the speech segment prior to quantization and hence, adjusting the quantizer levels accordingly. The adjustment of the quantizer levels is equivalent to designing the quantizer for unit variance and normalizing the input signal before quantization. This is called forward adaptation. A forward adaptive quantizer block diagram is shown in Figure 3.4. Assuming the speech is stationary during K samples, the *rms* is given by:

$$\sigma_x = \sqrt{\frac{1}{K}\sum_{n=1}^{K} x(n)^2} \tag{3.25}$$

where the speech samples in the block are represented by $x(n)$ and mean is assumed to be zero. However, the choice of block length K is very important because the probability density function of the normalized input signal can be affected by K. As K increases the probability density of the normalized speech signal changes from Gaussian ($K \leq 128$) to Laplacian ($K > 512$) [6]. This method requires the transmission of the speech block variances to the de-quantizer for correct signal amplitude adjustment. In order to make the normalization and de-normalization compatible, a quantized version of the speech *rms*, σ_x, is used at both the quantizer and the de-quantizer.

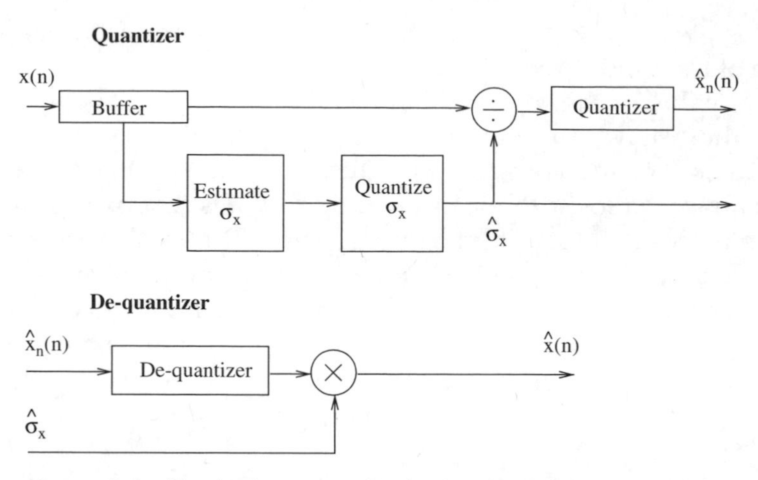

Figure 3.4 Block diagram of a forward adaptive quantizer

Another adaptation scheme which does not require transmission of the speech variance to the de-quantizer is called backward adaptation. Here, before quantizing each sample, the *rms* of the input signal is estimated from N previously quantized samples. Thus, the normalizing factor for the nth sample is:

$$\sigma_x(n) = \sqrt{\frac{a_1}{N}\sum_{i=1}^{N}\hat{x}^2(n-i)} \tag{3.26}$$

where $\hat{x}$ represents the quantized values of the past samples and a_1 is a tuning factor [6].

It has been shown [7] that for a band-limited stationary zero-mean Gaussian input, as the period N increases, the obtained signal to noise ratio tends to an asymptotic maximum. However, N must be such that the power of the signal is fairly constant during the samples of estimation. On average, the backward adaptive quantizer has 3–5 dB more signal to noise ratio compared with a logarithmic quantizer. A block diagram of a backward adaptive quantizer is shown in Figure 3.5.

An adaptation scheme called *one word memory* [8] has also been suggested. It looks at only one previously quantized sample and either expands or compresses the quantizer intervals as shown in Figure 3.6. Thus at the $(n+1)^{th}$ sample the value of the quantizer step size Δ is:

$$\Delta_{n+1} = \Delta_n M_i(|\hat{x}(n)|) \tag{3.27}$$

where, M_i is one of i fixed coefficients corresponding to quantizer levels which control the expansion–compression processes.

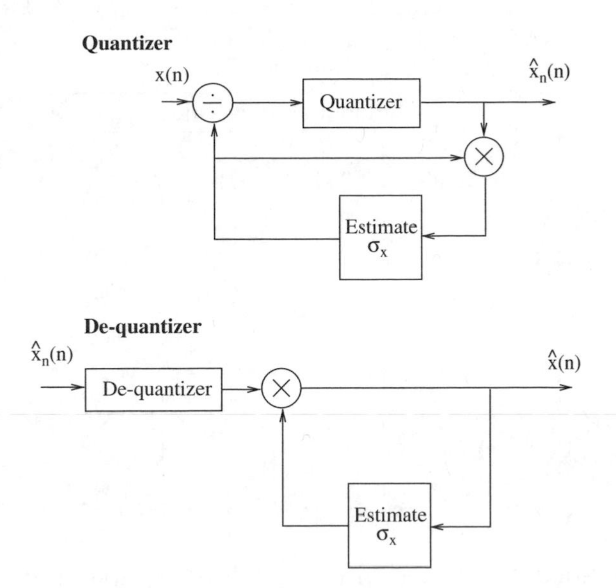

Figure 3.5 Block diagram of a backward adaptive quantizer

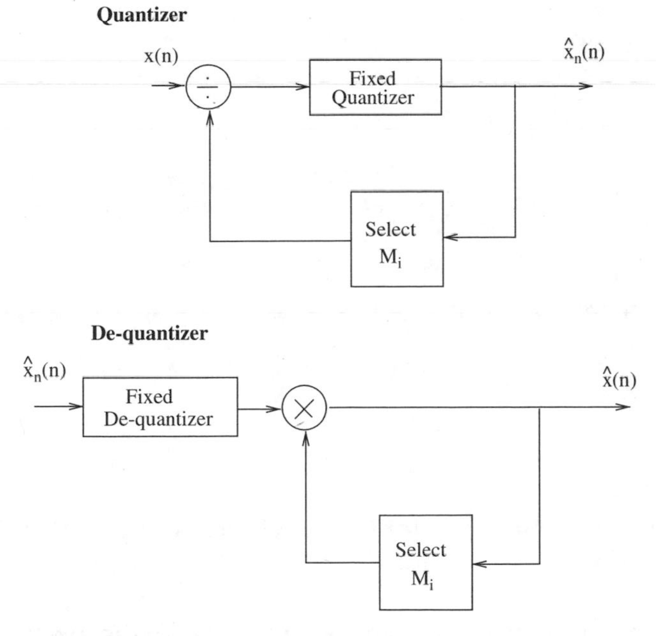

Figure 3.6 Block diagram of one word memory (Jayant) quantizer

For large quantized previous samples, multiplier values are greater than one and for small previously quantized samples multiplier values are less than one. A typical set of step size multiplier values for 2, 3 and 4 bit quantizers are shown in Table 3.2.

Table 3.2 Step size multiplier values for 2, 3, and 4 bit quantizers [9]

Adaptation multiplier values			
Previous o/p levels	2 bit	3 bit	4 bit
L1	0.60	0.85	0.80
L2	2.20	1.00	0.80
L3		1.00	0.80
L4		1.50	0.80
L5			1.20
L6			1.60
L7			2.00
L8			2.40

The recommended step size multiplier values [9] do not, in general, constitute critical target values. As can be seen from Table 3.2 [9], the middle values are fairly constant. What is critical, however, is that the step size increase should be more rapid than its decrease. This is very important for preventing quantizer overload.

3.3.6 Differential Quantizer

In a differential quantizer, the final quantized signal, $r(n)$ is the difference between the input samples $x(n)$ and their estimates $x_p(n)$.

$$r(n) = x(n) - x_p(n) \tag{3.28}$$

and

$$x_p(n) = \sum_{k=1}^{p} \hat{x}(n-k)a_k \tag{3.29}$$

where a_k is the weighting used for the previously quantized $(n-k)^{th}$ sample and p is the number of previously quantized samples considered in the estimation process.

The reason for this preprocessing stage to form the prediction residual (prediction error signal) before quantization is that, in speech signals, there is a strong correlation between adjacent samples and, hence, by removing some of the redundancies that speech signals possess, the signal variance is reduced before quantization. This reduces the quantization noise by employing a smaller quantizer step size Δ. Block diagrams of typical adaptive differential quantizers are shown in Figures 3.7 and 3.8.

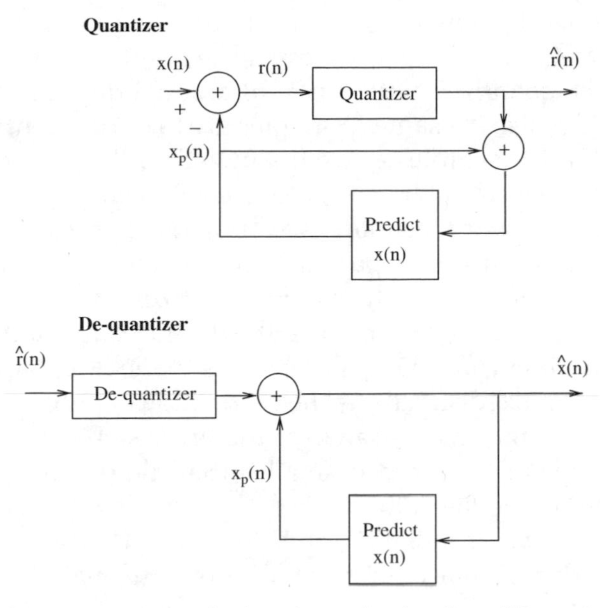

Figure 3.7 Block diagram of a backward adaptive differential quantizer

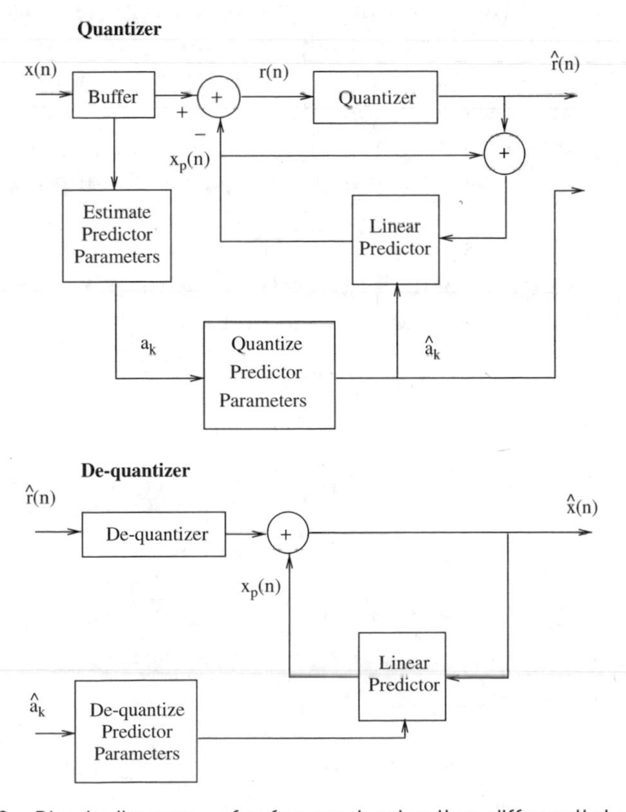

Figure 3.8 Block diagram of a forward adaptive differential quantizer

In order to show the advantage of a differential quantizer over a nondifferential quantizer, consider the following example: Assume that K input samples are to be quantized with a nondifferential quantizer with a total of $K.B_1$ bits. Consider also the same K samples are to be differentially quantized, in which case K error samples e_i are quantized to B_2 bits/sample accuracy. In a differential quantizer, the weighting coefficients a_k can be calculated using backward or forward techniques as shown in Figures 3.7 and 3.8. When backward estimation of the a_k parameters is used, the quantizer does not need to send extra information to the de-quantizer. However, in the case of forward estimation of the a_k parameters, the differential quantizer would also require $K.B_3$ bits to transmit the a_k parameters to the de-quantizer for correct recovery of the quantized signal. As the correlation between the input speech samples is usually high, the variance of the error signal to be quantized by the differential quantizer is much smaller than that of the original speech samples. Therefore, for the same accuracy of quantization, $B_2 < B_1$ and in general $B_3 \ll B_2$ which means $K.B_1 > K(B_2 + B_3)$. This shows that the main advantage of a differential over a nondifferential quantizer is due to the reduction in the speech dynamic range to be quantized.

The performance of a differential quantizer can be approximately defined by its prediction gain (the amount of signal reduction before quantization) and the performance of the residual error quantizer. Assuming that the same type of quantizer is used for both the differential and nondifferential quantization schemes, the difference in performance will depend on the accuracy of the predictor. For simplicity, if we assume a predictor depth of 1, and $\hat{x}(n-1) \simeq x(n-1)$ the residual error signal is obtained as,

$$r(n) = x(n) - ax(n-1) \tag{3.30}$$

where a is the weighting coefficient used on the previous sample to predict the current sample. The squared error is then given by,

$$r^2(n) = [x(n) - ax(n-1)]^2 \tag{3.31}$$

or,

$$r^2(n) = x^2(n) + a^2x^2(n-1) - 2ax(n)x(n-1) \tag{3.32}$$

Assuming, a is updated every N samples,

$$\sum_{n=1}^{N} r^2(n) = \sum_{n=1}^{N} x^2(n) + \sum_{n=1}^{N} a^2x^2(n-1) - 2a\sum_{n=1}^{N} x(n)x(n-1) \tag{3.33}$$

which can simply be written as,

$$\sigma_r^2 = \sigma_x^2 + a^2\sigma_x^2 - 2a\sum_{n=1}^{N} x(n)x(n-1) \tag{3.34}$$

Substituting $\rho = \frac{\sum_{n=1}^{N} x(n)x(n-1)}{\sum_{n=1}^{N} x^2(n)}$ (first order normalized autocorrelation coefficient) in (3.34) gives,

$$\sigma_r^2 = \sigma_x^2 + a^2\sigma_x^2 - 2a\sigma_x^2\rho \tag{3.35}$$

The prediction gain G_p is then found as,

$$G_p = \frac{\sigma_x^2}{\sigma_r^2} = \frac{1}{1 + a^2 - 2a\rho} \tag{3.36}$$

To maximize the prediction gain, the denominator of equation (3.36) should be minimized with respect to a, hence,

$$\frac{\partial(1 + a^2 - 2a\rho)}{\partial a} = 0 = (0 + 2a - 2\rho) \tag{3.37}$$

which gives,

$$a = \rho \tag{3.38}$$

Substituting $a = \rho$ in (3.36)

$$G_p = \frac{1}{1 + \rho^2 - 2\rho\rho} = \frac{1}{1 - \rho^2} \tag{3.39}$$

The above result shows that if the correlation between the adjacent samples is high, then a differential quantizer will perform significantly better than a nondifferential quantizer. In fact, if the signal to be quantized is a nonvarying DC signal, where $\rho = 1$, the gain of the prediction process will be infinite, i.e. no residual error will be left and, hence, no residual information will need to be transmitted. A typical ρ for speech is between 0.8 and 0.9 which may result in 4–7 dB signal reduction before quantization, hence achieving significant increase in quantization performance.

3.4 Vector Quantization

When a set of discrete-time amplitude values is quantized jointly as a single vector, the process is known as vector quantization (VQ), also known as block quantization or pattern-matching quantization. A block diagram of a simple vector quantizer is shown in Figure 3.9.

If we assume $\mathbf{x} = [x_1, x_2, \ldots, x_N]^T$ is an N dimensional vector with real-valued, continuous-amplitude (short or float representation is assumed to be continuous amplitude) randomly varying components $x_k, 1 \leq k \leq N$ (the

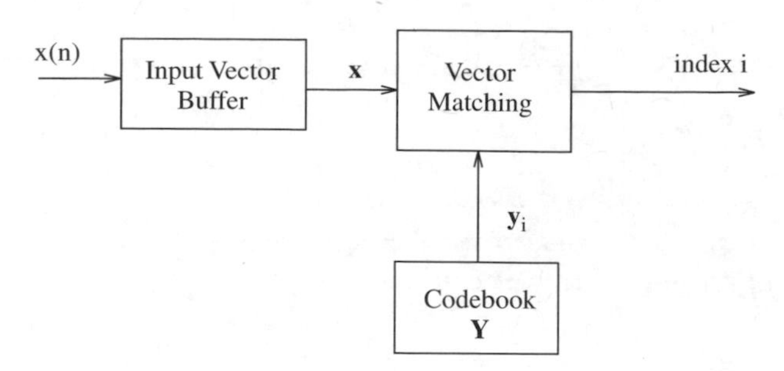

Figure 3.9 Block diagram of a simple vector quantizer

superscript T denotes transpose in vector quantization), this vector is matched with another real-valued, discrete-amplitude, N dimensional vector $\mathbf{y}$. Hence, $\mathbf{x}$ is quantized as $\mathbf{y}$, and $\mathbf{y}$ is used to represent $\mathbf{x}$. Usually, $\mathbf{y}$ is chosen from a finite set of values $\mathbf{Y} = \mathbf{y}_i, 1 \leq i \leq L$, where $\mathbf{y}_i = [y_{i1}, y_{i2}, \ldots, y_{iN}]^T$. The set $\mathbf{Y}$ is called the codebook or reference templates where L is the size of the codebook, and $\mathbf{y}_i$ are the codebook vectors. The size of the codebook may be considered to be equivalent to the number of levels in a scalar quantizer. In order to design such a codebook, N dimensional space is partitioned into L regions or cells $C_i, 1 \leq i \leq L$ and a vector $\mathbf{y}_i$ is associated with each cell C_i. The quantizer then assigns the codebook vector $\mathbf{y}_i$ if $\mathbf{x}$ is in C_i,

$$q(\mathbf{x}) = \mathbf{y}_i \quad \textit{if } \mathbf{x} \in C_i \tag{3.40}$$

The codebook design process is also known as training or populating the codebook. Figure 3.10 shows an example of the partitioning of a two-dimensional space ($N = 2$) for the purpose of vector quantization. The filled region enclosed by the bold lines is the cell C_i. During vector quantization, any input vector $\mathbf{x}$ that lies in the cell C_i is quantized as $\mathbf{y}_i$. The other codebook vectors corresponding to the other cells are shown by dots.

If the vector dimension, N, equals one vector quantization reduces to scalar quantization. Scalar quantization has the special property that whilst cells may have different sizes (step sizes) they all have the same shape. In vector quantization, however, cells may have different shapes which gives vector quantization an advantage over scalar quantization.

When $\mathbf{x}$ is quantized as $\mathbf{y}$, a quantization error results and, to measure the performance of a specific codebook, an overall distortion measure D is defined as,

$$D = \frac{1}{M} \sum_{i=1}^{M} d_i[\mathbf{x}, \mathbf{y}] \tag{3.41}$$

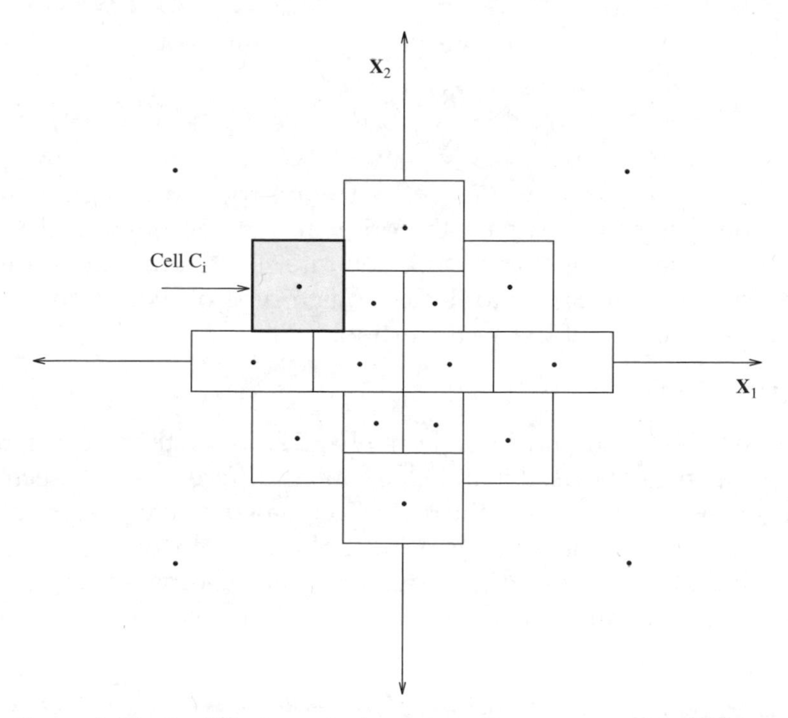

Figure 3.10 Partitioning of a two-dimensional space into 18 cells

where $d_i[\mathbf{x}, \mathbf{y}]$ is the distortion due to the ith vector in the database given by,

$$d_i[\mathbf{x}, \mathbf{y}] = \frac{1}{N} \sum_{k=1}^{N} d[x_{ik}, y_{mk}] \tag{3.42}$$

where M is the number of vectors in the database and $\mathbf{y}_m$ is the quantized version of $\mathbf{x}_i$. For transmission purposes, each vector $\mathbf{y}_i$ is encoded using a codeword of binary digits of length B_i bits. The transmission rate T is given by,

$$T = BF_c \quad \textit{bits/second} \tag{3.43}$$

where,

$$B = \frac{1}{M} \sum_{i=1}^{M} B_i \quad \textit{bits/vector} \tag{3.44}$$

is the average codeword length (usually $B = B_i$), B_i is the number of bits used to encode vector $\mathbf{y}_i$ and F_c is the number of codewords transmitted per second. The average number of bits per vector dimension (sample) is,

$$R = \frac{B}{N} \quad bits/sample \tag{3.45}$$

When designing a compression system, one tries to design a quantizer in which the distortion between the original and the quantized vectors is minimized for a given digital transmission rate. Therefore, during the design of a quantizer it is important to decide which type of distortion measure is likely to minimize the subjective distortion.

3.4.1 Distortion Measures

A distortion measure should be subjectively relevant, so that the differences in distortion values can be used to indicate similar differences in speech quality. However, a few dB decrease in the distortion may be quite perceptible by the ear in one case but not in another. Whilst objective distortion measures are necessary and useful tools in the design of speech coding systems, decisions on the direction for improving coder performance should be made using subjective quality testing.

Mean Squared Error

The most common distortion measure is the mean squared error (MSE) defined as,

$$d[\mathbf{x}, \mathbf{y}] = \frac{1}{N}(\mathbf{x} - \mathbf{y})(\mathbf{x} - \mathbf{y})^T = \frac{1}{N}\sum_{k=1}^{N}[x_k - y_k]^2 \tag{3.46}$$

The popularity of the MSE is due to its simplicity.

Weighted Mean Squared Error

In the mean squared error method, it is assumed that the distortion contributed by each element of the vector $\mathbf{x}$ is weighted equally. In general, unequal weights can be introduced to render contributions of certain elements to the distortion more important than others. Hence, a general weighted mean squared error is defined by,

$$d_w[\mathbf{x}, \mathbf{y}] = (\mathbf{x} - \mathbf{y})W(\mathbf{x} - \mathbf{y})^T \tag{3.47}$$

where W is a positive weighting matrix.

Perceptually Determined Distortion Measures

For high bit rates and hence small distortions, reasonable distortion measures, including the two mentioned above, perform well with similar performances.

Furthermore, they correlate well with subjective judgements of speech quality. However, as the bit rate decreases and distortion increases, simple distortion measures may not be related to the subjective quality of speech. Since the main application of vector quantization is expected to be at low bit-rates, it is very important to develop and use distortion measures that are better correlated with human auditory behaviour. A number of perceptually based distortion measures have been developed [10, 11, 12]. Since the main aim is to produce the highest speech quality possible at a given bit rate, it is essential to use a distortion measure that correlates well with human perception.

3.4.2 Codebook Design

When designing an L level codebook, N dimensional space is partitioned into L cells C_i, $1 \leq i \leq L$, and each cell C_i is assigned a vector $\mathbf{y}_i$. The quantizer chooses the codebook vector $\mathbf{y}_i$ if $\mathbf{x}$ is in C_i. To optimize a quantizer, the distortion in equation (3.41) is minimized over all L levels. There are two necessary conditions for optimality. The first condition is that the optimum quantizer finds a matching vector for every input vector by minimizing the distortion criterion. That is, the quantizer chooses the codebook vector that results in the minimum distortion with respect to $\mathbf{x}$ [13].

$$q(\mathbf{x}) = \mathbf{y}_i \quad \text{if } d[\mathbf{x}, \mathbf{y}_i] \leq d[\mathbf{x}, \mathbf{y}_j], \quad j \neq i, \quad 1 \leq j \leq L. \tag{3.48}$$

The second necessary condition for optimality is that each codebook vector $\mathbf{y}_i$ is optimized to give the minimum average distortion in cell C_i.

$$D_i = E\{[d(\mathbf{x}, \mathbf{y}_i)| \ \mathbf{x} \epsilon C_i]\} = \int_{\mathbf{x} \epsilon C_i} d[\mathbf{x}, \mathbf{y}_i] p(x) dx \tag{3.49}$$

where $p(x)$ is the probability density function of vectors that result in the quantized vector $\mathbf{y}_i$ in cell (cluster) C_i.

Vector $\mathbf{y}_i$ is called the centroid of the cell C_i. Optimization of the centroid of a particular cell depends on the definition of the distortion measure. For either the mean squared error or the weighted mean squared error, distortion in each cell is minimized by,

$$y_{in} = \frac{1}{M_i} \sum_{k=1}^{M_i} x_{kn} \quad \mathbf{x} \epsilon C_i \tag{3.50}$$

where y_{in} $\{n = 1, 2, \ldots, N\}$ is the nth element of the centroid $\mathbf{y}_i$ of the cluster C_i. That is, $\mathbf{y}_i$ is simply the sample mean of all the training vectors M_i contained in cell C_i. One of the most popular methods for codebook design is an iterative clustering algorithm known as the K-means algorithm [13] (also known as

Lloyd's algorithm [14]). The algorithm divides the set of training vectors into L clusters C_i in such a way that the two necessary conditions for optimality are satisfied.

K-means Algorithm

Given that m is the iteration index and C_{i_m} is the ith cluster at iteration m with $\mathbf{y}_{i_m}$ its centroid:

1. Initialization: Set $m = 0$ and choose a set of initial codebook vectors $\mathbf{y}_{i_0}$, $1 \leq i \leq L$.
2. Classification: Partition the set of training vectors $\mathbf{x}_n$, $1 \leq n \leq M$, into the clusters C_i by the nearest neighbour rule,

$$\mathbf{x} \epsilon C_{i_m} \quad \textit{if} \quad d[\mathbf{x}, \mathbf{y}_{i_m}] \leq d[\mathbf{x}, \mathbf{y}_{j_m}] \quad \textit{for all} \quad j \neq i.$$

3. Codebook updating: $m \to m + 1$. Update the codebook vector of every cluster by computing the centroid of training vectors in each cluster.
4. Termination test: If the decrease in the overall distortion at iteration m relative to $m - 1$ is below a certain threshold, stop; otherwise go to step 2.

Any other reasonable termination test may be used for step 4.

The above algorithm converges to a local optimum [14, 15]. Furthermore, any such solution is in general not unique [16]. Global optimality may be achieved approximately by initializing the codebook vectors to different values and repeating the above algorithm for several sets of initializations and then choosing the codebook that results in the minimum overall distortion.

3.4.3 Codebook Types

Vector quantization can offer substantial performance over scalar quantization at very low bit-rates. However, these advantages are obtained at considerable computational and storage costs. In order to compromise between the computation and storage costs, and quantizer performance, a number of codebook types have been developed. Some codebooks are precomputed and do not change while being used; others may be updated during quantization. Here, we will briefly explain some of the widely-used codebooks in speech coding.

Full Search Codebook

A full search codebook is one where during the quantization process each input vector is compared against all of the candidate vectors in the codebook. This process is called full search or exhaustive search. The computation and storage requirements of a typical full search codebook can be calculated as follows. If each vector in a full search codebook is represented by $B = RN$ *bits* for transmission, then the number of vectors in the codebook is given by,

$$L = 2^B = 2^{RN} \tag{3.51}$$

where N is the vector dimension in the codebook. In many applications, computing the absolute value of the quantization error may not be necessary as the main concern is to select the best performing vector. So a relative performance rather than the absolute error is required. It is therefore possible to compute the similarity rather than the difference between the input vector and the codebook vectors. Therefore, assuming that the cross-correlation of the input vector with each of the codebook candidates is computed and the one resulting in the highest cross-correlation value is selected as the quantized value of the input vector, the computation cost (assuming that all the vectors are normalized, as differences in the energy levels will give misleading cross-correlation values) is given by,

$$Com_{fs} = N2^{RN} \quad multiply - add\ per\ input\ vector \tag{3.52}$$

From this, we can also calculate the storage required for the codebook vectors as,

$$M_{fs} = NL = N2^B = N2^{RN} \quad locations \tag{3.53}$$

It can be seen from the above expressions that the computation and storage requirements of a full search codebook are exponentially related to the number of bits in the codewords.

For a 16-bit fixed point processor the storage M_{fs} in bytes is given by $2 \times M_{fs}$ and for a 32-bit floating point implementation, storage is $4 \times M_{fs}$. In general, the storage is defined by the required number of words each corresponding to a location. For example if $N = 10$ and $R = 1$ the number of codebook vectors L will be $2^{NR} = 1024$. The number of multiply–add operations needed will be $N2^{RN} = 10 \times 1024 = 10\,240$ per input vector. Assuming a sampling frequency of 8 kHz, the number of vectors per second will be $8000/10 = 800$. Therefore, the computation cost will be $800 \times 10\,240 = 8.192 \times 10^6$ multiply–add per second. The storage requirement will be $N2^{RN} = 10 \times 2^{10} = 10\,240$ words (locations).

Using the K-means algorithm, a full search codebook can be optimized (trained) in two possible ways.

- **Method 1:** The process starts with two initial vectors which may be chosen randomly or calculated as centroids of the two halves of the large training database. The K-means algorithm is used to optimize the initial vectors. After the optimization of each of the two initial vectors $\mathbf{v}_1 = [v_{11}, v_{12}, v_{13}, \ldots, v_{1N}]$ and $\mathbf{v}_2 = [v_{21}, v_{22}, v_{23}, \ldots, v_{2N}]$ with dimensions N, each is split into two further vectors as,

$$\mathbf{v}_3 = \mathbf{v}_1 - \varepsilon_1, \quad \mathbf{v}_4 = \mathbf{v}_1 + \varepsilon_1, \quad \mathbf{v}_5 = \mathbf{v}_2 - \varepsilon_2, \quad \mathbf{v}_6 = \mathbf{v}_2 + \varepsilon_2,$$

where $\varepsilon_1 = [e_{11}, e_{12}, e_{13}, \ldots, e_{1N}]$ and $\varepsilon_2 = [e_{21}, e_{22}, e_{23}, \ldots, e_{2N}]$. In most cases $\varepsilon_1 = \varepsilon_2$.

The vectors from the second stage are again optimized using the K-means algorithm and split into further vectors and so on until the number of optimized vectors is equal to the desired number. The optimization process can also be terminated by comparing the overall quantization noise performance of the codebook against a threshold.

During the optimization of a full search codebook using the above method, it is important to check that all of the optimized vectors are in the densely-populated areas and do not diverge into outer areas where their use will be wasted. In such cases the perturbation vector ε is modified to change the direction of the resultant vector.

- **Method 2:** The second method of optimization starts with randomly-selected vectors from the training database. The number of initial vectors is larger than the final desired number of vectors in the codebook. Using the K-means algorithm these vectors are optimized. After the first optimization process, the least used vectors are discarded from the codebook. The remaining vectors are then optimized and the least used vectors are again discarded from the optimized codebook. This process continues until the final size of the codebook is reached. Here, the number of vectors discarded at each stage and the number of optimization iterations may vary with the application but the initial size of the codebook should at least be 1.5 times the final size and the number of discarding stages should not be fewer than five or six. The number of vectors discarded in each stage should be reduced to increase the accuracy of optimization.

Binary Search Codebook

Binary search [17], known in the pattern recognition literature as hierarchical clustering [14], is a method for partitioning space in such a way that the search for the minimum distortion code-vector is proportional to $\log_2 L$ rather than L. In speech coding literature, binary search codebooks are also called tree codebooks or tree search codebooks.

In a binary search codebook, N dimensional space is first divided into two regions (using the K-means algorithm with two initial vectors), then each of the two regions is further divided into two subregions, and so on, until the space is divided into L regions or cells. Here, L is restricted to be a power

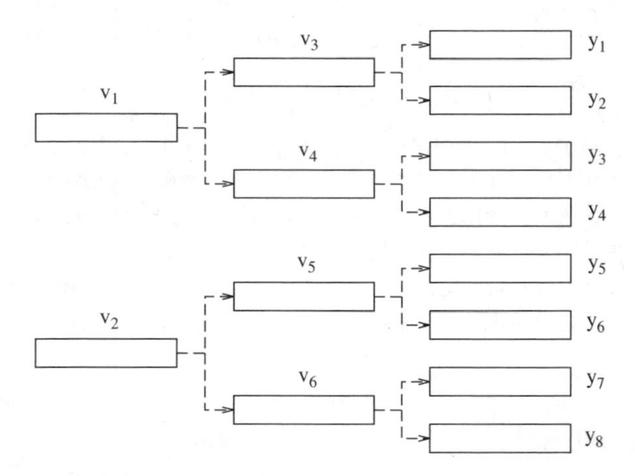

Figure 3.11 Binary splitting into eight cells

of 2, $L = 2^B$, where B is an integer number of bits. Each region is associated with a centroid. Figure 3.11 shows the division of space into $L = 8$ cells. At the first binary division $\mathbf{v}_1$ and $\mathbf{v}_2$ are calculated as the centroids of the two halves of the total space to be covered. At the second binary division four centroids are calculated as $\mathbf{v}_3$ to $\mathbf{v}_6$. The centroids of the regions after the third binary division are the actual codebook vectors $\mathbf{y}_i$. An input vector $\mathbf{x}$ is quantized, searching the tree along a path that gives the minimum distortion at each node in the path. Again assuming N multiply–adds for each distortion computation, the computation cost will be,

$$Com_{bs} = 2N \log_2 L = 2NB \quad multiply - add\ per\ input\ vector \tag{3.54}$$

At each stage, the input vector is compared against only two candidates. This makes the computation cost a linear function of the number of bits in the codewords.

The total storage cost, on the other hand, has gone up significantly,

$$M_{bs} = 2N(L-1) \quad locations \tag{3.55}$$

or,

$$M_{bs} = N \sum_{i=1}^{B} 2^i \quad locations \tag{3.56}$$

A tree search codebook need not be a binary search codebook. In other words the number of splitting stages may be less than the number of bits, B, in the codeword. In this case, each vector from the previous stage may point to more than two vectors in the current stage. This can be seen as a compromise between the extreme cases of low computation cost with high storage (binary codebook) and high computation cost with low storage requirement (full search codebook).

During the training of a binary codebook, at each stage of splitting using the K-means algorithm and method 1, the resultant optimum codebooks are stored. The database is also split into sections represented by each of the resultant vectors. When the vectors are further split, each new pair of vectors is optimized using the section of the database represented by their mother vector. This process continues until the final size codebook is reached and optimized.

Cascaded Codebooks

The major advantage of a binary search codebook is the substantial decrease in its computational cost, relative to a full search codebook, with a relatively small decrease in performance. However, the storage required for a binary search codebook relative to a full search codebook is nearly doubled. Cascaded vector quantization is a method intended to reduce storage as well as computational costs [18, 13]. A two-stage cascaded vector quantization is shown in Figure 3.12. Cascaded vector quantization consists of a sequence of vector quantization stages, each operating on the error signal of the previous stage. The input vector **x** is first quantized using a B_1 bit L_1 level vector

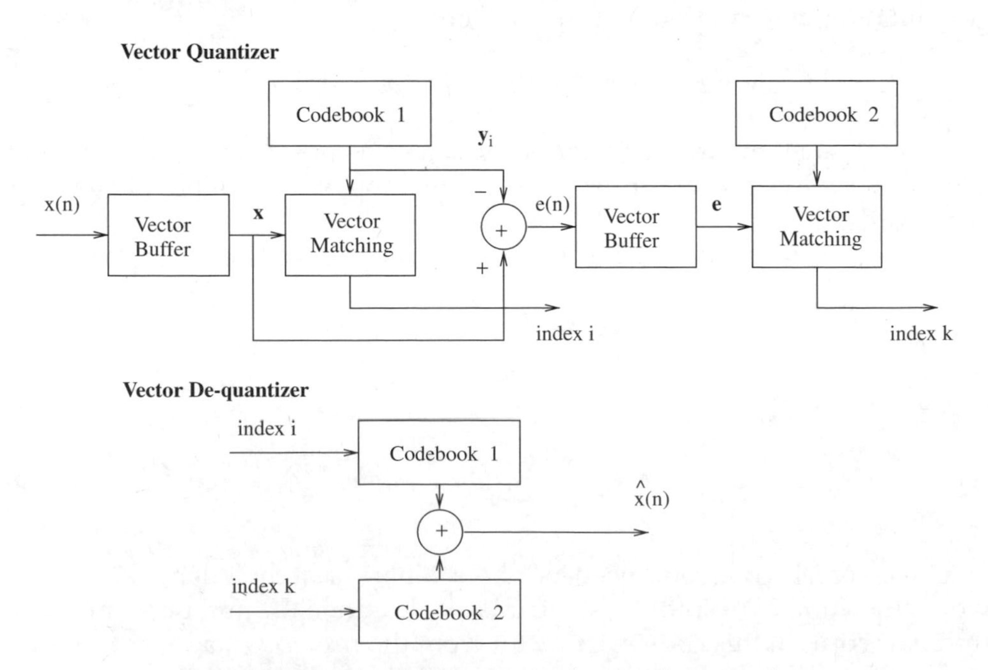

Figure 3.12 A two-stage cascaded vector quantizer

quantizer and the resulting error signal is then used in the input to a B_2 bit L_2 level second vector quantizer. The sum of the two quantized vectors results in the quantized value of the input vector **x**.

The computation and storage costs for a k -stage cascaded vector quantization are respectively,

$$Com_{cc} = N(L_1 + L_2 + \ldots + L_k) \quad \textit{multiply} - \textit{add per input vector} \tag{3.57}$$

$$M_{cc} = N(L_1 + L_2 + \ldots + L_k) \quad \textit{locations} \tag{3.58}$$

Assuming $L_1 = 2^{B_1}$, $L_2 = 2^{B_2}$ and $L_k = 2^{B_k}$ and the total number of bits per input vector $B = B_1 + B_2 \ldots + B_k$, we can see that the number of candidate vectors searched in a cascaded codebook for each input vector is less than in a full search codebook,

$$\sum_{n=1}^{k} 2^{B_n} < 2^B \quad \textit{if} \quad B = \sum_{n=1}^{k} B_n \quad \textit{and} \;\; k > 1 \tag{3.59}$$

We can also see that the storage of a cascaded codebook is less than that required by a binary codebook,

$$N\left(\sum_{n=1}^{k} 2^{B_n}\right) < N\left(\sum_{i=1}^{B} 2^i\right) \quad \textit{for} \;\; k > 1 \tag{3.60}$$

Given the condition that the total number of bits used at various stages of a cascaded codebook is B, both computation and storage requirements reduce with an increase in the number of stages.

Split Codebooks

In all of the above codebook types an N dimensional input vector is directly matched with N dimensional codebook entries. In a split vector quantization scheme, an N dimensional input vector is first split into P parts where $P > 1$. For each part of the split vector a separate codebook is used and each part may be vector quantized independently of the other parts using B_p bits. Assuming a vector is split into P equal parts and vector quantized using B_p bits for each part, the computation and storage requirements can be calculated as follows:

$$Com_{ss} = \frac{N}{P}(L_1 + L_2 + \ldots + L_P) \quad \textit{multiply} - \textit{add per input vector} \tag{3.61}$$

where $L_p = 2^{B_p}$ for $p = 1, 2, \ldots, P$. Similarly, the storage is given by:

$$M_{ss} = \frac{N}{P}(L_1 + L_2 + \ldots + L_P) \quad \textit{locations} \tag{3.62}$$

The usefulness of a split vector quantization is in its flexibility in choosing the dimension of each split part and in the allocation of the overall bits per input vector to these parts according to the perceptual importance of the vector elements contained in each split part.

Gain Shape Codebooks

In the earlier discussion of scalar quantization, it was mentioned that the variance of the input speech signal affected the performance of the quantizer. This is also true in the case of a vector quantizer. For example, if the input signal variance is fixed at a certain value, all of the codebook entries will have the same variance and differ only in the shape of vector elements. In addition, if we assume that the same number of shape combinations is repeated with another variance level at the input, the number of codebook entries would have to be doubled to cover the vector shapes at two different energy levels. Therefore, if the input vectors have a large dynamic range, the required codebook size may be too large for practical implementation in both computation and storage. This problem can be overcome by using the same idea that is used in scalar quantization: each input vector is normalized to a certain variance level (usually unity), and then its unit variance shape is vector quantized using a shape codebook containing candidate vectors with unity variance. The original variance of the input vector is separately quantized and transmitted to the de-quantizer for correct scaling. This process is called gain-shape vector quantization. A block diagram of a gain-shape vector quantizer is shown in Figure 3.13. The gain of the input vector is usually calculated and quantized using a scalar quantizer either before or during the search of the shape codebook.

If the gain of the input vector is to be calculated and quantized before finding its shape then the quantized gain is calculated as:

$$\hat{\sigma}_x = Q\left[\sqrt{\frac{1}{N}\sum_{i=1}^{N} x_i^2}\right] \tag{3.63}$$

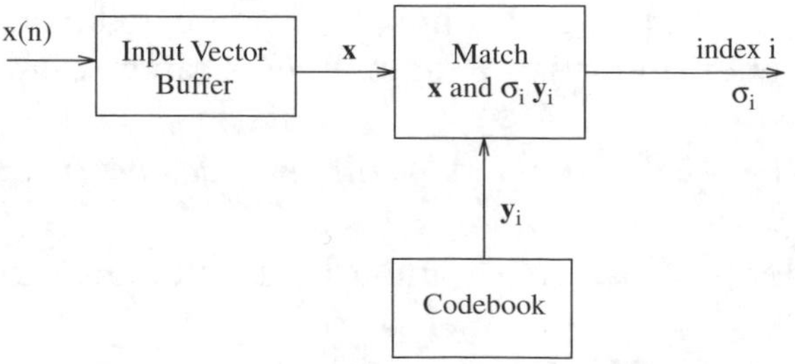

Figure 3.13 Gain-shape vector quantizer

where

$$Q[.]$$

denotes quantization operation. The shape codebook is then searched and the codebook vector which minimizes the expression,

$$D_k = \sum_{i=1}^{N} (x_i - \hat{\sigma}_x y_{ki})^2 \quad k = 1, 2, \ldots, L \tag{3.64}$$

is chosen for transmission. This search scheme, called open loop, is not optimum. Better performance can be achieved with a closed loop scheme where the shape is first found and then the corresponding gain is quantized before computing the final error. Here, we assume an optimum gain σ_k to be used for each of the L shape codebook entries and compute the corresponding distortion D_k as:

$$D_k = \sum_{i=1}^{N} (x_i - \sigma_k y_{ki})^2 \quad k = 1, 2, \ldots, L \tag{3.65}$$

We wish to find a vector $\mathbf{y}_k$ from the shape codebook with a gain value of σ_k such that the corresponding distortion D_k is minimized. However, we have two unknowns, namely, $\mathbf{y}_k$ and σ_k. To find σ_k in terms of $\mathbf{y}_k$ we differentiate (3.65) with respect to σ_k and set it to zero for minimum error gain. This gives the following σ_k for the codebook vector $\mathbf{y}_k$ in relation to an input vector $\mathbf{x}$,

$$\sigma_k = \frac{\sum_{i=1}^{N} (x_i y_{ki})}{\sum_{i=1}^{N} y_{ki}^2} \tag{3.66}$$

If we substitute (3.66) into (3.65) we can write the distortion D_k independently of σ_k as,

$$D_k = \sum_{i=1}^{N} (x_i)^2 - \frac{\left(\sum_{i=1}^{N} x_i y_{ki} \right)^2}{\sum_{i=1}^{N} y_{ki}^2} \quad k = 1, 2, \ldots, L \tag{3.67}$$

The first term of D_k in equation (3.67) does not change with k, and hence it is not computed during the search of the shape. The shape is found by maximizing only the second term in (3.67). During the codebook search process, the most likely shape values are found by maximizing the second term in equation (3.67). Then, corresponding gain values given by (3.66) are computed and quantized. Finally, each shape vector scaled by its quantized gain is compared with the input vector. This whole process can be simplified with only a small increase in the quantization error by computing the second part of equation (3.67) for all k to select the best shape vector without quantizing its gain (assuming that gain quantization noise will not, in general, render other vectors more favourable). In this case only one shape vector is considered which does not require further comparisons after the gain quantization process.

Adaptive Codebooks

The above discussed codebooks do not vary with time. Therefore, it is extremely important to train these codebooks for optimal performance with varying time and hence varying input vector characteristics. One way of making a codebook track the input vector characteristics with time is to make the codebook adaptive. As in the case of an adaptive scalar quantizer, the adaptation of a codebook can be achieved using either forward or backward schemes.

In a forward adaptive vector quantizer, the codebook is updated with respect to the input vectors before the quantization process, which requires some side information to be transmitted to the de-quantizer for compatible adaptation necessary for correct recovery of the signal.

In the case of a backward adaptive quantizer, the codebook is updated by the appropriately transformed most recent quantizer output vectors. In this case, no side information is needed since the same update process can be performed at the de-quantizer using the previously recovered vectors.

An adaptive codebook is usually used in cascade with other (generally, fixed) codebooks, which provide the initial vectors to the adaptive codebook as well as helping to speed up adaptation when significant signal variations occur. An adaptive codebook in a two-stage cascaded vector quantizer is shown in Figure 3.14. The first stage can be an adaptive codebook followed by a fixed second stage codebook. The adaptive codebooks used in these configurations are called *predictor codebooks* and the whole process is called *predictive* or *differential vector quantization*.

3.4.4 Training, Testing and Codebook Robustness

An important part of the codebook design is the training process used to populate the codebook. The training process simply optimizes a codebook for given training data by calculating the centroids of the cells. Because the K-means algorithm is not guaranteed to result in a codebook that is globally optimum, it is often suggested that one repeats the algorithm with a number of different initial sets of codebook vectors [19].

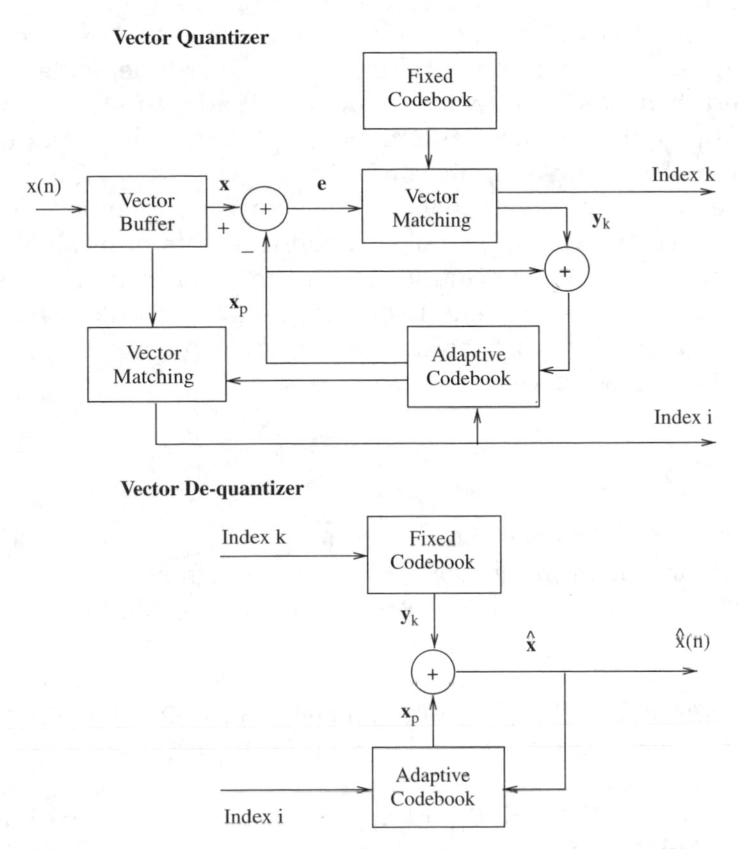

Figure 3.14 Adaptive vector quantizer in a cascaded setup

After designing a codebook to match a given set of training data, it is important to test the performance of that codebook on data that was not used in the training. Testing only on the training data will always give better performance than the codebook will actually give in practice.

The robustness of a codebook can be measured by measuring its performance on data whose distribution is different from that of the training data. In practice, one cannot usually predict all of the situations under which a quantizer will be used and so the distribution of the actual data may be different from that of the training data. There are two major types of variation that affect the design and operational performance of a codebook: input signal variability and digital transmission channel errors.

Signal variability can be further classified as speaker variability and environmental variability. Speaker variability covers the changes in the input signal due to a change in the speaker's voice and may, for example, be due to multiple speakers or the health conditions of each speaker. Environmental variability, on the other hand, refers to the background noise level and type. For a given bit rate and speaker, a speaker-dependent codebook performs better than a speaker-independent codebook. One method of maximizing the performance of a codebook is to design a speaker-independent codebook initially and then, as the system is used, have it adapt to the speech of new speakers [20]. In such a system, automatic adaptation to the background noise environment of the speaker is also possible.

As in the case of a scalar quantizer transmission channel errors affects the performance of a vector quantizer. Channel errors translate directly into distortion at the output, depending on the channel error rate. In general, vector quantization systems tend to be less robust to random channel errors than scalar quantizers, as a single bit error can cause all of the values represented by that vector to be in error.

3.5 Summary

Many quantization schemes have been designed and deployed in practice. With the advancement in the DSP technology which allowed more processing power as well as storage, vector quantization techniques have become widespread. Vector quantization schemes are very effective in reducing the bit rate of the signal that is being quantized at the expense of increased implementation complexity. It is however crucial that the codebooks are trained to match the incoming signal. As the training processes are usually applied off-line they can be allowed to run for a long time so that the best codebooks are obtained. In parallel with significant advances in the DSP technology, the implementation cost of various codebooks has been optimized by developing intelligent search algorithms as well as different types of codebook.

Bibliography

[1] L. Rabiner and R. Schafer (1978) *Digital Processing of Speech Signals*. Englewood Cliffs, NJ: Prentice-Hall

[2] K. W. Cattermole (1973) *Principles of Pulse Code Modulation*. London: Illiffe

[3] P. F. Panter and W. Dite (1951) 'Quantization distortion in PCM with non-uniform spacing of levels', in *Proc. IRE*, 39:44–8.

[4] J. Max (1960) 'Quantising for minimum distortion', in *IRE Trans. on Information Theory*, 6:7–12.

[5] M. D. Paez and T. H. Glisson (1972) 'Minimum mean-squared-error quantization in speech PCM and DPCM systems', in *IEEE Trans. on Communications*, 20(4):225–30.

[6] P. Noll (1974) 'Adaptive quantizing in speech coding systems', in *IEEE Int. Zurich Seminar on Digital Comm.*, pp. B3.1–6, March.

[7] R. W. Stroh (1970) 'Optimum and adaptive DPCM', Ph.D. thesis, Polytechnic Inst. of Brooklyn, USA.

[8] N. S. Jayant (1974) 'Adaptive quantization with a one-bit memory', in *Bell Sys. Technical Journal*, 52.

[9] N. S. Jayant (1974) 'Digital coding of speech waveforms: PCM, DPCM and DM quantizers', in *IEEE Proc.*, 62(5):611–632.

[10] N. S. Jayant and P. Noll (1984) *Digital Coding of Waveforms: Principles and applications to speech and video*. New Jersey: Prentice-Hall

[11] M. Schroeder and B. Atal (1979) 'Predictive coding of speech signals and subjective error criteria', in *IEEE Trans. on Acoust., Speech and Signal Processing*, 27:247–54.

[12] V. Viswanathan *et al.* (1983) 'Objective speech quality evaluation of medium band and narrow band real-time speech coders', in *Proc. of Int. Conf. on Acoust., Speech and Signal Processing*, pp. 543–6.

[13] J. Makhoul, S. Roucos, and H. Gish (1985) 'Vector quantisation in speech coding', in *Proc. of IEEE*, 23:1551–88.

[14] M. R. Anderberg (1973) *Cluster Analysis for Applications*, p. 22. Academic Press

[15] Y. Linde, A. Buzo, and R. Gray (1980) 'An algorithm for vector quantiser design', in *IEEE Trans. on Communications*, 28(1):84–95.

[16] R. Gray and E. Karnin (1982) 'Multiple local in vector quantization', in *IEEE Trans. on Information Theory*, 28:256–61.

[17] A. Buzo, AH Gray Jr., R. M. Gray, and J. D. Markel (1980) 'Speech coding based upon vector quantisation', in *IEEE Trans. on Acoust., Speech and Signal Processing*, 28(5):562–74.

[18] S. Roucos, R. Schwartz, and J. Makhoul (1982) 'Vector quantization for very low bit rate coding of speech', in *Proc. of Globecom*, pp. 1074–8.

[19] R. Gray (1984) 'Vector quantization', in *IEEE ASSP Magazine*, 1:4–28.

[20] D. B. Paul (1983) 'An 800 bps adaptive vector quantisation vocoder using a perceptual distance measure', in *Proc. of Int. Conf. on Acoust., Speech and Signal Processing*, pp. 73–6.

4

Speech Signal Analysis and Modelling

4.1 Introduction

The speech signal has been studied for various reasons and applications by many researchers for many years. Some studies broke down the speech signal into its smallest portions called phonemes. Here, we will describe the speech signal in terms of its general characteristics. Speech signals can be classified into *voiced* or *unvoiced*. A voiced speech segment is known by its relatively high energy content but, more importantly, it contains periodicity which is called the *pitch* of voiced speech. The unvoiced part of speech on the other hand looks more like random noise with no periodicity. However, there are some parts of speech that are neither voiced nor unvoiced, but a mixture of the two. These are usually called the transition regions, where there is a change either from voiced to unvoiced or unvoiced to voiced. The amplitude versus time plots of typical voiced and unvoiced speech are shown in Figure 4.1 (Note: The unvoiced sound has been amplified five times).

In some speech coding schemes the frequency domain representation of the speech signal is necessary. For this purpose, the short-time Fourier transform is very useful. The short-time spectral transformation is also important to look at a segment of the speech signal and determine features that are not obvious from the time domain representation.

4.2 Short-Time Spectral Analysis

The short-time Fourier transform plays a fundamental role in frequency domain analysis of the speech signal. It is used to represent the time-varying properties of the speech waveform in the frequency domain. A

Digital Speech. A. Kondoz
 ISBN 0-470-87007-9 (HB)

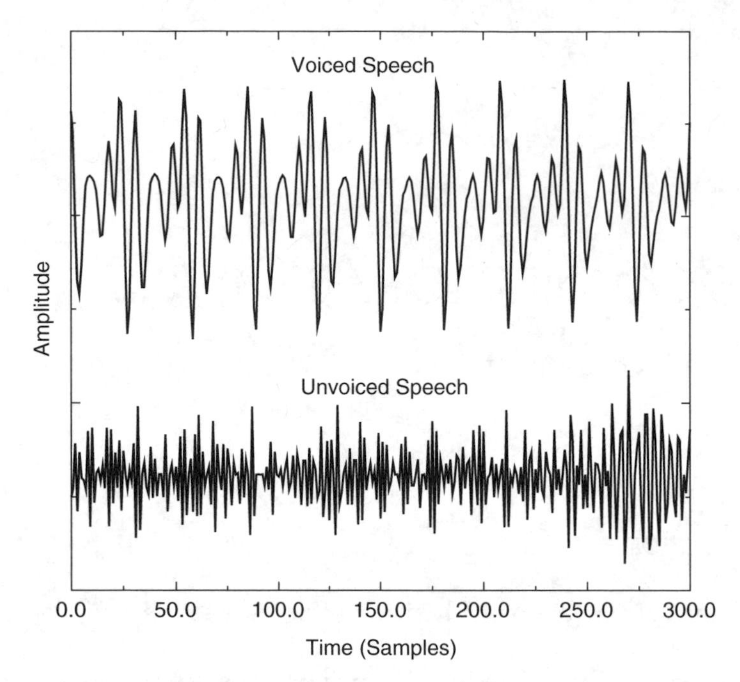

Figure 4.1 Voiced and unvoiced speech waveforms (unvoiced amplified by 5)

useful definition of the time-dependent Fourier transform is [1],

$$S_k(e^{j\omega}) = \sum_{n=-\infty}^{\infty} w(k-n)s(n)e^{-j\omega n} \tag{4.1}$$

where $w(k-n)$ is a real window sequence used to isolate the portion of the input signal that will be analysed at a particular time index, k. During the analysis of speech signals, the shape and length of the window can affect the frequency representation of speech (or any other signal). Various types of window have been studied by researchers, producing window shapes and characteristics suitable for various applications. In the following, a brief description of windowing and its effects on the short-time Fourier representation are given.

4.2.1 Role of Windows

The window, $w(n)$, determines the portion of the speech signal that is to be processed by zeroing out the signal outside the region of interest. The ideal window frequency response has a very narrow main lobe which increases the resolution and no side lobes (or frequency leakage). Since such a window is not possible in practice, a compromise is usually selected for each specific application. There are many possible windows (e.g. Rectangular, Bartlett,

Hamming, Hanning, Blackman, Kaiser, etc.), some of which are defined as follows:

Rectangular:

$$w(n) = \begin{cases} 1 & ; \quad 0 \le n \le N-1 \\ 0 & ; \quad otherwise \end{cases} \tag{4.2}$$

Bartlett:

$$w(n) = \begin{cases} \frac{2n}{N-1} & ; \quad 0 \le n \le \frac{N-1}{2} \\ 2 - \frac{2n}{N-1} & ; \quad \frac{N-1}{2} \le n \le N-1 \\ 0 & ; \quad otherwise \end{cases} \tag{4.3}$$

Hamming:

$$w(n) = \begin{cases} 0.54 - 0.46 \cos\left(2\pi \frac{n}{N-1}\right) & ; \quad 0 \le n \le N-1 \\ 0 & ; \quad otherwise \end{cases} \tag{4.4}$$

Hanning:

$$w(n) = \begin{cases} 0.5 - 0.5 \cos\left(2\pi \frac{n}{N-1}\right) & ; \quad 0 \le n \le N-1 \\ 0 & ; \quad otherwise \end{cases} \tag{4.5}$$

Blackman:

$$w(n) = \begin{cases} 0.42 - 0.5 \cos\left(2\pi \frac{n}{N-1}\right) + 0.08 \cos\left(2\pi \frac{2n}{N-1}\right) & ; \quad 0 \le n \le N-1 \\ 0 & ; \quad otherwise \end{cases} \tag{4.6}$$

Kaiser:

$$w(n) = \begin{cases} \frac{I_0\left(\beta\sqrt{1 - \left(\frac{2n}{N-1} - 1\right)^2}\right)}{I_0(\beta)} & ; \quad 0 \le n \le N-1 \\ 0 & ; \quad otherwise \end{cases} \tag{4.7}$$

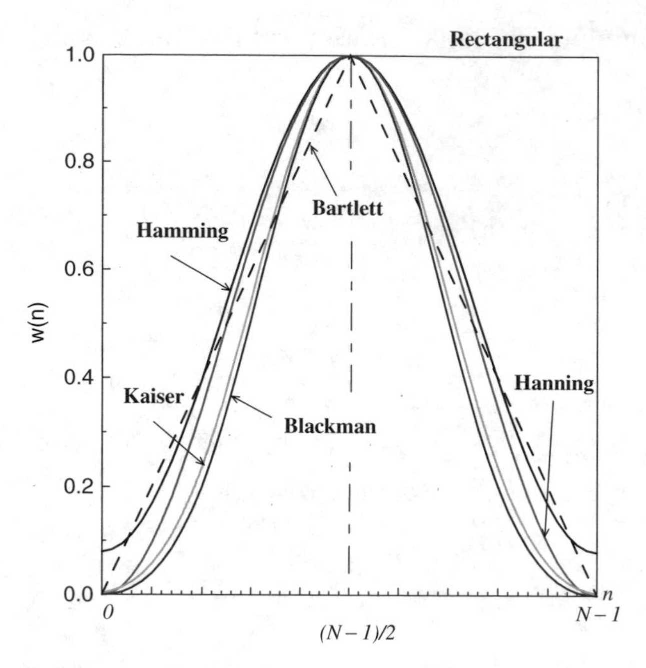

Figure 4.2 Time plots of various window functions

where I_0 is a zero order Bessel function given by,

$$I_0(\beta) = \sum_{k=0}^{\infty} \frac{\frac{\beta}{2}^{2k}}{(k!)^2} \tag{4.8}$$

The time and frequency domain shapes of these window functions are illustrated in Figures 4.2 and 4.3 respectively. As can be seen in Figure 4.3, the rectangular window has the highest frequency resolution, as it has the narrowest main lobe, but the largest frequency leakage. On the other hand, the Blackman window has the lowest resolution and the smallest frequency leakage. The effect of these windows on the time-dependent Fourier representation of speech can be illustrated by discussing the properties of two representative windows, e.g. the rectangular window and the Hamming window.

The effects of using the Hamming and rectangular windows for speech spectral analysis are shown in Figures 4.4, 4.5 and 4.6. In each figure, plots (a) and (b) show the windowed signal $s(n)w(k-n)$ and log magnitude of the Fourier transform, $S_k(\omega)$, respectively, of the rectangular window. Similarly, plots (c) and (d) show the windowed signal and log magnitude spectrum of the Hamming window. In Figure 4.4, the results for a window duration of 220 samples (27.5 ms for 8 kHz sampling rate) for a section of voiced speech is shown. When compared, the periodicity of the signal is clearly seen

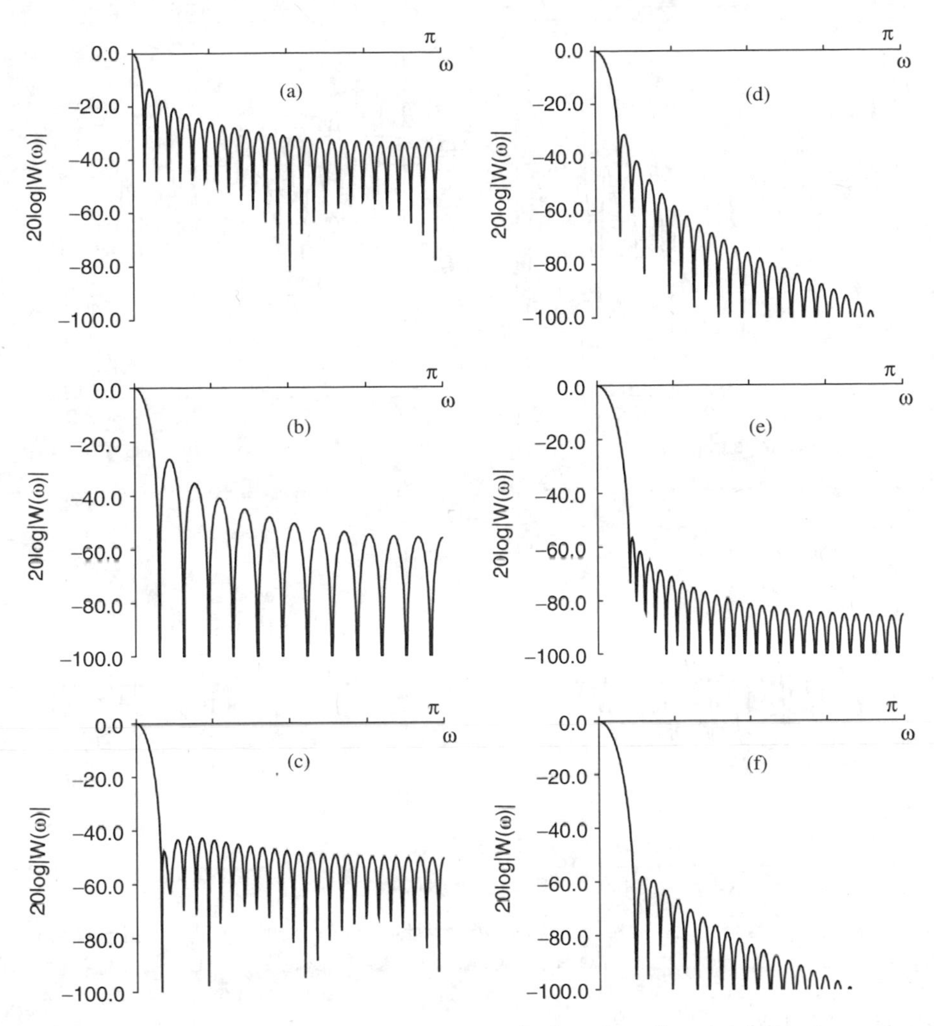

Figure 4.3 Frequency responses of various window functions: (a) Rectangular, (b) Bartlet, (c) Hamming, (d) Hanning, (e) Kaiser $\beta = 7.8$, and (f) Blackman

in both Figures 4.4b and 4.4d. However the harmonics peaks at multiples of the fundamental frequency are narrower and sharper in the rectangular windowed speech. Also noticeable in Figures 4.4b and 4.4d is the *formant* (speech sample) structure which consists of a strong first formant peak at about 500 Hz and three broader peaks at about 1350 Hz, 2300 Hz and 3400 Hz, as well as a tendency to fall off at higher frequencies due to the low-pass nature of the glottal pulse spectrum.

Although Figures 4.4b (rectangular window) and 4.4d (Hamming window) show considerable overall similarity in terms of the pitch harmonics, formant structure, and gross spectral shape, the pitch harmonics of Figure 4.4b are sharper, due to the greater frequency resolution of the rectangular window

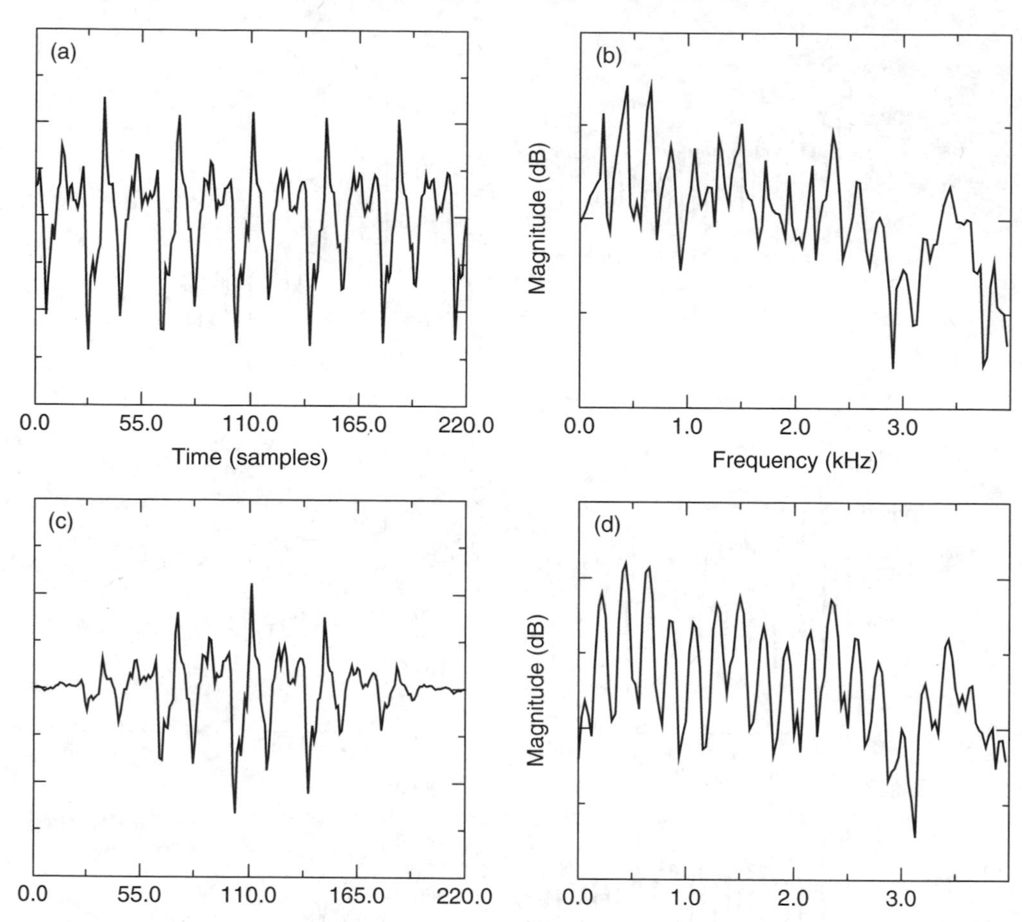

Figure 4.4 Effects of window types on voiced speech with a 220 sample window length: (a) and (b) show the time and frequency plots of speech using a rectangular window, and (c) and (d) show the time and frequency plots of speech using a Hamming window

relative to that produced by the Hamming window. However the high frequency leakage produced by the larger side lobes makes rectangular windowed speech look more noisy. This undesirable high frequency leakage between adjacent harmonics tends to offset the benefits of the flat time domain response (greater frequency resolution) of the rectangular window. As a result, rectangular windows are not usually used in speech spectral analysis.

The effect of windowing unvoiced speech is shown in Figure 4.5. Again the spectra are slowly varying with a series of sharp peaks and valleys. The noisy appearance of the spectrum (for both windows) however, is due to the random nature of unvoiced speech. Although the signal itself is random, again the Hamming window produces a smoother spectrum than the rectangular window.

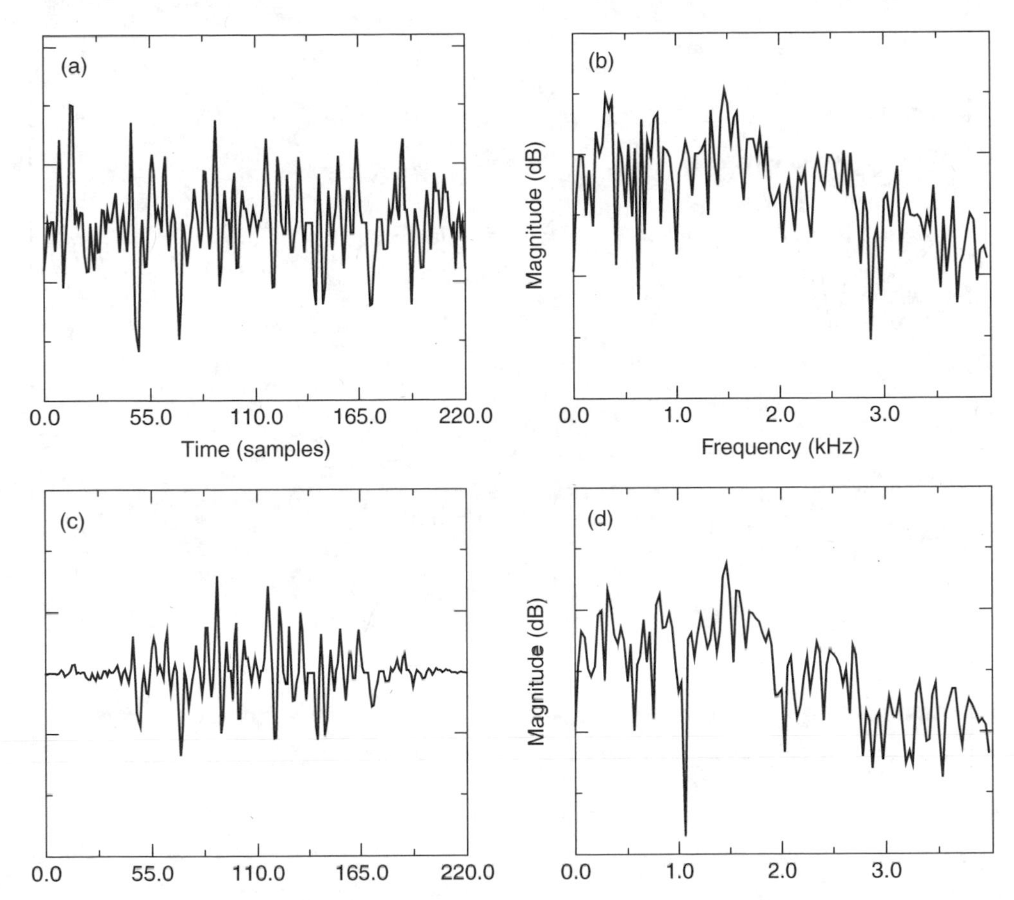

Figure 4.5 Effects of window types on unvoiced speech with a 220 sample window length: (a) and (b) show the time and frequency plots of speech using a rectangular window, and (c) and (d) show the time and frequency plots of speech using a Hamming window

In order to see the effect of varying the window length, consider the example in Figure 4.6 where a block of 40 sample (5 ms) long voiced speech is shown. In this case, the time domain speech $s(n)\, w(k-n)$ shown in Figures 4.6a and 4.6c do not show the signal periodicity accurately. This is also true for the signal spectra shown in Figures 4.6b and 4.6d. When compared with Figure 4.4, the spectra of Figure 4.6 show only a few rather broad peaks at about 500, 1350, 2300, and 3400 Hz corresponding to the formants of the speech contained within the window.

The effects of Hamming and rectangular windowing are still visible in the spectra of Figures 4.6b and 4.6d. If windows of 5 ms duration were to be positioned at the beginning and end of the 27.5 ms interval they would show different spectral characteristics. Therefore, good temporal resolution requires a short window while good frequency resolution of speech requires

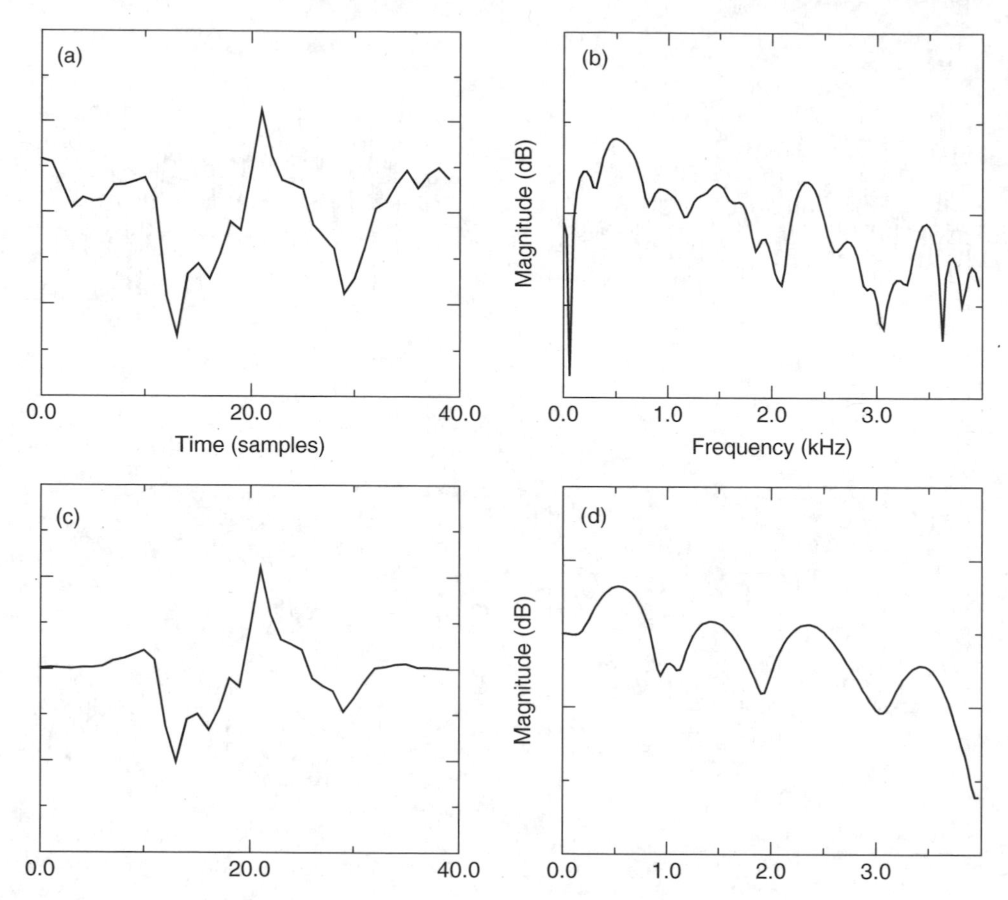

Figure 4.6 Effects of window types on voiced speech with a 40 sample window length: (a) and (b) show the time and frequency plots of speech using a rectangular window, and (c) and (d) show the time and frequency plots of speech using a Hamming window

a longer window (with a narrower main lobe). Since the attenuation of windows is essentially independent of the window duration, increasing the length, N, simply decreases the bandwidth (main lobe). If N is small, e.g. 40 samples, the short-time energy will change very rapidly. If N is too large on the other hand, e.g. on the order of several pitch periods, the short-time energy will be averaged over a long time, and hence will not adequately reflect the changing properties of the speech signal. This implies that there is no satisfactory value N which can be chosen because the duration of a pitch period varies from about 16 samples for a high pitched female or a child, up to 160 samples for a very low pitched male voice. Therefore, in practice a compromise is made by setting a suitable practical value for N between 120 and 240 samples (i.e. 15–30 ms duration). The size of the window is also determined by practical factors. That is, when speech is analysed, some form of parametric information is extracted for transmission, which would require a higher bit rate for a smaller window size (more frequent update rate). In addition, during the analysis of speech, it is necessary to have a window length which will represent the harmonic structure fairly accurately (i.e. to have more than one or two pitch periods in each window).

4.3 Linear Predictive Modelling of Speech Signals

One of the most powerful speech analysis methods is the method of linear prediction analysis [2, 3], or LPC analysis as it is commonly called. In LPC analysis, the short-term correlations between speech samples (formants) are modelled and removed by a very efficient short order filter. As LPC is a short term prediction process, the latest speech coders also call it a short term predictor (STP). Another equally powerful and related method is pitch prediction [4, 5]. In pitch prediction, the long-term correlation of speech samples are modelled. In the following sections these linear prediction techniques will be examined and discussed.

4.3.1 Source Filter Model of Speech Production

Before parameters can be extracted from a speech signal, it is necessary to have a theoretical model for our analysis. In speech processing, the source-filter model of speech production is generally used as a means of analysis. A simplified block diagram of this model [1] is shown in Figure 4.7. In this model the driving input, or excitation signal, is modelled as either an impulse train (for voiced speech) or random noise (for unvoiced speech). The combined spectral contributions of the glottal flow, the vocal tract, and the radiation of the lips are represented by a time-varying digital filter with a

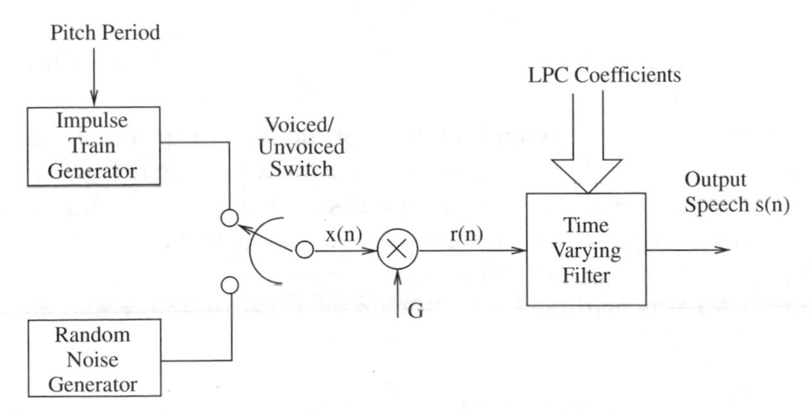

Figure 4.7 Block diagram of the simplified source-filter model of speech production

steady state system function as given by,

$$H(z) = \frac{S(z)}{X(z)} \tag{4.9}$$

$$= \frac{G\left(1 - \sum_{j=1}^{M} b_j z^{-j}\right)}{1 - \sum_{i=1}^{N} a_i z^{-i}} \tag{4.10}$$

In (4.10), both poles and zeros exist in the transfer function. However, if the order of the denominator is high enough, $H(z)$ can be approximated by an all-pole model as given by,

$$H(z) = \frac{G}{1 - \sum_{j=1}^{p} a_j z^{-j}} = \frac{G}{A(z)} \tag{4.11}$$

where,

$$A(z) = 1 - \sum_{j=1}^{p} a_j z^{-j} \tag{4.12}$$

Transforming equation (4.11) into the sampled time domain we obtain,

$$s(n) = Gx(n) + \sum_{j=1}^{p} a_j s(n-j) \tag{4.13}$$

Equation (4.13) is the well-known LPC difference equation which states that the value of the present output, $s(n)$, may be determined by summing the present input, $Gx(n)$, and a weighted sum of the past output samples. Hence, in LPC analysis the problem can be stated as follows: given the measurements of the signal, $s(n)$, determine the parameters a_j, $j = 1, \ldots, p$ which minimize $Gx(n)$. The resulting parameters are then assumed to be the parameters of our model system transfer function $H(z)$.

If α_j represents the estimates of a_j, the error or residual is given by,

$$e(n) = s(n) - \sum_{j=1}^{p} \alpha_j s(n-j) \tag{4.14}$$

It is now possible to determine the estimates by minimizing the mean squared error, i.e.

$$E\{e^2(n)\} = E\left\{\left[s(n) - \sum_{j=1}^{p} \alpha_j s(n-j)\right]^2\right\} \tag{4.15}$$

Setting the partial derivatives of the above with respect to α_j to zero for $j = 1, \ldots, p$ we get,

$$E\left\{\left[s(n) - \sum_{j=1}^{p} \alpha_j s(n-j)\right] s(n-i)\right\} = 0, \quad \text{for } i = 1, \ldots, p \tag{4.16}$$

That is, $e(n)$ is orthogonal to $s(n-i)$ for $i = 1, \ldots, p$. Equation (4.16) can be rearranged to give,

$$\sum_{j=1}^{p} \alpha_j \phi_n(i,j) = \phi_n(i,0), \quad \text{for } i = 1, \ldots, p \tag{4.17}$$

where

$$\phi_n(i,j) = E\{s(n-i)s(n-j)\} \tag{4.18}$$

In the derivation of equation (4.17), a major assumption is that the signal of our model is stationary. For speech, this is obviously untrue over a long duration. However, for short segments of speech the assumption that it is stationary is reasonable. Consequently, our expectations in equation (4.18) are replaced by finite summations over a short length of speech samples.

In this section the equation for LPC analysis was derived from the Least Mean Square approach. An equally valid result can be obtained using the Maximum Likelihood method and other formulations [6]. An interesting aspect of LPC analysis is that it applies not only to speech processing, but also to a wide range of other fields such as control and radar. However, it is in speech processing that LPC analysis has been perhaps the most successful, as it allows very accurate representation of speech with a small set of parameters.

4.3.2 Solutions to LPC Analysis

As mentioned above, in order to model the time-varying nature of the speech signal whilst staying within the constraint of our LPC analysis, i.e. a stationary signal, it is necessary to limit our analysis to short blocks of speech. This is achieved by replacing the expectations of equation (4.17) by summations over finite limits, i.e.

$$\begin{aligned}\phi_n(i,j) &= E\{s(n-i)s(n-j)\}\\ &= \sum_m s_n(m-i)s_n(m-j), \quad \text{for } i=1,\ldots,p, \quad j=0,\ldots,p \end{aligned} \tag{4.19}$$

There are two approaches to interpret equation (4.19), and these lead to two methods, namely the Autocorrelation and Covariance methods [1, 7].

The Autocorrelation Method

For the Autocorrelation Method (AM), the waveform segment, $s_n(m)$, is assumed to be zero outside the interval $0 \le m \le N-1$ where N is the length of the sample sequence. Since for $N \le m \le N+p$ we are trying to predict zero sample values (which are not actually zero) the prediction error for these samples will not be zero. Similarly, the beginning of the current frame will be affected by the same inaccuracy incurred in the previous frame. The limits for equation (4.19) can be expressed as,

$$\phi_n(i,j) = \sum_{m=0}^{N-1-|(i-j)|} s_n(m)s_n(m+|i-j|), \quad 1 \le i \le p, \quad 0 \le j \le p \tag{4.20}$$

Equation (4.20) can be reduced to the short-time autocorrelation function given by,

$$\phi_n(i,j) = R_n(|\, i-j\,|), \quad \text{for} \quad i=1,\ldots,p \quad j=0,\ldots,p \tag{4.21}$$

where,

$$R_n(j) = \sum_{m=0}^{N-1-j} s_n(m)s_n(m+j) \tag{4.22}$$

Using the AM, equation (4.17) can be expressed as

$$\sum_{j=1}^{p} \alpha_j R_n(|\, i-j\,|) = R_n(i), \quad 1 \le i \le p \tag{4.23}$$

or in matrix form by,

$$\begin{bmatrix} R_n(0) & R_n(1) & . & R_n(p-1) \\ R_n(1) & . & . & R_n(p-2) \\ \vdots & \vdots & \vdots & \vdots \\ R_n(p-1) & . & . & R_n(0) \end{bmatrix} \begin{bmatrix} \alpha_1 \\ \alpha_2 \\ \vdots \\ \alpha_p \end{bmatrix} = \begin{bmatrix} R_n(1) \\ R_n(2) \\ \vdots \\ R_n(p) \end{bmatrix}$$

The above matrix has the property that it is symmetrical and all the elements along a given diagonal are equal, i.e. it is a Toeplitz matrix. Equation (4.23) can be solved by inversion of the $p \times p$ matrix, however this is not usually performed as computational errors, such as finite precision, tend to accumulate. By exploiting the Toeplitz characteristic however, very efficient recursive procedures have been devised. The most widely used is perhaps Durbin's algorithm, which is a recursive process as follows:

$$E_n^{(0)} = R_n(0) \tag{4.24}$$

$$k_i = \left[R_n(i) - \sum_{j=1}^{i-1} \alpha_j^{i-1} R_n(i-j)\right] \Big/ E_n^{(i-1)} \quad 1 \le i \le p \tag{4.25}$$

$$\alpha_i^{(i)} = k_i \tag{4.26}$$

$$\alpha_j^{(i)} = \alpha_j^{(i-1)} - k_i \alpha_{i-j}^{(i-1)} \quad 1 \le j \le i-1 \tag{4.27}$$

$$E_n^{(i)} = (1 - k_i^2) E_n^{(i-1)} \tag{4.28}$$

After solving equations (4.25) to (4.28) recursively for $i = 1, 2, \ldots, p$, the α_j are given by,

$$\alpha_j = \alpha_j^{(p)} \quad 1 \le j \le p \tag{4.29}$$

Consider an example where the order, $p = 2$,

$$\begin{bmatrix} R_n(0) & R_n(1) \\ R_n(1) & R_n(0) \end{bmatrix} \begin{bmatrix} \alpha_1 \\ \alpha_2 \end{bmatrix} = \begin{bmatrix} R_n(1) \\ R_n(2) \end{bmatrix}$$

Then, for $i = 1$,

$$E_n^{(0)} = R_n(0)$$

$$k_1 = \frac{R_n(1)}{R_n(0)}$$

$$\alpha_1^{(1)} = \frac{R_n(1)}{R_n(0)}$$

$$E_n^{(1)} = \left[1 - \frac{R_n^2(1)}{R_n^2(0)}\right]$$

$$R_n(0) = \frac{R_n^2(0) - R_n^2(1)}{R_n(0)}$$

and for $i = 2$,

$$k_2 = [R_n(2) - \alpha_1^{(1)} R_n(1)]/E_n^{(1)} = \frac{R_n(2)R_n(0) - R_n^2(1)}{R_n^2(0) - R_n^2(1)}$$

$$\alpha_2^{(2)} = k_2$$

$$\alpha_1^{(2)} = \alpha_1^{(1)} - k_2\alpha_1^{(1)} = \frac{R_n(1)R_n(0) - R_n(1)R_n(2)}{R_n^2(0) - R_n^2(1)}$$

and, from this,

$$\alpha_1 = \alpha_1^{(2)}, \quad \text{and} \quad \alpha_2 = \alpha_2^{(2)}.$$

The Covariance Method

For the Covariance Method (CM), the opposite approach to the AM is taken. Here the interval over which the mean squared error is computed is fixed, i.e.

$$E = \sum_{m=0}^{N-1} e_n^2(m) \tag{4.30}$$

Equation (4.19) can be written as,

$$\phi_n(i,j) = \sum_{m=0}^{N-1} s_n(m-i)s_n(m-j), \quad 1 \leq i \leq p,\ 0 \leq j \leq p \tag{4.31}$$

Changing the summation index,

$$\phi_n(i,j) = \sum_{m=-i}^{N-i-1} s_n(m)s_n(m+i-j), \quad 1 \leq i \leq p,\ 0 \leq j \leq p \tag{4.32}$$

The expression given by equation (4.32) is slightly different to equation (4.20) used in the AM as it requires the use of samples in the interval $-p \leq m \leq N-1$. In effect, equation (4.31) is not a true autocorrelation function, but rather the cross-correlation between two very similar but not identical, finite-length sampled sequences. Using equation (4.31), our original LPC equation (4.17) can be expressed as,

$$\sum_{j=1}^{p} \alpha_j \phi_n(i,j) = \phi_n(i,0), \quad 1 \leq i \leq p \tag{4.33}$$

or in matrix form,

$$\begin{bmatrix} \phi_n(1,1) & \phi_n(1,2) & \cdot & \phi_n(1,p) \\ \phi_n(2,1) & \cdot & \cdot & \phi_n(2,p) \\ \vdots & \vdots & \vdots & \vdots \\ \phi_n(p,1) & \cdot & \cdot & \phi_n(p,p) \end{bmatrix} \begin{bmatrix} \alpha_1 \\ \alpha_2 \\ \vdots \\ \alpha_p \end{bmatrix} = \begin{bmatrix} \phi_n(1,0) \\ \phi_n(2,0) \\ \vdots \\ \phi_n(p,0) \end{bmatrix}$$

A solution to equation (4.33) is not as straightforward as for the equivalent AM. This is because the covariance matrix, $\phi_n(i,j) = \phi_n(j,i)$, but the $p \times p$ matrix ϕ is not Toeplitz. However, efficient matrix inversion solutions such as Cholesky decomposition can be applied where ϕ is expressed as [1]:

$$\phi = VDV^T \tag{4.34}$$

V is a lower triangular matrix whose main diagonal elements are 1s and D is a diagonal matrix. The elements of the V and D matrices are determined from equation (4.34) as follows:

$$\phi_n(i,j) = \sum_{m=1}^{j} V_{im} d_m V_{jm} \quad 1 \leq j \leq i-1 \tag{4.35}$$

or equivalently,

$$V_{ij} d_j = \phi_n(i,j) - \sum_{m=1}^{j-1} V_{im} d_m V_{jm} \quad 1 \leq j \leq i-1 \tag{4.36}$$

and for the diagonal elements of D,

$$\phi_n(i,i) = \sum_{m=1}^{i} V_{im} d_m V_{im} \tag{4.37}$$

or,

$$d_i = \phi_n(i,i) - \sum_{m=1}^{i-1} V_{im}^2 d_m \quad \textit{for } i \geq 2 \tag{4.38}$$

and,

$$d_1 = \phi_n(1,1) \tag{4.39}$$

Lattice Methods

As shown in the previous sections, the solution to the LPC equation involves two basic steps: (i) computation of a matrix of correlation values, $\phi_n(i,j)$, and (ii) solution of a set of linear equations. Although the two steps are already very efficient, another class of autocorrelation based methods, called Lattice Methods (LM), have been developed which combine the two steps to compute the LPC parameters. The basic idea behind the LM is that knowledge of the forward and backward prediction errors are incorporated during the calculation of the intermediate stages of the predictor parameters. A major incentive for using the LM is that the computed parameters are guaranteed to form a stable filter, a feature which neither the AM nor the CM possess.

Consider the ith stage of Durbin's algorithm where the set of coefficients $\alpha_j^{(i)}, j = 1, 2, \ldots, i$ are the optimum linear prediction coefficients of an ith order filter. The inverse filter $A(z)$ based on these i optimum coefficients will be,

$$A^{(i)}(z) = 1 - \sum_{j=1}^{i} \alpha_j^{(i)} z^{-j} \tag{4.40}$$

and the prediction error $e_n^{(i)}(m)$ (or for simplicity $e^{(i)}(m)$) will be,

$$e^{(i)}(m) = s(m) - \sum_{j=1}^{i} \alpha_j^{(i)} s(m-j) \tag{4.41}$$

In a z-transform notation, the above equation becomes,

$$E^{(i)}(z) = A^{(i)}(z) S(z) \tag{4.42}$$

By combining (4.27) and (4.40) we have

$$A^{(i)}(z) = 1 - \sum_{j=1}^{i-1} [\alpha_j^{(i-1)} - k_i \alpha_{i-j}^{(i-1)}] z^{-j} - \alpha_i^{(i)} z^{-i} \tag{4.43}$$

but $\alpha_i^{(i)} = k_i$ and hence,

$$A^{(i)}(z) = 1 - \sum_{j=1}^{i-1} \alpha_j^{(i-1)} z^{-j} + k_i \sum_{j=1}^{i-1} \alpha_{i-j}^{(i-1)} z^{-j} - k_i z^{-i} \tag{4.44}$$

$$A^{(i)}(z) = A^{(i-1)}(z) - k_i \left[z^{-i} - \sum_{j=1}^{i-1} \alpha_{i-j}^{(i-1)} z^{-j} \right] \tag{4.45}$$

Using (4.42) and (4.45),

$$E^{(i)}(z) = A^{(i-1)}(z)S(z) - k_i \left[z^{-i} - \sum_{j=1}^{i-1} \alpha_{i-j}^{(i-1)} z^{-j} \right] S(z) \tag{4.46}$$

The first term in equation (4.46) represents the prediction error for an $(i-1)^{th}$ order predictor employing the $s(m-1), s(m-2), \ldots, s(m-i+1)$ samples to predict $s(m)$. Since the output of the filter $\sum_{j=1}^{i-1} \alpha_{i-j}^{(i-1)} z^{-j}$ operating on $S(z)$ is $\alpha_{i-1}s_{m-1} + \alpha_{i-2}s_{m-2} + \ldots + \alpha_1 s_{m-i+1}$ the second term of equation (4.46) represents the backward prediction error of the same predictor attempting to predict $s(m-i)$ from the i samples $s(m-i+j)$, $j = 1, 2, 3, \ldots, i$ that follow $s(m-i)$. The prediction error sequence $e_m^{(i)}$ can therefore be expressed in terms of the forward and backward error sequences as,

$$e^{(i)}(m) = e^{(i-1)}(m) - k_i b^{(i-1)}(m-1) \tag{4.47}$$

It can also be shown that the ith stage backward prediction error $b^{(i)}(m)$ can be expressed as,

$$b^{(i)}(m) = b^{(i-1)}(m-1) - k_i e^{(i-1)}(m) \tag{4.48}$$

Equations (4.47) and (4.48) provide the forward and backward prediction error sequences for an ith order filter, in terms of the corresponding errors of a $(i-1)^{th}$ order filter. Note that

$$e^{(0)}(m) = n^{(0)}(m) = s(m) \tag{4.49}$$

i.e, the zero order filter error equals the original input. Furthermore, the k_i parameters can be directly computed from the forward and backward prediction errors as [8],

$$k_i = \frac{\sum_{m=0}^{N-1} e^{(i-1)}(m) b^{(i-1)}(m-1)}{\sqrt{\sum_{m=0}^{N-1} [e^{(i-1)}(m)]^2 \times \sum_{m=0}^{N-1} [b^{(i-1)}(m-1)]^2}} \tag{4.50}$$

without using the prediction coefficients α_j. From the above expression, it is clear that the k_i parameters represent the normalized cross-correlation function between the forward and backward error sequences. It is for this reason the k_i parameters are known as the partial correlation (PARCOR) coefficients [8].

A popular lattice implementation of LPC analysis is that developed by Burg [3]. Burg derived the k_i parameters by minimizing the sum of the mean squared forward and backward prediction errors, i.e.

$$\hat{E}^{(i)} = \sum_{m=0}^{N-1} \left[(e^{(i)}(m))^2 + (b^{(i)}(m))^2 \right] \tag{4.51}$$

$\hat{E}^{(i)}$ is differentiated with respect to k_i and then is set to zero to give,

$$k_i = \frac{2 \sum_{m=0}^{N-1} e^{(i-1)}(m) b^{(i-1)}(m-1)}{\sum_{m=0}^{N-1} [e^{(i-1)}(m)]^2 + \sum_{m=0}^{N-1} [b^{(i-1)}(m-1)]^2} \tag{4.52}$$

It can also be shown [8] that the above equation results in k_i parameters, $-1 \leq k_i \leq 1$. Burg's algorithm operates as follows [3]:

1. Set $e^{(0)}(m) = s(m) = b^{(0)}(m)$
2. Compute $k_1 = \alpha_1^{(1)}$
3. Determine $e^{(1)}(m)$ and $b^{(1)}(m)$ using (4.47) and (4.48)
4. Set $i = 2$
5. Find $k_i = \alpha_i^{(i)}$ using (4.52)
6. Find $\alpha_j^{(i)}$ for $j = 1, 2, \ldots, i-1$ using (4.27)
7. find $e^{(i)}(m)$ and $b^{(i)}(m)$ using (4.47) and (4.48)
8. Set $i = i + 1$
9. If $i \leq p$ go to step 5
10. End

4.3.3 Practical Implementation of the LPC Analysis

In the practical implementation of the LPC analysis, several important groups of factors need to be addressed. The first group comprises the performance, efficiency and stability factors, which are not too dissimilar for all three methods, although the LM is preferred in real-time systems where guaranteed stability is very important. However, with careful choice of windowing and fine precision arithmetic, the AM and CM are equivalent to the LM for stability. As quantization is usually applied to the coefficients, stability can always be maintained to some extent. The second group involves the choice of the filter order, p, and the analysis frame size, N. Speech is usually sampled at 8 kHz, thus giving a 4 kHz spectrum for analysis. Within the 4 kHz spectrum, the maximum number of formants displayed is usually four, thus indicating

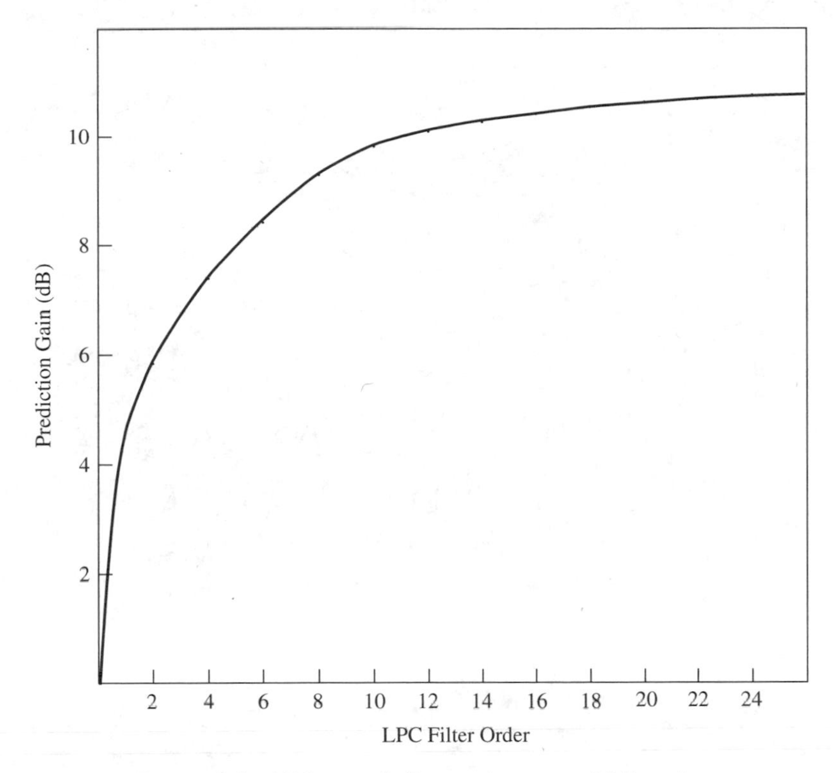

Figure 4.8 LPC prediction gain versus LPC order

that the filter order needs to be at least eight. Usually, a 10-pole filter is used so that formant resonances and general spectral shape is modelled accurately. (However, much higher order filters have been used in elaborate schemes such as the proposed ITU 16 kb/s G.728 standard [9], where a 50-order LPC filter is used!). The LPC filter prediction gain versus the order of the filter is shown in Figure 4.8 (note that the plot in this figure was obtained over a small number of speech samples and it is only indicative, the LPC gain may vary from sample to sample), and in Figure 4.9 the spectral envelope of various filter orders is illustrated.

As for the frame size, the stationarity constraint applies, thus it is necessary for us to choose a size which will conform to this. This usually implies a frame size of 16–32 ms. Another related factor is the partitioning points of the analysis frame. The position of the analysis window may affect significantly the performance of the LPC analysis. In order to reduce the effect of window positioning some common preprocessing stages include the use of pre-emphasis of the signal prior to the LPC analysis, and the use of overlapping windowed frames. Overlapping frames try to overcome some of the block-edge effects of the frame-based LPC analysis. The amount of overlap is typically around 10–20 % of the frame size. Interpolation of the

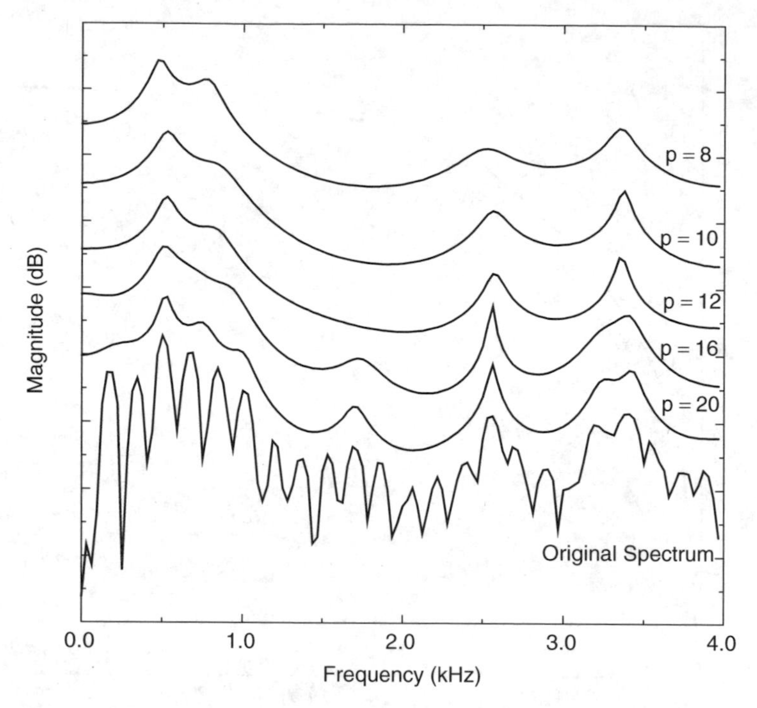

Figure 4.9 LPC magnitude responses for various LPC orders

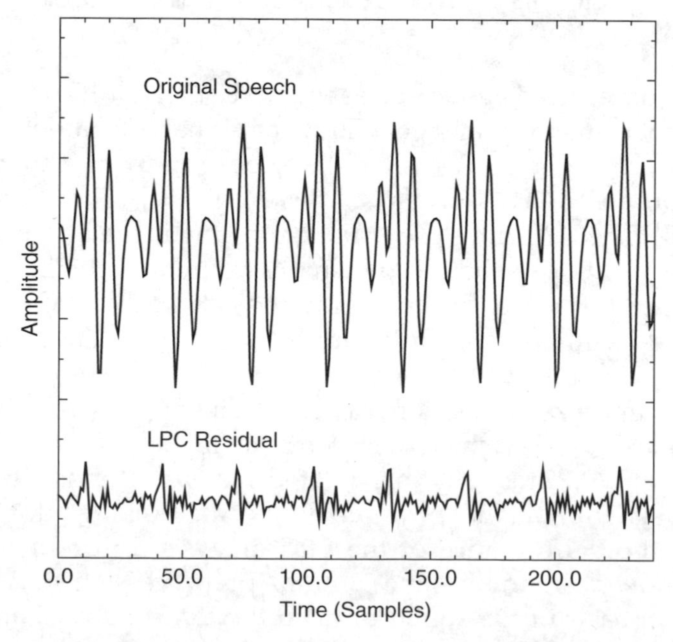

Figure 4.10 Waveform plots of original and LPC inverse-filtered speech

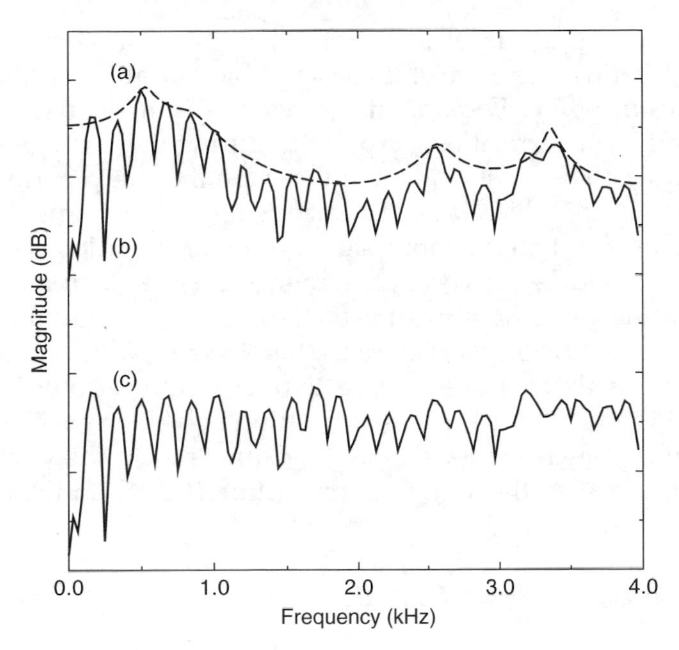

Figure 4.11 (a) the original speech spectral envelope, (b) the original speech spectrum and (c) the LPC residual spectrum

LPC coefficients from one frame to the next is also commonly applied to smooth out transitional effects.

After the LPC inverse filtering, the resultant signal, $e(n)$, should have a much lower spectral variation than the original, $s(n)$. This is illustrated in Figures 4.10 and 4.11 where the time and frequency domain representation of a typical frame of $s(n)$ and $e(n)$ are shown. Clearly, the error signal spectrum is much flatter. This result is not surprising since LPC can be viewed as a method of short-time spectrum estimation.

Also illustrated in Figure 4.11 is the frequency response or spectral envelope of the LPC filter. A feature that can be observed is that the LPC spectral envelope matches the signal spectrum much better in the spectral peaks than the spectral valleys. This can be expected as our model transfer function, $H(z)$, has poles only to model the formant peaks and no zeros to model the spectral valleys.

4.4 Pitch Prediction

4.4.1 Periodicity in Speech Signals

In the previous section, the ability of LPC analysis to remove the adjacent or neighbouring sample correlations present in speech was described. As observed, this was equivalent to removing the spectral envelope in the signal spectrum. However, as can be seen from Figure 4.11, after the LPC analysis and inverse filtering there are still considerable variations in the spectrum, i.e. it is far from white. Looking at the residual signal in Figure 4.10, it is clear that long-term correlations, especially during voiced regions, still exist between samples. The most evident of these are the sharp periodic pulses of the excitation signal, which is hardly surprising as our original source-filter model assumes this type of input signal. This also explains why the LPC analysis, which models our vocal tract, cannot adequately remove them.

To remove the periodic structure of the residual or excitation signal, a second stage of prediction is required. The objective of this second stage is again to spectrally flatten our signal, i.e. to remove the periodic fine structure. But unlike LPC analysis, it exploits correlation between speech samples that are one or more 'pitch' periods away. For this reason, the pitch prediction (filter) is usually called the long-term prediction (LTP) and the filter delay is called the lag.

4.4.2 Pitch Predictor (Filter) Formulation

Before discussing methods of pitch or long-term prediction, it is perhaps worth considering what our objectives are. Our aim is to model the long-term correlation left in the speech residual signal after LPC inverse filtering (or in the original speech signal) such that when the model parameters are used in a filter, it will remove the long-term correlation as much as possible, or spectrally flatten our signal. There are no obvious reasons why we must use the LPC residual and not the original signal to model the long-term correlation in the speech signal, as long as the effects of the formants are taken into account during the determination of the long-term delay (pitch) in our model. Indeed, in Atal's original formulations in APC [10] (and in other APC-related schemes), the pitch predictor was applied before the LPC. The order in which they are combined is not too critical if the combination is carefully optimized, e.g. block edge effects must be carefully compensated to avoid 'clicking' type distortions. It is worth noting that the prediction gain of the combined system will always be less than the sum of the gains in systems employing the pitch and LPC filters in isolation. This is because in reality the vocal tract and excitation are not completely separable as assumed in our model, but are interconnected. The pitch filter can be interpreted as

$$P(z) = \frac{1}{1 - \sum_{j=-I}^{I} b_j z^{-(j+T)}} \tag{4.53}$$

where T is the 'pitch period', and b_j are the 'pitch gain' coefficients which reflect the amount of correlation between the distant samples. Referring

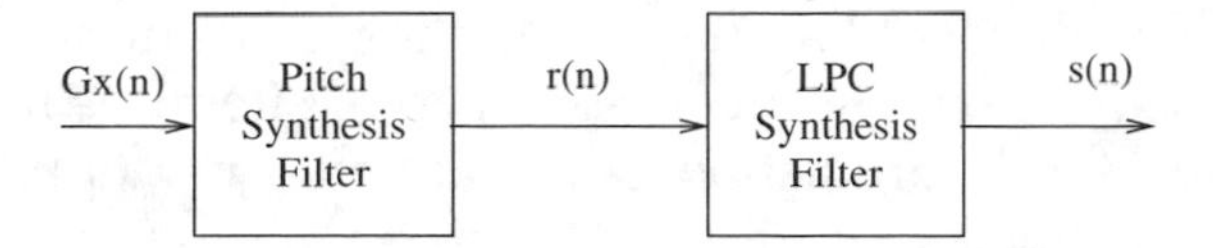

Figure 4.12 A typical pitch-LPC formulation model

to Figure 4.12, the combined analysis model can be represented by a time domain difference equation,

$$s(n) = Gx(n) + \sum_{j=-I}^{I} b_j r(n - T - j) + \sum_{j=1}^{p} a_j s(n - j) \tag{4.54}$$

where $r(n)$ is the past excitation (LPC residual) signal. Following a similar procedure to that of LPC analysis, our goal is to determine estimates $(\beta_j, \tau, \alpha_j)$ of the model parameters (b_j, T, a_j). Then, the prediction error is given by $(e(n) = Gx(n))$,

$$e(n) = s(n) - \sum_{j=-I}^{I} \beta_j r(n - \tau - j) - \sum_{j=1}^{p} \alpha_j s(n - j) \tag{4.55}$$

The mean squared error solution to equation (4.55) is not as straightforward as for the LPC analysis due to the presence of the delay factor τ. In order to overcome this hurdle two suboptimal approaches can be taken:

- **One-Shot Optimization:** If one assumes that the pitch spectrum information of the residual $r(n)$ is close to the pitch spectrum information of the input speech $s(n)$, then we can solve for α_j as before and use the residual from the LPC inverse filter to determine (β_j, τ). Thus during the first iteration, the LPC coefficients are estimated to minimize the intermediate residual energy. The pitch filter coefficients are then found using this intermediate residual signal. This procedure can be considered to be near optimal provided the long-term lag, τ, is greater than the analysis frame size, i.e. $\tau > N$.
- **Iterative Sequential Approach:** An analysis similar to the One-Shot method described above is first performed. During subsequent iterations, the LPC is re-optimized with the previously-determined pitch filter coefficients [11]. Also, the pitch filter parameters are recalculated based on the newly-formed intermediate residual. This iteration process can be continued until a certain threshold is reached or for a fixed number of iterations.

For practical reasons, the one-shot method is usually preferred as it only requires one iteration. In the iterative sequential method the main difficulty is in setting a suitable threshold for the termination of the iteration run. Overall, it is substantially more complicated. However, the iterative method has been reported to give a better prediction gain and better perceptual performance [5]. This is usually achieved with a shifting of the LPC prediction gain to the pitch prediction gain. Here, only the one-shot method is considered as follows:

By removing the LPC effect in equation (4.55), we obtain,

$$e(n) = r(n) - \sum_{j=-I}^{I} \beta_j r(n - \tau - j) \tag{4.56}$$

The estimates can now be determined by minimizing the mean squared error, i.e.

$$E = E\{e^2(n)\} = E\left\{\left[r(n) - \sum_{j=-I}^{I} \beta_j r(n - \tau - j)\right]^2\right\} \tag{4.57}$$

Replacing the expectation with finite summations, we get

$$E = \sum_m e_n^2(m) = \sum_m \left[r_n(m) - \sum_{j=-I}^{I} \beta_j r_n(m - \tau - j)\right]^2 \tag{4.58}$$

By setting $\partial E/\partial \beta_j$ to zero, we obtain

$$\sum_{j=-I}^{I} \beta_j V(i,j) = R(\tau + i) \quad -I \le i \le I \tag{4.59}$$

which can be written in matrix form as,

$$\begin{bmatrix} V(-I,-I) & \cdots & V(-I,I) \\ \vdots & \vdots & \vdots \\ V(I,-I) & \cdots & V(I,I) \end{bmatrix} \begin{bmatrix} \beta_{-I} \\ \vdots \\ \beta_I \end{bmatrix} = \begin{bmatrix} R(\tau - I) \\ \vdots \\ R(\tau + I) \end{bmatrix}$$

where,

$$R(\tau + i) = \sum_{m=0}^{N-1} r(m - \tau - i) r(m) \tag{4.60}$$

$$V(i,j) = \sum_{m=0}^{N-1} r(m - \tau - i) r(m - \tau - j), \ -I \le i \le I, \ -I \le j \le I \tag{4.61}$$

The β_j coefficients can now be solved by inverting $V(i, j)$, e.g. using Cholesky decomposition. In the above formulation, a 'fix-up' may be used to ensure that the filter so formed is stable, e.g. by adding a small noise source into the formulation, the matrix inversion to obtain $[V(i, j)]^{-1}$ can be made more reliably. However, a stable pitch filter is not a pre-condition for the pitch analysis as rapid transitions are sometimes desired.

In the above formulation it is assumed that the pitch lag, τ, has already been found and that $\beta_j = \beta_{j,\tau}$. In order to determine τ, various pitch measurement algorithms can be used. These include the Autocorrelation [12], average magnitude difference function (AMDF) [13], Cepstrum [14] and Maximum Likelihood [15]. These methods exhibit different characteristics especially with a noisy input signal.

As the preceding analysis to determine β_j has shown, pitch analysis is performed on a block containing N samples. However, the size of our window in which the block is taken is required to be considerably longer than the analysis frame length, N. This is because our pitch value, τ, can vary between a minimum, τ_{min}, of around 16 samples to a maximum, τ_{max}, of around 160 samples. Therefore, our ideal analysis window is significantly greater in length ($N + \tau_{max} \geq 200$ samples) such that it contains more than one complete pitch period.

For simplicity, consider a 1-tap pitch filter, i.e. ($I = 0$),

$$P_1(z) = \frac{1}{1 - \beta z^{-\tau}} \tag{4.62}$$

Thus,

$$\beta = \frac{R(\tau)}{V(0,0)} \tag{4.63}$$

$$= \frac{\sum_{m=0}^{N-1} r(m)r(m-\tau)}{\sum_{m=0}^{N-1} r^2(m-\tau)} \qquad \tau_{min} \leq \tau \leq \tau_{max} \tag{4.64}$$

Substituting this into equation (4.58),

$$E = \sum_{m=0}^{N-1} r^2(m) - \frac{\left[\sum_{m=0}^{N-1} r(m)r(m-\tau)\right]^2}{\sum_{m=0}^{N-1} r^2(m-\tau)} \tag{4.65}$$

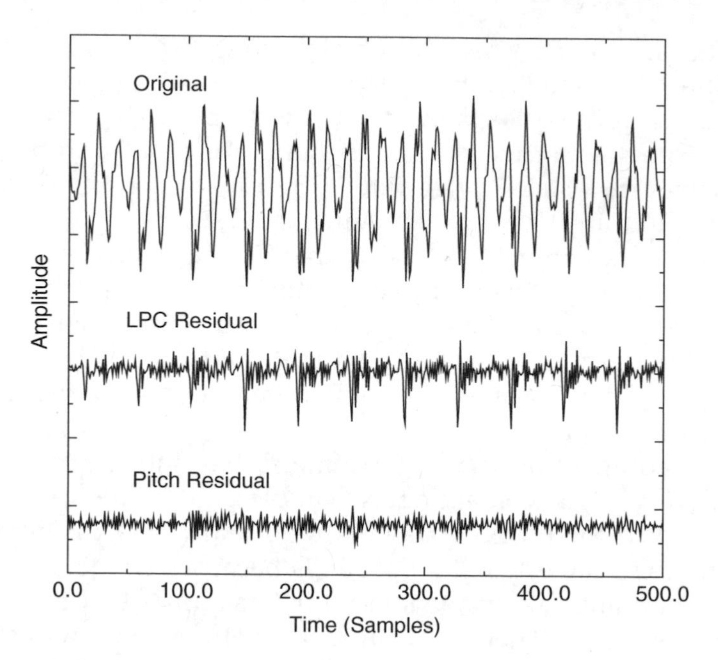

Figure 4.13 Time domain plots of original, LPC and pitch residuals

In order to determine the optimum τ, values of the lags are tested between τ_{min} and τ_{max}, and the lag which minimizes the error E is the optimal value. Having found τ, the gain β can be found. A plot of the LPC residual and the signal after pitch inverse filtering is shown in Figure 4.13. It is clear that the pitch residual (secondary excitation) no longer possesses the sharp pulse-like characteristics of the residual, i.e. it looks much whiter than the LPC residual. Similar formulations can also be given for multiple-tap pitch filters.

A typical plot of τ and β for a block of voiced, unvoiced and transitional speech is shown in Figures 4.14 and 4.15. As can be observed, during voiced regions (refer to the steady regions in Figure 4.14), β stays close to unity, whereas during transitional regions β fluctuates significantly.

As well as a single-tap filter, the three-tap pitch filter given by equation (4.66) is commonly used. Here $I = 1$ which forms the pitch prediction based on three past samples at $\tau - 1, \tau, \tau + 1$.

$$P_3(z) = \frac{1}{1 - \sum_{j=-1}^{1} \beta_j z^{-(j+\tau)}} \tag{4.66}$$

A multiple-tap pitch filter tends to provide better performance than the single-tap, but with increased complexity and larger capacity requirement for the extra two filter taps β_{-1} and β_1.

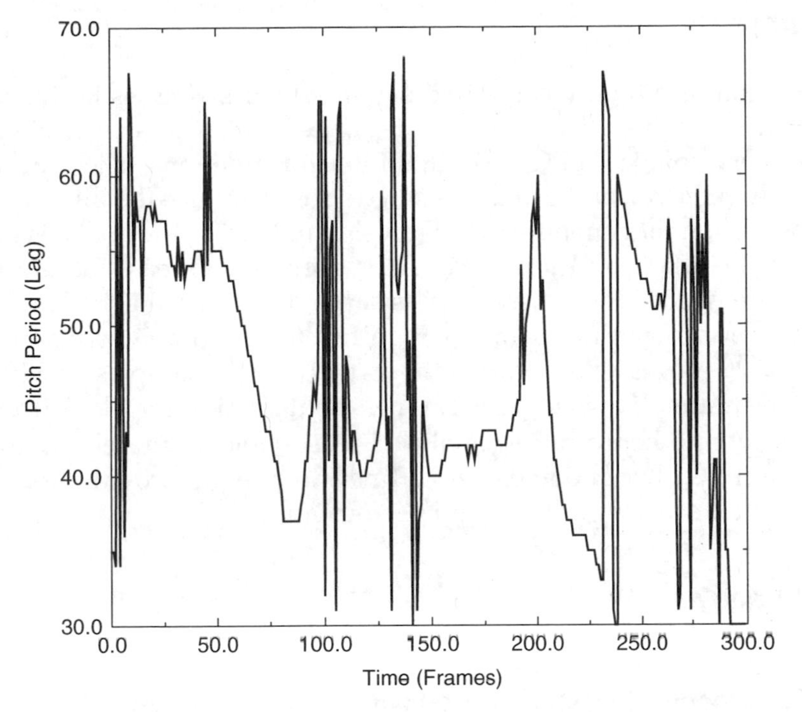

Figure 4.14 Pitch lag variation with time

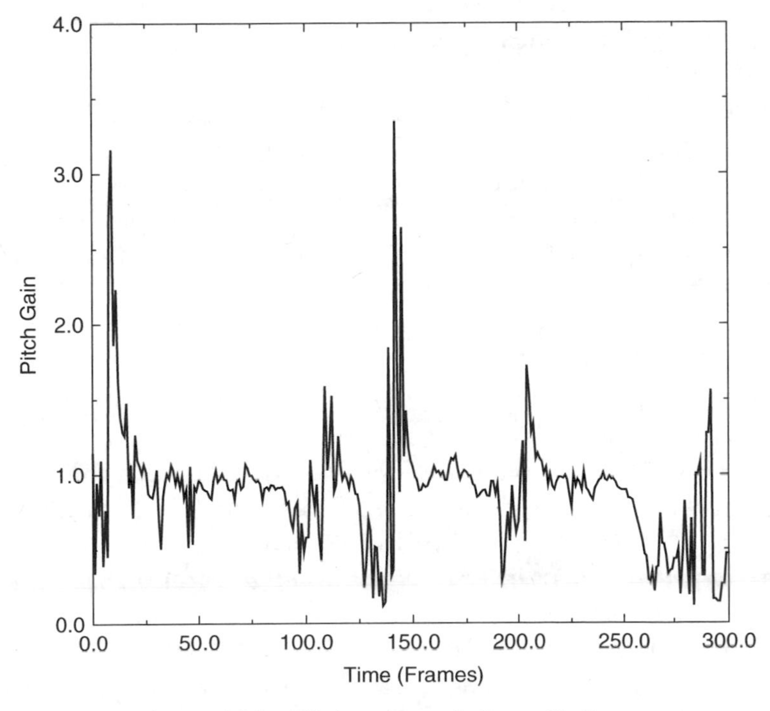

Figure 4.15 Pitch gain variation with time

4.5 Summary

Speech signal is a highly-correlated signal which possesses both short- and long-term similarities. These similarities or redundancies can easily be modelled by very compact LPC and pitch filter formulations. The redundancies are usually removed at the analysis stage so as to reduce the bit rate required for transmitting the remaining residual signal. During the analysis of speech to obtain the short- and long-term filter coefficients, reasonable lengths of samples are needed, which introduces some delay into the analysis process. A typical block length of samples required for good analysis performance is around 20–30 ms which corresponds to 160–240 samples at 8 kHz sampling. The assumption is that the samples contained in the block do not vary significantly and hence can be analysed reasonable accurately. A 10th-order short-term LPC filter updated every 20 ms and a single-order long-term pitch filter updated every 5 ms give good performance.

Bibliography

[1] L. Rabiner and R. Schafer (1978) *Digital Processing of Speech Signals*. Englewood Cliffs, NJ: Prentice-Hall

[2] B. Atal and M. Schroeder (1970) 'Adaptive predictive coding of speech signals', in *Bell Sys. Technical Journal*, pp. 1973–87. October 1970.

[3] J. Makhoul (1975) 'Linear prediction: A tutorial review', in *Proc. of IEEE*, 63:561–80.

[4] C. McGonegal, L. R. Rabiner and A. E. Rosenberg (1977) 'A subjective evaluation of pitch detection methods using LPC synthesised speech', in *IEEE Trans. on Acoust., Speech and Signal Processing*, 25(3):221–9.

[5] R. P. Ramachandran and P. Kabal (1989) 'Pitch prediction filters in speech coding', in *IEEE Trans. On Acoust., Speech and Signal Processing*, 37:467–78.

[6] M. Srinath and P. Rajasekaran (1979) *An Introduction to Statistical Signal Processing with Applications*. John Wiley & Sons Ltd

[7] S. Saito and K. Nakata (1985) *Fundamentals of Speech Signal Processing*, Chapter 9. Academic Press

[8] J. Makhoul (1977) 'Stable and efficient lattice methods for linear prediction', in *IEEE Trans. on Acoust., Speech and Signal Processing*, 25:423–8.

[9] J. H. Chen (1990) 'High quality 16 kbit/s speech coding with a one-way delay less than 2 ms', in *Proc. of Int. Conf. on Acoust., Speech and Signal Processing*, pp. 453–6.

[10] M. Schroeder and B. Atal (1979) 'Predictive coding of speech signals and subjective error criteria', in *IEEE Trans. on Acoust., Speech and Signal Processing*, 27:247–54.

[11] P. Kabal and R. P. Ramachandran (1988) 'Joint solutions for formant and pitch predictors in speech processing', in *Proc. of Int. Conf. on Acoust., Speech and Signal Processing*, pp. 315–18.

[12] L. Rabiner (1977) 'On the use of autocorrelation analysis for pitch detection', in *IEEE Trans. on Acoust., Speech and Signal Processing*, 25(1):24–33.

[13] E. Chilton (1990) 'Factors affecting the quality of linear predictive coding of speech at low bit-rates', Ph.D. thesis, CCSR, University of Surrey, UK.

[14] A. M. Noll (1967) 'Cepstrum pitch determination', in *Journal of the Acoustic Soc. of America*, 41:293–309.

[15] J. Wise, J. R. Caprio, and T. W. Parks (1976) 'Maximum likelihood pitch estimation', in *IEEE Trans. on Acoust., Speech and Signal Processing*, 24:418–23.

5

Efficient LPC Quantization Methods

5.1 Introduction

Linear predictive coding is a very powerful analysis technique and is used in many speech processing systems. In speech coding and synthesis systems, the analysis techniques for obtaining the LP coefficients (LPC), e.g. autocorrelation, covariance, lattice, and the quantization of the LPC are very important aspects of LPC analysis as minimization of coding capacity is the ultimate aim in these applications. The main objective of the quantization procedure is to code the LPC with as few bits as possible without introducing audible spectral distortion. Whilst perfect reconstruction is not possible, subjective transparency is achievable. Quantization of the LPC is usually performed by transforming the LPC to other forms which enables predictive coding and allows an easy filter stability check. The most popular LPC transformation is the use of Line Spectrum Pairs (LSP), related to the Line Spectral Frequency (LSF) representation of the LPC [1, 2]. In this chapter, the LSF representation of the LPC will be described, followed by various LPC quantization schemes using LSF transformation.

5.2 Alternative Representation of LPC

As was shown in Chapter 4, the LPC filter is given by

$$H(z) = \frac{1}{1 + \sum_{i=1}^{p} \alpha_i z^{-i}} \tag{5.1}$$

where p is the order of LPC filter.

Digital Speech. A. Kondoz
 ISBN 0-470-87007-9 (HB)

The α_i coefficients are the direct form of LPC. The filter $H(z)$ is stable if it is minimum phase, i.e. all the roots of the equation (5.1) are within the unit circle. If α_i were quantized directly, the stability of the filter $H(z)$ is not easily guaranteed as the roots of equation (5.1) are not usually computed to check for stability. Thus a more useful parameter, the PARCOR (partial correlation) coefficients, k_i, are usually used for quantization. The distribution plots of PARCOR parameters for a 10th-order LPC filter are shown in Figure 5.1. The forward and backward transformation are given below [3].

LPC to PARCOR:

$$a_j^p = \alpha_j \quad 1 \le j \le p$$

$$\begin{aligned} &\text{For } i = p, p-1, \ldots, 1 \\ &a_j^{i-1} = (a_j^i + a_i^i a_{i-j}^i)/(1 - k_i^2), \qquad 1 \le j \le i-1 \\ &k_{i-1} = a_{i-1}^{i-1} \end{aligned} \tag{5.2}$$

PARCOR to LPC:

$$\begin{aligned} &\text{For } i = 1, 2, \ldots, p \\ &\quad a_i^i = k_i \\ &\quad a_j^i = a_j^{i-1} - k_i a_{i-j}^{i-1}, \qquad 1 \le j \le i-1 \end{aligned} \tag{5.3}$$

$$\alpha_j = a_j^p, \quad 1 \le j \le p$$

The LPC filter is stable if $|k_i| \le 1.0$. Although k_i can easily be checked for stability, they are not suitable for quantization because they possess a nonflat spectral sensitivity, i.e. values of k_i near unity require more quantization accuracy than those away from unity. Thus, nonlinear functions of k_i are required, with the Log-Area Ratio (LAR) and inverse sine (IS) functions being the most widely used [4]. For LAR and IS, the forward and backward transformation are given below:

PARCOR to LAR:

$$g_i = \log\left(\frac{1-k_i}{1+k_i}\right), \qquad 1 \le i \le p \tag{5.4}$$

LAR to PARCOR:

$$k_i = \left(\frac{1-10^{g_i}}{1+10^{g_i}}\right), \qquad 1 \le i \le p \tag{5.5}$$

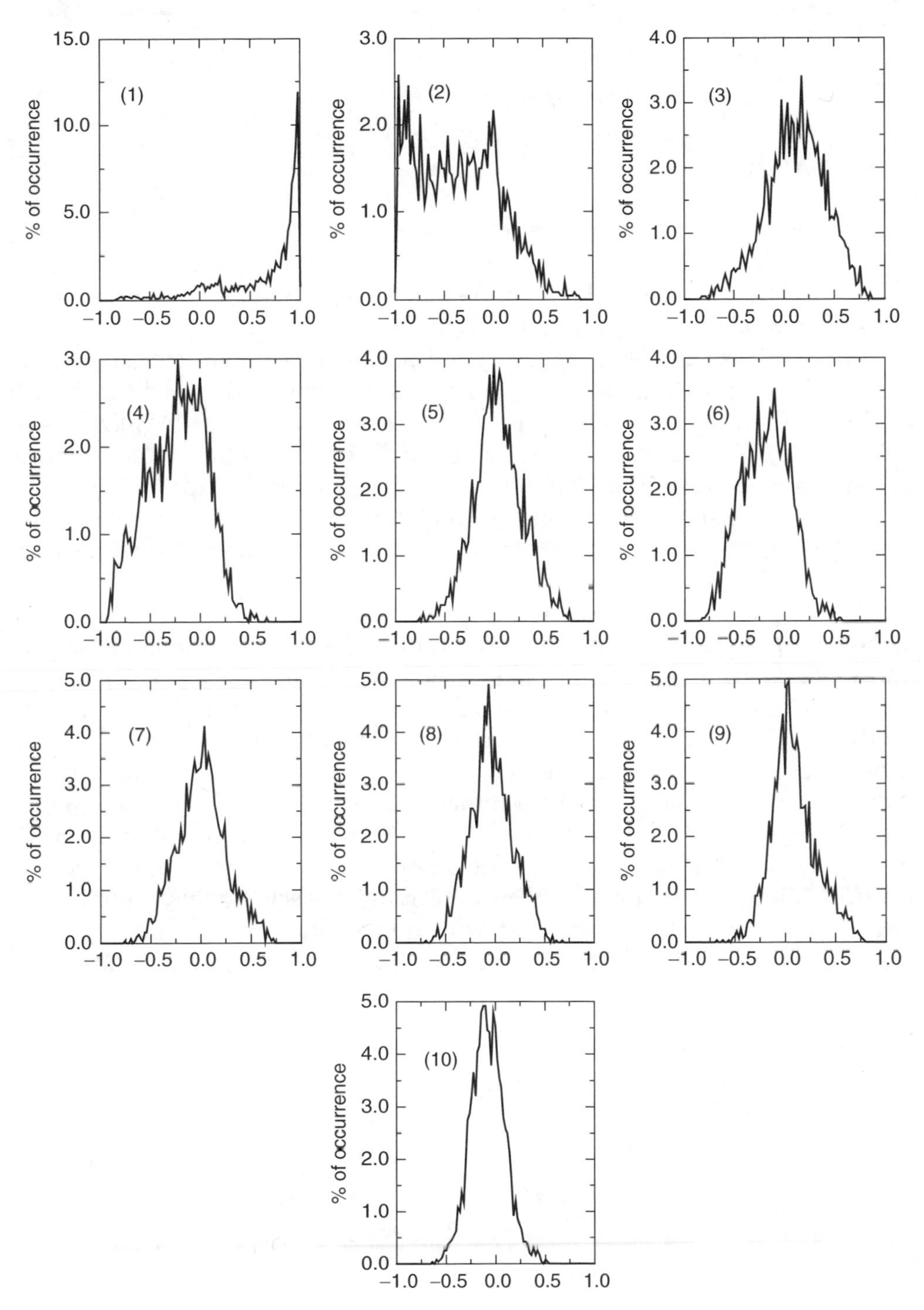

Figure 5.1 The distribution plots of PARCOR LPC

PARCOR to IS:

$$s_i = \sin^{-1}(k_i), \quad 1 \le i \le p \tag{5.6}$$

IS to PARCOR:

$$k_i = \sin(s_i), \quad 1 \le i \le p \tag{5.7}$$

The distribution plots of LAR and IS parameters for a 10th order LPC filter are shown in Figures 5.2 and 5.3 respectively.

Although it is possible to design good performance quantizers using the LAR and IS representations, the frame-to-frame correlation of LPC (which evidently exists for slowly-varying parts of speech) is not highlighted in either LAR or IS representations, i.e. it is difficult to predict frame-to-frame parameter values. Thus, not all the redundancies are fully exploitable.

In view of the shortcomings of LAR and IS representation, the line spectral pairs (LSP) or frequencies (LSF) representations of LPC have been investigated [2]. The concept of LSF was introduced by Itakura, but it remained almost dormant until its usefulness was re-examined in the latest speech coding standards. LSFs encode speech spectral information in the frequency domain and have been found to be capable of improving the coding efficiency by more than other transformation techniques, especially when incorporated into predictive quantization schemes. For use in conventional scalar quantization, it has been shown by Cox [4] and others that LSF is not significantly better than LAR or IS, but it does have other properties which are desirable, as will be discussed in later sections. The fact that LSF representation is in the frequency domain means that quantization can easily incorporate spectral features known to be important in perceiving speech signals. In addition, LSFs lend themselves to frame-to-frame interpolation with smooth spectral changes because of their intimate relationship with format frequencies.

5.3 LPC to LSF Transformation

An all-pole digital filter for speech synthesis, $H(z)$, can be derived from linear predictive analysis and is given by

$$H(z) = 1/A_p(z) \tag{5.8}$$

where,

$$A_p(z) = 1 + \sum_{k=1}^{p} \alpha_k z^{-k} \tag{5.9}$$

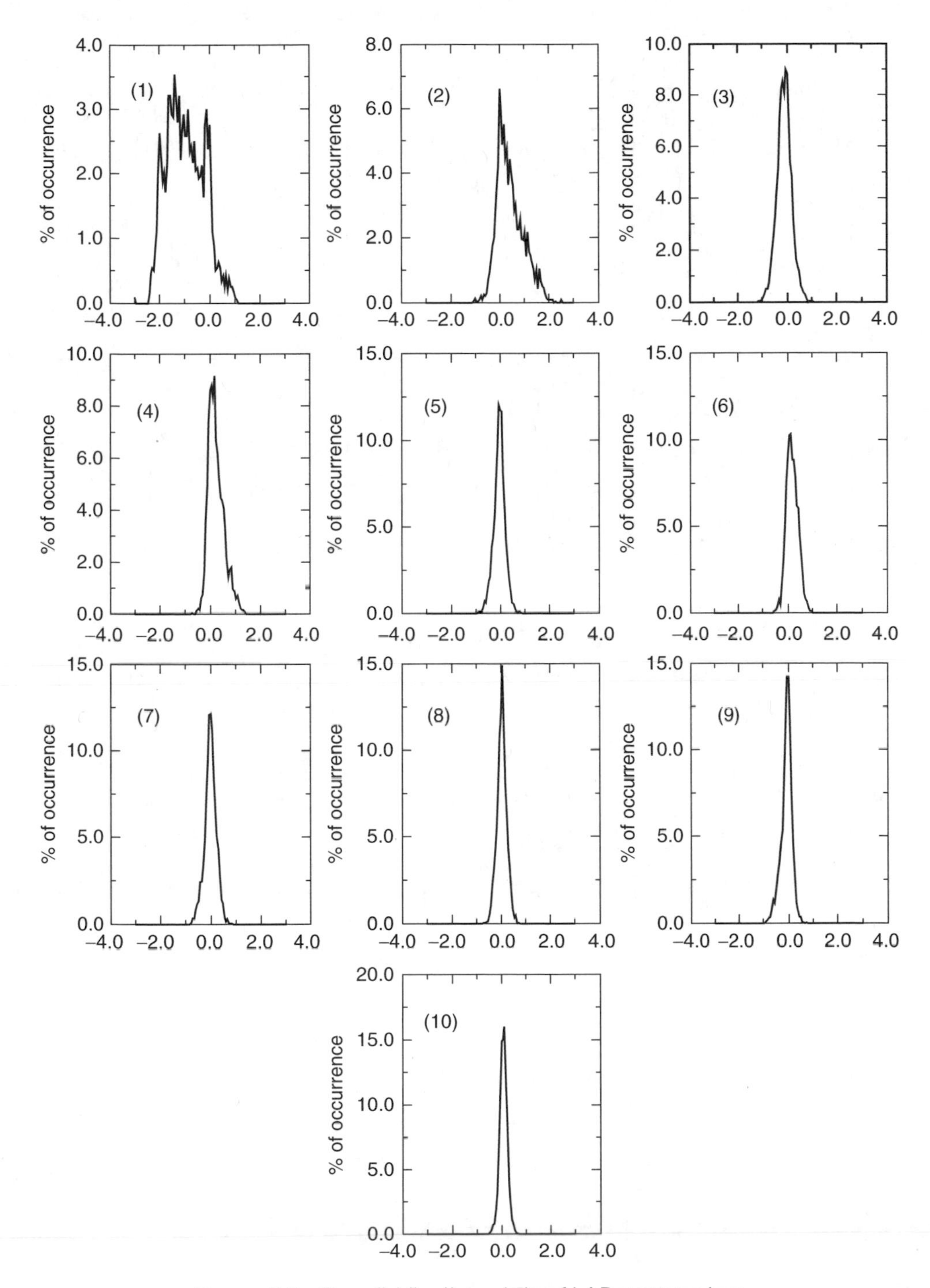

Figure 5.2 The distribution plots of LAR parameters

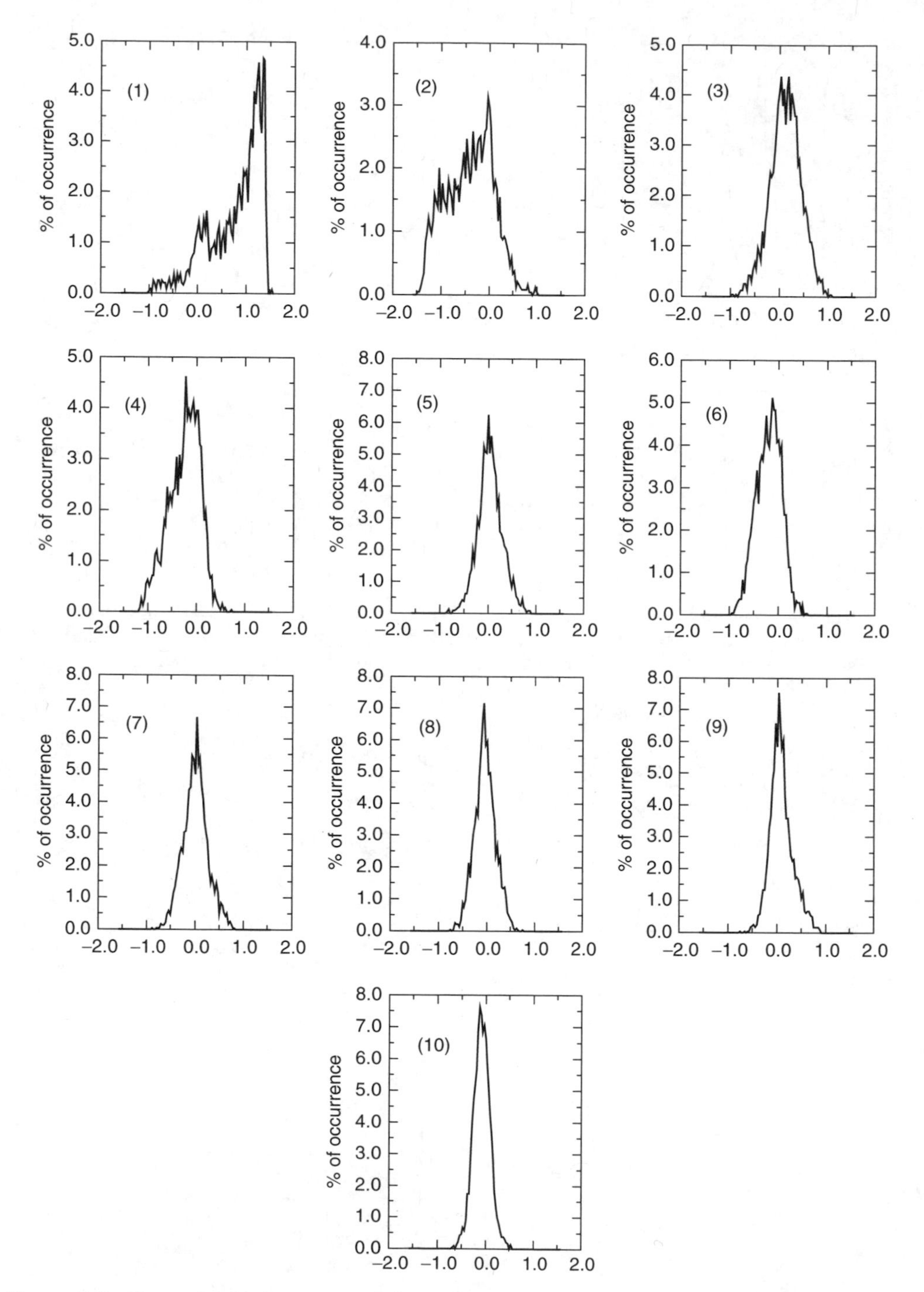

Figure 5.3 The distribution plots of inverse sine parameters (horizontal axis is in radians)

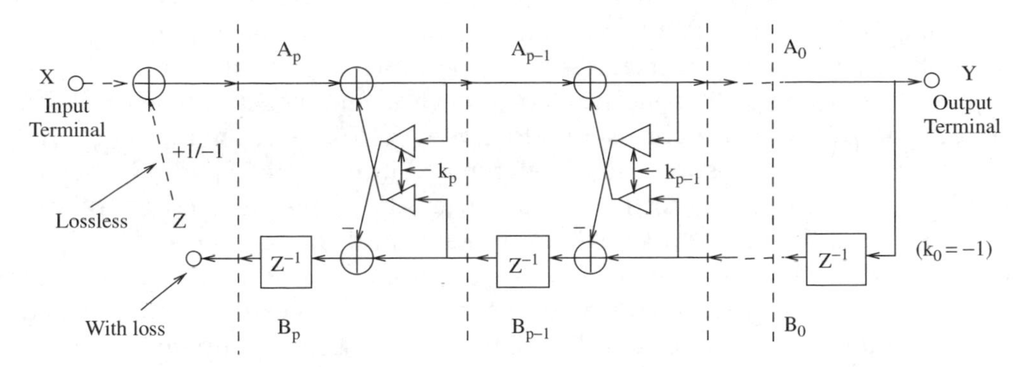

Figure 5.4 PARCOR structure of LPC synthesis

The PARCOR representation is an equivalent version and its digital form is as shown in Figure 5.4, where,

$$A_{p-1}(z) = A_p(z) + k_p B_{p-1}(z) \tag{5.10}$$

$$B_p(z) = z^{-1}[B_{p-1}(z) - k_p A_{p-1}(z)] \tag{5.11}$$

where $A_0(z) = 1$ and $B_0(z) = z^{-1}$, and

$$B_p(z) = z^{-(p+1)} A_p(z^{-1}) \tag{5.12}$$

The PARCOR representation as shown in Figure 5.4 is stable for $|k_i| < 1$ for all i. In Figure 5.4, the transfer function, TF, from X to Y is $H_p(z)$, and from Y to Z is $B_p(z)$, therefore the TF from X to Z is given by equation (5.13) where $R_p(z)$ is the ratio filter,

$$R_p = B_p(z)/A_p(z) \tag{5.13}$$

The PARCOR synthesis process can be viewed as sound wave propagation through a lossless acoustic tube, consisting of p sections of equal length but nonuniform cross sections. The acoustic tube is open at the terminal corresponding to the lips and each section is numbered from the lips. Mismatching between the adjacent sections p and $(p+1)$ causes wave propagation reflection. The reflection coefficients are equal to the pth PARCOR coefficient k_p. Section $p+1$, which corresponds to the glottis, is terminated by a matched impedance. The excitation signal applied to the glottis drives the acoustic tube.

In PARCOR analysis, the boundary condition at the glottis is impedance-matched. Now consider a pair of artificial boundary conditions where the acoustic tube is completely closed or open at the glottis. These conditions correspond to $k_{p+1} = 1$ and $k_{p+1} = -1$, a pair of extreme values for the artificially-extended PARCOR coefficients which correspond to perfectly lossless tubes. The value Q of each resonance becomes infinite and the spectrum of distributed energy is concentrated in several line spectra. The feedback conditions for $k_{p+1} = -1$ correspond to a perfect closure at the input (glottis) and for $k_{p+1} = 1$ correspond to an opening to infinite free space. To derive the line spectra or line spectrum frequencies (LSF), we proceed as follows (it is assumed that the PARCOR filter is stable and the order is even). $A_p(z)$ may be decomposed to a set of two transfer functions, one having an even symmetry and the other having an odd symmetry. This can be accomplished by taking a difference and sum between $A_p(z)$ and its conjugate functions. Hence the transfer functions with $k_{p+1} = \pm 1$ are denoted by $P_{p+1}(z)$ and $Q_{p+1}(z)$.

$$\begin{aligned} &\text{For } k_{p+1} = 1, \quad P_{p+1}(z) = A_p(z) - B_p(z) \quad \text{(Difference filter)} \\ &\text{For } k_{p+1} = -1, \quad Q_{p+1}(z) = A_p(z) + B_p(z) \quad \text{(Sum filter)} \end{aligned} \tag{5.14}$$

$$\Rightarrow A_p(z) = \frac{1}{2}[P_{p+1}(z) + Q_{p+1}(z)] \tag{5.15}$$

Substituting equation (5.12) into (5.14),

$$\begin{aligned} P_{p+1}(z) &= A_p(z) - z^{-(p+1)} A_p(z^{-1}) \\ &= 1 + (\alpha_1 - \alpha_p) z^{-1} + \ldots + (\alpha_p - \alpha_1) z^{-p} - z^{-(p+1)} \\ &= z^{-(p+1)} \prod_{i=0}^{p+1} (z + a_i) \end{aligned} \tag{5.16}$$

where a_i is generally complex. Similarly,

$$Q_{p+1}(z) = z^{-(p+1)} \prod_{i=0}^{p+1} (z + b_i) \tag{5.17}$$

As we know that two roots exist ($k_{p+1} = \pm 1$), the order of $P_{p+1}(z)$ and $Q_{p+1}(z)$ can be reduced, i.e.

$$\begin{aligned} P'(z) &= \frac{P_{p+1}(z)}{(1 - z)} \\ &= A_0 z^p + A_1 z^{(p-1)} + \ldots + A_p \end{aligned} \tag{5.18}$$

and,

$$Q'(z) = \frac{Q_{p+1}(z)}{(1+z)} \tag{5.19}$$
$$= B_0 z^p + B_1 z^{(p-1)} + \ldots + B_p$$

where,

$$A_0 = 1 \tag{5.20}$$
$$B_0 = 1 \tag{5.21}$$
$$A_k = (\alpha_k - \alpha_{p+1-k}) + A_{k-1} \tag{5.22}$$
$$B_k = (\alpha_k + \alpha_{p+1-k}) - B_{k-1} \tag{5.23}$$
$$\text{for } k = 1, \ldots, p$$

The LSFs are the angular positions of the roots of $P'(z)$ and $Q'(z)$ with $0 \leq \omega_i \leq \pi$. The roots occur in complex conjugate pairs and have the following properties:

1. All roots of $P'(z)$ and $Q'(z)$ lie on the unit circle.
2. The roots of $Q'(z)$ and $P'(z)$ alternate with each other on the unit circle, i.e. the following is always satisfied, $0 \leq \omega_{q,0} < \omega_{p,0} < \omega_{q,1} < \omega_{p,1} \ldots, \leq \pi$.

5.3.1 Complex Root Method

The roots of equation (5.18) can be solved using complex arithmetic. This will give complex conjugate roots on the unit circle and the frequencies are then given by the inverse tangent of the roots. This method is obviously very complex as it involves solving two polynomials of pth order using complex arithmetic. Also, as it uses an iteration procedure for determining the roots, the time required for this method is not deterministic which is undesirable for real-time implementations.

5.3.2 Real Root Method

As the coefficients of $P'(z)$ and $Q'(z)$ are symmetrical the order of equation (5.18) can be reduced to $p/2$.

$$P'(z) = A_0 z^p + A_1^{p-1} + \ldots + A_1 z^1 + A_0 \tag{5.24}$$
$$= z^{p/2}[A_0(z^{p/2} + z^{-p/2}) + A_1(z^{(p/2-1)} + z^{-(p/2-1)}) + \ldots + A_{p/2}]$$

Similarly,

$$Q'(z) = B_0 z^p + B_1^{p-1} + \ldots + B_1 z^1 + B_0 \tag{5.25}$$
$$= z^{p/2}[B_0(z^{p/2} + z^{-p/2}) + B_1(z^{(p/2-1)} + z^{-(p/2-1)}) + \ldots + B_{p/2}]$$

As all roots are on the unit circle, we can evaluate equation (5.24) on the unit circle only.

$$\text{Let } z = e^{j\omega} \text{ then } z^1 + z^{-1} = 2\cos(\omega) \tag{5.26}$$

$$P'(z) = 2e^{jp\omega/2}\left[A_0 \cos\left(\frac{p}{2}\omega\right) + A_1 \cos\left(\frac{p-2}{2}\omega\right) + \ldots + \frac{1}{2}A_{p/2}\right] \tag{5.27}$$

$$Q'(z) = 2e^{jp\omega/2}\left[B_0 \cos\left(\frac{p}{2}\omega\right) + B_1 \cos\left(\frac{p-2}{2}\omega\right) + \ldots + \frac{1}{2}B_{p/2}\right] \tag{5.28}$$

By making the substitution $x = \cos(\omega)$, equations (5.27) and (5.28) can be solved for x. For example, with $p = 10$, the following is obtained:

$$P'_{10}(x) = 16A_0x^5 + 8A_1x^4 + (4A_2 - 20A_0)x^3 + (2A_3 - 8A_1)x^2$$
$$+(5A_0 - 3A_2 + A_4)x + (A_1 - A_3 + 0.5A_5) \tag{5.29}$$

and similarly,

$$Q'_{10}(x) = 16B_0x^5 + 8B_1x^4 + (4B_2 - 20B_0)x^3 + (2B_3 - 8B_1)x^2$$
$$+(5B_0 - 3B_2 + B_4)x + (B_1 - B_3 + 0.5B_5) \tag{5.30}$$

The LSFs are then given by:

$$LSF(i) = \frac{\cos^{-1}(x_i)}{2\pi T}, \quad \text{for } 1 \le i \le p \tag{5.31}$$

The distribution plots of LSFs for a 10th order LPC filter are shown in Figure 5.5 and a typical LSF plot is shown in Figure 5.6, where the first half is active speech and the second half is silence. Notice that during silent regions the frequencies are evenly spread between 0 and $f_s/2$ where f_s is the sampling frequency. This method is obviously considerably simpler than the complex root method, but it still suffers from indeterministic computation time. However, a faster root search can be accomplished by noting that the change from one LSF vector to the next is not too drastic in most cases. Thus by using the previous values as the starting estimates of the roots, the number of iterations required per root is considerably reduced, e.g. typically from 5 to 10 iterations.

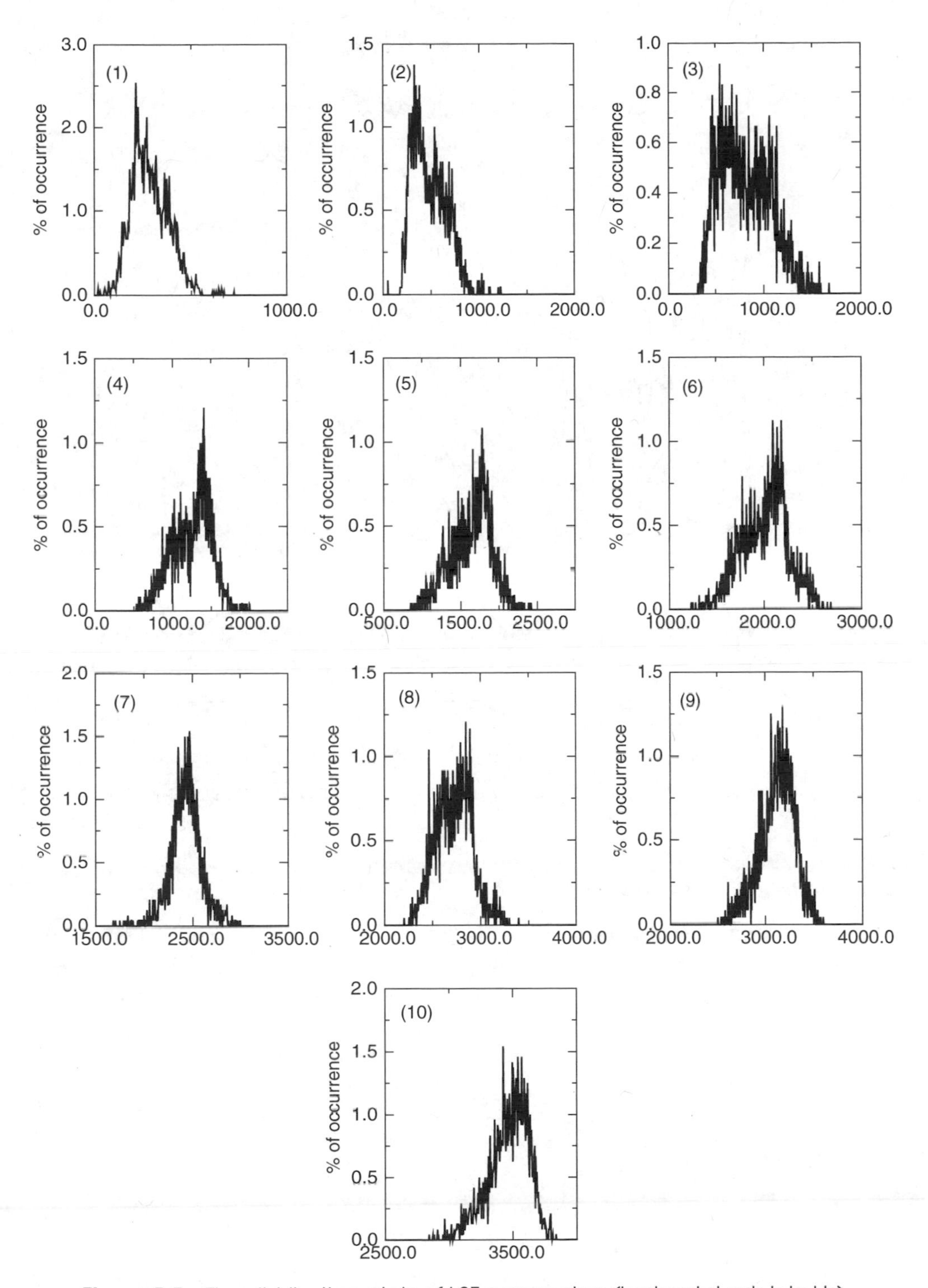

Figure 5.5 The distribution plots of LSF parameters (horizontal axis is in Hz)

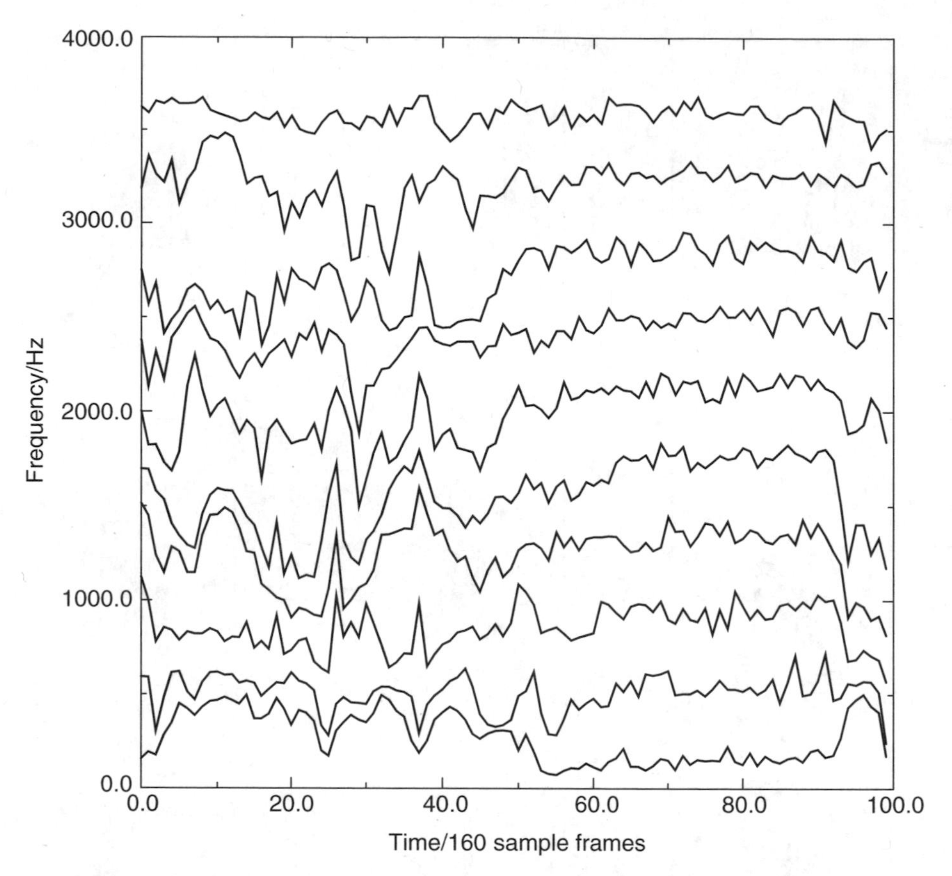

Figure 5.6 Typical LSF trajectories for voiced and unvoiced speech

5.3.3 Ratio Filter Method

The expression for the ratio filter is given by equation (5.32). The phase response, $\phi(kf_s)$, of the ratio filter is given by equation (5.34). The frequency corresponding to a multiple of $-\pi$ and -2π radians are the lower and upper line spectra of the LSF [5].

$$R_p(z) = \frac{z^{-(n+1)} A_p(z^{-1})}{A_p(z)} \tag{5.32}$$

where,

$$A_p(z) = 1 - \sum_{i=1}^{n} \beta_i z^{-i} \tag{5.33}$$

and $\beta_i = -\alpha_i$ where α_i are the LPC.

$$\phi(kf_s) = -(n+1)(2\pi Tkf_s) - 2\tan^{-1}\left\{\frac{\sum_{i=1}^{n}\beta_i \sin(2\pi iTkf_s)}{1-\sum_{i=1}^{n}\beta_i \cos(2\pi iTkf_s)}\right\} \tag{5.34}$$

where T is the sampling period, f_s is the frequency step, and $k = 1, 2, 3, \ldots, K_{max}$.

By performing a Discrete Fourier Transform (DFT) on the coefficient sequence, A_k and B_k, ω_i can be solved as the zero-valued frequencies of a power spectrum. A typical plot, showing the partial minima of the spectrum, is shown in Figure 5.7.

If the spectrum were to be obtained directly, it would involve an enormous number of computations. Fortunately, a number of computation reductions can be made. The aim is to find the partial minima of the response, thus

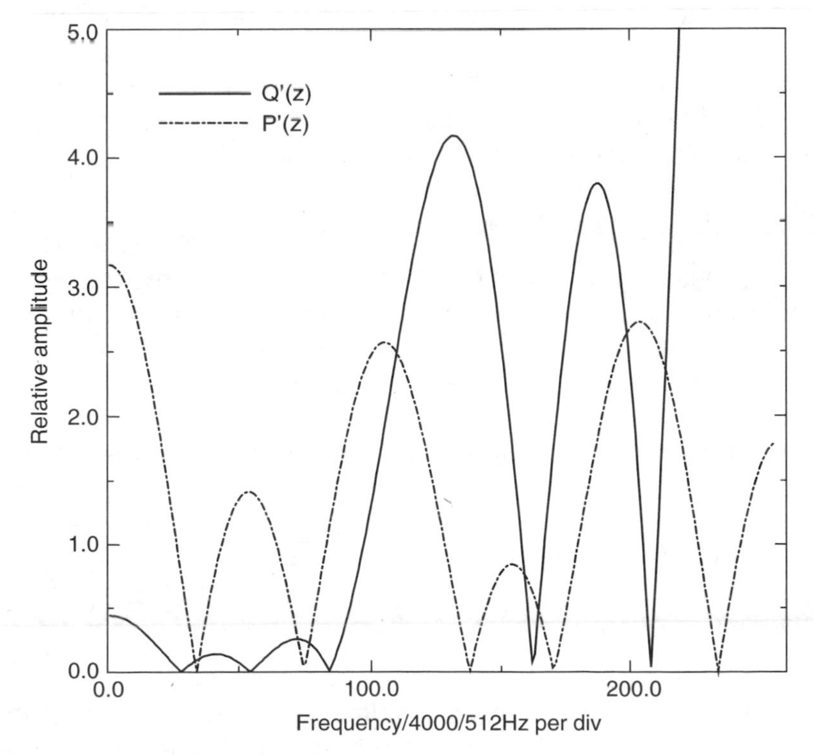

Figure 5.7 Zero frequency plot for one frame of the DFT–LSF method

the absolute values of the response are not critical; only the locations of the minima are vital. The spectrum is given by equation (5.35) where $\mathbf{P}$ is the spectrum, $\mathbf{W}$ is the $L \times L$ DFT kernel, and $\mathbf{S}$ is the input sequence. L is the size of the transform.

$$\begin{bmatrix} - \\ \mathbf{P} \\ - \end{bmatrix} = \begin{bmatrix} - & - & - \\ - & \mathbf{W} & - \\ - & - & - \end{bmatrix} \begin{bmatrix} - \\ \mathbf{S} \\ - \end{bmatrix} \tag{5.35}$$

As the input sequences A_k and B_k are real, we can move them from the start to the middle of $\mathbf{S}$ with zeros elsewhere. This will produce an even spectrum which means that only $f_s/2$ terms need to be computed. Also, the spectrum will be real, thus only the cosine-terms in the kernel require computing. Since the sequences A_k and B_k are even, only half of the values need to be computed, i.e. A_0 to $A_{p/2-1}$ and $1/2A_{p/2}$, and similarly for B_k. With these savings the number of multiply–adds is reduced to $p/2 + 1$ per spectrum point. The cosine terms are fixed for a particular transform size, therefore they can be pre-computed and stored in a lookup table.

Once the spectrum is found the partial minima need to be located and this involves computationally expensive comparisons. As the LSF are naturally ordered, i.e. the frequencies alternate between $Q(z)$ and $P(z)$, they can be located in an efficient manner. The first $Q(z)$ LSF starts at the origin, then the first $P(z)$ LSF starts from the previous $Q(z)$ LSF location. Once the first $P(z)$ LSF is found the second $Q(z)$ LSF is located, starting from the previous $P(z)$ location. This alternation is repeated until all LSFs are found. Thus in total only one pass of the frequency range is made instead of two.

5.3.4 Chebyshev Series Method

Another step-wise method which requires no prior storage or calculation of trigonometric functions is the Chebyshev Series Method [6]. By expanding equation (5.24) with the Chebyshev polynomial set, the mapping $x = \cos(\omega)$ maps the upper semicircle in the z-plane to the real interval $[+1, -1]$. Therefore, all the roots x_i lie between -1 and $+1$, with the root corresponding to the lowest frequency LSF being the one nearest to $+1$. Thus the basic task is similar to the DFT method, i.e. we isolate the roots of $P'(z)$ and $Q'(z)$ by searching incrementally for intervals in which the sign changes which is refined by successive bisections of the root interval.

5.3.5 Adaptive Sequential LMS Method

All of the previously described methods for deriving the LSF parameters required the intermediate step of calculating the LPC before proceeding to the computation of the LSF parameters. However, using a Least Mean

Squares adaptive method [7] the LSF parameters can be computed directly from the speech samples themselves. The LMS algorithm aims to minimize the mean-square value of the PARCOR lattice filter output, and thus flatten its frequency spectrum by a 'noisy steepest-descent' procedure which uses the squared value of a single output sample to approximate the mean-square value. Thus the algorithm begins the sequential estimation using evenly-distributed estimated LSFs and, as each sample of speech is processed, a new LSF vector estimate is obtained. Depending on the adaptation rate required, the algorithm converges to the correct value after around 100 samples of input.

The LMS method is very attractive because it requires no LPC analysis. However, as it is a 'learning' type algorithm, it is susceptible to 'out-lier' input samples, i.e. samples which are different in character to the majority of speech samples. The effect of these unusual inputs is to throw the algorithm off its convergence curve; if this occurs at the end of a frame there will be no time for correction before the final values are used.

5.4 LSF to LPC Transformation

There are two methods for the inverse transformation, neither of which is as computationally intensive as the forward transformation. The two methods are equivalent but the LPC synthesis method is perhaps more easily visualized.

5.4.1 Direct Expansion Method

In all of the LPC to LSF methods above the aim is to find the roots of equation (5.16), i.e. a_i and b_i. Having found these roots using any of the methods, the LPC, α_i, can be simply found by multiplying out the product terms of equation (5.16), i.e.

$$
\begin{aligned}
P_{p+1}(z) &= z^{-(p+1)}[P'(z)(1-z)] \qquad (5.36)\\
&= z^{-(p+1)}[(1-z)(z-r_0)(z-r_0^*)\ldots(z-r_{p/2})(z-r_{p/2}^*)]\\
&= z^{-(p+1)}[(1-z)(z^2-2u_0z+t_0)\ldots(z^2-2u_{p/2}z+t_{p/2})]\\
&= S_0+S_1z^{-1}+\ldots+S_pz^{-p}+S_{p+1}z^{-(p+1)} \qquad (5.37)
\end{aligned}
$$

Similarly,

$$
Q_{p+1}(z) = T_0+T_1z^{-1}+\ldots+T_pz^{-p}+T_{p+1}z^{-(p+1)} \qquad (5.38)
$$

where,

$$\begin{aligned} r_i &= u_i + jv_i \quad \text{and} \quad r_i^* = u_i - jv_i \\ \Rightarrow \quad r_i + r_i^* &= 2u_i \quad \text{and} \quad r_i \times r_i^* = u_i^2 + v_i^2 = t_i \end{aligned} \tag{5.39}$$

Equating the terms of equations (5.37) and (5.16),

$$S_0 = 1 \tag{5.40}$$

$$T_0 = 1 \tag{5.41}$$

$$S_{p+1} = -1 \tag{5.42}$$

$$T_{p+1} = 1 \tag{5.43}$$

$$\alpha_i = \frac{1}{2}(T_i + S_i) \tag{5.44}$$

$$\alpha_{p+1-i} = \frac{1}{2}(T_i - S_i) \tag{5.45}$$

$$\text{for } i = 1, \ldots, P/2$$

5.4.2 LPC Synthesis Filter Method

An LPC synthesis can be constructed directly using the LSF coefficients. The filter is derived from the following,

$$\begin{aligned} H(z) &= 1/A_p(z) = 1/[1 + (A_p(z) - 1)] \\ &= \frac{1}{1 + 1/2[(P_{p+1}(z) - 1) + (Q_{p+1}(z) - 1)]} \end{aligned} \tag{5.46}$$

i.e.

$$A_p(z) - 1 = 1/2[(P_{p+1}(z) - 1) + (Q_{p+1}(z) - 1)] \tag{5.47}$$

$$\begin{aligned} &= 1/2 \left\{ (1 - z) \prod_{i=1}^{p/2} (1 - 2\cos\omega_i z + z^2) - 1 \right. \\ &\left. + (1 + z) \prod_{i=1}^{p/2} (1 - 2\cos\theta_i z + z^2) - 1 \right\} \end{aligned} \tag{5.48}$$

Let $u_i = -2\cos\omega_i, v_i = -2\cos\theta_i$

where w_i and θ_i are the even and odd number LSFs given by $LSF(i)2\pi T$.

$$A_p(z) - 1 = 1/2\left\{\prod_{i=1}^{p/2}(1 + u_i z + z^2)\right. \tag{5.49}$$

$$\left. -z\prod_{i=1}^{p/2}(1 + u_i z + z^2) - 1\right\}$$

$$+1/2\left\{\prod_{i=1}^{p/2}(1 + v_i z + z^2)\right.$$

$$\left. -z\prod_{i=1}^{p/2}(1 + v_i z + z^2) - 1\right\} \tag{5.50}$$

$$= z/2\left\{(u_1 + z) - \prod_{j=1}^{p/2}(1 + u_j z + z^2)\right.$$

$$\left. + \sum_{i=1}^{p/2-1}(u_{i+1} + z)\prod_{j=1}^{i}(1 + u_j z + z^2)\right\}$$

$$+z/2\left\{(v_1 + z) - \prod_{j=1}^{p/2}(1 + v_j z + z^2)\right.$$

$$\left. + \sum_{i=1}^{p/2-1}(v_{i+1} + z)\prod_{j=1}^{i}(1 + v_j z + z^2)\right\} \tag{5.51}$$

An 8th order inverse filter is shown in Figure 5.8. The LPC are simply the impulse response of the filter.

5.5 Properties of LSFs

A very important LSF property, as mentioned earlier, is the natural ordering of its parameters. This ordering property was already used to good effect in speeding up the LPC to LSF transformation procedure. The ordering property indicates that the LSFs within a frame, and from frame to frame, are correlated. In order to illustrate the intra-frame correlation property of the

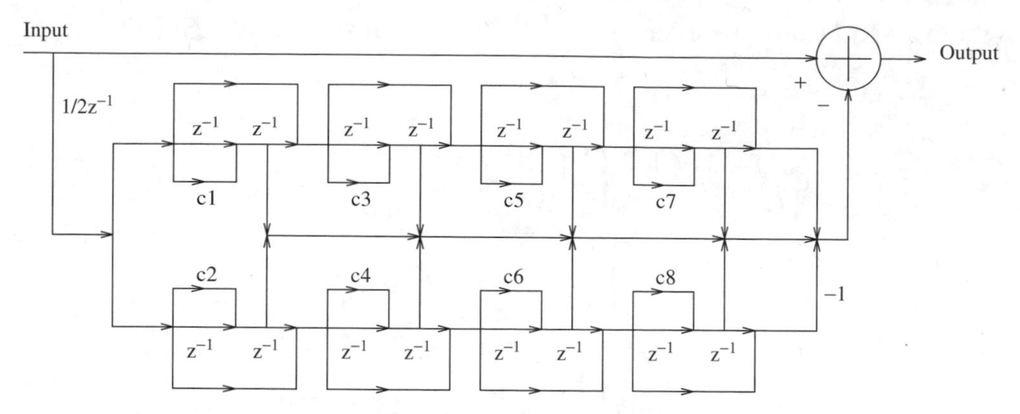

Figure 5.8 Practical scheme of LSF inverse filter ($c_i = -2\cos\omega_i$, for even i, and $c_i = -2\cos\theta_i$, for odd i)

Table 5.1 Experimental conditions for estimating $\mathbf{\Omega}$ and $\mathbf{\Psi}$

Sampling Frequency	8 kHz
Frame Update	10 ms
Window	20 ms Hamming
Analysis order	10
Number of Frames	6000

LSF vector, ω, Table 5.2 presents the matrix $\mathbf{\Omega} = \{\phi_{i,j}\}$ where,

$$\phi_{i,j} = \omega_{n,i} \times \omega_{n,j}, \quad i = 1, 2, \ldots, p, \; j = 1, 2, \ldots, p \tag{5.52}$$

for the experimental conditions according to Table 5.1. The relatively high correlation between neighbouring LSFs is clear. Similarly, to illustrate the inter-frame correlation of the LSF parameters, Table 5.3 presents the matrix $\mathbf{\Psi} = \{\phi_{i,k}\}$ where,

$$\phi_{i,k} = \omega_{n,i} \times \omega_{n-k,i}, \quad i = 1, 2, \ldots, p, \; k = 1, 2, \ldots, p \tag{5.53}$$

From Tables 5.2 and 5.3, it is clear that there is a strong correlation between the LSFs of adjacent frames as well as neighbouring parameters in the same frame. Therefore, any compression algorithm that effectively makes use of these correlations can result in improved performance over those that do not incorporate this correlation property.

Table 5.2 Intra-frame correlation coefficients $\mathbf{\Omega}$

	j									
i	1	2	3	4	5	6	7	8	9	10
1	1.00	0.65	−0.30	−0.35	−0.41	−0.49	−0.39	−0.40	−0.36	−0.20
2	0.65	1.00	0.28	0.11	−0.07	−0.13	−0.07	−0.05	−0.06	−0.07
3	−0.30	0.28	1.00	0.72	0.50	0.53	0.46	0.54	0.39	0.28
4	−0.35	0.11	0.72	1.00	0.72	0.62	0.46	0.42	0.45	0.21
5	−0.41	−0.07	0.50	0.72	1.00	0.79	0.52	0.47	0.34	0.26
6	−0.49	−0.13	0.53	0.62	0.79	1.00	0.71	0.61	0.49	0.28
7	−0.39	−0.07	0.46	0.46	0.52	0.71	1.00	0.73	0.58	0.41
8	−0.40	−0.05	0.54	0.42	0.47	0.61	0.73	1.00	0.58	0.46
9	−0.36	−0.06	0.39	0.45	0.34	0.49	0.58	0.58	1.00	0.41
10	−0.20	−0.07	0.28	0.21	0.26	0.28	0.41	0.46	0.41	1.00

Table 5.3 Inter-frame correlation coefficients $\mathbf{\Psi}$

	k									
i	1	2	3	4	5	6	7	8	9	10
1	0.93	0.84	0.76	0.68	0.61	0.55	0.50	0.45	0.41	0.36
2	0.89	0.75	0.63	0.54	0.46	0.38	0.32	0.27	0.22	0.18
3	0.92	0.80	0.70	0.60	0.51	0.43	0.36	0.30	0.24	0.20
4	0.92	0.82	0.73	0.64	0.56	0.49	0.43	0.37	0.32	0.27
5	0.95	0.88	0.81	0.74	0.67	0.61	0.54	0.48	0.43	0.37
6	0.94	0.85	0.77	0.69	0.62	0.56	0.49	0.44	0.38	0.33
7	0.93	0.83	0.75	0.66	0.58	0.50	0.43	0.37	0.31	0.26
8	0.91	0.81	0.72	0.64	0.56	0.49	0.43	0.37	0.32	0.28
9	0.87	0.73	0.64	0.55	0.48	0.42	0.37	0.33	0.29	0.25
10	0.82	0.66	0.57	0.50	0.44	0.38	0.34	0.30	0.27	0.24

5.6 LSF Quantization

Most modern speech coders make use of LPC modelling during speech processing. Although some coders use a backward-adaptive LPC filter [8], most speech coders extract the LPC parameters from the input speech at regular intervals, transform them into the LSF domain, and quantize them for transmission to the decoder.

Low distortion LSF quantization is essential for the overall quality of decoded speech, and the number of bits allocated to LSFs usually takes a significant proportion of the overall bit rate, up to over 50% for very low bit-rate speech coders. Therefore the overall success of a given speech coding scheme depends greatly on the quality of the LSF quantizer used.

Scalar schemes can be used, as they present very low complexity and storage requirements. However they cannot make use of the high intra-frame correlation exhibited by LSF vectors and, hence, they are very rarely used due to their poor performance. Vector quantization (VQ) schemes can be used to exploit intra-frame correlations. VQ exploits the redundancies in the LSF vector well and can provide high quality quantization for a relatively limited number of bits per frame of speech. As a result, they are widely used in modern speech coders. The following sections investigate the use of VQ for LSF quantization and ways of maximizing the performance of such schemes in several coder configurations.

5.6.1 Distortion Measures

In order to achieve good performance quantization of LSF parameters, it is necessary to have a way of linking the quantization error to the distortion in perceptual quality. Due to the complex relationship that exists between a set of LSF coefficients and the frequency response of the corresponding LPC filter, using a Mean-Square Error (MSE) measurement may not lead to an optimal performance of the quantizer.

A widely-used technique for computing the distortion that exists between the original set of LSFs and their quantized version is the Log Spectral Distortion measure. However a Weighted Mean-Square Error (WMSE) measurement may also lead to good results if an appropriate weighting function is used.

5.6.2 Spectral Distortion

The mean square log spectral distortion, which will be referred to simply as spectral distortion (SD), is defined as:

$$sd = \frac{1}{\pi} \int_0^{\pi} [10\, log_{10} S(w) - 10\, log_{10} S'(w)]^2 \tag{5.54}$$

where $S(w)$ and $S'(w)$ are the frequency responses of the LPC filter derived from the original and quantized LSFs, respectively. $S(w)$ can therefore be defined as:

$$S(w) = 1/ \mid A(w) \mid^2 \tag{5.55}$$

which leads to,

$$S(w) = 1/\mid 1 - \sum_{k=1}^{p} a_k e^{-jwk} \mid^2 \tag{5.56}$$

where a_k are the LPC coefficients. This can be evaluated using an N-point Fourier Transform, giving the following expression:

$$SD = \frac{1}{N/2} \sum_{k=0}^{N/2-1} \left[10\, log_{10} \mid A'(k) \mid^2 - 10\, log_{10} \mid A(k) \mid^2 \right]^2 \qquad (5.57)$$

Moreover, it is common practice to restrict the computation of the distortion to a limited portion of the spectrum, typically the 125–3100 Hz band. The reason is that the portions of the spectrum below 125 Hz and above 3100 Hz usually have perceptually little impact but may significantly affect the computed spectral distortion, due to the use of the *log* function.

5.6.3 Average Spectral Distortion and Outliers

The spectral distortion (SD) measure gives a good indication of the perceptual difference between two sets of LSFs. The overall distortion caused by a quantization scheme can be computed by simply averaging the SD obtained over a large sequence of LSF vectors. It is commonly accepted that an average SD below 1 dB is necessary for an LSF quantizer to be transparent, i.e. not to add any audible distortion to synthesized speech. However, the average SD (*aveSD*) is not sufficient to determine the performance of a quantizer. The human ear is very sensitive to occasional large quantization errors. Therefore it is also important that the number of times the quantizer gives a large distortion is kept to a minimum. It is customary to use the percentage of input vectors giving spectral distortions above 2 dB and 4 dB as a quality measure. These measures are referred to as outliers at 2 dB and 4 dB, respectively.

The set of requirements usually considered necessary to achieve good quality speech is [9]:

- Average spectral distortion less than 1 dB
- Fewer than 2 % outliers at 2 dB
- No outliers at 4 dB

These three parameters need to be considered when evaluating the performance of an LSF quantizer. However an optimization has to be carried out to achieve the best overall performance for a given bit rate, i.e. accepting a larger average spectral distortion in return for fewer outliers.

5.6.4 MSE Weighting Techniques

Although spectral distortion is a fairly accurate representation of how quantization noise in the LSF is perceived, its high computational complexity limits its use. In order to compare two sets of LSFs, two fairly large fast Fourier Transforms (FFT) need to be computed and a logarithm must then be computed for every bin of each FFT output. This is, of course, not a problem when estimating the performance of a quantizer off-line, but severely limits its use in a real-time coder.

On the other hand, simple MSE techniques have much lower complexity and can easily be implemented in real-time coders. However the basic MSE methods do not take into account the different perceptual effect of each of the LSFs, and this may lead to poor performance of the quantizer. One simple way to reduce this problem is to introduce an appropriate weighting function in the calculation of the MSE (WMSE). The WMSE between the LSF vector $\mathbf{f}$ and the candidate vector $\hat{\mathbf{f}}$ (frequencies are in Hz) is given by:

$$d(\mathbf{f},\hat{\mathbf{f}}) = (\mathbf{f}-\hat{\mathbf{f}})^T W(\mathbf{f}-\hat{\mathbf{f}}) \tag{5.58}$$

where W is a positive diagonal matrix. This is equivalent to:

$$d(\mathbf{f},\hat{\mathbf{f}}) = \sum_{n=1}^{p} w_n (f_n - \hat{f}_n)^2 \tag{5.59}$$

where $\mathbf{w}$ is a positive weighting vector.

The weighting vector renders contributions of certain elements more important than others in the summation process. The weighting vector is usually a function of the original LSF vector, and therefore needs to be computed only once per quantization (i.e once for every frame). A correctly chosen weighting function will improve the perceptual quality of the quantization but finding a suitable weighting function is difficult, as it needs to be related to perceptual quality. Various weighting functions have been investigated in the literature and the most popular ones are presented here.

Paliwal-Atal

This LSF weighting method is based on the frequency response of the original LPC filter [9]. The weights are calculated as:

$$w_n = c_n \, [P(f_n)]^{\tau} \tag{5.60}$$

where $P(f_n)$ is the LPC power spectrum associated with the original set of LSFs, f_n is the nth LSF. τ is a constant used to determine the relative importance of the LSF and is experimentally set to 0.3. Finally, the fact that the human ear cannot resolve high frequencies very well is used in introducing the factor c_n, which reduces the influence of the last two LSFs in the summation.

$$c_n = \begin{cases} 1.0 & \text{for } 1 \le \text{n} \le 8 \\ 0.8 & \text{for n} = 9 \\ 0.4 & \text{for n} = 10 \end{cases} \tag{5.61}$$

EFR weighting

This weighting function is used in the GSM Enhanced Full Rate standard (EFR) [10]. The weights are calculated as follows:

$$w_n = \begin{cases} 3.347 - \dfrac{1.547}{450} d_n & \text{for } d_n \leq 450 \\ 1.8 - \dfrac{0.8}{1050} (d_n - 450) & \text{otherwise} \end{cases} \tag{5.62}$$

where

$$d_n = f_{n+1} - f_{n-1} \tag{5.63}$$

and f_n is the nth LSF, $f_0 = 0$ and $f_{11} = 4000$.

LSF inverse distance

This method is based on the principle that the peaks in the LPC filter are located where two consecutive LSFs are close to each other. The weighting is given by:

$$w_n = \frac{4000}{(f_n - f_{n-1})} + \frac{4000}{(f_{n+1} - f_n)} \tag{5.64}$$

Group Delays

This weighting is based on the group delay of the LPC filter and is defined as [11]:

$$w_n = \begin{cases} u(f_n)\sqrt{\dfrac{D_n}{D_{max}}} & 1.375 \leq D_n \leq D_{max} \\ u(f_n)\dfrac{D_n}{\sqrt{1.375\, D_{max}}} & D_n < 1.375 \end{cases} \tag{5.65}$$

where

$$u(f_n) = \begin{cases} 1 & f_n < 1000 \\ 1 - \dfrac{0.5}{3000}(f_n - 1000) & 1000 \leq f_n \leq 4000 \end{cases} \tag{5.66}$$

D_n is the group delay of the LPC filter at the frequency f_n in milliseconds whilst D_{max} is the maximum group delay, experimentally found to be around 20 ms.

The group delays of the filter are larger at the formant frequencies, therefore the weighting will be higher for these frequencies. The factor $u(f_n)$ simply reduces the weights for the higher frequencies to take into account the

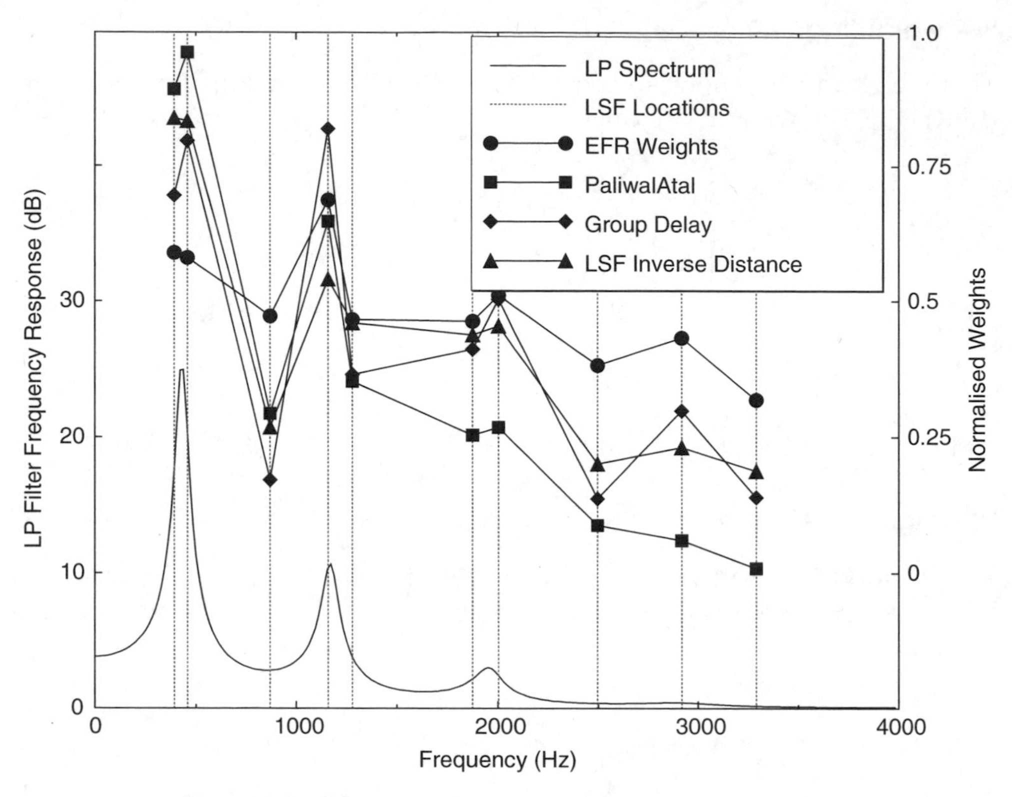

Figure 5.9 Example of various LSF weighting functions

lesser sensitivity of the ear to the spectral distortion above 1000 Hz. The relationship between the position of the LSF and the peaks in the LPC spectrum is illustrated in Figure 5.9. It can be seen that a peak in the LPC spectrum usually corresponds to a pair of LSFs close to each other, which justifies in particular the LSF inverse distance weighting method. The result of each of the weighting functions described above is also plotted at each LSF location, after normalization. It can be seen that although all have similar overall characteristics close to the peaks, they vary significantly in the importance they place on the LSFs situated in the valleys and the higher frequencies.

5.7 Codebook Structures

In order to obtain good quality speech with a low bit-rate speech coder, it is necessary for the LSF quantizer to fulfil the requirements on the spectral distortion described in Section 5.6.3. This is usually achieved using a vector quantizer in order to maximize the quantization efficiency, and such a system typically requires 20 to 25 bits to represent a set of 10 LSF parameters with the required accuracy. Such a large number of bits precludes the use of straightforward vector quantization of the LSF vector, as the complexity and storage requirements of such a system would be far too great to be implemented on any reasonably priced device. Therefore, alternative suboptimal methods have to be used, which add structure to the codebook in order to reduce implementation costs. The two most common methods are split vector quantization (SVQ) and multi-stage vector quantization (MSVQ).

5.7.1 Split Vector Quantization

Direct quantization of a set of LSF parameters with a typical vector quantizer of 25 bits, would require a codebook with 10^{25} entries, which is not practical from both the search complexity and memory point of view. An alternative method is to use SVQ, where the 10-element LSF vector is split into a number of smaller subvectors, each quantized independently using a small number of bits. Since the complexity and storage requirements of a full-search vector quantizer are exponential functions of the number of bits used to represent the input vector, SVQ requires only a fraction of the complexity required by a full search VQ.

In an SVQ system, an input vector $\mathbf{f}$ is represented by a vector $\hat{\mathbf{f}}$ given by:

$$\hat{\mathbf{f}} = \{\{y_0^{i_0}(0), .., y_0^{i_0}(N_0)\}, .., \{y_{K-1}^{i_{K-1}}(0), .., y_{K-1}^{i_{K-1}}(N_{K-1})\}\} \tag{5.67}$$

where K is the number of subvectors, each of length N_k, $y_k^{i_k}(n)$ is the nth element from the kth codebook, and i_k is the codebook index for the kth subvector. Obviously K and N_k are chosen so that the sum of N_k for $k = 0, 1, \ldots, K-1$ is equal to the length of the input LSF vector.

Splitting the 10-element LSF vector can be performed in various ways and some classic configurations are illustrated in Table 5.4. The split usually takes into account some of the perceptual properties of the LSF vector, such as the fact that lower frequency LSFs are usually more sensitive to distortion than higher frequency ones. Therefore, a $\{4, 6\}$ split would be preferred to a $\{5, 5\}$ split for instance. The configurations shown here have been chosen so that they all have the same bit-rate of 24 bits. Complexity (in multiply–adds) and memory storage (in words) for the typical SVQ configurations are presented in Table 5.5. It can be seen that although the direct VQ approach is extremely complex, the SVQ configurations are all practical. Even the most complex one requires only 40 960 multiply–adds per input vector, which translates to only 2 MIPS if performed at a 20 ms update rate.

However, there are several drawbacks which relate to the efficiency of SVQ quantization:

Table 5.4 Typical examples of SVQ LSF quantizers (24 bits/frame)

Sub-vectors	Elements per subvector	Bit allocation
2	5,5	12,12
3	3,3,4	8,8,8
4	3,2,2,3	6,6,6,6
5	2,2,2,2,2	5,5,5,5,4

Table 5.5 Complexity and memory requirements for various SVQ schemes

Sub-vectors	Split	Bits	Complexity	Memory storage
1	10	24	1.67×10^8	1.67×10^8
2	5,5	12,12	40 960	40 960
3	3,3,4	8,8,8	2560	2560
4	3,2,2,3	6,6,6,6	640	640
5	2,2,2,2,2	5,5,5,5,4	288	288

- The correlations between subvectors are not exploited. Therefore only a fraction of the intra-frame correlation is used. In particular, a pair of LSFs close to a peak in the spectrum may be split into two different subvectors and, although there is a correlation between them, they are quantized independently. As a result the quantization efficiency decreases greatly as the size of the subvectors reduces.
- Some combinations of subvectors do not respect the ordering of the LSF, or lead to neighbouring LSFs being too close to each other. As there is a minimum spacing limit that a pair of adjacent LSFs are allowed to have, this means that certain SVQ vector combinations will never be used, which is a waste of bandwidth. This can however be alleviated to some extent. For example, once the first subvector has been quantized, a simple transformation such as an offset shift can be applied to the vectors that violate the minimum distance in the second codebook, so as to make them usable. However this is difficult to include in the training process, and the resulting quantizer may not be optimal.
- The number of bits allocated to each subvector is fixed. The effect of the weighting function will therefore be limited to within one subvector. If a subvector contains only LSFs of relatively small importance, they will still use all the bits allocated to this subvector, whereas a classic VQ would effectively shift some of that bandwidth towards the more important LSFs, through the weighting function. This effectively reduces the use of the weighting function to the LSF within a given subvector and lowers the overall quantization efficiency of an SVQ quantizer.

5.7.2 Multi-Stage Vector Quantization

In an Multi-Stage Vector Quantizer (MSVQ), the input vector is quantized as a sum of vectors from a number of codebooks. Each of these codebooks can therefore be of relatively small size, making the storage requirements reasonable. That is, an input vector $\mathbf{f}$ is represented by a vector $\hat{\mathbf{f}}$ given by:

$$\hat{\mathbf{f}} = y_0^{i_0} + y_1^{i_1} + \ldots + y_{K-1}^{i_{K-1}} \tag{5.68}$$

where K is the number of stages and i_k is the codebook index for the kth stage.

It can easily be seen that SVQ systems are a particular type of MSVQ system, where the codebook vectors for a given stage contain nonzero elements only in the locations corresponding to the SVQ subvectors. This is illustrated in the following example, where it is easily seen that an SVQ codebook can be mapped onto an MSVQ codebook.

$$\begin{array}{lcllllllll} y_0^{i_0} & = & \{y_0^{i_0}(0) & y_0^{i_0}(1) & 0 & 0 & \ldots & \ldots & 0 & 0\} \\ y_1^{i_1} & = & \{0 & 0 & y_1^{i_1}(2) & y_1^{i_1}(3) & \ldots & \ldots & 0 & 0\} \\ \vdots & = & \{0 & 0 & 0 & 0 & \vdots & \vdots & 0 & 0\} \\ y_{K-1}^{i_{K-1}} & = & \{0 & 0 & 0 & 0 & \ldots & \ldots & y_{K-1}^{i_{K-1}}(p-2) & y_{K-1}^{i_{K-1}}(p-1)\} \end{array}$$

This obviously implies that an MSVQ system will have a performance at least equivalent to that of an SVQ system and probably much higher, as the SVQ imposes a strong constraint on the structure of the codebook. On the other hand, complexity and memory requirements for the MSVQ will be higher, i.e. the sparse structure of the SVQ codebook significantly reduces the storage requirement and a sequential search for each subvector is equivalent to an exhaustive search, which is not the case for MSVQ. Examples of typical bit allocations for MSVQ codebooks are illustrated in Table 5.6, including the bit allocation for the 2.4 kb/s MELP coder [12].

Table 5.6 Typical examples of MSVQ LSF quantizers (24–25 bits/frame)

Stages	Bit allocation	Total number of bits
2	12,12	24
3	8,8,8	24
4	6,6,6,6	24
4	7,6,6,6	25
5	5,5,5,5,4	24

5.7.3 Search strategies for MSVQ

The usual search strategy for an SVQ codebook is straightforward: a full search (FS) for each of the subvectors is applied. The structure of an MSVQ quantizer, however, allows different types of search strategy depending on the desired complexity. The simplest of the searches is the sequential search (SS). In this search, the input vector $\mathbf{f}$ is first approximated by the i_0^{th} vector from the first codebook Y_0 which minimizes:

$$d(\mathbf{f}, \hat{\mathbf{f}}) = \sum_{n=1}^{p} w_n \left(f_n - (y_0^{i_0})_n\right)^2 \tag{5.69}$$

The index for the first codebook i_0 is then fixed and the quantization error $\mathbf{f} - \mathbf{y}_0^{i_0}$ is then quantized using the i_1^{th} vector from the second codebook Y_1 which minimizes:

$$d(\mathbf{f}, \hat{\mathbf{f}}) = \sum_{n=1}^{p} w_n \left((f_n - (y_0^{i_0})_n) - (y_1^{i_1})_n\right)^2 \tag{5.70}$$

This process is repeated for each stage in the codebook. The complexity of this search is the sum of the complexity of a full search through each codebook, given by,

$$C = N \sum_{k=1}^{K} 2^{B_k} \tag{5.71}$$

where K is the number of stages, each with B_k bits, and N is the length of the input vector. This search is, however, nonoptimal as there is no guarantee that the set of codebook vectors giving the lowest intermediate distortion will also result in the best overall distortion. A better way to ensure that the best performance is obtained is simply to perform a full search on all codebooks jointly. That is, every combination of codebook vectors $\hat{\mathbf{x}} = \mathbf{y}_0^{i_0} + \mathbf{y}_1^{i_1} + \ldots + \mathbf{y}_{K-1}^{i_{K-1}}$ is tested against the original input vector. This guarantees optimal quantization, but at the cost of a very high complexity, given by,

$$C = N\, 2^{\sum_{k=1}^{K} B_k} \tag{5.72}$$

This complexity is equal to that of a direct vector quantization of the LSF, which is far too high for practical applications. The only advantage of the full-search MSVQ over a standard full search vector quantizer is the reduced storage requirement. However, it is possible to obtain most of the

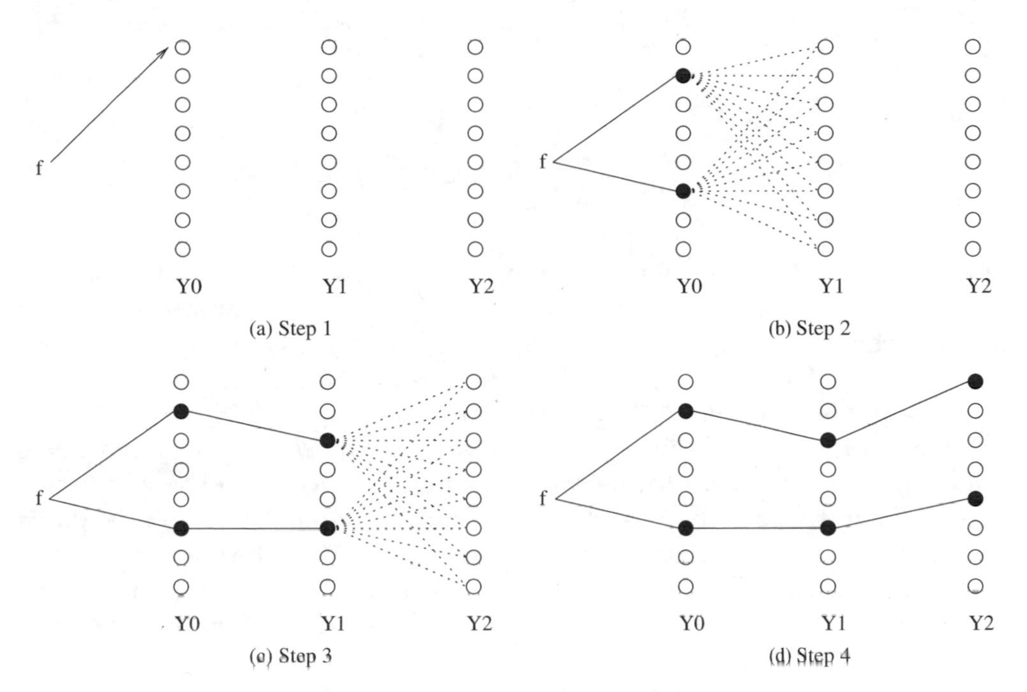

Figure 5.10 Steps in an M-best search

advantages of the full search over the sequential search, while still maintaining a reasonable level of complexity, by using a tree-search algorithm (TS), such as an M-best tree search.

An M-best tree search operates by exploring a certain number, M, of paths in the quantizer tree. Starting with the first codebook, the M code-vectors giving the lowest distortion when compared with the input are kept, as well as the M quantization error vectors resulting from these vectors. The second codebook is then searched M times, once for each of these error vectors, and the M paths which achieve the lowest overall distortion are kept. This procedure is performed for each stage of the codebook. Finally, for the last stage, the path giving the lowest overall distortion is selected. This process is illustrated in Figure 5.10. For this example, M has been set to 2 and the codebook consists of three stages of 3 bits each. In Figure 5.10a, the first codebook $Y0$ is searched to find the M vectors that best match the original LSF vector. In Figure 5.10b, the second codebook is searched to best match the difference between the original input LSFs and the selected vector from the first codebook. This is performed for each of the M selected vectors in the first codebook. The M best paths are selected for the next stage. Figure 5.10c shows the same process repeated for the third and final codebook. Finally, Figure 5.10d shows the final M best paths. Out of these, the path with the lowest overall distortion is selected. Experiments show that such a tree search can give performance close to that of a full search even with a small value of

M (i.e. 8–16). The complexity of this search is given by:

$$C = N\left(2^{B_0} + M\sum_{k=2}^{K} 2^{B_k}\right) \tag{5.73}$$

Obviously for $M = 1$, this equates to the complexity of the sequential search. It can be seen that the M factor does not apply to the complexity of the first codebook search. This can be exploited in designing the structure of the codebook. For example, if we have three stages for a total of 25 bits, it is significantly less complex to have a $\{9, 8, 8\}$ structure than a $\{8, 9, 8\}$ structure, whereas storage is the same and performance is expected to be similar. One interesting improvement to the M-best search strategy is to use a complex perceptual measure in the final stage only, to select which of the M final paths are the best. Since this computation only needs to be performed M times, it is possible to use much more complex distortion measures than the WMSE normally used. It is also possible to compute this measure on only a subset of the M best final paths, i.e. the ones which give the lowest WMSE. This procedure significantly enhances the performance of the quantizer, partly solving the problem that the WMSE is not such a good distortion measure compared to the SD for example.

5.7.4 MSVQ Codebook Training

The basic codebook training algorithms usually cater for single-stage codebooks. It is however possible to adapt the algorithm for MSVQ codebook training. The most basic technique is called *sequential optimization*. In this method, the codebook for stage 1 of the MSVQ is first designed. The quantization errors for the training database are then computed and the codebook for stage 2 is trained over the error vectors. This is then repeated for each stage, until reaching the final codebook.

However sequential optimization does not provide the best performance, as each codebook is optimized as if it was the last stage of the MSVQ quantizer. A better alternative is *iterative sequential optimization*, where an initial codebook is chosen for each stage. Each codebook is then optimized by assuming all the other stages to be fixed and known, i.e. the quantization error is computed using all the other stages except the current one, and training is used to obtain an updated version of the current codebook. This process can then be repeated until all of the codebooks have converged.

It is also possible to jointly optimize all codebooks using *simultaneous joint codebook design*. This method gives slightly better results than the previous methods but has a high computational cost, which is described in [13].

5.8 MSVQ Performance Analysis

In order to compare the relative performance of various MSVQs, quantizers have been trained using the same training database, which has the following characteristics:

- MIRS and FLAT filtered speech in various languages are included
- Only speech-active regions are included
- LSFs are extracted with an update rate of 20 ms, over a 200 sample Hamming window
- A bandwidth expansion factor of 0.994 is applied to the LP coefficients prior to LSF conversion
- 50 000 sets of LSF coefficients are included

The speech database used is rather small to produce quantizers with good performance in real-life applications. Typically, a speech database of over 1 000 000 LSF vectors is used for training codebooks for actual applications. However, for the purpose of comparing performances of various quantization schemes, the smaller speech database is adequate in providing indicative results. Additionally, it significantly reduces the time required to train the quantizers, which is prohibitive for the bigger database (several weeks of computing are usually required for typical codebook training with the larger database).

5.8.1 Codebook Structures

For a given bit rate, MSVQ and SVQ codebooks can differ in the number of stages and in the vector splits. The actual structure of the quantizer affects complexity and memory storage, as discussed earlier, but also affects performance. Typically, the more structure imposed on the codebooks, the lower the complexity and storage, but also the poorer the performance. All of the SVQ and MSVQ quantizers have been trained using 24 bits, for various numbers of stages, from 2 to 5. The configurations used are shown in Table 5.7. The results are plotted in Figure 5.11. As expected, the performance is directly linked to the amount of structure present in the codebook.

5.8.2 Search Techniques

In order to compare the performance of various types of searches available for a given codebook, an MSVQ codebook of 21 bits, using three stages of 7 bits each, has been trained. It uses no prediction and the search algorithm used during training was a sequential search (SS). The performance of the codebook was then measured using SS, FS, and TS with values of M from 2 to 32. The WMSE, average SD, and number of outliers at 2 dB are plotted

Table 5.7 MSVQ and SVQ structures for Figure 5.11

Stages	MSVQ	SVQ	
	bit allocation	Bit allocation	Vector split
2	12,12	12,12	5,5
3	8,8,8	8,8,8	3,3,4
4	6,6,6,6	6,6,6,6	3,2,2,3
5	5,5,5,5,4	5,5,5,5,4	2,2,2,2,2

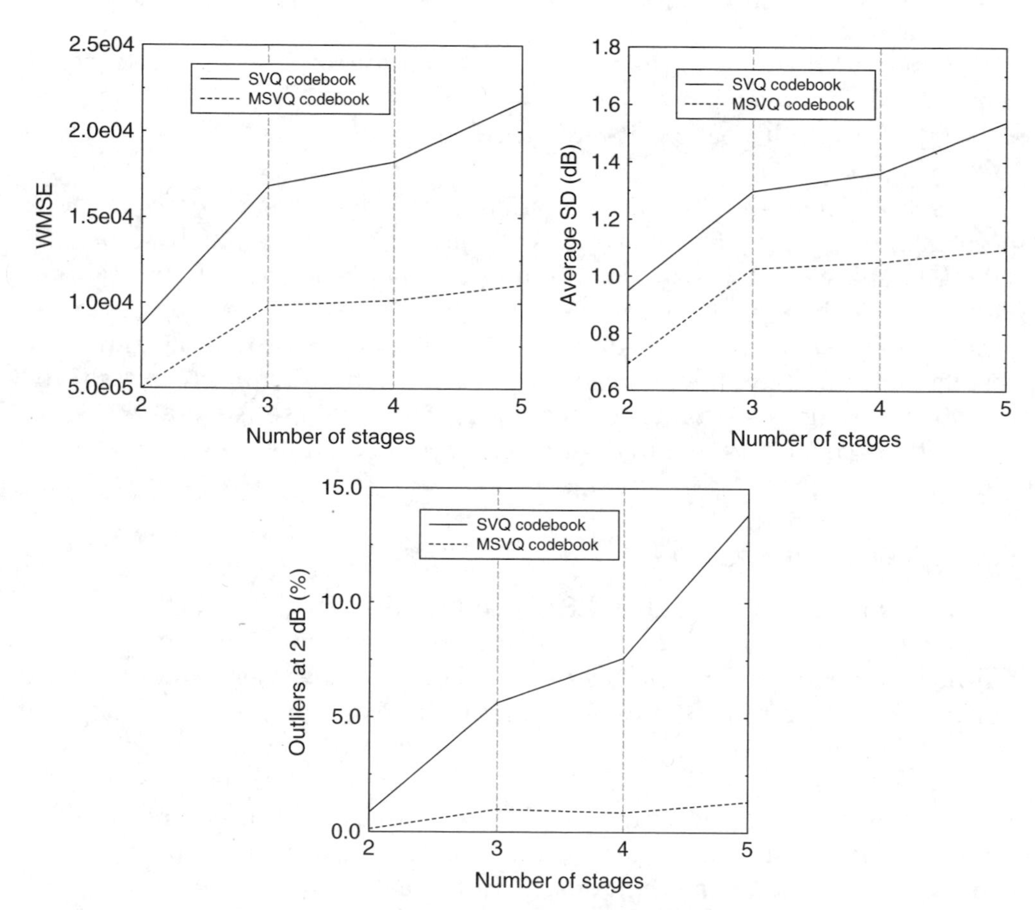

Figure 5.11 Performance comparison of various SVQ and MSVQ codebook structures

in Figure 5.12. Outliers at 4 dB have not been plotted, as they are zero for all cases. The advantage of a TS over both SS and FS is evident in these graphs. For M greater than or equal to eight, the performance of the TS is very close to that of the FS, at a much reduced complexity. It is also significantly better than that of SS, for a relatively small increase in complexity. The complexity

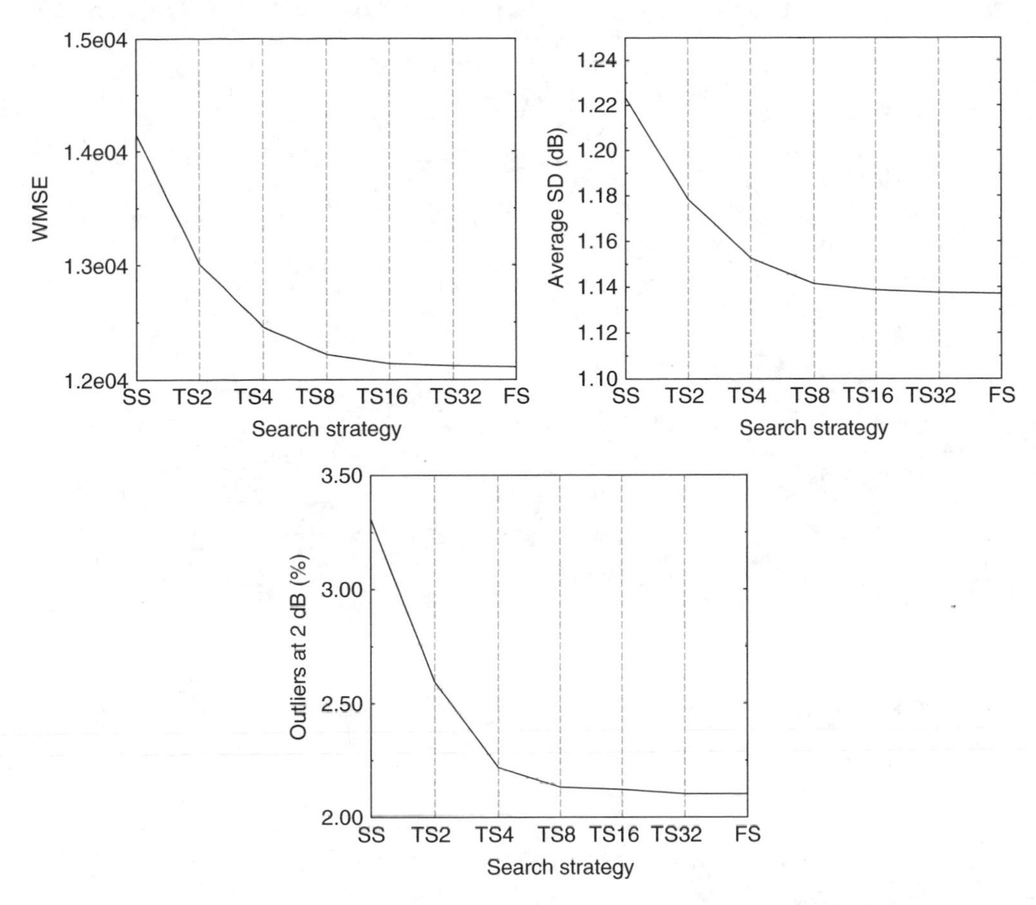

Figure 5.12 Performance comparison of various search techniques

in multiply–adds per input vector is given in Table 5.8. It is to be noted that, in the test, codebooks have been trained using the SS algorithm. Therefore, they are only optimal for an SS search. Better performance for the TS and FS cases can be obtained by using the same search in the training as the one used during the operation of the quantizer. This is illustrated in Figure 5.13, where WMSE, average SD and outliers at 2 dB are plotted for the original codebook and the retrained codebooks, for SS and TS with values of M ranging from 2 to 32. Due to the very high complexity of the FS, it was not possible to fully retrain the codebook using FS, although the results are expected to be similar to that of TS with $M = 32$.

5.8.3 Perceptual Weighting Techniques

Several weighting techniques were described in Section 5.6.4. A good weighting technique should give a distortion measure which is well correlated with the spectral distortion measure, which is our reference here. For testing, we

Table 5.8 Complexity of various search strategies for a $\{7,7,7\}$ MSVQ codebook

Search type	Complexity
SS	3840
TS: $M = 2$	6400
TS: $M = 4$	11 520
TS: $M = 8$	21 760
TS: $M = 16$	42 240
TS: $M = 32$	83 200
FS	20 971 520

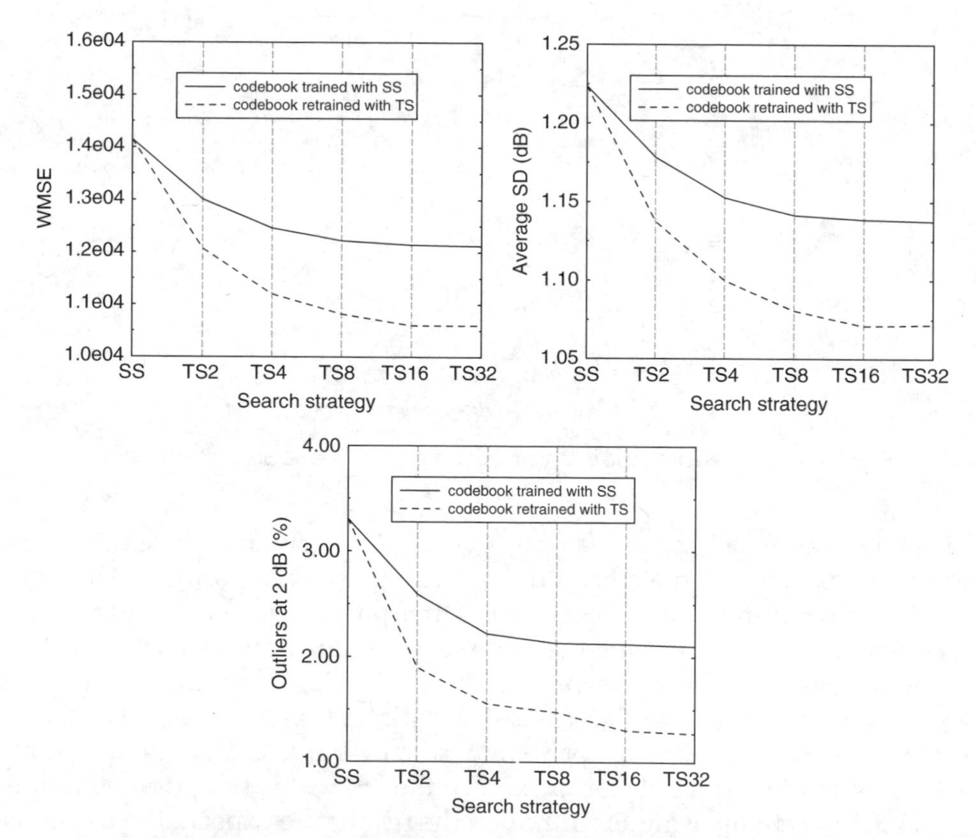

Figure 5.13 Performance comparison with and without codebook reoptimization

use a 21-bit MSVQ codebook with three stages of 7 bits each which has been trained using the weighting methods listed below:

- W1: no weighting (all weights are equal to 1.0)
- W2: EFR weighting method

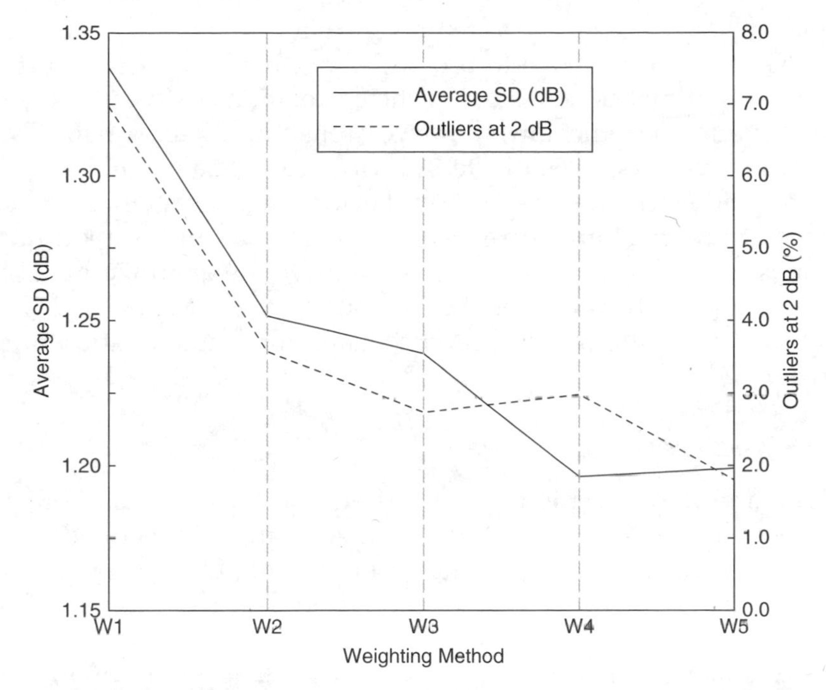

Figure 5.14 Performance comparison of various weighting functions

- W3: LSF inverse distance method
- W4: Paliwal–Atal method
- W5: Group delays method

The results are shown in Figure 5.14. Only average SD and the number of outliers at 2 dB are shown, as there are no outliers at 4 dB in any of the cases. Figure 5.14 clearly shows the performance gain given by the use of weighting over the simple MSE method. As indicated, a well-chosen weighting method can give a reduction of up to 0.15 dB in average SD and up to 5 % in the number of outliers. The figure also shows that some weighting methods clearly outperform others. The weighting technique used in EFR and the LSF inverse distance method are not the best. Better results are obtained with the Paliwal–Atal method, however the best of all is the group delays method. The main advantage of the group delays method is in the reduction in the number of outliers at 2 dB (from 3 % to 1.8 %), whereas its average SD is virtually identical to that of the Paliwal–Atal method.

5.9 Inter-frame Correlation

When quantizing LSF parameters, or any other parameter for that matter, a good quantization scheme must make use of all the redundancies in the parameters to be quantized, so as to maximize the efficiency of the quantizer. The MSVQ was shown above to provide better performance than the SVQ, mostly because it makes better use of the correlations between the elements of an LSF vector, i.e. the intra-frame correlations as shown in Table 5.2. However, LSF vectors are extracted at a typical update rate of 20 ms and speech characteristics often remain similar for longer than 20 ms. Therefore, successive LSF vectors are correlated (see Table 5.3) and a good quantizer should make use of these similarities to improve the quantization accuracy. The inter-frame correlation can be exploited in various ways, the most popular ones being the use of a predictor and joint quantization of several sets of LSFs.

5.9.1 LSF Prediction

A popular approach to exploiting the inter-frame correlations of LSF vectors is the use of prediction. Instead of quantizing an LSF vector directly, the difference between a predicted vector and the actual LSF vector is quantized and transmitted. If the predictor is good, then the residual signal should be easier to quantize than the original LSF vector. It is common practice to remove the long-term mean of each LSF before applying prediction. The residual LSF is given by:

$$r_n = f_n - \tilde{f}_n \tag{5.74}$$

where $\tilde{f}_n$ is the prediction vector. The decoded LSF vector is then given by,

$$\hat{f}_n = \hat{r}_n + \tilde{f}_n \tag{5.75}$$

where $\hat{r}_n$ is the quantized value of r_n.

This obviously implies that the decoder should have knowledge of $\tilde{f}_n$. Therefore, the prediction used should be a function of some parameters available at both the decoder and the encoder. One of the simplest predictors assumes that a set of LSFs can be predicted using the previous quantized set of LSFs, scaled by a weighting factor,

$$\tilde{f}_n^k = \alpha_n \hat{f}_n^{k-1} \tag{5.76}$$

This will be referred to as an LSF differential quantizer (LSF-DQ). The computation of the prediction gain is made difficult by the fact that knowledge of the quantizer (codebook) is necessary to compute the prediction. One way around this problem is to assume that the final quantizer will be quite good and therefore $\hat{f}_n^{k-1}$ can be approximated by f_n^{k-1} in the equation above. The optimal factors α_n can then be determined by maximizing the prediction gain over a speech database. In order to increase the prediction accuracy, higher order predictors can also be used. The prediction then becomes a weighted sum of the LSF vectors for a given number of past frames. This increases the performance of the predictor, at the small expense of slightly higher memory requirements for storing the past values. Unfortunately this scheme has a major drawback: the decoder must have correct knowledge of the prediction used at the encoder. If a channel error occurs and corrupts the bitstream for one frame, then the decoded LSF will be corrupted. Since the decoded erroneous LSFs will be used for prediction, the LSF for the next frame will also be corrupted and the error will then propagate indefinitely.

A better approach is to generate the prediction from the decoded codebook entries, rather than the decoded LSFs which will limit error propagation. Such predictors are called moving average (MA) predictors. A first-order MA predictor is given by,

$$\tilde{f}_n^k = \alpha_n \hat{r}_n^{k-1} \tag{5.77}$$

The decoded vector is then given by,

$$\hat{f}_n^k = \hat{r}_n^k + \alpha_n \hat{r}_n^{k-1} \tag{5.78}$$

Therefore if an error occurs, the only frames affected will be the frame where the error occurs and the N following frames, where N is the order of the predictor. For a first-order MA predictor, only one extra frame will be affected compared with a quantizer not using prediction. Intuitively, an MA predictor will not be as efficient as a DQ predictor, but its error resilience capabilities are very significant. This makes the MA predictor a better choice for the majority of applications.

Assuming all α_n are chosen equal to a constant α, the prediction gains of the DQ and MA predictors are plotted against α in Figure 5.15, for an update rate of 20 ms. Experiments show that forcing all α_n to be equal does not significantly reduce the prediction gain over the ideal case.

Figure 5.15 shows that the DQ predictor can achieve a gain of up to 5 dB with an α value of 0.8, whereas the MA predictor can achieve 3 dB for α around 0.65. The MA predictor is not as efficient as the DQ predictor, but still provides a useful prediction gain, which in turn can help improve the overall performance of the quantizer. The prediction gains for both DQ and MA predictors depend on the LSF update rate which directly affects correlation between adjacent sets of parameters. A faster update rate will give a higher prediction gain, as consecutive sets of LSFs will be more correlated, and it will usually be achieved with a higher value of α. For example, an update rate of 10 ms gives an optimal α of around 0.8 for the MA predictor. In the following sections, only the MA predictor will be considered as the DQ predictor is not

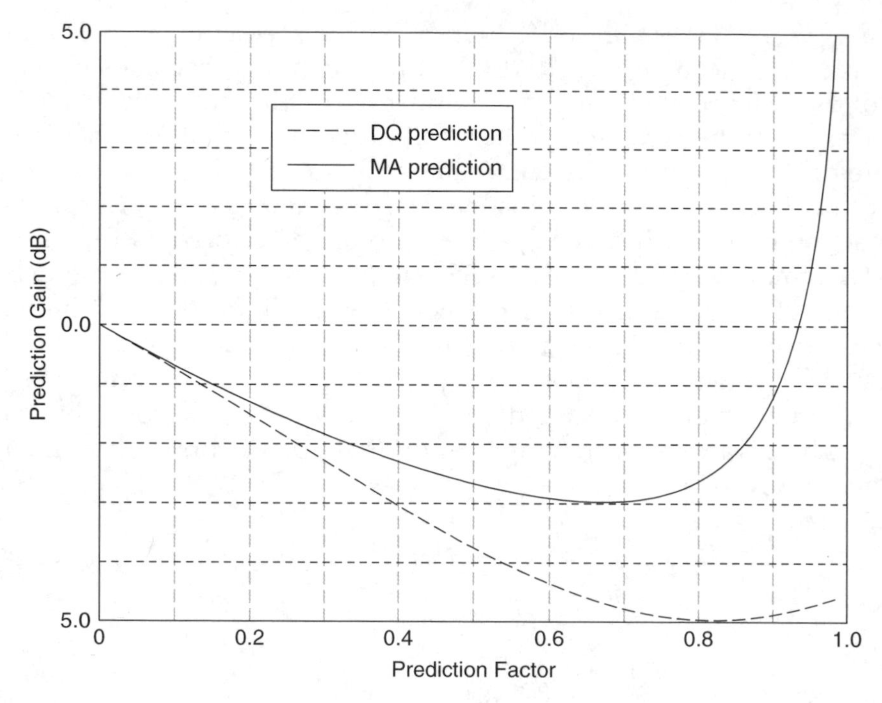

Figure 5.15 Prediction gain of first order MA vs DQ Predictors (20 ms update rate)

suitable for a general purpose coder (with possible channel errors). However, a DQ predictor may be applicable in cases where virtually no channel errors are encountered, such as voice storage applications.

5.9.2 Prediction Order

MA prediction has been presented above for the case of a first-order predictor. It is also possible to have a higher order MA predictor, where the prediction is a weighted sum of the quantized residuals received in N previous frames. An Nth-order predictor would exploit correlations between the current frame n and the frames $n-1$, $n-2$, $\ldots$, $n-N$. As a result, its performance is expected to be greater than that of a first-order MA predictor. However, the drawback of the improved performance is greater sensitivity to channel errors as an error on one set of parameters will corrupt $N+1$ frames of speech. The optimal order of an MA predictor requires a trade-off between prediction gain and error resilience. In order to estimate the optimal order to be used in most applications, where channel errors are expected, the optimal prediction factors have been derived for various orders. This was achieved by computing the prediction gain for all possible combinations of the prediction parameters, in steps of 0.05, over a database of 30 000 LSF, extracted at 20 ms intervals with all silences removed. The results are shown in Table 5.9. It can

Table 5.9 Prediction gain vs MA predictor order

Order	Optimum prediction parameters	Prediction gain (dB)
1	0.65	2.97
2	0.85,0.43	4.13
3	0.85,0.60,0.35	4.61
4	0.9,0.7,0.45,0.2	4.84

be seen from Table 5.9 that the increase in prediction gain when increasing the order of the MA predictor from 1 to 2 is 1.16 dB. An increase in prediction gain of 1.87 dB can be achieved by increasing the order from 1 to 4, whereas the increase from order 0 (no prediction) to 1 is nearly 3 dB. Although higher order predictors help to increase the prediction, the degradation in speech quality due to channel errors is expected. If the order is 1, 40 ms of speech are corrupted. With proper error concealment techniques, it is usually possible to limit the distortion caused by the loss of LPC for 40 ms to an acceptable level. However, for higher prediction orders, 60 ms or more are lost and the speech degradation caused by such a loss is usually difficult to recover, affecting the overall speech quality significantly. Therefore, for most applications with a 20 ms parameters update rate which involve a noisy channel, it is better to use a first-order MA prediction. In case of shorter update rates, or very low bit error conditions, higher order prediction can be used to improve the MA prediction performance. In the following discussion, only first-order MA prediction is considered.

5.9.3 Prediction Factor Estimation

Figure 5.15 shows that the best prediction gain for a first order MA is achieved with a value of 0.65. Therefore it would be reasonable to assume that a prediction factor of 0.65 will give the best performance in a first-order MA quantizer. Indeed, such a value is used in some speech coders such as EFR [10]. However, this value has been derived using the assumption that the original residual r_n^k is close enough to the quantized residual $\hat{r}_n^k$ that it can be used instead to obtain the curve shown in Figure 5.15. In a practical quantizer, there is no guarantee that this assumption will be true. Therefore the only way to determine the optimal prediction factor is by training quantizers with various prediction factors and comparing their performances. Various first-order MA quantizers have been trained for values of α ranging from 0.3 to 0.7 in 0.05 steps, for a 20 ms update rate. An MSVQ quantizer comprising three stages of 8 bits each has been selected to quantize the residual, as it provides good performance. The performance of these quantizers is plotted in Figure 5.16, together with the performance of the quantizer without prediction. It can be seen that the best overall performance of the quantizer is achieved for a value

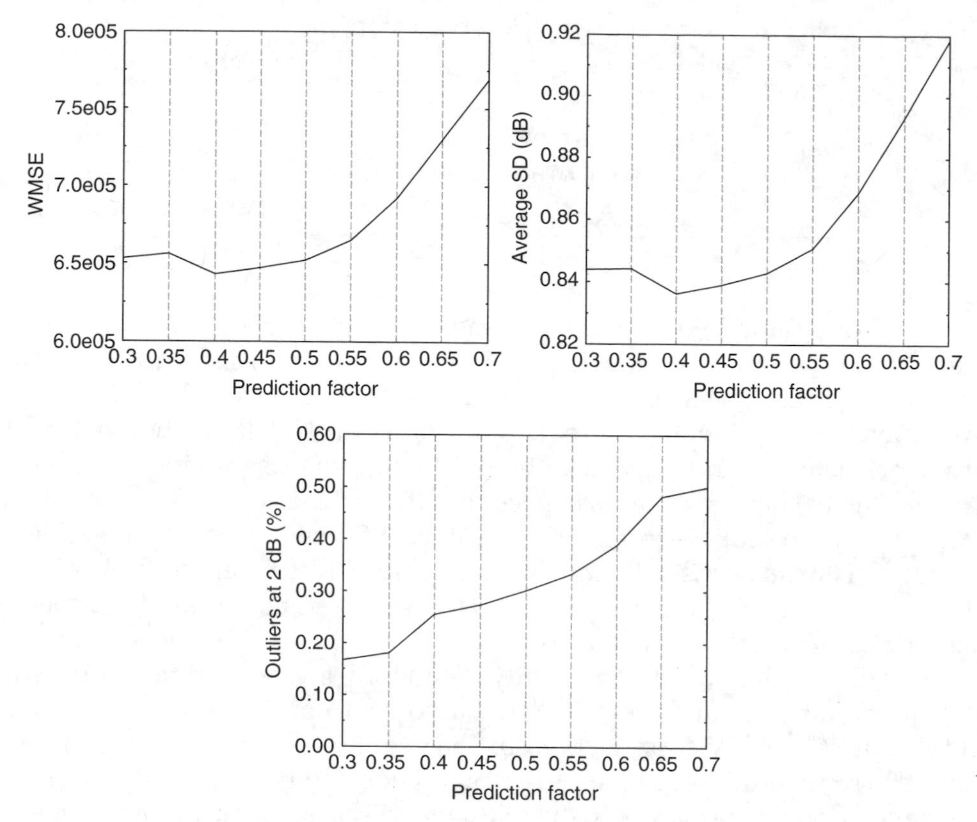

Figure 5.16 Performance of a MA-MSVQ Quantizer vs Prediction Factor

of α around 0.4, and not 0.65 as would be expected from Figure 5.15. WMSE and average SD are lowest for 0.4, although outliers at 2 dB are slightly lower at 0.3 than at 0.4. This is to be expected if the predictor does not work well, at speech transitions for example. However this is not a problem as it is possible to bias the training process towards producing fewer outliers with only a small increase in the average SD. Moreover, the performance obtained with 0.4 is significantly better than that of 0.65. This clearly shows that the 'intuitive' way of determining the prediction factor may not be correct.

5.9.4 Performance Evaluation of MA Prediction

In order to compare the performance of quantizers with and without MA prediction, several codebooks have been trained with $\alpha = 0.4$ for various bit rates, using a 20 ms update rate. In order to make comparisons with the previous graphs easier, the quantizers all have 3-stage MSVQ codebooks. The bit rates range from 20 to 26 bits, and the codebook structures are detailed in Table 5.10. The search algorithm is a tree search with a depth of 32. The overall performances are shown in Figure 5.17. The gain provided by the MA

Table 5.10 MSVQ bit allocation for Figure 5.17

Bits	Bit allocation
20	7,7,6
21	7,7,7
22	8,7,7
23	8,8,7
24	8,8,8
25	9,8,8
26	9,9,8

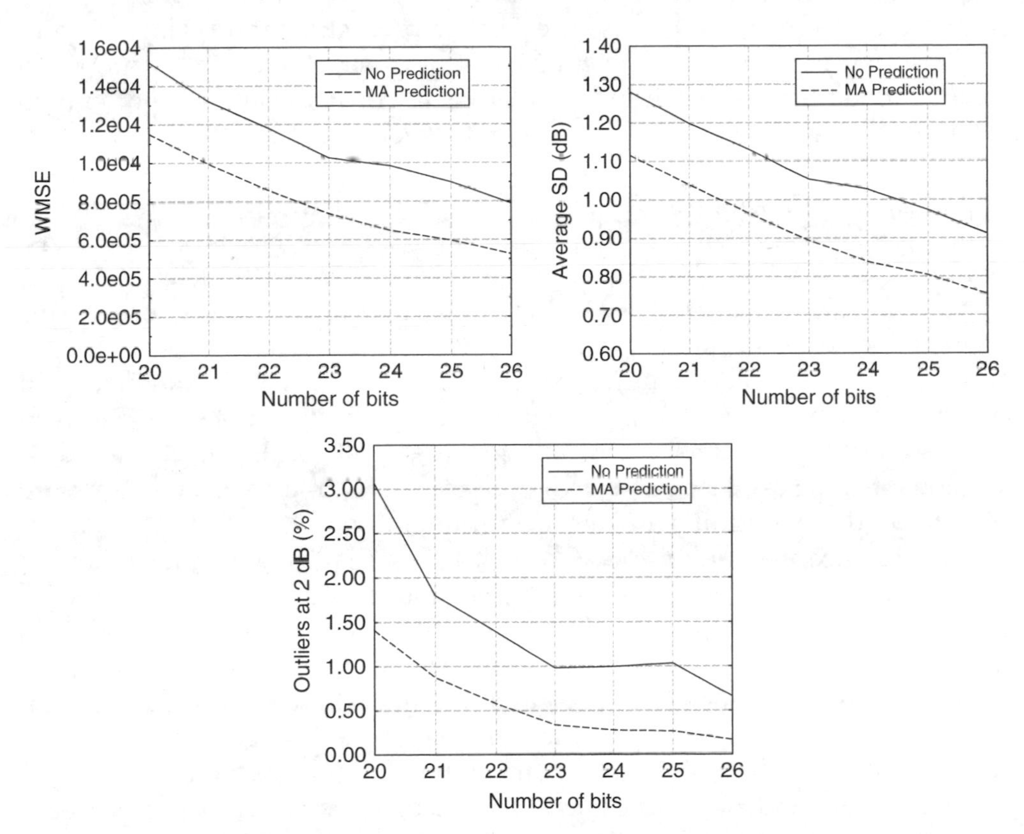

Figure 5.17 MA prediction vs no prediction at various bit rates

prediction is evident from the graphs, where similar performance is obtained for the MA-MSVQ with around 3 bits less than for the MSVQ without prediction. This 3 bits advantage is present for all performance measures. Therefore it is possible to achieve a saving of 10–15 % in bit rate by using MA prediction with an MSVQ quantizer, on top of the bit reduction already obtained by using MSVQ instead of SVQ. The only cost of the MA prediction is a slightly increased sensitivity to channel errors. However, during testing of coders using such schemes, this extra sensitivity did not turn out to be a significant problem as the prediction order is limited to one.

5.9.5 Joint Quantization of LSFs

Prediction is an efficient way of removing correlation from two or more neighbouring sets of parameters. However it is a one-way only process where information from frame $k-1$ is used in the prediction and quantization of frame k, but information from frame k is not used for the prediction and quantization of frame $k-1$. Indeed it is assumed that frame k is not known when quantizing frame $k-1$, in order to keep the delay to a minimum. However, in some applications it is worth accepting a slight increase in delay and using a quantization scheme which makes use of the extra redundancies. A simple way of achieving this is to jointly quantize several sets of parameters. For example, a 1.2 kb/s version of the SB-LPC coder jointly quantizes three sets of parameters extracted at 20 ms intervals, giving a 60 ms frame size. This enables the coder to quantize the three sets of parameters jointly, making the best use of the redundancies existing between them. This quantizer will be referred to as JQ-MSVQ, and the large frame composed of several speech frames will be referred to as a meta-frame. JQ-MSVQ is also used in a 4 kb/s version of the SB-LPC, where two sets of LSF extracted every 10 ms are quantized jointly, forming a 20 ms meta-frame.

One significant issue with a JQ quantizer is that of weighting. Various weighting functions have been discussed above and they can be used to provide weights for each individual set of LSFs. However, all sets of LSFs in a meta-frame are not usually of equal importance. For example, at a speech onset, the first set can be in a nonspeech region, whereas the other sets can be in a speech-active region. Therefore the weight vector should ideally take this into consideration, so as to maximize the quantization efficiency for the important sets and not waste bits on a set of LSFs which will have very little influence on the speech quality. This can be achieved by including a bias based on the relative energies of the speech for each set of LSFs and multiplying the weights for the nonspeech LSFs by a factor smaller than one. A value of 0.1 has been found to give good performance. It is risky to use a smaller value, as problems can arise from interpolation at the decoder if the 'not so important' set of LSFs is too poorly quantized.

Table 5.11 Comparative performance of JQ-MSVQ and MA-MSVQ

	JQ-MSVQ	MA-MSVQ	MA-MSVQ
Number of bits	44	15	18
Number of bits per 60 ms	44	45	54
Stages	8	3	3
Bit allocation	6,6,6,6,6,6,6,2	5,5,5	6,6,6
M	32	32	32
Complexity (per 60 ms)	374 400	62 400	124 800
Memory	13 560	960	1920
WMSE	1.541 e-04	2.574 e-04	1.594 e-04
Average SD (dB)	1.2576	1.6383	1.3053
Outliers at 2 dB (%)	4.6563	17.2014	4.3119
Outliers at 4 dB (%)	0.0	0.1159	0.0185

The performance of JQ-MSVQ is illustrated in Table 5.11. The LSF quantizer used in the 1.2 kb/s coder, which quantizes three sets of LSFs jointly using 44 bits in an 8-stage JQ-MSVQ quantizer, is compared against a classic MA-MSVQ quantizer of similar bit rate and one of similar performance. Complexity and memory requirements are also indicated. In this example, LSFs are extracted every 20 ms. The results clearly show the advantage of JQ-MSVQ over MA-MSVQ in terms of performance. JQ-MSVQ has the same performance at 44 bits as MA-MSVQ at 54 bits, and is far superior to the MA-MSVQ at 45 bits. Complexity is higher for the JQ-MSVQ, but this may be reduced by lowering the depth of the tree search M. Memory requirements are also higher for the JQ-MSVQ, but again they can be reduced by adding more structure to the codebook (more stages of smaller sizes) and accepting a slight reduction in performance. Overall, JQ-MSVQ is very effective at providing reasonable LSF quantization at very low bit-rates. At 1.2 kb/s, only 72 bits are available every 60 ms for quantizing all speech parameters. Assuming that the gain, pitch and voicing are quantized using 28 bits, only 44 bits are left for the spectral parameters. As shown in Table 5.11, an MA-MSVQ quantizer would not work well under those circumstances, giving significantly degraded speech quality with over 17 % outliers at 2 dB. However the use of JQ-MSVQ quantization makes a 1.2 kb/s coder a practical possibility, with only 4.6 % outliers at 2 dB.

5.9.6 Use of MA Prediction in Joint Quantization

When using JQ-MSVQ, the redundancies between the jointly quantized sets of LSFs are exploited. Using MA prediction within the meta-frame will not therefore achieve any more gain. Indeed, a JQ codebook using MA from one set to the next can be transformed to a JQ codebook using no MA, by simply adding the prediction to the predicted set. Therefore, MA prediction is only useful if it uses correlation with a previously transmitted set of LSFs, i.e. from one meta-frame to another. This means that the distance between the predicted sets and the sets used to compute the prediction is usually larger than in a nonJQ case, thereby reducing the efficiency of the prediction. Moreover, a channel error on JQ-MSVQ quantizers using a first-order MA will affect two meta-frames, instead of just two speech frames for a nonJQ quantizer. For the 1.2 kb/s configuration with a 60 ms meta-frame described above, this means an error will affect 120 ms of speech instead of 40 ms for a nonJQ quantizer at the same update rate. As a result errors will have a much greater impact on speech quality, as it is usually possible to limit the effect of the loss of 40 ms of speech, but not the loss of 120 ms of speech. Therefore MA prediction for JQ quantizers is mostly useful when the meta-frames are relatively small.

For example, two sets of LSFs obtained at a 10 ms update rate can be jointly quantized in a 20 ms meta-frame. The 20 ms meta-frame is small enough that the MA prediction will give good prediction gain, while keeping the error propagation down to a manageable level. For the prediction to be optimal, it is better to predict both sets of LSFs in the meta-frame with the last set of LSFs of the previous meta-frame. This way the time difference between the predicting and predicted frames is kept to a minimum. However the optimal prediction factors for both sets will not be the same, as the first set will be more correlated with the predicting set than the second set. Experiments using a 10 ms update rate with two sets forming a meta-frame indicated that, for this configuration, prediction factors of $\{0.5, 0.4\}$ are suitable, i.e. the first set is predicted with a factor of 0.5 and the second set with 0.4. The quantizer jointly quantizes two sets of LSFs extracted every 10 ms using 36 bits. Codebooks are organized in six stages of 6 bits each, using the group delays weighting method and a tree search of depth 32. The results are shown in Table 5.12. They show that the MA prediction gives a large performance gain over the nonMA case, and that using both MA and JQ together allows two sets of LSFs to be quantized accurately with only 36 bits, i.e. only 18 bits per set. The performance gain given by the MA predictor in the JQ case is consistent with that observed in the nonJQ case.

5.10 Improved LSF Estimation Through Anti-Aliasing Filtering

When estimating speech model parameters at about 50 Hz over a 20–30 ms analysis window, speech is assumed to be locally stationary [14] within this analysis window. When closely investigated however, one can see that speech has considerable variation even within the analysis window. Speech parameters in general, and LSFs in particular, may contain high frequency

Table 5.12 Performance comparison of JQ-MSVQ and MA-JQ-MSVQ

	JQ-MSVQ	MA-JQ-MSVQ
Number of bits per 20 ms	36	
Number of bits per set	18	
Stages	6	
Bit allocation	6,6,6,6,6,6	
M	32	
WMSE	2.227 e-04	1.622 e-04
Average SD (dB)	1.0926	0.9335
Outliers at 2 dB (%)	1.8135	0.5627
Outliers at 4 dB (%)	0.0052	0.0052

variations which violate the Nyquist sampling criterion. Therefore the use of an anti-aliasing filter with cut-off frequency adequate for the chosen LSF sampling rate may be used to alleviate possible spectral overlapping of the LSFs. It is confirmed [15] that this method offers an advantage over the classic LSF extraction methods; during quantization, bit-saving and significant reduction in the percentage of outliers have been possible.

5.10.1 LSF Extraction

Al-Naimi investigated the speech stationarity assumption over the analysis window with regard to LSF vector extraction by calculating LSF vectors at every sample [15]. The centre of the analysis window is shifted by one sample at a time, leading to an LSF vector extraction rate of 8 kHz. Evolution of each LSF parameter over time, also referred to as an LSF track, is then produced from the over-sampled LSF vectors. Decimation without any filtering of the LSF tracks at a given LSF vector transmission rate (i.e. 20, 10 or 5 ms) should produce exactly the same LSF vectors as the classic methods. It is therefore clear that LSF track frequency variations greater than half of the LSF computation rate (frequency) will cause problems during the decimation process i.e. by introducing aliasing distortion. Note that the LSF computation rate need not be same as the frame transmission rate.

In order to measure the amount of aliasing introduced, the following test was used:

1. Ten LPC parameters were calculated for every sample using Hamming windowing over 200 samples and bandwidth expanded by 15 Hz, then converted to LSFs.
2. The evolution of each LSF track f_i over time was taken and FFT transformed. The logarithmic magnitude of the FFT spectrum is shown in Figure 5.18.

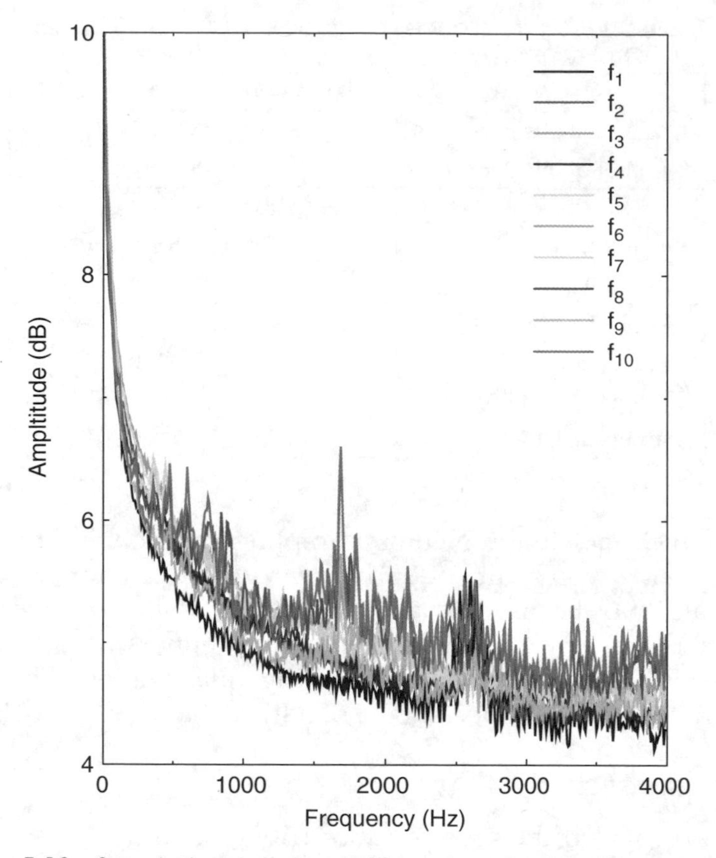

Figure 5.18 Spectral variations of LSF tracks calculated every sample

In Figure 5.18, we can see that most of the LSF tracks' spectra have a substantial amount of their energy in the low frequency band (below 100 Hz). However, if a coder calculates its LSF vectors every 20 ms (i.e. 50 Hz sampling), for example, then all the energy in the band greater than 25 Hz will be a source of spectral overlapping, producing inaccurate LSF parameters. In order to identify the source of these high frequencies, two further tests were carried out.

- **Window position test**
  A synthetic speech segment was prepared by repeating a whole pitch cycle from a voiced speech segment. Using this as the input, LSF vectors were calculated every sample, as before. The results showed that the LSF tracks were not affected by the positioning of the window. Therefore the conclusion was that as long as the window size is of sufficient length and the speech contained in the window is stationary, the window position will not be the source of the high frequency components evident in the spectra of the LSF tracks.

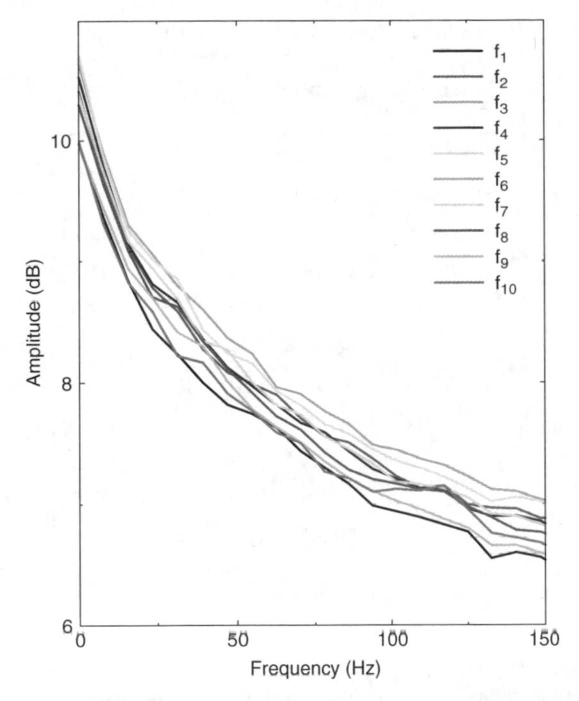

Figure 5.19 Low frequency region of the plots in Figure 5.18 expanded

- **The wide-sense stationary assumption of speech**
  In general, a signal $s(t)$ is said to be wide-sense stationary (WSS) if the expectation, $E\{s(t)s(t+\tau)\}$, is independent of time t and only dependent on the time difference τ. In the window position test, the LSF tracks do not contain high frequencies indicating that, for the synthetic speech file, the WSS assumption is valid. In reality, speech is changing in characteristics during the analysis frame. Therefore, the stationary assumption of our speech segment within the analysis window is not strictly correct and this is why high frequency variations are evident in the spectra of the LSF tracks.

Table 5.13 shows the percentage of energies for three different bands computed over four male and four female speakers each uttering eight seconds from the NTT speech database. The band below 25 Hz corresponds to a 20 ms LSF vector transmission rate whereas a band below 50 Hz corresponds to a 10 ms LSF vector transmission rate. Even though more than 92 % of the energy is present in the band below 25 Hz, the remaining 8 % of the energy is enough to produce higher LSF parameter variations in some specific speech sections (Note that these figures are average over 32 seconds of speech and instantaneous variations are much larger.) Therefore, following from the discussion

Table 5.13 Percentage of energy contained in frequency bands: A<25 Hz, 25 Hz≤B<50 Hz, 50 Hz≤C<100 Hz and D≥100 Hz

LSF parameters	Frequency band			
	A	B	C	D
f_1	94.52	4.24	1.07	0.17
f_2	95.44	3.61	0.83	0.12
f_3	96.67	2.71	0.54	0.08
f_4	96.81	2.56	0.54	0.09
f_5	98.10	1.51	0.33	0.05
f_6	97.46	1.99	0.45	0.10
f_7	96.36	2.88	0.64	0.12
f_8	95.54	3.28	0.71	0.47
f_9	94.64	4.41	0.98	0.24
f_{10}	92.72	3.97	1.13	2.18

above, a low pass filtering as a preprocessing stage prior to decimation has been proposed [15] to alleviate the possible spectral overlapping distortion.

Of course one may question the use of low-pass filtering when the same can be achieved by increasing the analysis window length with overlapping. Increasing the analysis window length, i.e. to greater than two and a half times the average pitch, would increase the frequency resolution, but in the time domain, the speech signal would have evolved considerably during a longer analysis window. Even though a large window may result in smoothed spectra, important details within the frame will not be modelled accurately. In addition, even if the window length was increased there would still be no guarantee that the high frequency components of the LSF tracks would not be present. Al-Naimi's proposal of the use of a low-pass filter with a cut-off frequency that is dependent on the LSF vector transmission rate, is therefore justifiable [15].

The following set-up has been used to show the effect of low-pass filtering over 8 seconds of speech [15]. First the LSF vectors **f** were extracted every frame from the tracks f_i which are formed by calculating the LSFs every sample. Next, filtering was applied in the frequency domain separately for each LSF track, f_i, with a cut-off frequency that is dependent on the LSF vector transmission rate and another set of LSFs $\mathbf{g} = g_1, g_2, g_3, \ldots, g_p$ were extracted. In order to avoid the rectangular windowing effect at the edges of the blocks, one large FFT transformation was used for whole of the 8 seconds. Figures 5.20–5.23 show a section of the variations of certain LSF tracks for both classic f_i and low-pass filtered g_i methods. It is evident in these figures that significant variations exist in the LSF tracks produced by the classic

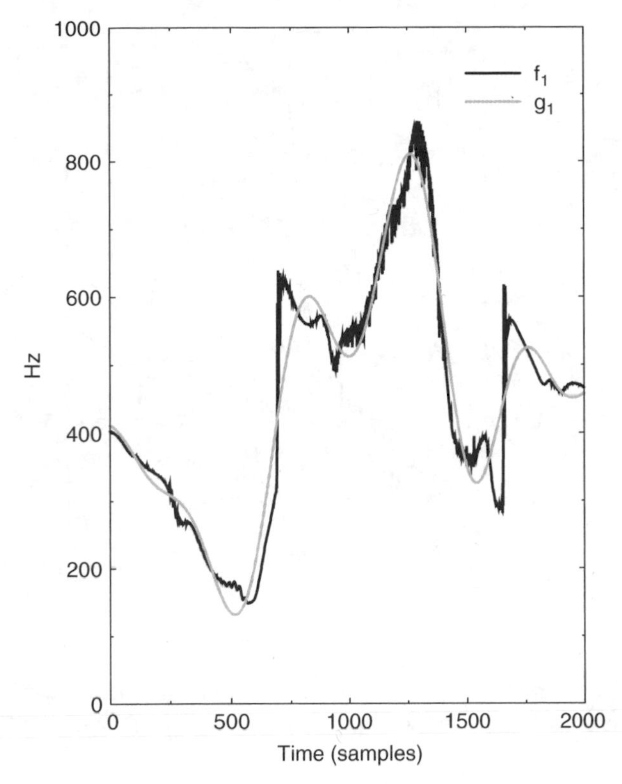

Figure 5.20 LSF tracks f_1 and g_1 variations over time

method due to the weak stationarity assumption within the analysis window, especially at transitions from voiced speech segments to unvoiced (offsets) and vice versa (onsets). The low-pass filtered method, on the other hand, produces smoother and more slowly evolving LSF tracks. The differences in the LSF tracks are more evident in the higher LSF parameters (f_7 and f_{10}) as shown in Figures 5.22 and 5.23.

Work in [16] showed that using a perceptually-smoothed power spectral envelope leads to a significant increase in subjective performance. Additionally, [17] showed that low-rate quantization is possible through smoothing the LSF parameter evolution. An informal listening test comparing both the classic, $\mathbf{f}$, and low-pass filtered, $\mathbf{g}$, LSF vectors used in a 4 kb/s SB-LPC coder showed no difference in speech quality. An advantage during quantization is therefore expected with regard to smoother evolution of the LSF tracks.

5.10.2 Advantages of Low-pass Filtering in Moving Average Prediction

Although using the unquantized LSF parameters for both the new and classical methods did not show any subjective quality difference, the new method is expected to produce better performance under predictive quantization.

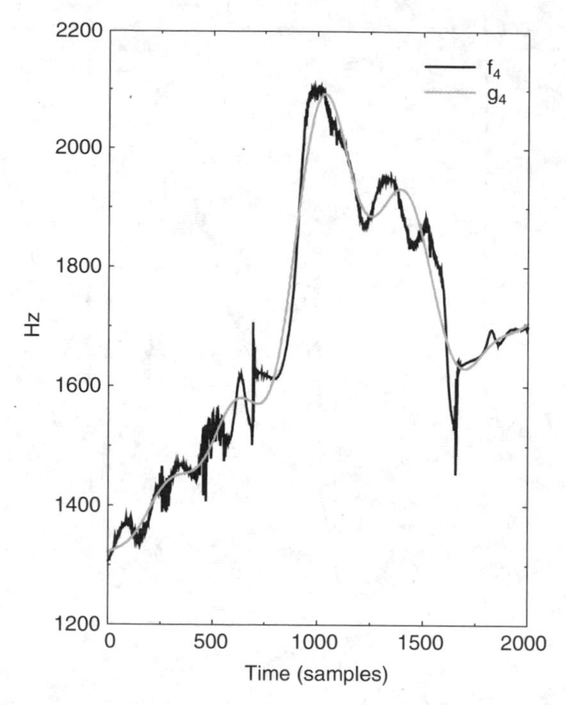

Figure 5.21 LSF tracks f_4 and g_4 variations over time

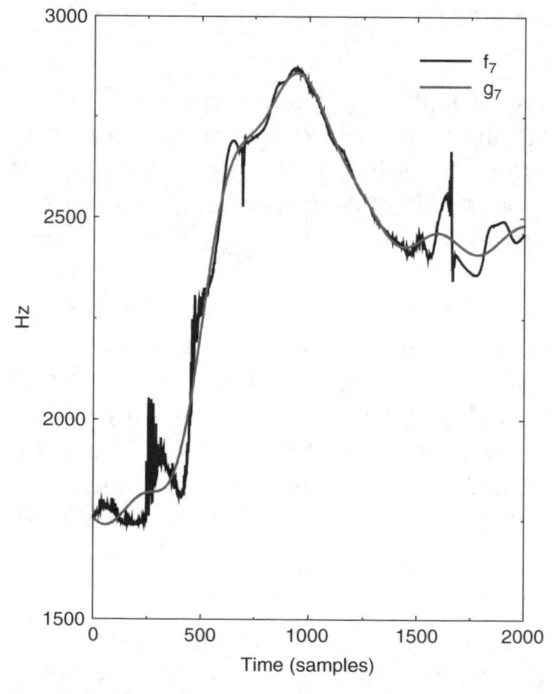

Figure 5.22 LSF tracks f_7 and g_7 variations over time

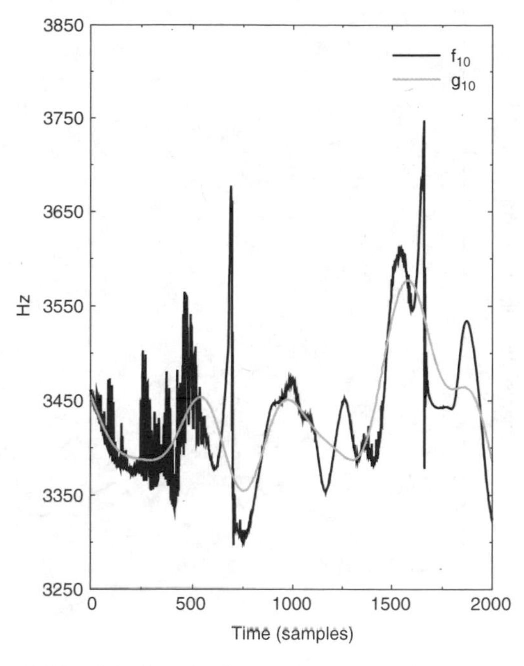

Figure 5.23 LSF tracks f_{10} and g_{10} variations over time

This advantage is shown in the following test. First, the classical method of LSF extraction is applied at various update rates. Next, the low-pass filtered method is used where LSFs are calculated at every sample. Each LSF track is then filtered with a low-pass filter which had its cut off frequency suitably selected to be half of the LSF transmission frequency. A subsampling is then applied to get the required number of LSF vectors. Finally, the variance for each set of LSF vectors is computed after a single-order MA prediction. According to the earlier observations, the new method is expected to produce smaller prediction residual with a greater prediction coefficient owing to its smoother evolution and hence higher correlation between successive sets. Figure 5.26 shows that for a 20 ms update rate, the variance of the LSF prediction residual is lower for the new method and the minimum variance (best prediction) occurs at a higher value of prediction coefficient which indicates that the new method produces LSF vectors that are more correlated. Figures 5.24–5.28 show similar results for various other LSF vector transmission rates. It can be seen that the variance of the LSF prediction residual is always less in the new method, regardless of the LSF vector rate. In order to quantify the amount of prediction achieved, prediction gain, P_g, is computed using,

$$P_g = \frac{x_0 - x_{\min}}{x_0} \times 100 \tag{5.79}$$

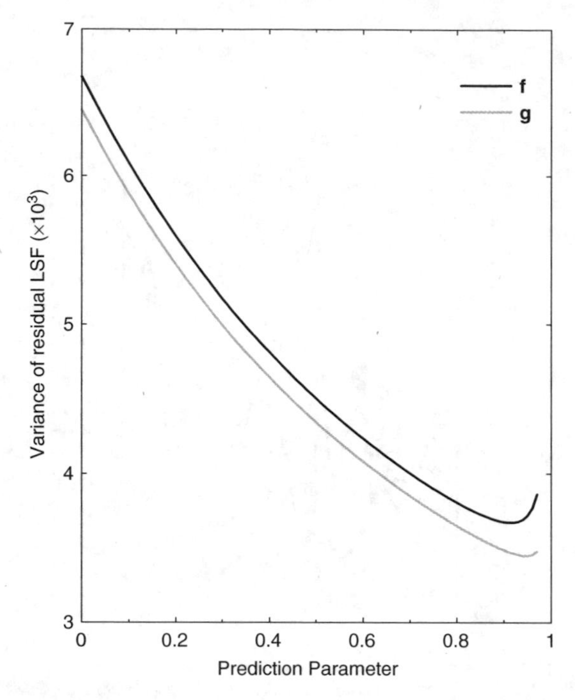

Figure 5.24 Moving average LSF prediction residual variance for 5 ms LSF vector rate

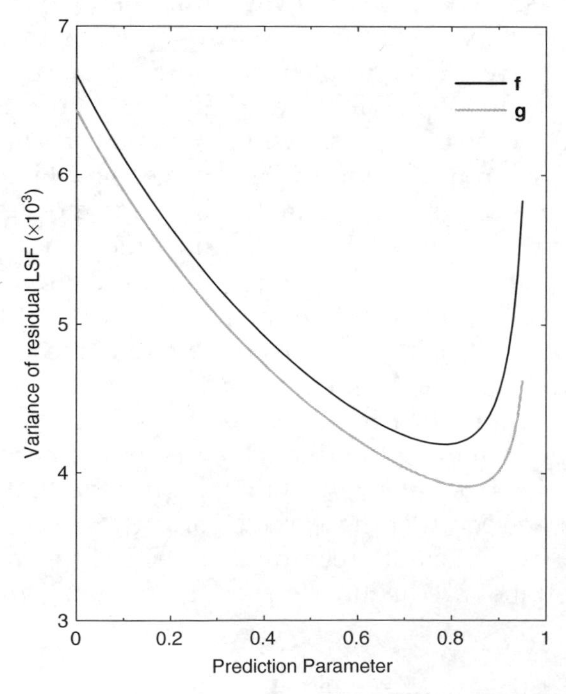

Figure 5.25 Moving average LSF prediction residual variance for 10 ms LSF vector rate

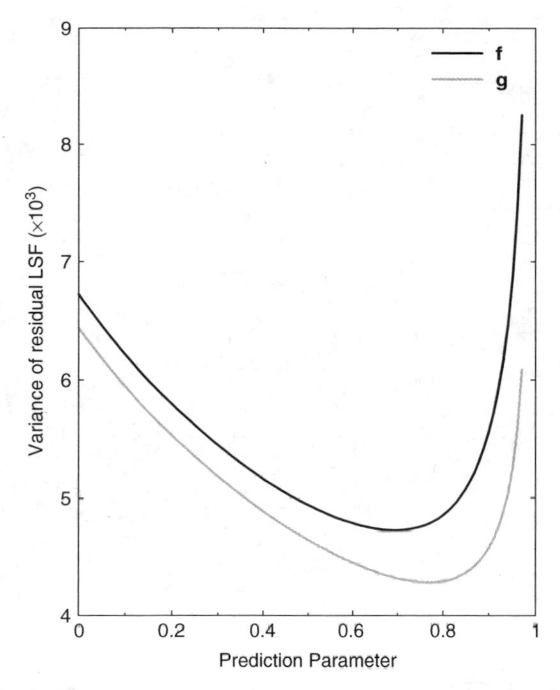

Figure 5.26 Moving average LSF prediction residual variance for 20 ms LSF vector rate

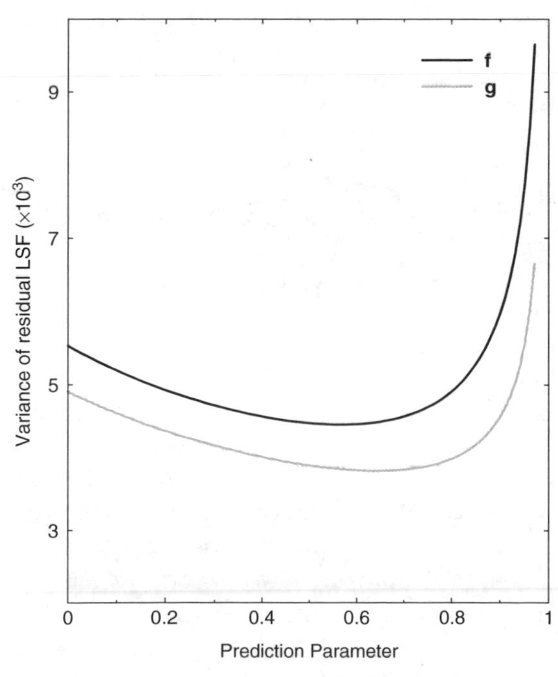

Figure 5.27 Moving average LSF prediction residual variance for 30 ms LSF vector rate

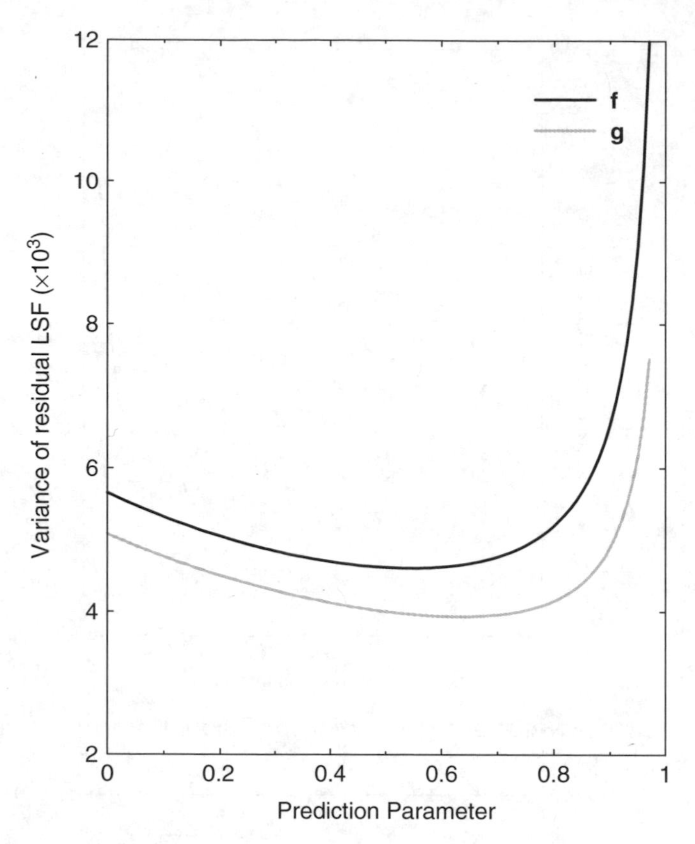

Figure 5.28 Moving average LSF prediction residual variance for 40 ms LSF vector rate

Table 5.14 Prediction gain for low-pass filtered and classic LSF extraction methods at various vector rates

	LSF vector transmission rate				
	40 ms	30 ms	20 ms	10 ms	5 ms
P_g for **g**	29.55	33.82	36.53	43.34	49.57
P_g for **f**	12.50	16.60	29.60	37.60	42.60

where x_0 is the variance of LSF prediction residual when the prediction factor is zero (the original LSF variance) and x_{min} is the minimum variance of LSF prediction residual computed by selecting the optimum prediction coefficient. Higher P_g is an indication of performance improvement that can be achieved through MA prediction before quantization. Table 5.14 shows the value of prediction gains at different LSF vector transmission rates. The

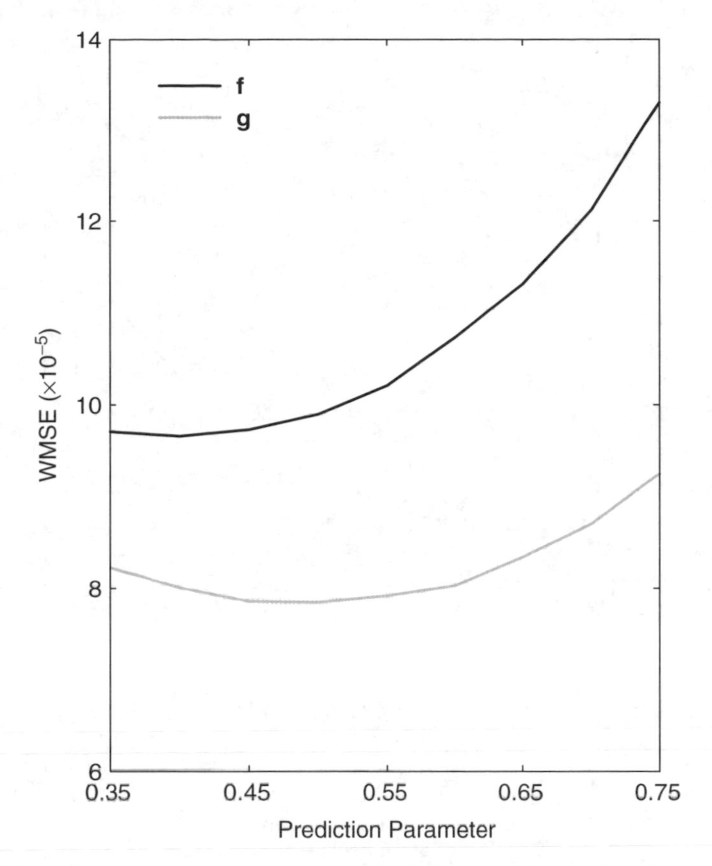

Figure 5.29 WMSE performance curves for a range of MA prediction parameters

new method always has a higher prediction gain compared to the classic extraction and, as expected, the difference between them becomes smaller for higher LSF update rates.

In the MA predictor used above, the prediction is a function of the unquantized LSF prediction residual from the preceding set. As shown before, when used in a quantizer such as MSVQ, the prediction will be a function of the quantized LSF prediction residual and, hence, it is expected to be different.

Figures 5.29–5.31 show the effect of quantizing the prediction residual on the moving average prediction coefficient. These results were obtained by varying the prediction coefficient in steps of 0.05 and training a multi-stage VQ with three stages of 7 bits each with M-best factor of eight. Table 5.15 gives a summary of the comparative performance results for the classic and new methods of LSF extraction where the low-pass filtered method significantly outperforms the classical method.

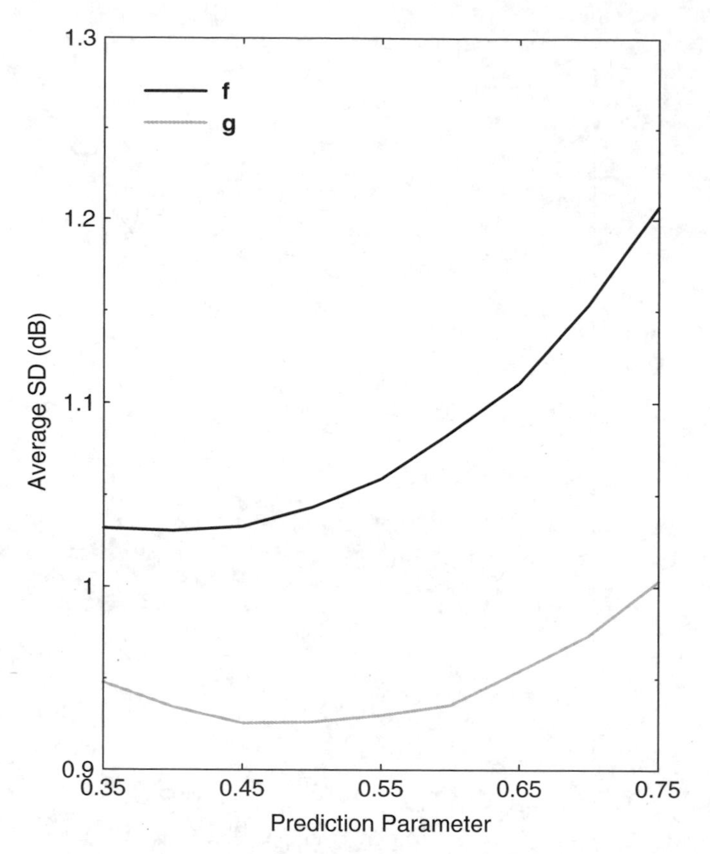

Figure 5.30 Average SD performance curves for a range of MA prediction parameters

Table 5.15 Performance comparison of the new and classical LSF extraction with quantization

	Prediction parameter	Average SD (dB)	2 dB outlier (%)	4 dB outlier (%)	WMSE
g	0.5	0.926	0.036	0	7.85E-05
f	0.4	1.031	0.23	0	9.66E-05

As the WMSE, average SD and percentage of 2 dB outliers is significantly lower for the new method, bit savings can be achieved whilst maintaining the same performance as the classic LSF VQ. The percentage of 4 dB outliers is not shown since it was zero in all cases. Other bit combinations in a three-stage MSVQ are shown in Table 5.16 where the new method has a clear advantage. Figures 5.32–5.35 present the results obtained for WMSE, average SD (ASD)

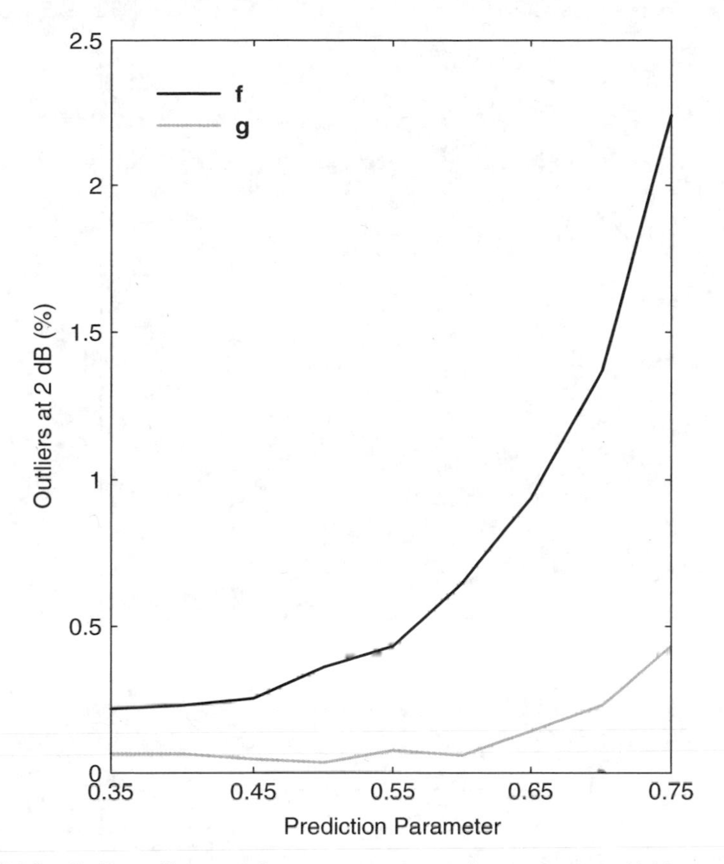

Figure 5.31 2 dB outliers performance curves for a range of MA prediction parameters

Table 5.16 Bit allocation for MA-MSVQ codebooks

Total bit allocation	Bits allocated per codebook stage
15	5,5,5
16	6,5,5
17	6,6,5
18	6,6,6
19	7,6,6
20	7,7,6
21	7,7,7
22	8,7,7
23	8,8,7
24	8,8,8

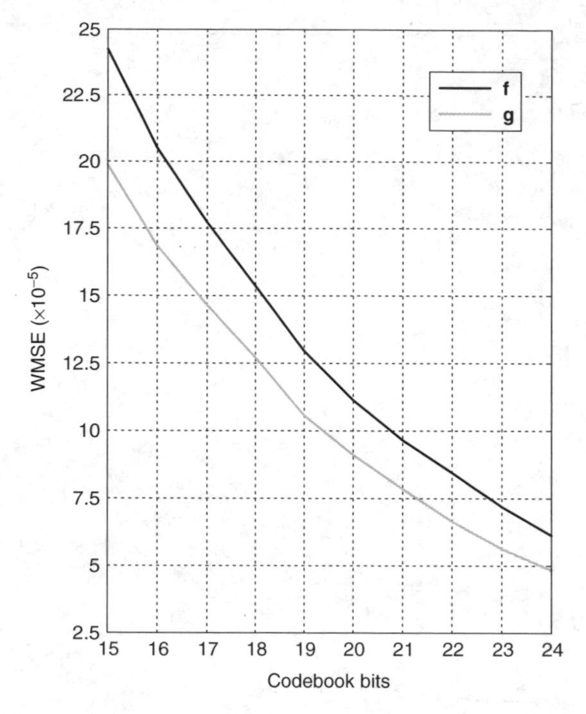

Figure 5.32 WMSE performance curves for a range of codebook bit allocations

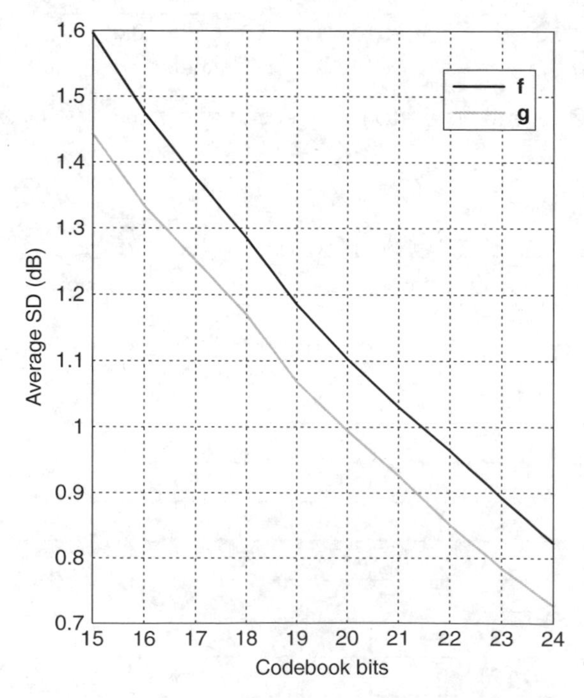

Figure 5.33 Average SD performance curves for a range of codebook bit allocations

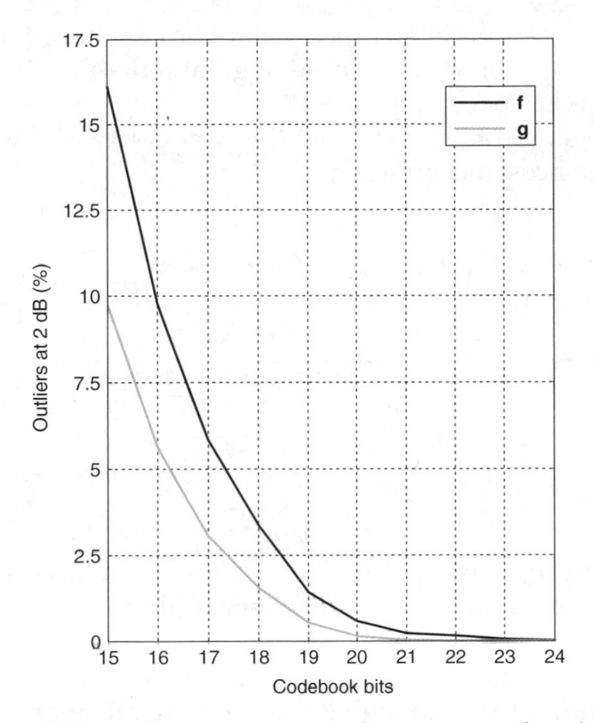

Figure 5.34 2 dB outliers performance curves for a range of codebook bit allocations

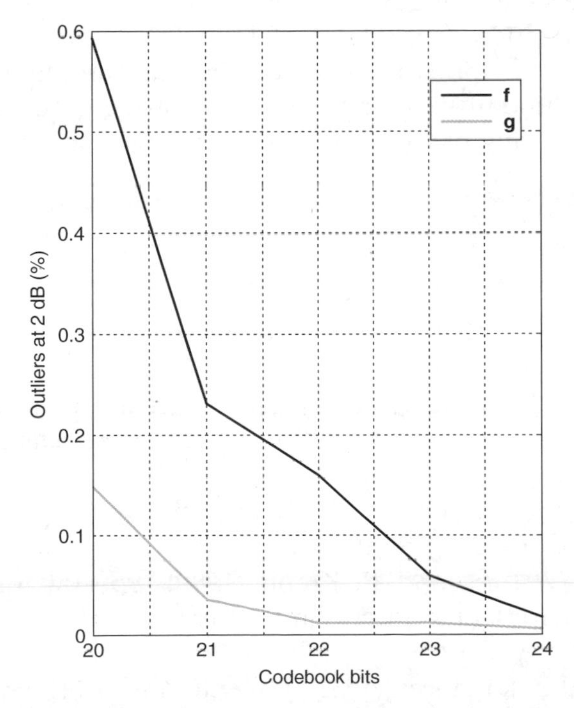

Figure 5.35 2 dB outliers performance curves for codebook bit allocations ranging from 20 to 24 bits

and percentage of 2 dB outliers respectively. The percentage of 4 dB outliers is shown in Table 5.17. For VQ bit allocation greater than 18 bits, the percentage of 4 dB outliers is zero.

Informal listening tests also showed the superiority of the new method at the expense of increased complexity.

Table 5.17 Percentage of 4 dB outliers

VQ bit allocation	15 bits	16 bits	17 bits	18 bits
g	0.0059	0.0059	0.0	0.0
f	0.0415	0.0119	0.0059	0.0

5.11 Summary

This chapter has presented the fundamental aspects of speech spectral representation via linear prediction. Accurate spectral representation of speech is crucial to the performance of low bit-rate speech coders, especially in sinusoidal coders where the simplified excitation model cannot compensate for shortcomings in the LPC modelling. Various quantization techniques have been investigated and have lead to the design of LSF quantization schemes optimized for specific configurations. These schemes offer solutions to LPC parameter quantization in the form of LSFs for several applications, with varying degrees of performance at a given bit rate and implementation complexity. A more fundamental approach to improving quantization by way of anti-aliasing filtering has also been presented with increased quantization performance.

Bibliography

[1] S. Saito and K. Nakata (1985) *Fundamentals of Speech Signal Processing*, Chapter 9. Academic Press

[2] F. Soong and B. H. Juang (1984) 'Line spectrum pairs and speech data compression', in *Proc. of Int. Conf. on Acoust., Speech and Signal Processing*, pp. 1.10.1–4.

[3] L. Rabiner and R. Schafer (1978) *Digital Processing of Speech Signals*. Englewood Cliffs, NJ: Prentice-Hall

[4] B. Atal, R. V. Cox, and P. Kroon (1989) 'Spectral quantization and interpolation for CELP coders', in *Proc. of Int. Conf. on Acoust., Speech and Signal Processing*, pp. 69–72.

[5] G. Kang and C. J. Francen (1984) 'Low bit rate speech encoders based on line spectrum frequencies (LSF)', *Tech. Rep. 8857*, U.S. Naval Lab. Report.

[6] P. Kabal and R. P. Ramachandran (1986) 'The computation of line spectral frequencies using Chebyshev polynomials', in *IEEE Trans. on Acoust., Speech and Signal Processing*, 34(6):1419–25.

[7] B. Cheetham (1987) 'Adaptive LSP filter', in *IEE Electronics Letters*, 23:89–90.

[8] ITU-T (1992) *Coding of speech at 16 kbit/s using low-delay code excited linear prediction*, ITU-T Rec. G.728.

[9] K. K. Paliwal and B. Atal (1993) 'Efficient vector quantisation of LPC parameters at 24 bits/frame', in *IEEE Trans. on Acoust., Speech and Signal Processing*, 1(1):3–14.

[10] ETSI (1998) *Digital cellular telecommunications system (phase 2); Enhanced full rate (EFR) speech transcoding*, GSM 06.60 v4.1.0 (ETS 301 245), June.

[11] F. Tzeng (1989) 'Analysis by synthesis linear predictive speech coding at 2.4 kb/s', in *Proc. of Globecom*, pp. 1253–7.

[12] A. McCree, K. Truong, E. B. George, T. P. Barnwell, and V. Viswanathan (1996) 'A 2.4 kbit/s MELP coder candidate for the new US Federal Standard', in *Int. Conf. on Acoust., Speech and Signal Processing*, 1:200–203.

[13] W. P. LeBlanc, B. Bhattacharya, S. A. Mahmoud, and V. Cuperman (1993) 'Efficient search and design procedures for robust multi-stage VQ of LPC parameters for 4 kb/s speech coding', in *IEEE Trans. on Speech and Audio Processing*, 1(3):373–85.

[14] J. Makhoul (1975) 'Linear prediction: A tutorial review', in *Proc. of IEEE*, 63:561–80.

[15] K. T. Al-Naimi (2002) 'Advanced speech processing and coding techniques', Ph.D. thesis, CCSR, University of Surrey, UK.

[16] W. B. Kleijn and J. Hagen (1995) 'Waveform interpolation for coding and synthesis', in *Speech coding and synthesis* by W. B. Kleijn and K. K. Paliwal (Eds), pp. 175–207. Amsterdam: Elsevier Science

[17] T. Eriksson, H-G Kang, and P. Hedelin (2000) 'Low-rate quantization of spectrum parameters', in *Int. Conf. on Acoust., Speech and Signal Processing*, pp. 1447–50.

6

Pitch Estimation and Voiced–Unvoiced Classification of Speech

6.1 Introduction

Low bit-rate speech coders, traditionally called vocoders, rely heavily on extracting the correct speech parameters from a given speech segment. The three main speech features are the spectral envelope, the pitch and the voiced–unvoiced classification. The spectral envelope is usually extracted by a standard autocorrelation method which results in a linear predictive (LP) parameters representation. However extracting the correct pitch and voicing classification is not as straightforward and may require a combination of methods.

When measuring the pitch, it is assumed that the voiced signals are formed by passing quasi-periodic excitation signals through the LPC filter. The duration between the pulses in the excitation signal is called the pitch period T_0 or fundamental frequency f_0. Correct estimation of the pitch is essential for good quality speech-coding. Incorrect estimation of the pitch period can seriously degrade the quality of synthesized speech. Pitch determination algorithms (PDAs) have been studied in both the time and frequency domains, and a comparison is discussed in [1, 2]. Traditionally, autocorrelation-based methods [3] and their variants [4, 5] have been intensively investigated and widely applied to various speech coders [6–11]. Frequency domain approaches [12–14], on the other hand, have become popular recently due to the growing interest in sinusoidal speech coders, such as the multi-band excitation (MBE) [13] and the sinusoidal transform coder (STC) [14], which conduct pitch determination based on a spectral synthesis (SS) method.

Digital Speech. A. Kondoz
 ISBN 0-470-87007-9 (HB)

In addition to correct pitch estimation, correct voiced–unvoiced estimation is also crucial for good quality speech synthesis. Traditional vocoders, which have been in use for many years, classify the input speech signal either as voiced or unvoiced. A voiced speech segment is known by its relatively high energy content but, more importantly, it contains periodicity. The unvoiced part of speech on the other hand looks more like random noise with no periodicity. However, there are some parts of speech that are neither voiced nor unvoiced, but a mixture of the two. These are usually called the transition regions, where there is a change either from voiced to unvoiced or unvoiced to voiced. In low bit-rate speech coding, correct classification of speech blocks (usually frames or subframes 20 ms long, or shorter) is very critical for good quality speech synthesis. If voiced speech is classified as unvoiced, the synthesized output will sound rough and less intelligible. If, on the other hand, unvoiced speech is classified as voiced, the synthesized speech will sound annoyingly metallic or robotic. In older versions of vocoders, a hard decision voicing was used and the transitions were classified into either fully voiced or fully unvoiced. In newer vocoders, such as sinusoidal based coders (IMBE, MELP, etc.), soft decision voicing is employed: a third class, in which both voiced and unvoiced exists together, has been defined. This mix of voiced and unvoiced decision is usually carried out in the frequency domain where voiced and unvoiced frequencies are appropriately selected to represent the mixed signal. As a result, better quality synthesized speech is produced. In this chapter we review some of the advanced techniques which are used in extracting the correct pitch and subsequently estimating the correct voicing in each speech segment.

6.2 Pitch Estimation Methods

The excitation model used in source-filter vocoders relies heavily on the correct determination of the pitch parameter. Incorrect pitch estimation may significantly degrade the speech quality, and in particular its intelligibility, by introducing artifacts into the synthetic speech. Moreover, other parameter estimations such as voicing and spectral amplitudes in vocoders often assume accurate pitch determination, and are severely affected by pitch errors. Therefore, the reliability of the pitch determination algorithm (PDA) used has a dramatic effect on the quality of the synthesized speech.

Pitch period is defined as the time interval between two consecutive voiced (periodic) excitation cycles. Although, this interval may vary from cycle to cycle, it usually evolves slowly, and therefore it can be estimated. Estimating the pitch period is generally easy for highly periodic sounds, but some speech segments do not exhibit such characteristics. In some parts of speech as well as having the pitch period varying the speech may contain a mixture of voiced (periodic) and unvoiced (random) signals which may cause estimation errors. Formant interaction can also be a problem as the speech may become highly resonant and this may cause incorrect pitch estimation. Onsets and offsets are also problem areas. Finally, large amounts of background noise present in the signal can also complicate the task of the PDA.

PDAs are generally classified in two main categories: time or frequency domain techniques. However in the last few years more complicated techniques which use both time and frequency domain characteristics of speech have been developed. These are summarized below.

6.2.1 Time-Domain PDAs

The most obvious feature of periodic signals is the similarity of the waveform at different times. The main principle of pitch detection algorithms (PDAs) which rely on time-domain waveform similarities is to find the pitch period by comparing the similarity between the original signal and its shifted version. If the shifted distance is equal to the pitch period, the two signal waveforms should have the greatest similarity. The majority of existing PDAs are based on this concept. Among them, the average magnitude difference function (AMDF) and the autocorrelation (AC) method are the two most widely used.

Average Magnitude Difference Method

A simple way to compare the current speech with its time-delayed version is to compute the average magnitude difference function (AMDF) [4] given by:

$$A(\tau) = \sum_{n=0}^{N-1} |s(n) - s(n-\tau)| \tag{6.1}$$

where τ is the lag. This function is computed over a given pre-determined range for τ and the value of τ minimizing $A(\tau)$ is selected as the pitch period. The value of N is typically 160 samples, corresponding to a 20 ms speech frame. A plot of the AMDF function against the speech signal is shown in Figure 6.1. The main advantage of the AMDF function is that it only requires additions and subtractions, making it very suitable for hardware implementation. However, current DSPs normally offer a one-cycle multiply–add instruction, making this less significant. The performance of the AMDF function is relatively poor and, in particular, it does not cater for variations in the energy of the speech.

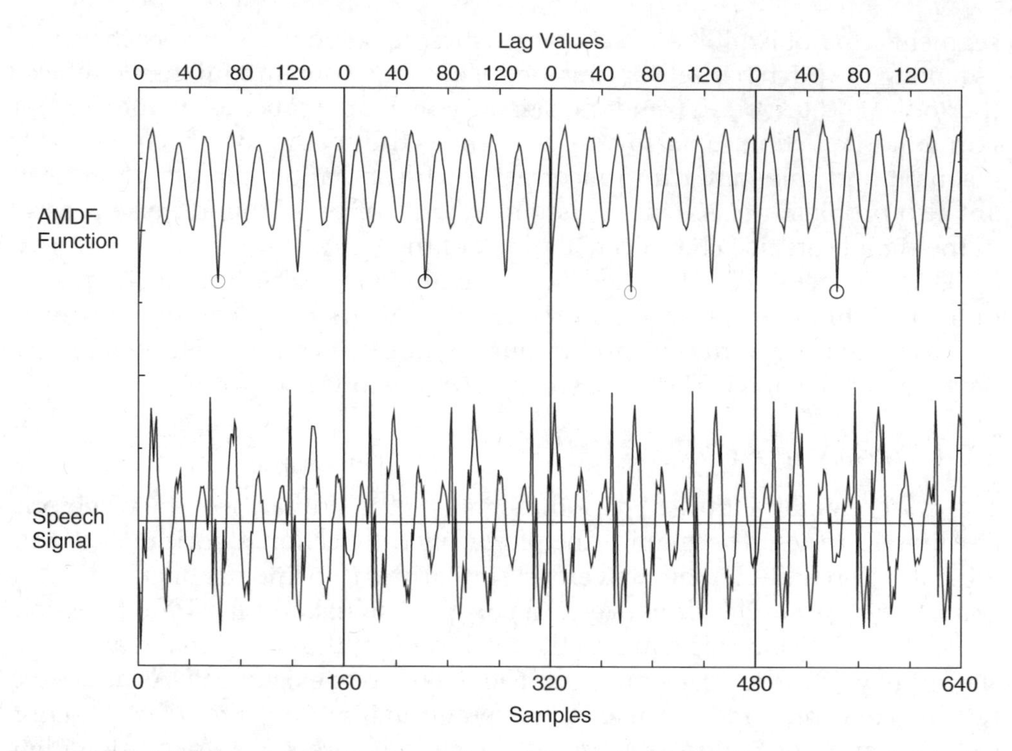

Figure 6.1 AMDF and speech signal: the minima of the AMDF corresponding to the pitch values are indicated by circles

Autocorrelation Method

The direct distance measurement is the most popular criterion examining the similarity between two waveforms; it can be expressed as,

$$E(\tau) = \frac{1}{N}\sum_{n=0}^{N-1}[s(n) - s(n+\tau)]^2 \tag{6.2}$$

Equation (6.2) assumes that the average signal level is fixed. However, at speech onsets and offsets this is not true, hence, the use of the normalized similarity criterion which considers the effect of nonstationarity of speech signals,

$$E(\tau) = \frac{1}{N}\sum_{n=0}^{N-1}[s(n) - \beta s(n+\tau)]^2 \tag{6.3}$$

where β is a scaling factor, or the pitch gain, controlling the changes in signal level. Under the assumption that the signal is stationary (i.e. $s(n) = s(n+\tau)$), the error criterion of equation (6.2) can be written as,

$$E(\tau) = [R(0) - R(\tau)] \tag{6.4}$$

where

$$R(\tau) = \sum_{n=0}^{N-1} s(n)s(n+\tau) \tag{6.5}$$

The minimization of the estimation error, $E(\tau)$, in equation (6.2) is equivalent to maximizing the autocorrelation (or cross-correlation) $R(\tau)$. The variable τ is called lag, or delay, and the pitch is equal to the value of τ which results in the maximum $R(\tau)$. Although the autocorrelation computation involves a large number of multiplications, it is very easy to implement these in real-time due to its regular form of computation, i.e. multiply–adds. With today's modern DSPs, multiply–add operations are very easily computed in one instruction. Another advantage of autocorrelation PDA is that it is phase-insensitive. Hence, it performs well in detecting the pitch of speech which may suffer some degree of phase distortion.

Nguyen generalized the direct similarity measure [15] as,

$$E(\tau) = \frac{1}{N}\left\{\sum_{n=0}^{N-1} |s(n) - s(n+\tau)|^k\right\}^{\frac{1}{k}} \tag{6.6}$$

where k is a constant. Although k can be arbitrary, Nguyen proved that k values of 1, 2 and 3 are appropriate. In his experimental investigation, he showed that 2 is the most appropriate value for speech signals, implying that the autocorrelation method is superior to the AMDF. A typical autocorrelation function is shown in Figure 6.2.

Speech in the long-term is a nonstationary signal and the direct similarity criterion may exhibit large errors, implying fewer similarities in positions where the shift is equal to the real pitch period. Figure 6.3b illustrates the direct autocorrelation function which indicates more similarities in the triple pitch period as the amplitude increases. The normalized similarity criterion of equation (6.3) is derived under the consideration of such a nonstationary process. Setting $\partial E(\tau, \beta)/\partial \beta = 0$ in equation (6.3) the optimum normalization coefficient (pitch gain) can be calculated using,

$$\beta = \frac{\sum_{n=0}^{N-1} s(n)s(n+\tau)}{\sum_{n=0}^{N-1} s^2(n+\tau)} \tag{6.7}$$

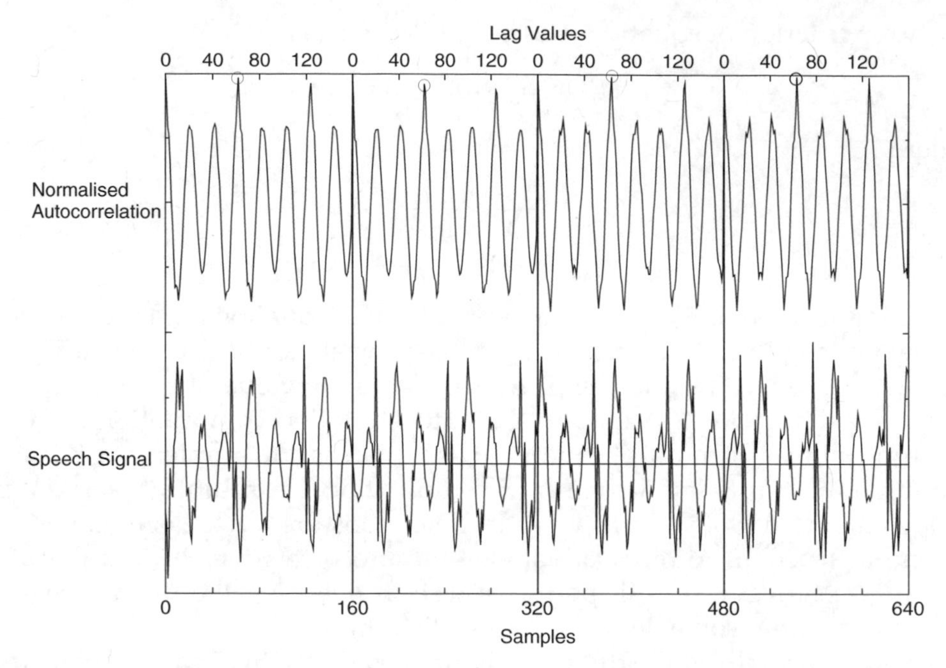

Figure 6.2 Autocorrelation and speech signal. The maxima of the autocorrelation function corresponding to the pitch values are indicated by circles

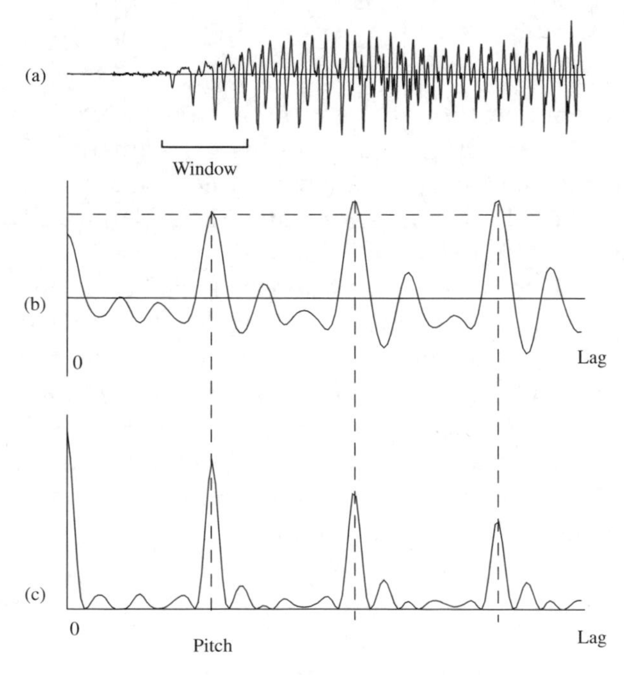

Figure 6.3 Autocorrelation of speech: (a) original speech, (b) direct autocorrelation function and (c) normalized autocorrelation function

By substituting the optimum gain back into the error function of equation (6.3), the pitch can be estimated by minimizing

$$E(\tau, \beta) = \sum_{n=0}^{N-1} s^2(n) - \frac{\left[\sum_{n=0}^{N-1} s(n)s(n+\tau)\right]^2}{\sum_{n=0}^{N-1} s^2(n+\tau)} \tag{6.8}$$

This is equivalent to maximizing the second term on the right hand side,

$$R_n^2(\tau) = \frac{\left[\sum_{n=0}^{N-1} s(n)s(n+\tau)\right]^2}{\sum_{n=0}^{N-1} s^2(n+\tau)} \tag{6.9}$$

Direct use of the above equation may result in some errors. This is because the square of the autocorrelation may result in a maximum even if the correlation is negative, forcing possible pitch-halving errors. In order to eliminate this problem, the square root of equation (6.9) is taken to remove the square from the correlation and, hence, eliminate the possibility of lags with negative correlation from being selected as the pitch. The final normalized autocorrelation function is therefore given by,

$$R_n(\tau) = \frac{\sum_{n=0}^{N-1} s(n)s(n+\tau)}{\sqrt{\sum_{n=0}^{N-1} s^2(n+\tau)}} \tag{6.10}$$

The normalized autocorrelation function, shown in Figure 6.3c, shows much better performance than the direct (un-normalized) autocorrelation method.

6.2.2 Frequency-Domain PDAs

Although most waveform similarity methods have their frequency domain equivalents, the frequency domain PDAs directly operate on the speech spectrum. The main frequency domain feature of a periodic signal is the harmonic structure, with the distance between harmonics being the fundamental frequency or the frequency equivalent of the pitch period. The main drawback of frequency-domain methods is their high computational complexity.

However, modern DSP techniques make the computational complexity of frequency-domain PDAs insignificant, making them very popular in sinusoidal coders. In the following, we briefly explain two frequency-domain PDAs.

Harmonic Peak Detection

An obvious way of determining the pitch in the frequency domain would be to extract the spectral peak at the fundamental frequency. This requires the first harmonic to be present, which cannot, in general, be expected because of the front-end filtering. A more practical method is to detect all of the harmonic peaks and then measure the fundamental frequency (pitch frequency) as either the common divisor of these harmonics or the spacing of the adjacent harmonics. This can be done using a comb filter given by

$$C(\omega, \omega_0) = \begin{cases} W(k\omega_0) & ; \quad \omega = k\omega_0, \ k = 1, 2, \ldots \frac{\Omega_m}{\omega_0} \\ 0 & ; \quad otherwise \end{cases} \tag{6.11}$$

and correlating it with the speech spectrum. The output of the correlation, $A_c(\omega_0)$, is the summation of weighted comb peaks as,

$$A_c(\omega_0) = \frac{\omega_0}{\Omega_m} \sum_{k=1}^{\Omega_m/\omega_0} S(k\omega_0)W(k\omega_0) \qquad \frac{2\pi}{\tau_{max}} \leq \omega_0 \leq \frac{2\pi}{\tau_{min}} \tag{6.12}$$

where Ω_m is the maximum frequency considered in the speech spectrum. If ω_0 is equal to the fundamental frequency, the comb response will match the harmonic peaks, and the maximum output will be obtained as shown in Figure 6.4. In order to obtain better subjective quality, a weighting coefficient can be applied to the individual teeth, normally decreasing weights with increasing frequency [16].

Spectrum Similarity

This method assumes that the spectrum is fully voiced and is composed only of a number of harmonics each located at multiples of the pitch frequency. A synthetic spectrum is reconstructed using this assumption for each possible pitch frequency candidate and is compared to the original spectrum. The pitch frequency leading to the best matching reconstructed spectrum is then selected [13] as the fundamental or pitch frequency. The speech spectrum is assumed to be composed of voiced harmonics only, located at multiples of the candidate pitch frequency ω_0. Therefore the synthetic spectrum $\hat{S}(m, \omega_0)$ is an approximation of the convolution of pulses located at multiples of the candidate pitch frequency ω_0, by the spectrum W of the window used on

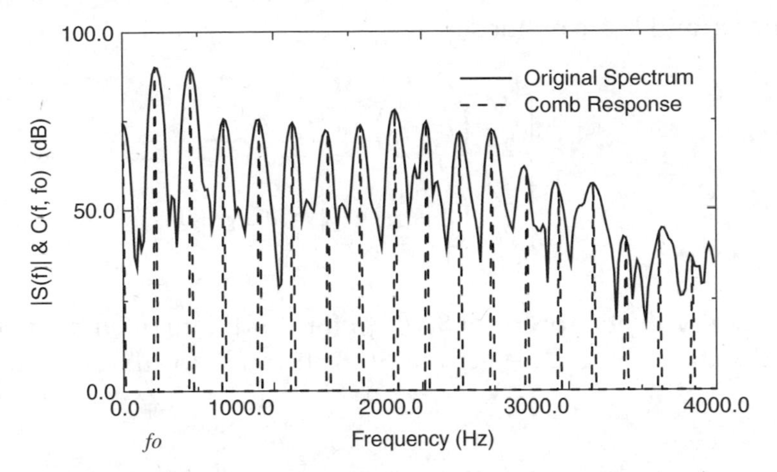

Figure 6.4 Harmonic peak matching method

the original speech prior to its Fourier transformation. The pulses are scaled by a factor $A_l(\omega_0)$ so as to provide the best possible match with the original spectrum. The synthetic spectrum $\hat{S}(m, \omega_0)$ is defined by:

$$\hat{S}(m,\omega_0) = \begin{cases} A_0(\omega_0)\, W\left(\frac{2\pi}{M}\, m\right) \\ A_1(\omega_0)\, W\left(\frac{2\pi}{M}\, m - \omega_0\right) \\ \vdots \\ A_l(\omega_0)\, W\left(\frac{2\pi}{M}\, m - l\,\omega_0\right) \\ \vdots \end{cases} \tag{6.13}$$

where M is the length of the DFT and $A_l(\omega_0)$ is defined as:

$$A_l(\omega_0) = \frac{\sum_{m=a_l}^{b_l} S(m) W\left(\frac{2\pi}{M}\, m - l\,\omega_0\right)}{\sum_{m=a_l}^{b_l} \left| W\left(\frac{2\pi}{M}\, m - l\,\omega_0\right)\right|^2} \tag{6.14}$$

$A_l(\omega_0)$ is such that the scaled harmonic lobe spectrum $A_l(\omega_0)W(\frac{2\pi}{M}m - l\omega_0)$ is the best possible match for $S(m)$, using an MSE criterion. The harmonic boundaries a_l and b_l are defined as:

$$a_l = \left\lceil \frac{M}{2\pi}\left(l - \frac{1}{2}\right)\omega_0 \right\rceil \tag{6.15}$$

$$b_l = \left\lfloor \frac{M}{2\pi}\left(l + \frac{1}{2}\right)\omega_0 \right\rfloor = a_{l+1} - 1 \tag{6.16}$$

Finally, the synthetic spectrum $\hat{S}(m, \omega_0)$ for candidate pitch frequency ω_0 is compared with the speech spectrum $S(m)$ through an MSE measure, given by:

$$E(\omega_0) = \sum_{m=0}^{M-1} \left(S(m) - \hat{S}(m, \omega_0)\right)^2 \tag{6.17}$$

The value of ω_0 minimizing $E(\omega_0)$ is then selected as the pitch frequency. Typical original and synthetic spectra with correct pitch are shown in Figure 6.5.

6.2.3 Time- and Frequency-Domain PDAs

Pitch Estimation using Spectral Autocorrelation

The time domain autocorrelation (temporal autocorrelation, or TA) has been used in various PDAs. Given a segment of speech signals $s(n)$, $0 \le n \le N-1$,

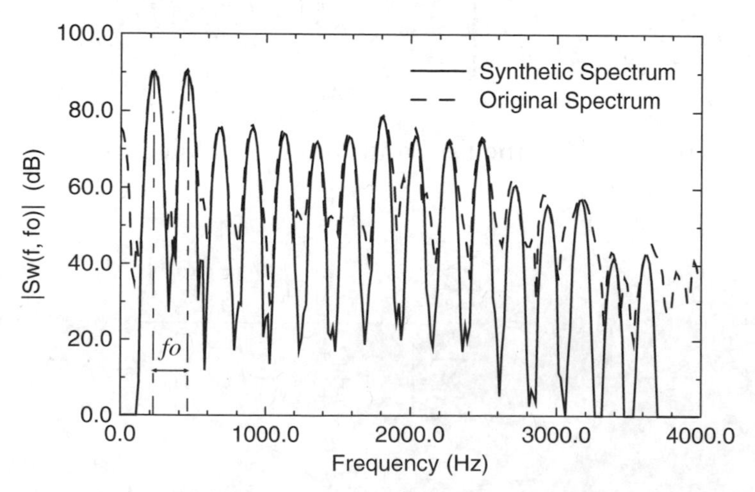

Figure 6.5 Original and synthesized speech spectra used in the spectrum-similarity PDA method

the normalized TA for a pitch candidate τ is given by

$$R_T(\tau) = \frac{\sum_{n=0}^{N-\tau-1} s(n)s(n+\tau)}{\sqrt{\sum_{n=0}^{N-\tau-1} s^2(n) \sum_{n=0}^{N-\tau-1} s^2(n+\tau)}} \tag{6.18}$$

which differs from the autocorrelation method discussed earlier in the limits of the summations (the earlier method was more like a cross-correlation). The TA has been widely used for PDAs due to its relatively good performance especially over noisy speech signals [2]. Autocorrelation can also be used in the frequency domain to bring out spectral similarities which are mainly due to the pitch frequency spacing of the harmonics. If the spectrum of windowed speech is given by $S(m) = A(m)e^{j\theta(m)}$ for $0 \le m \le M-1$, where $A(m)$ and $\theta(m)$ are the magnitude and phase of the normalized spectral autocorrelation (SA), $R_S(\tau)$ can be defined as

$$R_S(\tau) = \frac{\sum_{m=0}^{\lfloor M/2 \rfloor - \omega_\tau} A_z(m) A_z(m+\omega_\tau)}{\sqrt{\sum_{m=0}^{\lfloor M/2 \rfloor - \omega_\tau} A_z^2(m) \sum_{m=0}^{\lfloor M/2 \rfloor - \omega_\tau} A_z^2(m+\omega_\tau)}}, \text{ for } T_0^{(l)} \le \tau \le T_0^{(u)} \tag{6.19}$$

where $\omega_\tau = \lfloor M/\tau \ + 0.5 \rfloor$, and $T_0^{(l)}$ and $T_0^{(u)}$ are the lower and upper limits for the pitch search. In equation (6.19), the zero-crossing spectrum $A_z(m)$ is given by

$$A_z(m) = A(m) - g\overline{A}(m) \tag{6.20}$$

where $\overline{A}(m)$ is the spectral envelope of $A(m)$. The envelope may be estimated using the peak-picking method [17, 18]. The magnitude spectrum, $A(m)$, is converted into the zero-crossing spectrum $A_z(m)$ to make it feasible for the autocorrelation defined in equation (6.19). The gain, g, is calculated as:

$$g = \sum_{m=0}^{\lfloor M/2 \rfloor} A(m)\overline{A}(m) / \sum_{m=0}^{\lfloor M/2 \rfloor} A(m)A(m) \tag{6.21}$$

In equation (6.20), the logarithmic spectrum can also be considered to obtain a zero-crossing spectrum. However, the SA with the logarithmic spectrum produces a high correlation ratio for large lags, τ, close to $T_0^{(u)}$ (small

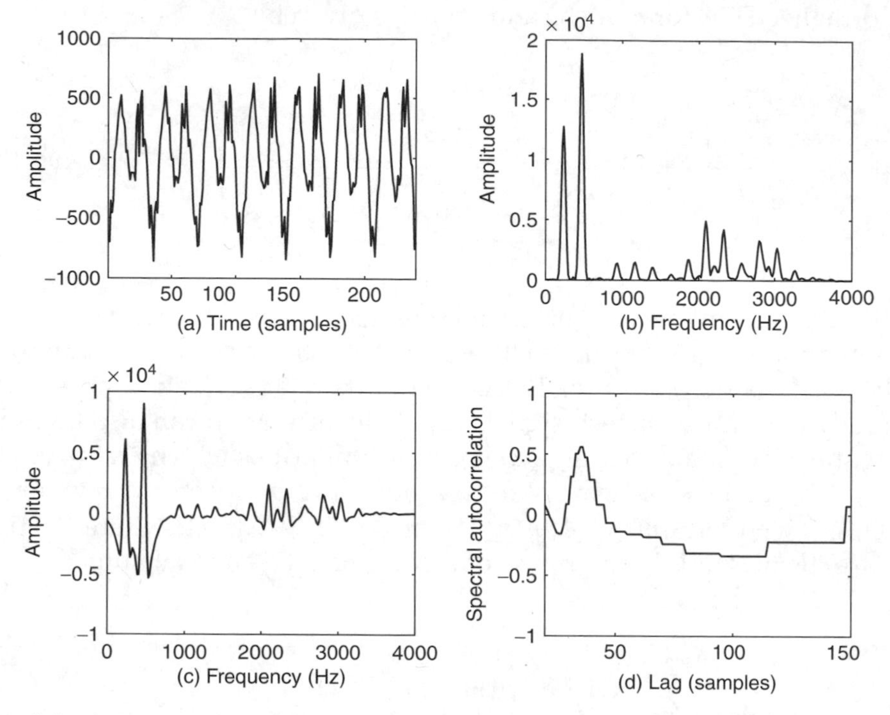

Figure 6.6 An example of (a) speech signal of $T_0 = 34$-sample ($F_s = 8\,\text{kHz}$), (b) magnitude spectrum, (c) zero-crossing spectrum, and (d) spectral autocorrelation

overlapping area) corresponding to very small ω_τ, i.e. $\lfloor M/T_0^{(u)} + 0.5 \rfloor \leq \omega_\tau << \lfloor M/(2T_0) + 0.5 \rfloor$. Thus, the linear magnitude spectrum is used instead of the logarithmic one.

Figure 6.6 shows an example illustrating the characteristics of SA. For a speech segment in Figure 6.6a, the magnitude and its zero-crossing spectra are shown in Figures 6.6b and 6.6c, respectively. Finally, the spectral autocorrelation is shown in Figure 6.6d, indicating a prominent peak at the pitch lag. The TA over a periodic signal produces high correlation for integer multiples of the pitch period T_0. This means that the spectral autocorrelation, $R_S(\tau)$ in equation (6.19), has peaks for the integer submultiples of T_0, i.e. $\tau = T_0/k$, for $1 \leq k \leq \lfloor T_0/T_0^{(l)} \rfloor$. Figure 6.7 shows an example featuring high SAs for pitch period submultiples. Thus, the TA-based PDA may result in detecting an unwanted pitch period multiple, and the SA-based PDA may result in pitch-halving. The pitch period multiple and submultiple problems can be compensated for by combining the two autocorrelation methods, TA and SA, in an advantages way. Hence, the spectro-temporal autocorrelation (STA) is defined as [19],

$$R_{ST}(\tau) = \alpha R_T(\tau) + (1 - \alpha) R_S(\tau) \tag{6.22}$$

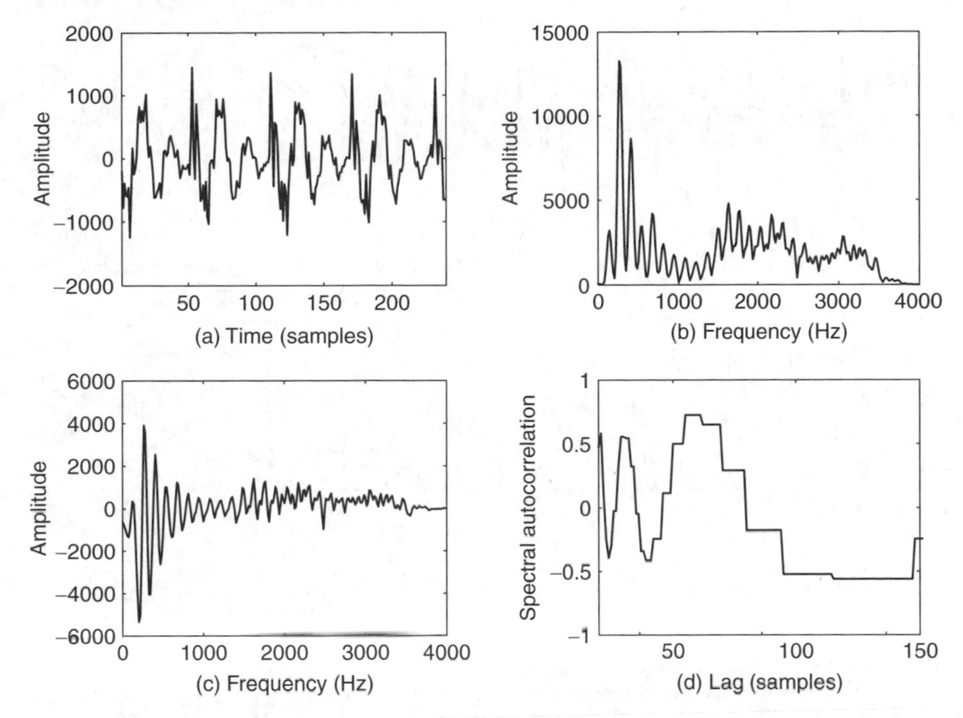

Figure 6.7 An example of (a) speech signal of $T_0 = 59$-sample ($F_s = 8\,\text{kHz}$), (b) magnitude spectrum, (c) zero-crossing spectrum, and (d) spectral autocorrelation

where α is a weighting factor, $0 \leq \alpha \leq 1$. The cases of $\alpha = 0$ and $\alpha = 1$ reduce the STA to SA and TA, respectively. The estimated pitch period $\hat{T}_0$ using the STA is the argument maximizing (6.22) as:

$$\hat{T}_0 = \arg\max_\tau \{R_{ST}(\tau)\} \tag{6.23}$$

Because of the dual relation between the temporal and the spectral autocorrelations, it is found that the STA has a useful property for pitch estimation. For a segment of periodic signal with a pitch period T_0, $T_0^{(l)} \leq T_0 \leq T_0^{(u)}$, $R_{ST}(\tau)$ in (6.22) has the strongest peak at $\tau = T_0$ compared with the integer multiple and submultiple periods of T_0, i.e. $\tau = pT_0$ and T_0/p for $2 \leq p \leq \lfloor T_0^{(u)}/T_0 \rfloor$ and $2 \leq p \leq \lfloor T_0/T_0^{(l)} \rfloor$. In (6.22), $R_S(\tau)$ and $R_T(\tau)$ terms suppress the undesirable high peaks for the multiples and submultiples of T_0 excluding $\tau = T_0$. Consequently, the STA for $\tau = T_0$ remains relatively more prominent compared with those for the rest.

The range of the pitch period can be split into three groups as high (short pitch period), mid, and low (long pitch period), based on the expected number of prominent peaks in TA and SA. The minimum pitch period producing a pitch period submultiple in SA is $2T_0^{(l)}$. The SA can only rarely produce pitch

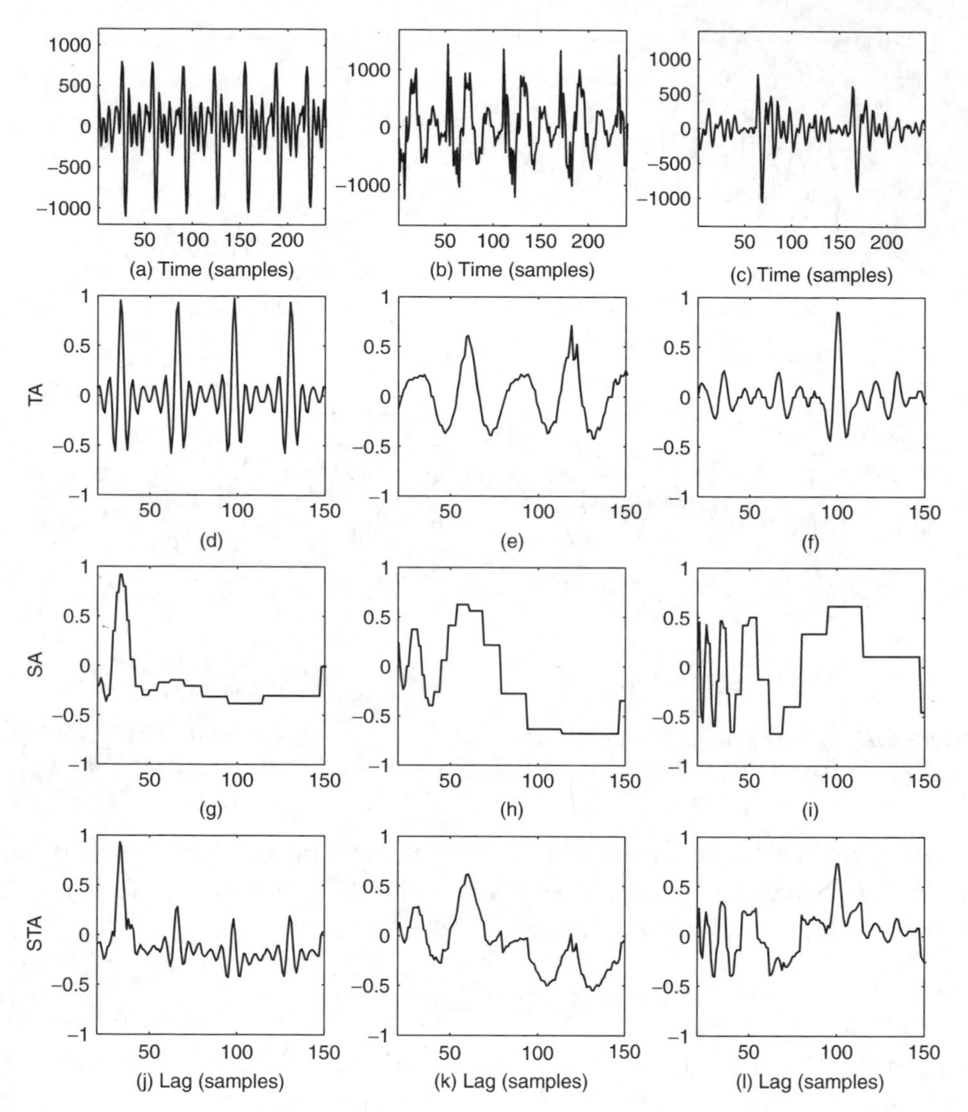

Figure 6.8 Comparison of TA, SA, and STA ($\alpha = 0.5$) for 32-sample (left column), 59-sample (middle column), and 100-sample (right column) speech signals

period submultiples for short pitch period signals, i.e $T_0^{(l)} \leq T_0 \leq 2T_0^{(l)} - 1$. On the other hand, in TA, the maximum pitch period generating the pitch period multiple is $T_0^{(u)}/2$. Thus, for $T_0^{(u)}/2+1 \leq T_0 \leq T_0^{(u)}$, TA can be relatively robust against pitch multiple errors. However, STA gives robust results for the whole pitch range by combining the two functions. Examples comparing the characteristics of TA, SA, and STA for speech signals with various pitch periods are shown in Figure 6.8. As can be seen from the figure, STA has a prominent peak regardless of the pitch period.

Spectral Synthesis–Spectral Autocorrelation PDA

The frequency spectrum of windowed speech can be decomposed into its spectral envelope and the fine structure spectra. The spectral envelope is the smoothed version of the speech spectrum. Spectral fine details (the excitation spectrum), on the other hand, exhibit harmonics for voiced components in which each harmonic typically has the shape of a sinc function corresponding to the applied window frequency response. Spectral synthesis (SS) methods [13, 14] determine the pitch so as to minimize the distortion between the original and synthesized spectra. The synthesized spectrum is generated by shifting the centre frequency of the sync function spectrum to harmonic frequencies.

In [14], McAulay's SS-based PDA, the metric for pitch determination is given by,

$$\Gamma(\tau) = \sum_{h=1}^{H(\tau)} \bar{A}\left(\frac{hK}{\tau}\right)\left\{\sum_{k=1}^{K/2} A(k)D\left(\frac{k}{K}-\frac{h}{\tau}\right) - \frac{1}{2}\bar{A}\left(\frac{hK}{\tau}\right)\right\} \tag{6.24}$$

where $H(x) = \lfloor x/2 \rfloor$ and $D(x) = \sin(2\pi x)/(2\pi x)$. The performance of the above PDA has been improved against the spectral formant effect by incorporating an energy-based metric $\varphi(\tau)$ [20, 17], given by,

$$\varphi(\tau) = \frac{\sum_{k=0}^{N} |d_\tau(k)|}{\sum_{k=0}^{N} |e_\tau(k)|} \tag{6.25}$$

where $e_\tau(n) = \sum_{k=0}^{\tau} s^2(n-\lfloor \tau/2 \rfloor + k)$ and $d_\tau(n) = 0.95 d_\tau(n-1) + e_\tau(n) - e_\tau(n-1)$ with $d_\tau(0) = 0$. The improved SS-based metric is defined as,

$$\Gamma_\varphi(\tau) = \frac{\Gamma(\tau)}{\varphi(\tau)} \tag{6.26}$$

in which $\Gamma_\varphi(\tau)$, if not positive, is bounded to a small positive value.

The SS-based PDA can be further improved by incorporating the spectral autocorrelation metric given in (6.19) to reduce pitch multiple effects which may occur in $\Gamma_\varphi(\tau)$. Hence, SS incorporating SA (called SS–SA [19]), is defined as:

$$\Gamma_{SA}(\tau) = \{\Gamma_\varphi(\tau)\}^\beta \left\{\frac{R_S(\tau)+1}{2}\right\}^{1-\beta} \tag{6.27}$$

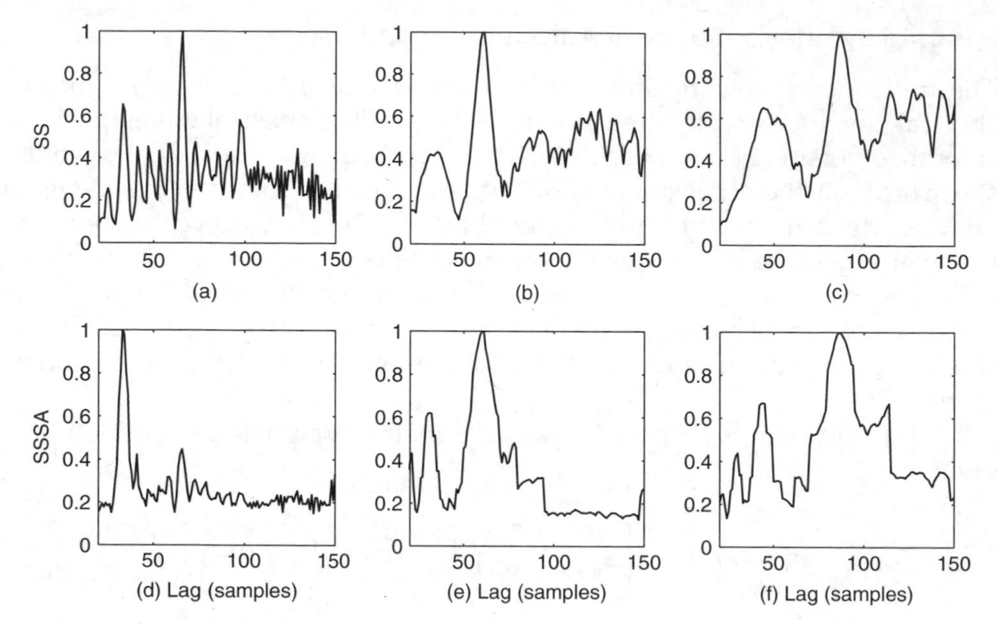

Figure 6.9 Comparison of SS and SS–SA ($\beta = 0.25$) for the high-, mid-, and low-pitched signals shown in Figures 6.8a–c

where β is a weighting factor, $0 \leq \beta \leq 1$, controlling the effect of SA. The SS–SA becomes SS when $\beta = 1$ and SA when $\beta = 0$. Examples examining the characteristics of $\Gamma_{\varphi}(\tau)$ and $\Gamma_{ST}(\tau)$, are shown in Figure 6.9; the measured value of each subfigure is normalized by each maximum value. The input speech signals used in Figure 6.9 are the same as the ones used in the analysis of the STA. For the high-pitched signals (short pitch periods) in Figure 6.8a, the lag corresponding to the pitch period double has a strong peak in Figure 6.9a, which seems to be even stronger than the peak at the correct pitch. The SS–SA alleviates this problem as shown in Figure 6.9d where the peak at the correct pitch lag becomes prominent in comparison with other peaks. For the mid- and low-pitched (long pitch period) signals in Figures 6.8b and 6.8c, the maximum peaks of $\Gamma_{\varphi}(\tau)$ and $\Gamma_{ST}(\tau)$ are relatively obvious as illustrated in Figures 6.9b, 6.9c, 6.9e, and 6.9f.

Comparison

An objective test was conducted to determine various tuning factors. The performance of the PDAs was measured in terms of pitch error rates (E_p). The speech test material, sampled at 8 kHz and filtered through the modified intermediate response system (MIRS) [21], was composed of 56 seconds each of male and female speech, each uttered by eight speakers. The reference pitch periods were manually marked for each 10 ms frame.

The range of the pitch search was limited to between 15 and 150 samples. Spectral analysis was conducted using a 240-sample Hamming window and a 256-point FFT with 16-sample zero padding. When computing the TA, a 240-sample rectangular or Hamming window was applied to the input signals. Pitch error decisions were checked in each frame by comparing the detected pitch period with the reference. A frame was classified as erroneous if the absolute difference between the reference and the detected pitch periods was more than 1 ms (8-sample) as in [1]. Extra processing, such as pitch tracking using the pitch history of the past frames, was not incorporated in order to evaluate only the main algorithmic contributions. Although the unvoiced speech regions were not taken into account, transitions were included in the performance evaluations as these regions are perceptually very important.

- **Analysis of the STA Weighting Factor**
  The effect of the STA rate α in terms of E_p is shown in Figure 6.10. The results show that the STA gives improved performance compared with the TA and the SA, corresponding to $\alpha = 1$ and $\alpha = 0$, respectively. The lowest E_p was obtained when $\alpha = 0.5$ for both the female and male speech samples.
- **Analysis of the SS–SA Weighting Factor**
  The weighting factor β of SS–SA in equation (6.27) was analysed by varying β between 0 and 1 (see Figure 6.11). As in STA, the SS–SA also shows much less E_p in comparison with those of the SS and the SA, corresponding to $\beta = 1$ and $\beta = 0$, respectively. Additionally, the lowest E_p values were

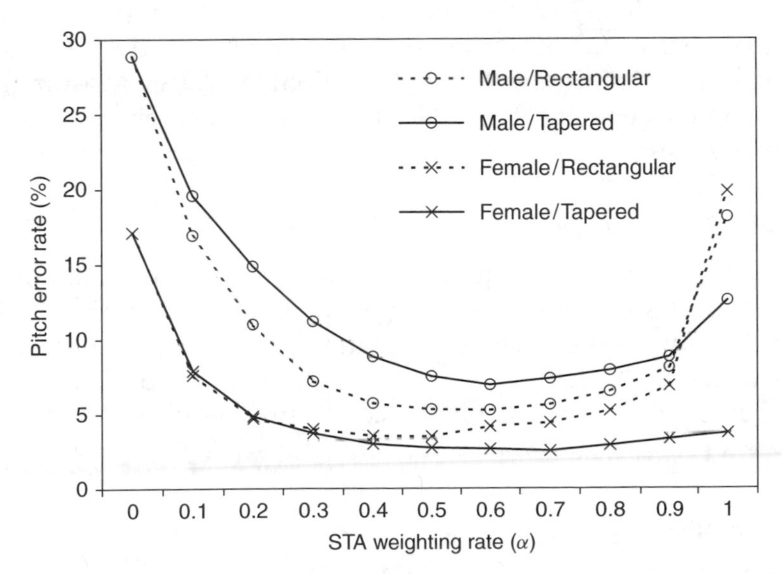

Figure 6.10 Analysis of the effect of the STA weighting factor α in terms of the pitch error rate; the formant weighting factor γ is 0.9

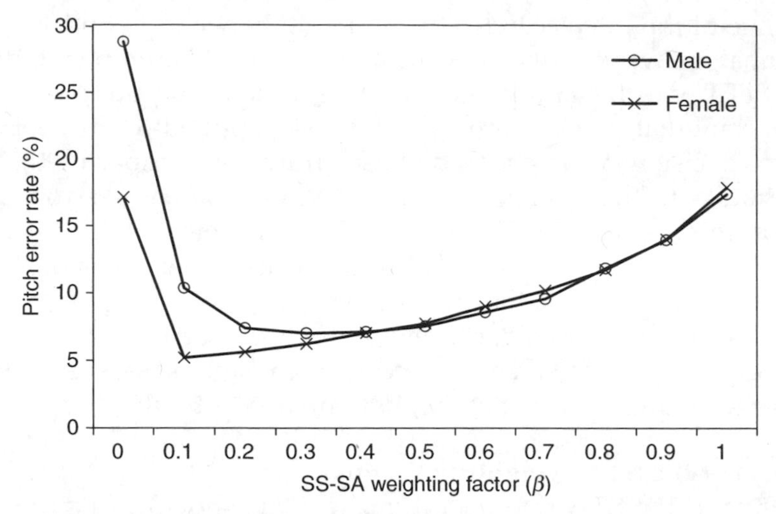

Figure 6.11 Analysis of the effect of the SS–SA weighting factor β in terms of the pitch error rate. Here, the formant weighting factor γ is 0.9

obtained when $\beta = 0.1$ for the female speech and $\beta = 0.3$ for the male speech, which means that the optimum β differs slightly depending on the pitch period of the signal. Higher performance can be achieved by weighting the SA more during shorter pitch period speech and less during longer pitch period speech.

Examples of pitch contours of the various PDAs are illustrated in Figures 6.12 and 6.13 in which the rectangular window is applied to the TA and STA. It shows that pitch errors in strongly-voiced regions are reduced considerably by the combination of time and frequency domain PDAs. Most of the errors were caused at speech onset and offset regions where irregular pitch pulse sequences are present.

6.2.4 Pre- and Post-processing Techniques

In addition to the main pitch determination processes described previously, there are several important pre- and post-processing techniques which can significantly improve the pitch determination performance. These techniques supplement the PDAs and are used before or after the pitch determination process. Hence, they are usually called preprocessing or post-processing stages.

Spectrum Flattening

Although the pitch of a voiced speech segment can be directly estimated from the original speech, the first formant frequency may affect the accuracy

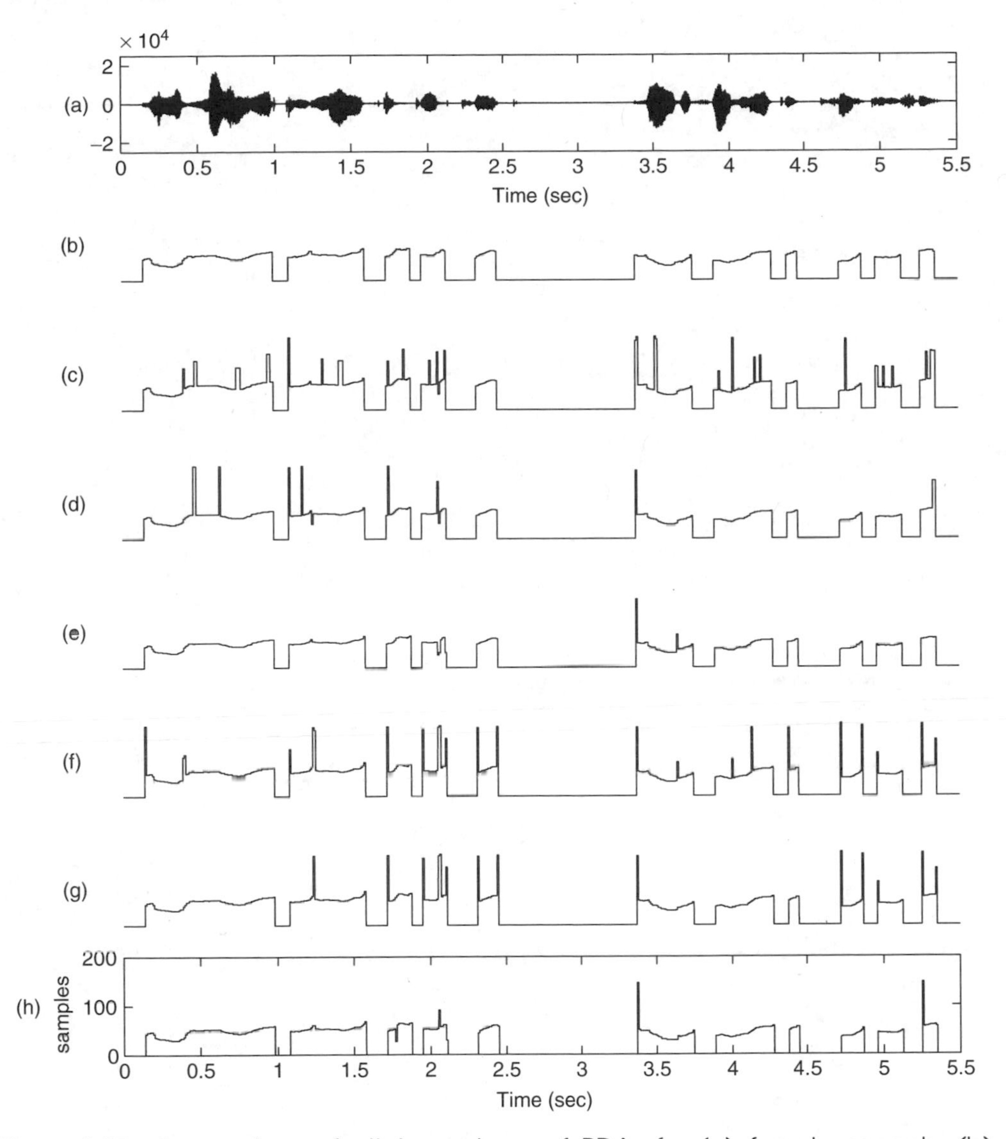

Figure 6.12 Comparison of pitch contours of PDAs for (a) female speech. (b) Reference; (c) TA, (d) WTA, (e) STA, (f) SS, (g) WSS, and (h) SS–SA-based PDAs

of the estimation. Several methods have been proposed to flatten the speech spectrum in order to avoid the formant interaction effect [20, 5, 22, 23]. The speech spectrum is first flattened by removing the formants (by either linear or nonlinear methods) before the pitch estimation process can begin.

The linear spectrum-flattening method uses the LPC inverse filter to remove the formants from the speech signal. The main drawback of this method is that for high-pitched speech, like that of females and children, the first complex

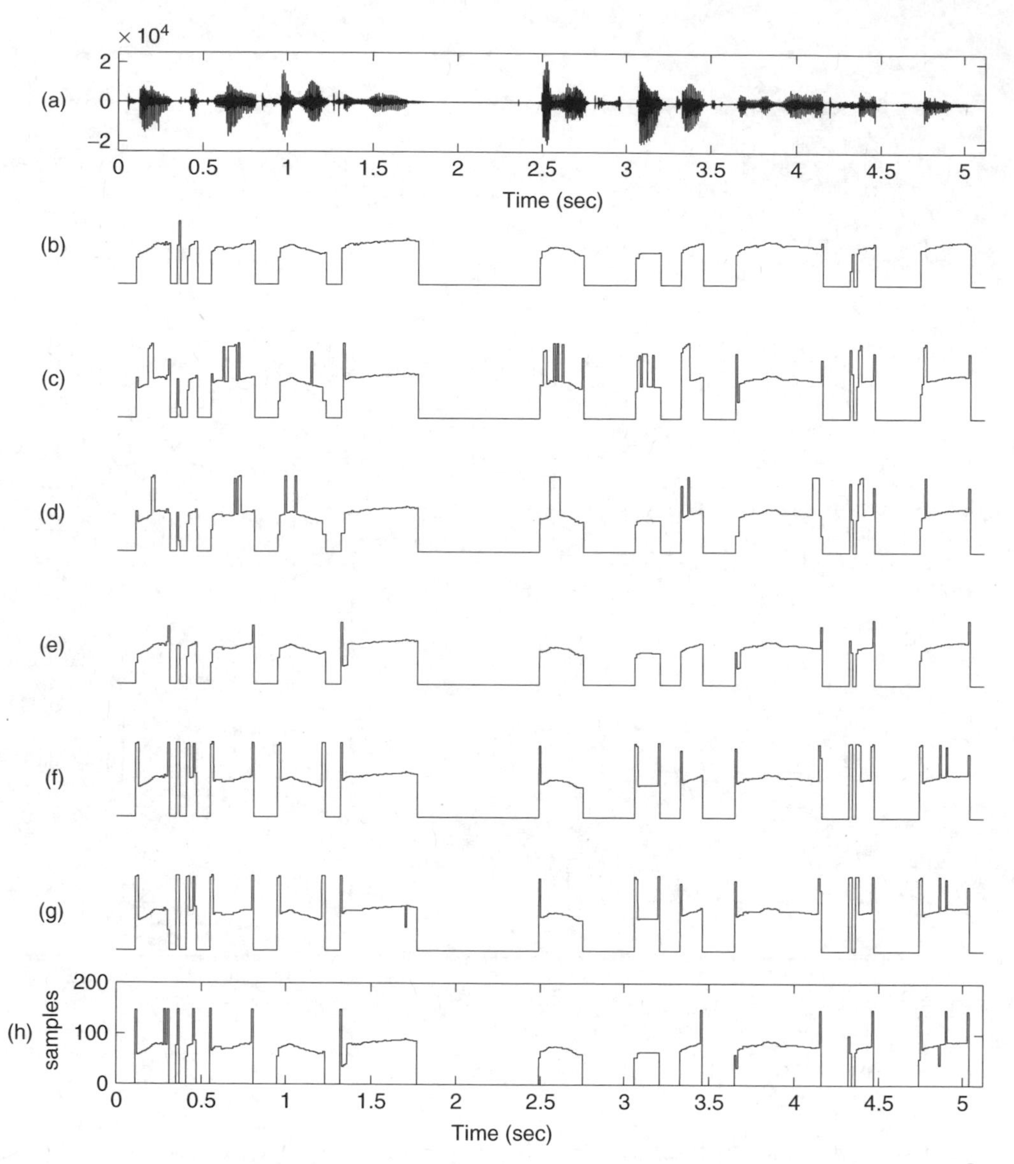

Figure 6.13 Comparison of pitch contours of PDAs for (a) male speech (b) Reference; (c) TA, (d) WTA, (e) STA, (f) SS, (g) WSS, and (h) SS-SA-based PDAs

zero of the inverse filter may be adjusted to become the first harmonic, and the second complex zero to the second harmonic. This may destroy the entire periodicity information in the residual signals [16, 1]. Thus, a formant weighting filter [24] is adopted as a preprocessor to control the de-emphasizing factor of the formants while keeping the harmonics structure. As described in the equation below, it is, effectively, a process of obtaining the

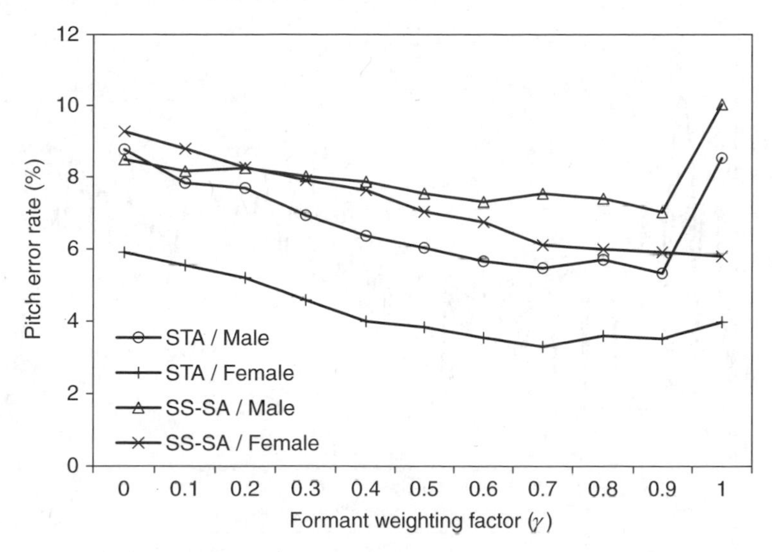

Figure 6.14 Analysis of the effect of the formant weighting factor γ in terms of the pitch error rate; α and β defined in the STA and SS–SA functions are 0.5 and 0.25, respectively, and the rectangular window is applied to TA calculation in the STA

intermediate signal between the original speech and its LPC residual signals.

$$S_f(z) = \frac{A(z)}{A(z/\gamma)} S(z) \tag{6.28}$$

where $S(z)$, $S_f(z)$, and $A(z)$ are the z-transform of the input speech signal $s(n)$, the formant-suppressed signal, and the inverse filter, respectively. The parameter γ is the formant weighting factor, $0 \leq \gamma \leq 1$. For the case of $\gamma = 1$, the filtered signal is identical to the original speech signal. On the other hand, $\gamma = 0$ makes the filtered signal equal to the LPC residual of $s(n)$. It can be seen that $S_f(z)$ is the intermediate spectrum between the original and residual spectra for $0 < \gamma < 1$. The effect of the formant weighing factor γ in equation (6.28) was observed over the STA and SS–SA-based PDAs and the results are shown in Figure 6.14. It can be seen that the value around $0.7 \sim 0.9$ gives improved performance.

The effect of the flattening filter is shown in Figure 6.15. The formant influence has been greatly reduced but not completely eliminated, while the harmonic structure is well-preserved. A better performance may be obtained by making the spectral-flattening factor a function of the average pitch (tracked pitch) as shown in Figure 6.16.

Nonlinear spectrum-flattening is usually achieved by centre-clipping the speech signal. The first centre-clipping PDA was proposed by Sondhi [22] in 1968 and various centre-clippers for autocorrelation PDAs were investigated by Rabiner [3] in 1976. The characteristics of three types of centre-clipping

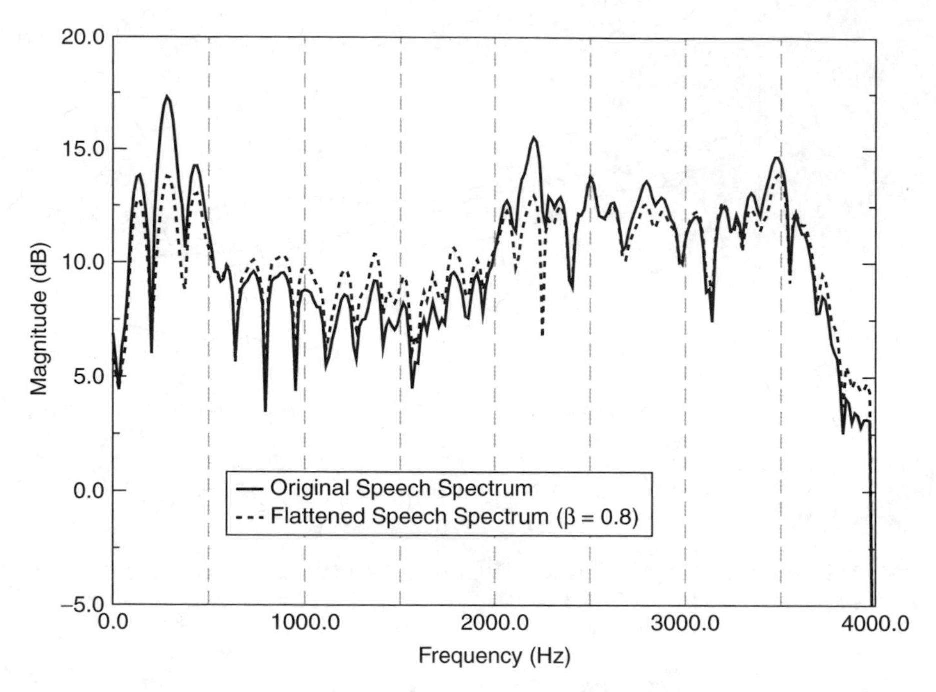

Figure 6.15 Influence of the spectrum-flattening filter

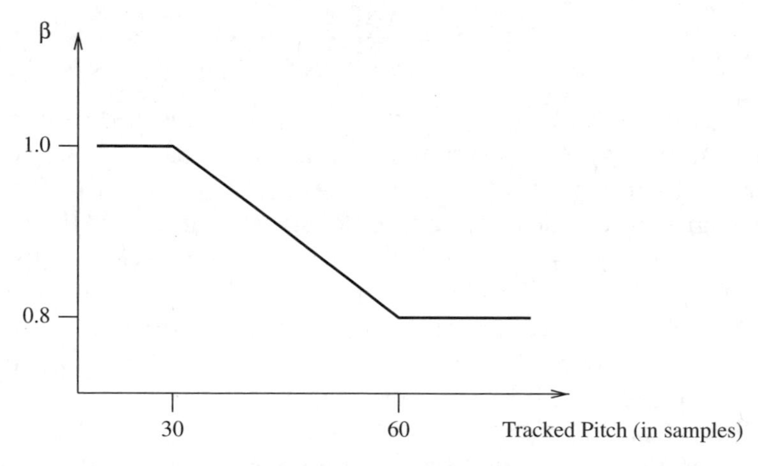

Figure 6.16 Value of the flattening factor β against tracked pitch

functions used for flattening the spectrum are shown in Figure 6.17 where,

$$y = clc(x) = \begin{cases} x + CL & ; \quad x \leq -CL \\ x - CL & ; \quad x \geq CL \\ 0 & ; \quad -CL < x < CL \end{cases} \tag{6.29}$$

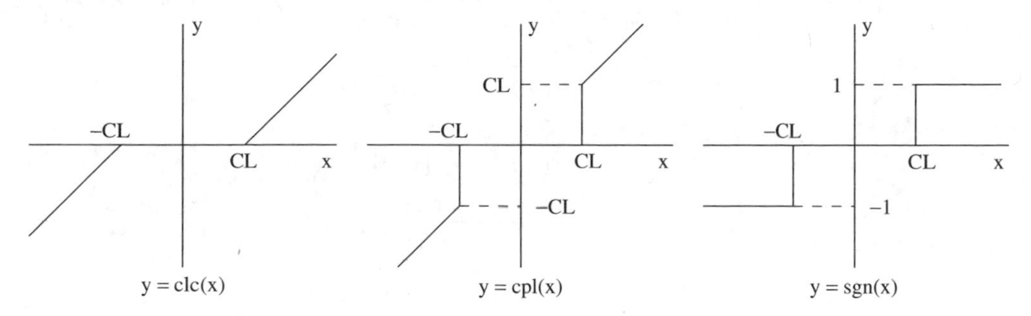

Figure 6.17 Clipper functions

$$y = clp(x) = \begin{cases} x & ; \quad -CL \geq x \geq CL \\ 0 & ; \quad -CL > x < CL \end{cases} \tag{6.30}$$

$$y = sgn(x) = \begin{cases} 1 & ; \quad x \geq CL \\ -1 & ; \quad -CL \geq x \\ 0 & ; \quad -CL > x < CL \end{cases} \tag{6.31}$$

The centre clipped signal $y(n)$ is generally defined as:

$$y(n) = f[s(n)] \tag{6.32}$$

The clipper function $f[.]$ can be any one of the functions in equations (6.29)–(6.31). For the autocorrelation method, the clipped autocorrelation function $R_c(\tau)$ is defined as:

$$R_c(\tau) = \sum_{n=0}^{N-1} y(n)y(n+\tau) = \sum_{n=0}^{N-1} f[s(n)]\, f[s(n+\tau)] \tag{6.33}$$

More generally, these two clipper functions can have any combination, e.g.

$$R_c(\tau) = \sum_{n=0}^{N-1} f_1[s(n)]\, f_2[s(n+\tau)] \tag{6.34}$$

A set of typical combinations of $f_1(n)$ and $f_2(n)$ are listed in Table 6.1. It has been shown that[3, 16]:

- For high-pitch speakers, the differences in performance scores between the various clipping combinations are small and probably insignificant.
- For low-pitch speakers, fairly significant differences in performance scores exist. Combination 1 in Table 6.1 gives the worst performance for all utterances in this class. Combinations 4, 5, and 6 (those involving one unprocessed component) are also poor in their overall performances.

Table 6.1 Combinations of clipper functions

Type	$f_1(n)$	$f_2(n)$
1	$s(n)$	$s(n)$
2	$clc[s(n)]$	$clc[s(n)]$
3	$clp[s(n)]$	$clp[s(n)]$
4	$s(n)$	$clc[s(n)]$
5	$s(n)$	$clps(n)$
6	$s(n)$	$sgns(n)$
7	$clcs(n)$	$sgns(n)$
8	$clps(n)$	$sgns(n)$
9	$clps(n)$	$clcs(n)$
10	$sgns(n)$	$sgns(n)$

- Differences in the performance of the remaining six combinations are not consistent, thus any one of these correlators can be applied for flattening the spectrum.
- The performance is improved if the nonlinear processing is performed before low-pass filtering. This applies especially to band-limited signals, where the weakly-attenuated waveform of the first formant is often the only information available for periodicity detection after low-pass filtering.

The key problem in centre clipping is the choice of the clipping threshold. Sondhi proposed a method based on a short time interval (5 ms), where the threshold for clipping was set at 30 % of the maximum absolute signal value within the block [22]. In Rabiner's method, the threshold was set to be a fixed percentage (68 %) of the smaller of the maximum absolute signal value over the first and last 10 ms of the analysis frame, which is normally 30 ms [25].

Pitch Tracking

The principle of pitch tracking is based on the continuity characteristic of pitch, i.e. once a voiced sound is established, its pitch varies within a limited narrow range. The pitch tracking principle can be used in two ways, one operating after the main pitch determination process as an error-checking function and the another within the main pitch determination process ensuring the estimation follows the correct route.

The first method of pitch tracking is also called pitch smoothing because it forces the pitch contour to be smooth. Pitch smoothing is a passive way of utilizing the continuity characteristic. The risk in using this method is that some abrupt changes in pitch are smoothed out, as there are occasional instances of dramatic change.

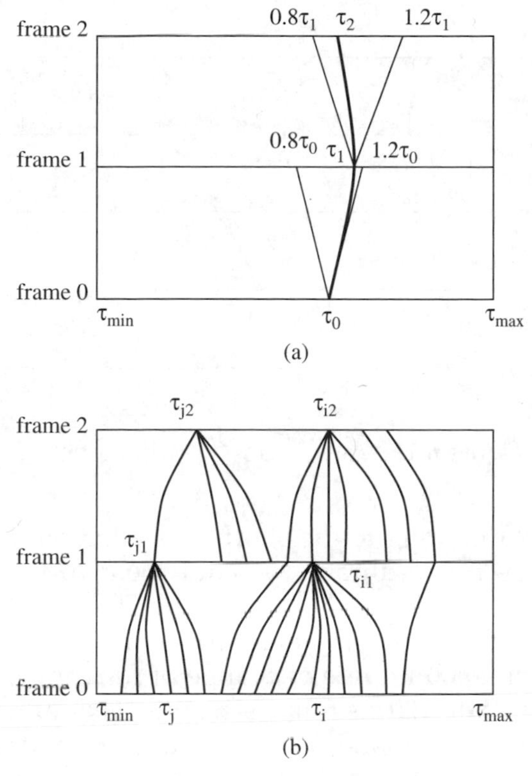

Figure 6.18 Forward pitch tracking: (a) Setting search range limits and (b) Possible tracked pitch candidates

The active way of using pitch tracking is to apply it at the beginning of the main processing. Thus, the pitch is not estimated in isolation but by considering the neighbouring frames. With this pitch-tracking method, the pitch is estimated as a minimum path error overall. Path error refers to an accumulated error for a number of adjacent frames, also called the path penalty. For instance, if a pitch path consists of τ_0, τ_1, τ_2 (see Figure 6.18), the path penalty is the accumulated error on the path given by:

$$E_{path} = E_0(\tau_0) + E_1(\tau_1) + E_2(\tau_2) \tag{6.35}$$

where $E_i(\tau_j)$ is the estimated error for candidate τ_j in the ith frame. Constraint conditions must be applied to the possible pitch paths so that the continuity characteristic can be maintained. Pitch-tracking constraints are as follows:

$$\begin{aligned}(1-\alpha)\tau_0 \leq \tau_1 \leq (1+\alpha)\tau_0 \\ (1-\alpha)\tau_1 \leq \tau_2 \leq (1+\alpha)\tau_1\end{aligned} \tag{6.36}$$

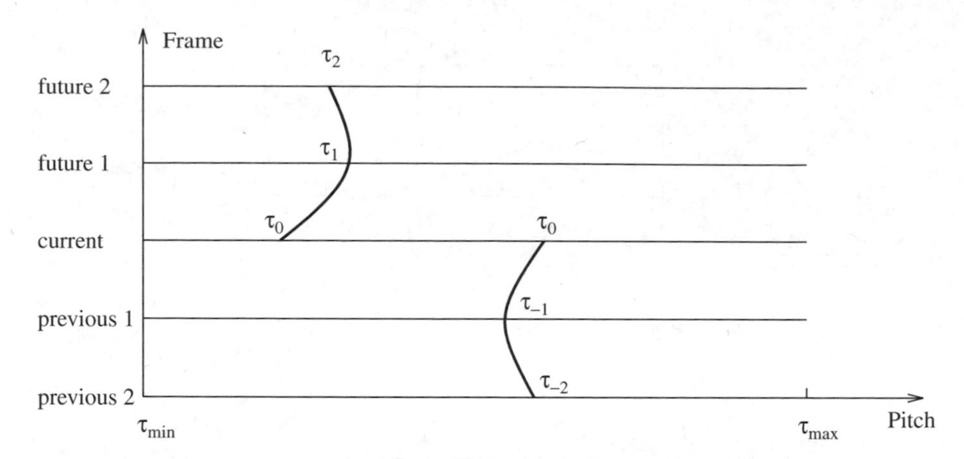

Figure 6.19 A two trace-tracking scheme

where α is chosen according to the short-time analysis (frame duration) or path step. Since the frame duration is equivalent to the interval between two consecutive pitch analysis blocks, as the frame gets larger, the next pitch could be expected to have more deviation. According to the data by Sundberg [26], the maximum rate of change of fundamental frequency is in the order of 1 %/ms. For a 20 ms frame size, the maximum frequency change would be 20 % which corresponds to a pitch range from $0.8\omega_0$ to $1.2\omega_0$. In the time domain, the corresponding range is $0.8\tau_0$ to $1.2\tau_0$, approximately, with $\alpha = 0.2$.

In order to take into account the effects of both continuous pitch and changing pitch, two pitch paths should be considered (see Figure 6.19). One traces the pitch from previous frames to the current frame, and the other traces the pitch from the current frame to incoming or future frames that forecast a new pitch trace. If the future path penalty is less, it is assumed that a new pitch trace is starting; if the path penalty with the previous frames is smaller, then the existing pitch is assumed to continue into the current frame. Since future pitch tracking requires the storage of the future frames, extra delay is unavoidable. The pitch-tracking procedure using three frames can be described as follows:

- **Forward Tracker**

  1. For each candidate pitch (τ_0) in the current frame, find $E_0(\tau_0)$.
  2. Find the joint minimum $E_1(\tau_1)+E_2(\tau_2)$ under the constraints of equation (6.36).
  3. Add $E_0(\tau_0)$ to the corresponding minimum $E_1(\tau_1) + E_2(\tau_2)$ to form each candidate's accumulated forward path error.
  4. Search for the minimum accumulated error and find the forward pitch.

- **Backward Tracker**

1. Add $E_{-2}(\tau_{-2})$ and $E_{-1}(\tau_{-1})$ corresponding to the pitch periods of the two previous frames to find the accumulated error up to the current frame.
2. Find the minimum $E_0(\tau_0)$ under the constraint that $0.8\tau_{-1} \leq \tau_0 \leq 1.2\tau_{-1}$.
3. Add $E_0(\tau_0)$ to the backward-accumulated error to find the backward pitch-tracking error.

Finally the forward and backward path errors are compared and the optimum pitch τ_{opt} is selected.

It can be seen from the above description that the forward pitch-tracking procedure is equivalent to a full search scheme for a given path boundary which makes it very complex to implement in real-time. Although dynamic programming techniques can reduce this procedure to a sequential search, it is still very complex and thus not widely used in practice. However, using the fact that the search ranges are heavily overlapped, a fast algorithm has been developed that reduces the computations significantly [27].

If we assume R_i is the search range in the next frame for the candidate pitch τ_i in the current frame and similarly, R_j defines the search range of τ_j, then according to the definition, we have

$$R_i = \{\tau_b^i,\ \tau_b^i + p,\ \tau_b^i + 2p,\ \tau_b^i + 3p, \ldots, \tau_e^i\} \tag{6.37}$$

$$R_j = \{\tau_b^j,\ \tau_b^j + p,\ \tau_b^j + 2p,\ \tau_b^j + 3p, \ldots, \tau_e^j\} \tag{6.38}$$

where τ_b^i and τ_e^i are the first and last pitch values in the range R_i and similarly τ_b^j and τ_e^j are the first and last pitch values in the range R_j, p is the pitch resolution or the step size and,

$$\tau_b^k = \begin{cases} \tau_{min} & \text{if } (1-\alpha)\tau_k < \tau_{min} \\ (1-\alpha)\tau_k & \text{otherwise} \end{cases}$$

$$\tau_e^k = \begin{cases} \tau_{max} & \text{if } (1+\alpha)\tau_k > \tau_{max} \\ (1+\alpha)\tau_k & \text{otherwise} \end{cases}$$

It can be shown that all possible search ranges can be divided into three main groups. The first and third cases, in Figures 6.20a and 6.20c respectively, have the same characteristic in that one range is fully overlapped by the next, i.e. $R_i \in R_j$. In these cases, it is not necessary to search through all these ranges

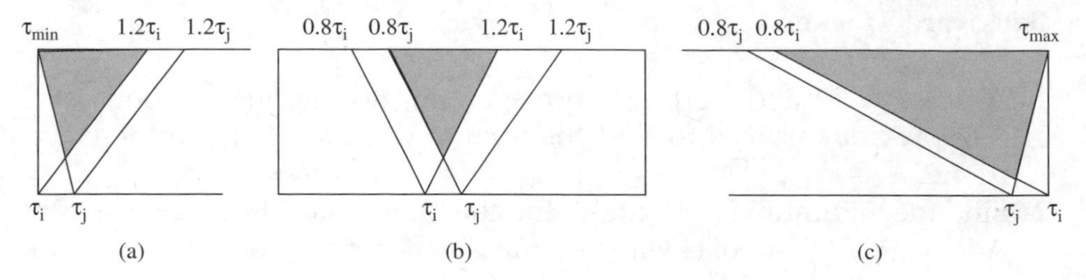

Figure 6.20 Three possible pitch-lag search ranges

separately. The smaller range is fully searched and then its minimum error value is compared with the new candidates in the next bigger range.

The second case (Figure 6.20b) is not as easy as the other two, because the ranges are only partially overlapping. Assuming the error function is as sketched in Figure 6.21a, the two search ranges should be fully searched to prevent the incorrect pitch being selected. However each partly-overlapped range, R_i, can be split into two fully overlapped subranges, $R'i$ and $R'i$, as shown in Figure 6.21b.

$$R'4 \in R'3 \in R'2 \in R'1 \tag{6.39}$$

$$R'1 \in R'2 \in R'3 \in R'4 \tag{6.40}$$

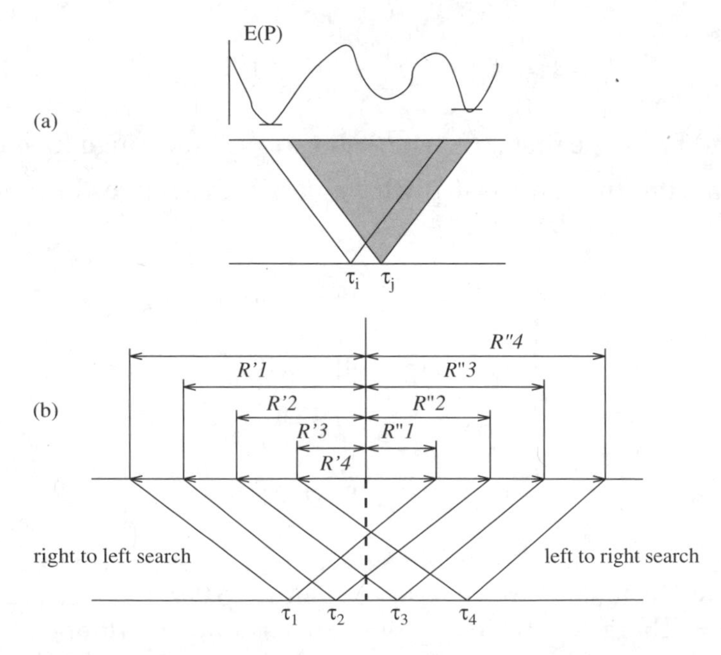

Figure 6.21 Partially-overlapped search schematics

The whole search, therefore, is divided into two procedures: subrange search and subrange comparison. Since these subranges are all fully overlapped, searching over the subranges only need be done twice, from left to right and from right to left. We start with the range $R''1$ and compare its minimum with the nonoverlapped part of $R''2$ and so on until all of the right hand side is completed. The same procedure is applied to the left hand side starting with $R'4$. Finally, the left hand and right hand side minima are compared and the overall minimum is selected. We can also see that the number of comparisons during the search is independent of the size of the pitch search ranges and is equal to three times the number of pitch candidates.

Multiple Pitch and Half Pitch Errors

Almost all PDAs have a peak detector which decides the pitch by the peak position. In time-domain methods for example, the peak to be detected is not only positioned at the correct pitch lag, but also at its integer multiples. Therefore it is possible that a multiple of the real pitch may be chosen. In order to find the desired peak among the peaks, a complicated procedure is normally needed. The basic idea for solving this problem includes two steps: picking the maximum peak; checking the submultiple positions to see if there is a comparable peak. However, since there is no fixed solution to this problem, tuned comparison thresholds are generally used.

For example, in the case of the cross-correlation pitch estimation method, the comparison is made by looking at the ratio $R(\tau_0/i)/R(\tau_0)$ where i is an integer, which produces pitch submultiples greater than or equal to the minimum expected pitch. The smallest submultiple which may produce a ratio greater than the set threshold is selected as the pitch.

In frequency-domain methods, such as the spectrum similarity method, a similar procedure can be applied. In this case, the average sum of the harmonics in the signal may be used in the comparison. At every submultiple, the average sum of harmonic magnitudes are computed by

$$A_v(\omega_k) = \frac{1}{L_k}\sum_{i=1}^{L_k} A(i\omega_k) \quad ; \quad k = 1, 2, 3, \ldots, n. \tag{6.41}$$

where L_k is the total number of harmonics in a 4 kHz speech bandwidth, $A(i\omega_k)$ are harmonic magnitudes and $\omega_k = \frac{2\pi}{\tau_0/k}$ is the fundamental frequency of the kth submultiple of the initial pitch. The ratio between the $A_v(\omega_k)$ of the smallest submultiple and the initial pitch, τ_0, is then computed and compared with a threshold which may vary for each submultiple. If this ratio is bigger than the corresponding threshold, then the smallest submultiple is selected as the pitch estimate. Otherwise the next largest submultiple is checked against the above procedure and it is selected as the pitch estimate if it satisfies the condition. This process continues until all submultiples have been tested against this condition. If none of the submultiples of the initial pitch satisfy the condition, then the initial τ_0 becomes the final pitch estimate.

In some cases, where the decision threshold is wrongly exceeded, a multiple or submultiple of the correct pitch may be selected. This may cause significant performance degradations. Therefore, when designing a vocoder that requires accurate pitch estimation, other measures which can reduce the effects of these occasional pitch errors should be considered.

6.3 Voiced–Unvoiced Classification

The voicing is another very important parameter which must be estimated correctly for good quality speech reproduction. In the old vocoders, a single (binary) voicing decision was made by classifying the frame (or half the frame) as either voiced or unvoiced. However, it is well known that the transitions are very important for good quality speech synthesis and most of the time the transitions are a mixture of voiced and unvoiced signals. Therefore a mixed decision of voicing has been developed and used in many of the latest vocoders. In the following, we review and discuss both binary (hard) and mix (soft) decision voicing.

6.3.1 Hard-Decision Voicing

Voiced and unvoiced sounds have very well-known characteristics which can be used to classify them reasonably correctly. Some of the most distinctive of these characteristics are discussed below.

Periodic Similarity

The most prominent characteristic that separates voiced speech from unvoiced speech is its regularity and fairly well-defined pitch. During voiced speech, samples in one pitch period look very similar to the samples in the adjacent pitch period. Hence, measuring the similarity between samples in consecutive pitch cycles can give a reasonably good idea if the speech is voiced or unvoiced. The measurement of similarity, Ps, can be computed by

$$Ps = \frac{\left[\sum_{i=1}^{N} s(i)s(i-T)\right]^2}{\sum_{i=1}^{N} s^2(i) \sum_{i=1}^{N} s^2(i-T)} \tag{6.42}$$

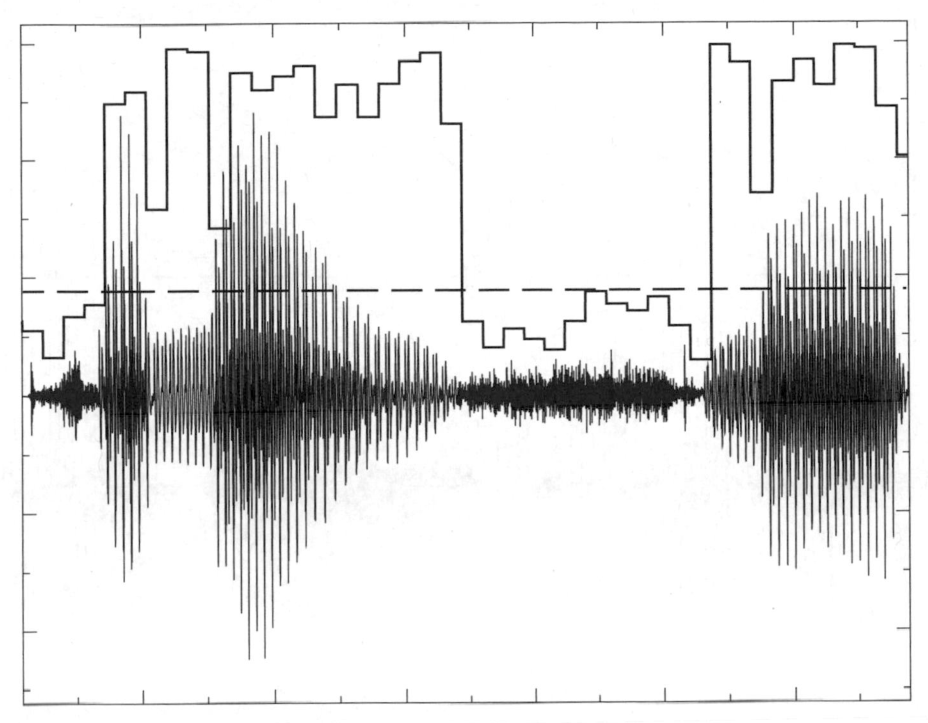

Figure 6.22 Original speech waveform and the corresponding pitch similarity plot with a possible voicing threshold of 0.5 (shown by the dashed line)

which has a value between 0 and 1, indicating no similarity and 100% similarity, respectively. Time plots of typical voiced and unvoiced speech against pitch similarity are shown in Figure 6.22. As can be seen from the figure the voiced parts of speech clearly have higher pitch similarity than the unvoiced parts. This is expected since two adjacent unvoiced speech segments do not possess noticeable similarities.

Peakiness of Speech

Periodic or voiced speech contains regular pulses which do not appear in unvoiced speech. This feature is described as peakiness of speech and it can be used to identify voiced speech when it has a relatively high value. In order to enhance the peakiness, the LPC residual can be used to compute its value.

$$Pk = \frac{\sqrt{\frac{1}{N}\sum_{i=1}^{N} r^2(i)}}{\frac{1}{N}\sum_{i=1}^{N} |r(i)|} \tag{6.43}$$

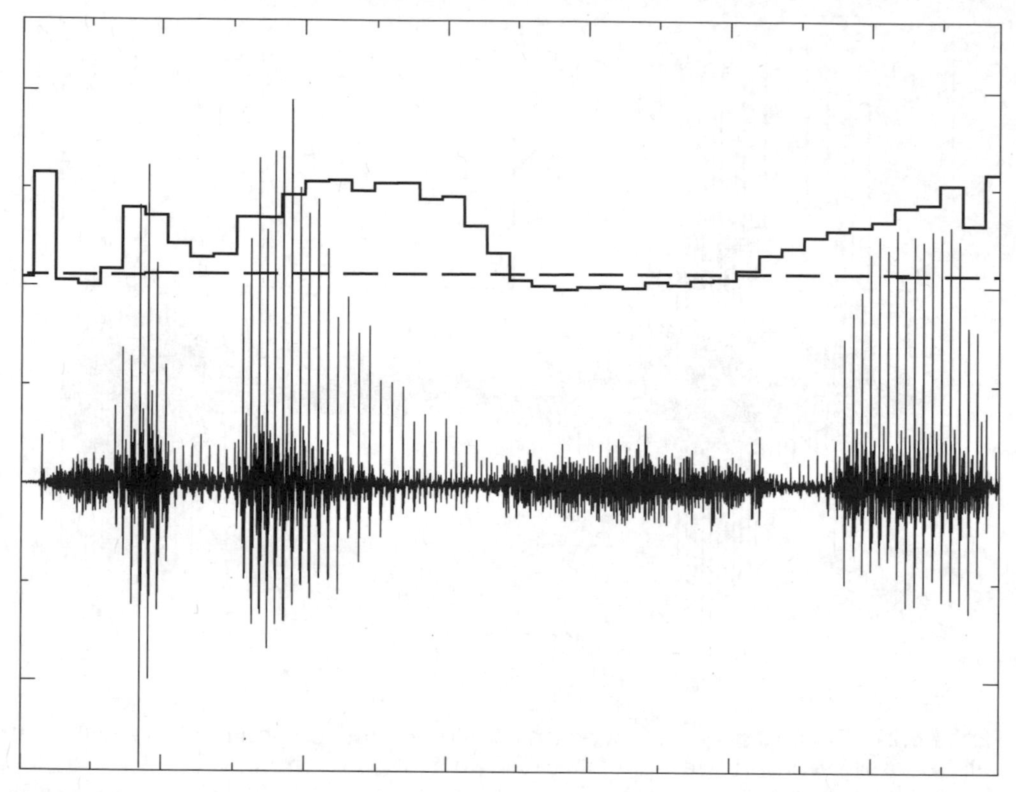

Figure 6.23 LPC residual and corresponding peakiness plots with a possible voicing threshold of 1.4 (shown by the dashed line)

where $r(i)$ is the LPC residual signal. Plots of LPC residual and the corresponding peakiness measure are shown in Figure 6.23. Although the voiced speech is clearly peaky, there are some unvoiced parts which contain a major spike. In these cases, the peakiness measure may incorrectly indicate the frame as voiced instead of unvoiced. In order to avoid this problem, a second peakiness measure can be computed by excluding the largest magnitude sample and its immediate neighbours from the computation. If the two peakiness measures are significantly different, then the frame is not really voiced but contains a spike.

Zero Crossing

Unvoiced speech has random characteristics, which means that the number of times the signal crosses the zero line (i.e. that the sign changes) is significantly higher than with the voiced part of speech, which has a much slower zero-

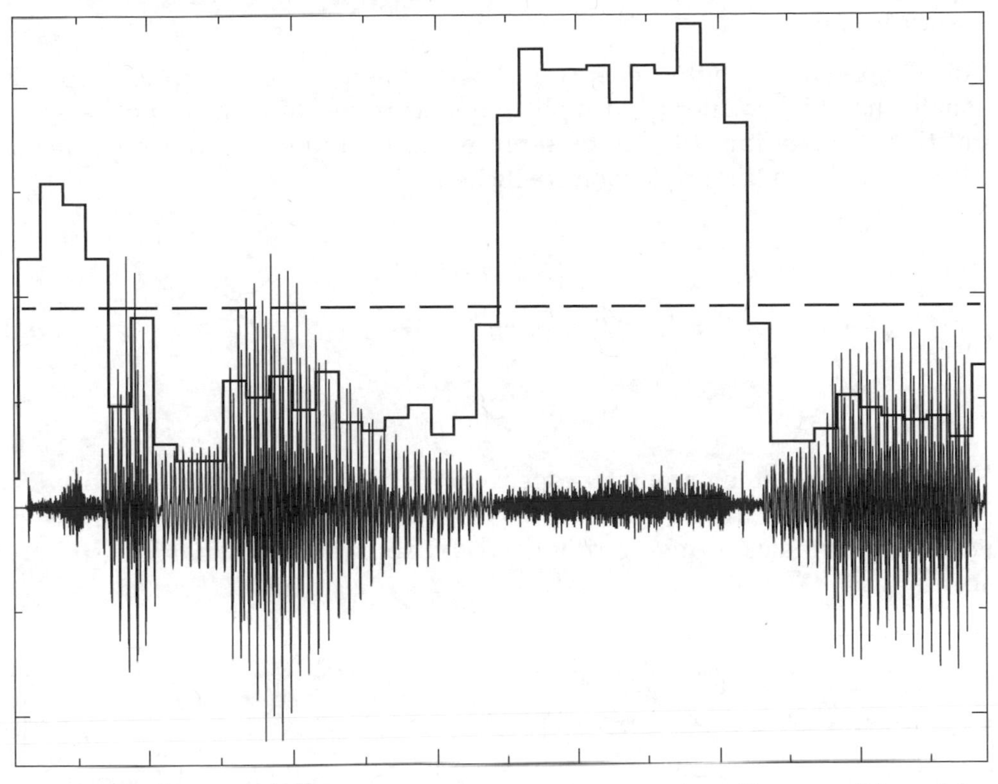

Figure 6.24 Speech waveform and its zero-crossing rate with a possible voicing threshold of 60 (shown by the dashed line)

crossing rate. The simple logic shown below can be used to compute the zero-crossing rate:

```
count=0;
for(i=1;i<N;i++)
  {
   if((data[i] x data[i-1]) < 0.0)
   count = count + 1
  }
 Zc=count
```

A speech waveform and its corresponding zero-crossing rate is shown in Figure 6.24. The zero-crossing rate also depends on the pitch of the signal (if voiced). For example, the zero-crossing rate of voiced female speech (with a short pitch period) is higher than that of voiced male speech (with a long pitch period). A small pitch weighting can be used to weight the decision threshold.

Spectrum Tilt

Voiced speech has higher energy in low frequencies and unvoiced speech usually has higher energy in high frequencies resulting in opposite spectral tilts. The spectral tilt can be represented by the first-order normalized autocorrelation or first reflection coefficient.

$$St = \frac{\sum_{i=1}^{N} s(i)s(i-1)}{\sum_{i=1}^{N} s^2(i)} \tag{6.44}$$

This is a very reliable parameter especially for plosive detection and to avoid individual spikes in low-level signals. As can be seen from Figure 6.25, its ability to indicate unvoiced and voiced sounds in general is also very accurate.

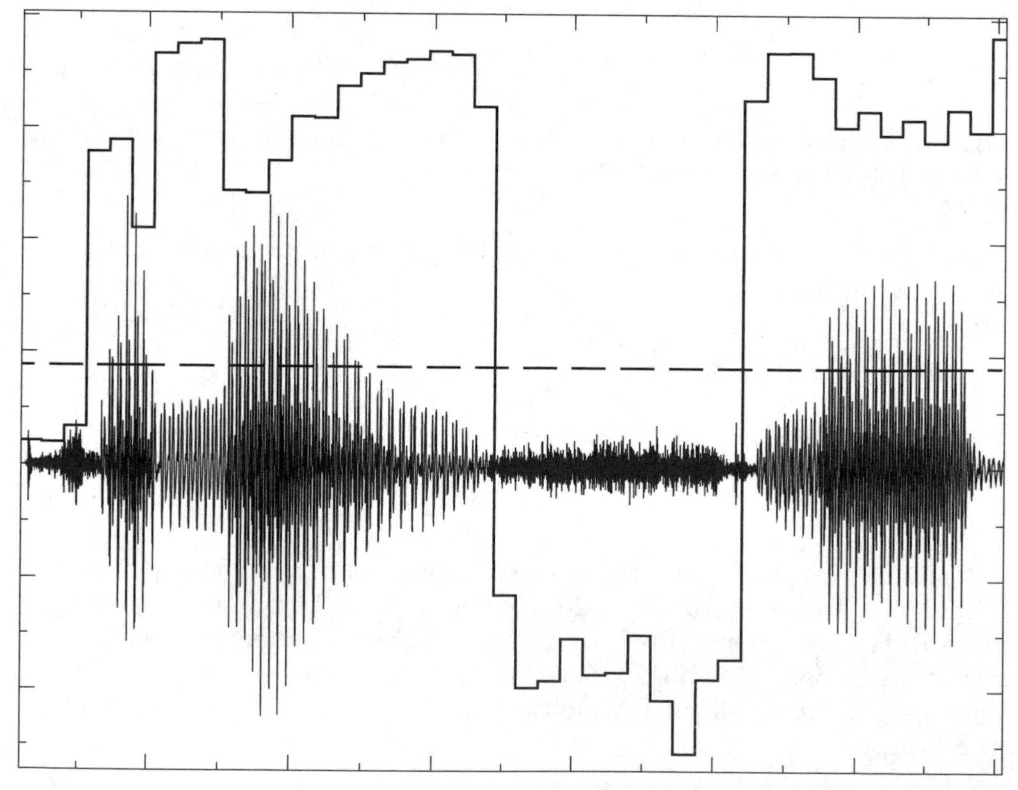

Figure 6.25 Speech waveform and its spectral tilt with a possible voicing threshold of 0.25 (shown by the dashed line)

Pre-emphasized Energy Ratio

Voiced and unvoiced speech can be discriminated by normalized pre-emphasized energy.

$$Pr = \frac{\sum_{i=1}^{N} |s(i) - s(i-1)|}{\sum_{i=1}^{N} |s(i)|} \qquad (6.45)$$

The variance of the difference between adjacent samples is usually much lower in voiced regions than in unvoiced regions. The first-order correlation of voiced samples is around 0.85 but that of unvoiced samples is nearly zero, which is a clear indication of the voiced–unvoiced discriminatory characteristic of this parameter. A speech waveform and its corresponding normalized pre-emphasized energy is shown in Figure 6.26.

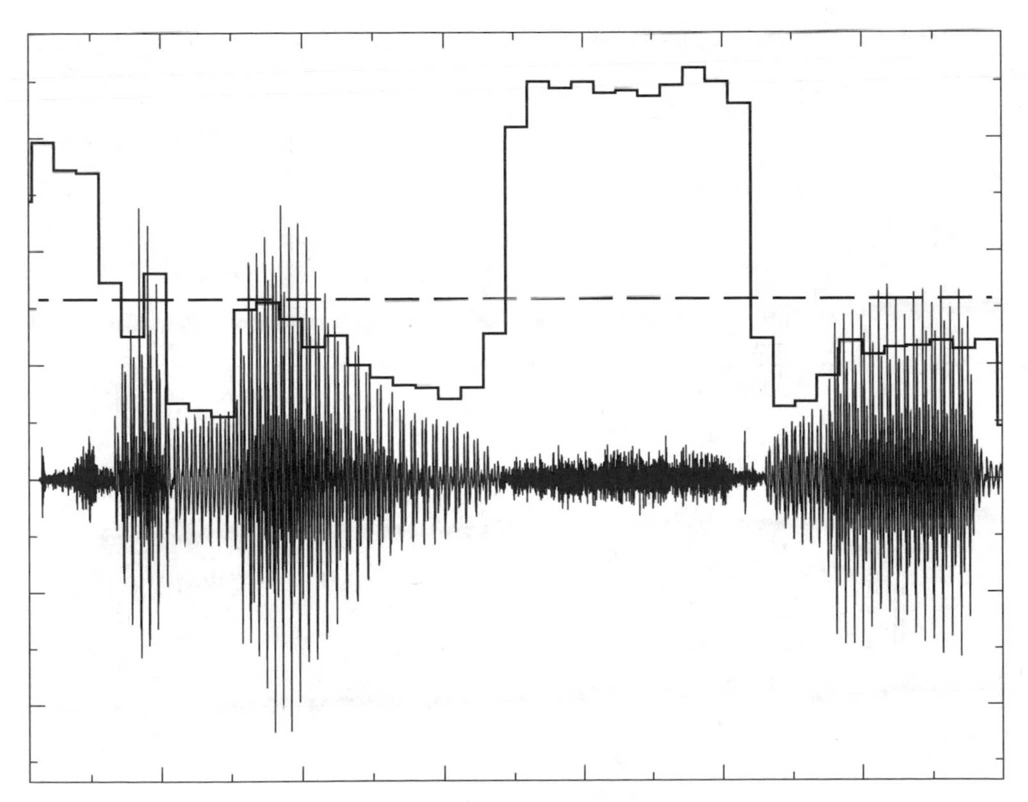

Figure 6.26 Speech waveform and its normalized pre-emphasized energy with a possible voicing threshold of 0.9 (shown by the dashed line)

Low-Band to Full-Band Energy Ratio

Voiced speech usually has a higher low-frequency energy than unvoiced speech. Therefore the energy ratio of the first 1 kHz to the full-band energy can give a good indication whether the speech is voiced. When voiced, the energy ratio is close to one and when unvoiced, since the low-band energy is significantly smaller, the ratio will be less than one.

$$LF = \frac{\sum_{i=1}^{N} s_{lpf}^2(i)}{\sum_{i=1}^{N} s^2(i)} \tag{6.46}$$

where $s_{lpf}(i)$ is low-pass filtered speech at 1 kHz. A speech waveform and its corresponding low-band to full-band energy ratio is shown in Figure 6.27.

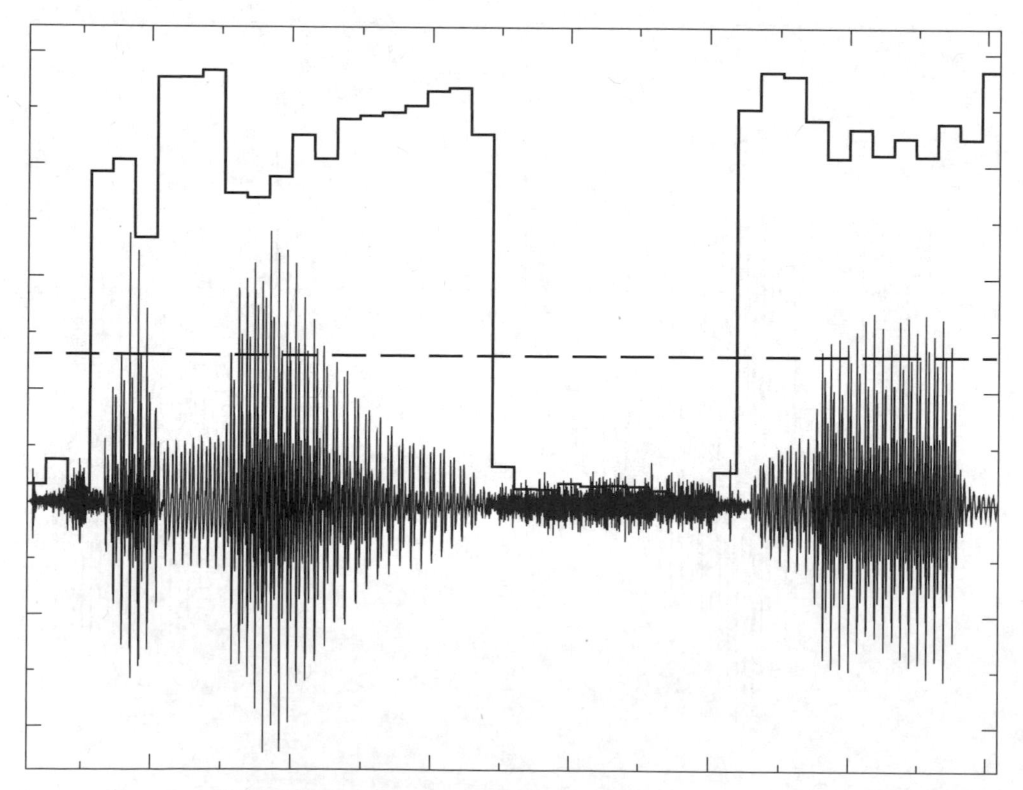

Figure 6.27 Speech waveform and its low-band to full-band energy ratio with a possible voicing threshold of 0.4 (shown by the dashed line)

Frame Energy

Voiced speech usually has a higher energy than unvoiced speech. However, the actual value of the energy in each frame also depends on the dynamic range of the signal. Therefore a more useful measure is to have a comparison of current frame energy with the tracked maximum and minimum energies. The voiced speech should ideally be closer to the maximum track energy and unvoiced speech should be closer to the minimum track energy (excluding silences). The maximum track energy must go up quickly and come down slowly and the minimum tracked energy must come down quickly and go up slowly.

$$E_{max}(n) = \begin{cases} \alpha E_{max}(n-1) + (1-\alpha)E_0 & ; \; if\ E_0 > E_{max}(n-1) \\ \gamma E_{max}(n-1) + (1-\gamma)E_0 & ; \; otherwise \end{cases} \tag{6.47}$$

where E_0 is the current frame energy and $E_{max}(n-1)$ is the previously tracked maximum energy. Typically $\alpha = 0.5$ and $\gamma = 0.98$ enables the maximum energy to go up fast and come down slowly.

$$E_{min}(n) = \begin{cases} \zeta E_{min}(n-1) + (1-\zeta)E_0 & ; \; if\ E_0 < E_{min}(n-1) \\ \beta E_{min}(n-1) + (1-\beta)E_0 & ; \; otherwise \end{cases} \tag{6.48}$$

where $E_{min}(n)$ and $E_{min}(n-1)$ are the current and previously tracked minimum energies. Typical values of $\zeta = 0.55$ and $\beta = 0.98$ are selected so that the minimum energy can come down fast and go up slowly. In addition to the above tracked maximum and minimum energies, the average energy of the speech signal may also be tracked by,

$$E_{av}(n) = 0.75E_{av}(n-1) + 0.25E_0 \tag{6.49}$$

The current frame energy, tracked average energy and tracked minimum energy will be low in the unvoiced regions. In the voiced regions, on the other hand, current frame energy will be close to the tracked maximum. A speech waveform with its corresponding maximum, minimum and average tracked energies, and the frame energy are shown in Figure 6.28. The following piece of logic can be used to indicate voiced or unvoiced,

$$\begin{aligned} & if((E_0 + th1 > E_{max}) || (E_0 > E_{ave})) \\ & \qquad Fe = voiced \\ & else\ \ if(E_0 < th2 + E_{min}) \\ & \qquad Fe = unvoiced \\ & else \\ & \qquad Fe = notsure \end{aligned} \tag{6.50}$$

where *th1* and *th2* are tuning tolerance thresholds.

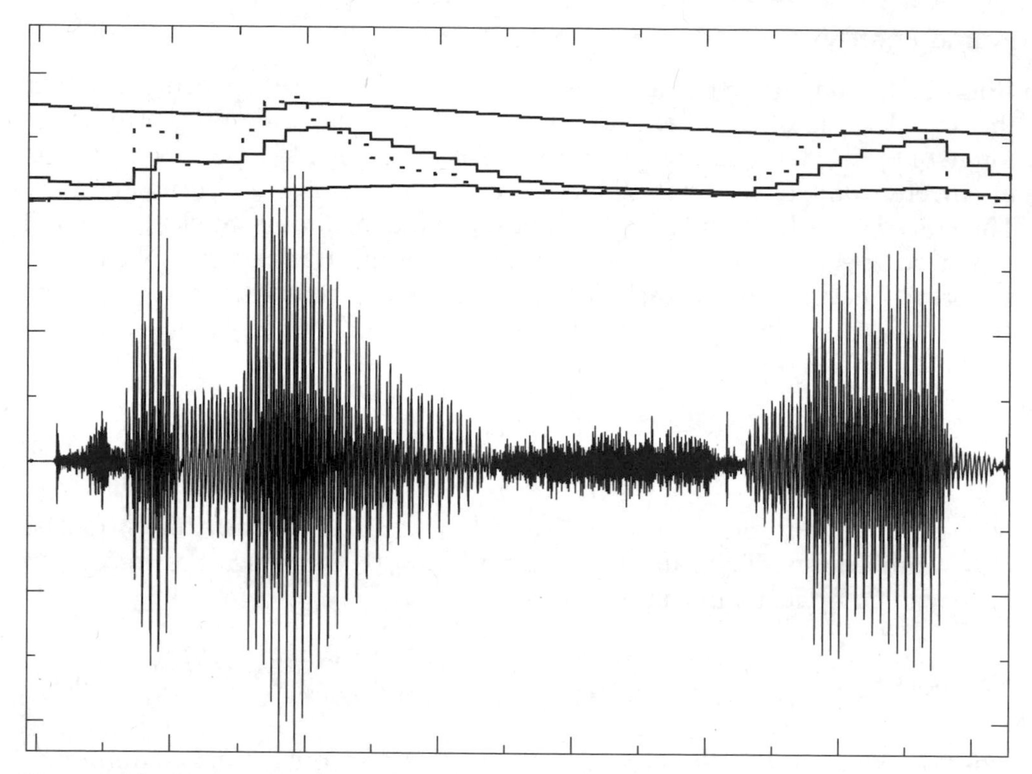

Figure 6.28 Speech waveform and (top) maximum, (middle) average, and (bottom) minimum tracked energies, and the frame energy (shown by the dotted line); energies have been shifted up

Decision-Making

Having computed the most useful voicing indicators, a combined decision has to be made. The simplest decision-making rule is to use a majority vote. A better decision rule could be to use a weighted combination of the voicing indicators. Two types of weighting can be applied to produce a combined decision. Different parameters have different degrees of reliability in indicating the correct voicing and the weighting could be used to reflect these variations in reliability. The weighting of parameters such as periodic similarity and spectral tilt can be higher to reflect their greater reliability. In addition, a second set of weightings can be used to reflect the difference of each parameter from the optimum decision threshold. For example, the variations of each parameter can be normalized to be ± 1 around the optimum threshold and these values can be used in a summation with the appropriate weights which reflect the importance of the corresponding parameter. The normalized parameters are given by,

$$Ps' = \begin{cases} (Ps - Th_{ps})/(Ps_{max} - Th_{ps}) & ; \quad if\ Ps > Th_{ps} \\ (Ps - Th_{ps})/(Th_{ps} - Ps_{min}) & ; \quad if\ Ps < Th_{ps} \end{cases} \tag{6.51}$$

$$Pk' = \begin{cases} (Pk - Th_{pk})/(Pk_{max} - Th_{pk}) & ; \quad if\ Pk > Th_{pk} \\ (Pk - Th_{pk})/(Th_{pk} - Pk_{min}) & ; \quad if\ Pk < Th_{pk} \end{cases} \tag{6.52}$$

$$Zc' = \begin{cases} (Th_{zc} - Zc)/(Th_{zc} - Zc_{min}) & ; \quad if\ Zc < Th_{zc} \\ (Th_{zc} - Zc)/(Zc_{max} - Th_{zc}) & ; \quad if\ Zc > Th_{zc} \end{cases} \tag{6.53}$$

$$St' = \begin{cases} (St - Th_{st})/(St_{max} - Th_{st}) & ; \quad if\ St > Th_{st} \\ (St - Th_{st})/(Th_{st} - St_{min}) & ; \quad if\ St < Th_{st} \end{cases} \tag{6.54}$$

$$LF' = \begin{cases} (LF - Th_{lf})/(LF_{max} - Th_{lf}) & ; \quad if\ LF > Th_{lf} \\ (LF - Th_{lf})/(Th_{lf} - LF_{min}) & ; \quad if\ LF < Th_{lf} \end{cases} \tag{6.55}$$

$$Pr' = \begin{cases} (Th_{pr} - Pr)/(Th_{pr} - Pr_{min}) & ; \quad if\ Pr < Th_{pr} \\ (Th_{pr} - Pr)/(Pr_{max} - Th_{pr}) & ; \quad if\ Pr > Th_{pr} \end{cases} \tag{6.56}$$

$$Fe' = \begin{cases} (E_0 - Th_v)/(E_{max} - Th_v) & ; \quad if\ \ voiced \\ (E_0 - Th_{uv})/(Th_{uv} - E_{min}) & ; \quad if\ \ unvoiced \\ 0 & ; \quad if\ \ not\ sure \end{cases} \tag{6.57}$$

where Th_{ps}, Th_{pk}, Th_{zc}, Th_{st}, Th_{lf} and Th_{pr} are fixed voicing thresholds for the pitch similarity, peakiness, zero crossing, spectral tilt, low-band to full-band energy ratio, and pre-emphasized energy ratio respectively, and Th_v and Th_{uv} are adaptive voiced and unvoiced thresholds used to compare the frame energy. The overall voicing indicator V is then computed by combining the contributions of all indicators.

$$V = w_1 Ps' + w_2 Pk' + w_3 Zc' + w_4 St' + w_5 LF' + w_6 Pr' + w_7 Fe' \tag{6.58}$$

The weights $w_1, \ldots, w_7$ are chosen according to the reliability of each indicator. The sign of the voicing V will indicate voiced when positive and unvoiced when negative. If V is close to zero it will indicate an unsure case, and the voicing of the previous frame could be used to increase reliability. Furthermore, in cases where $V = \pm\delta$ where δ has a small value (indicating an unsure case), individual voicing parameters can be checked to see if one or more of them has a clear indication of voiced or unvoiced. This can be achieved by selecting two further thresholds for each parameter, one indicating voiced and the other unvoiced. These thresholds must be selected by carrying out long simulations. Typically Ps can be above 0.7 for voiced and below 0.3 for

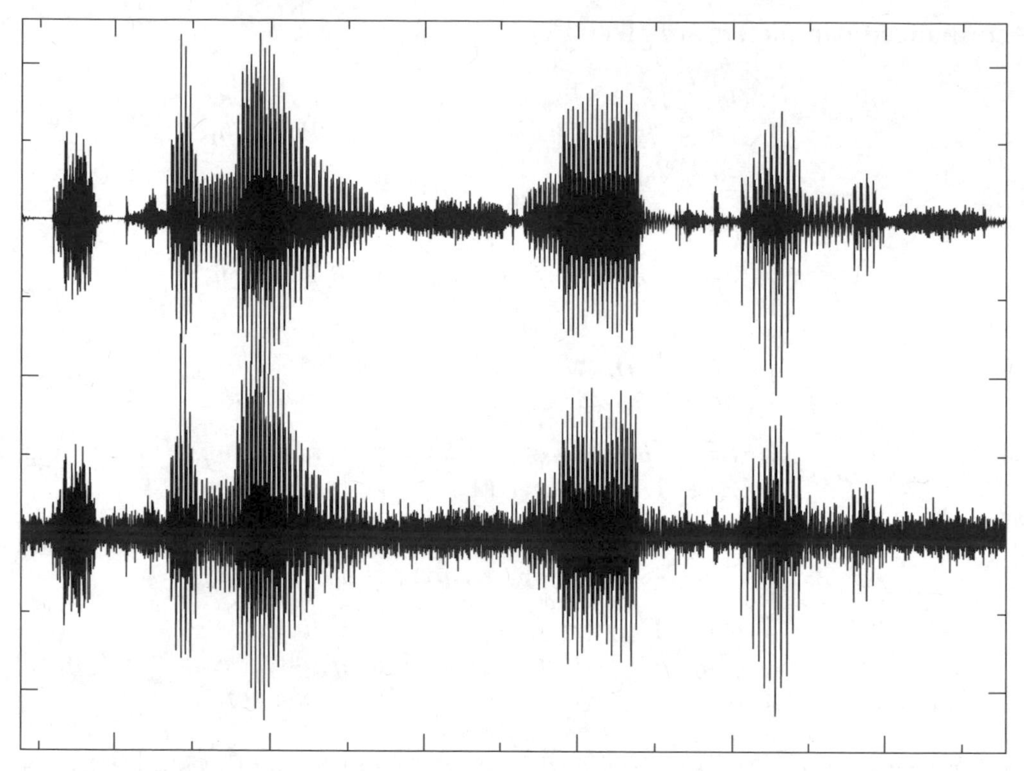

Figure 6.29 Clean (top) and 10 dB SNR noisy (bottom) speech waveforms

unvoiced. The *St* can have values above 0.6 for voiced and below 0.2 for unvoiced. Similarly *Zc* can have values below 40 and above 90 out of 160 for voiced and unvoiced respectively.

The above hard-decision voicing method works very well with clean background speech signals. However when speech is mixed with background noise, the set thresholds may not be valid anymore. Hence a more careful decision-making logic needs to be employed. Waveforms of original speech and 10 dB SNR heavy vehicle noise are shown in Figure 6.29. As can be seen from the figure, most unvoiced and some voiced sounds have been submerged in the noise, making it very difficult to see them. Under noisy conditions, voicing parameters are expected to differ considerably. The variations of three voicing parameters (spectrum tilt, pre-emphasized energy ratio and pitch similarity) are shown in Figure 6.30.

When there is a transition from voiced to unvoiced or unvoiced to voiced, even during clean speech conditions, a frame can be mistakenly declared as voiced or unvoiced since both voiced and unvoiced exist together in that frame. It is therefore necessary to refine the voicing decision further by introducing a mixed frame type in addition to completely voiced and unvoiced frames. The all-important question is what proportion of the frame

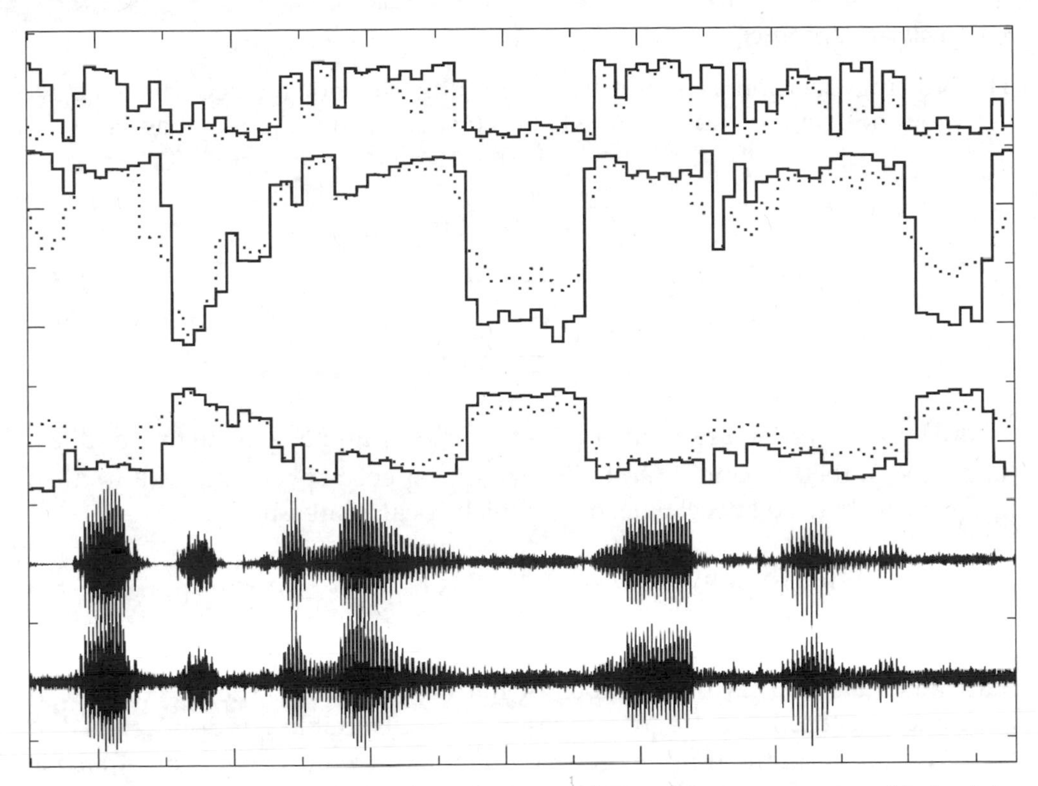

Figure 6.30 Reading from the top: *St*, *Pr*, and *Ps* voicing parameters (dotted for noisy speech), and the original and noisy speech waveforms

will be voiced and unvoiced? This leads to an adaptive mixed-voicing decision process which has been used in MBE, MELP, etc.

6.3.2 Soft-Decision Voicing

Although fully voiced and fully unvoiced frames can be identified in the time domain by using the voicing parameters discussed above, in the case of noisy speech this becomes more difficult and more mistakes are made. In order to avoid this problem and to deal with the mixed frames in one process, a frequency-domain voicing-decision process is more appropriate. The mixed voicing-decision process usually makes use of the harmonic and random structures of voiced and unvoiced sounds in the frequency domain. For example, in MBE-based coders, a synthetic spectrum (constructed by using the measured pitch of the frame) tests the degree of match with the original spectrum. Better-matched frequencies are declared voiced and the rest are classified as unvoiced. In the case of MELP, the input frame is first split into subbands and the long-term correlation in each band is measured to classify the band as voiced (high correlation) or unvoiced (low correlation).

MBE Mixed Voicing

The voicing decision is made by examining the normalized distance D_k between the original and estimated speech spectra in frequency bands,

$$D_k = \frac{\sum_{m=a_k}^{b_k} |S(m) - \hat{S}(m, \omega_0)|^2}{\sum_{m=a_k}^{b_k} |S(m)|^2} \tag{6.59}$$

where ω_0 is the refined fundamental frequency, a_k and b_k are the first and last harmonic in the kth band, $S(m)$ is the original speech spectrum, and $\hat{S}(m, \omega_0)$ is the reconstructed speech spectrum which is calculated using:

$$\hat{S}(m, \omega_0) = A_l(\omega_0)W(m) \quad 1 \leq l \leq L, \quad \lceil a_l \rceil \leq m < \lceil b_l \rceil \tag{6.60}$$

where $a_l = (l - 0.5)\omega_0$, $b_l = (l + 0.5)\omega_0$, $\lceil . \rceil$ means the nearest integer greater than or equal to, L is the number of harmonics within the 4 kHz speech bandwidth, $W(m)$ is the frequency response of a suitable window centred at the lth harmonic of the fundamental frequency (see Figure 6.31) and $A_l(\omega_0)$ is

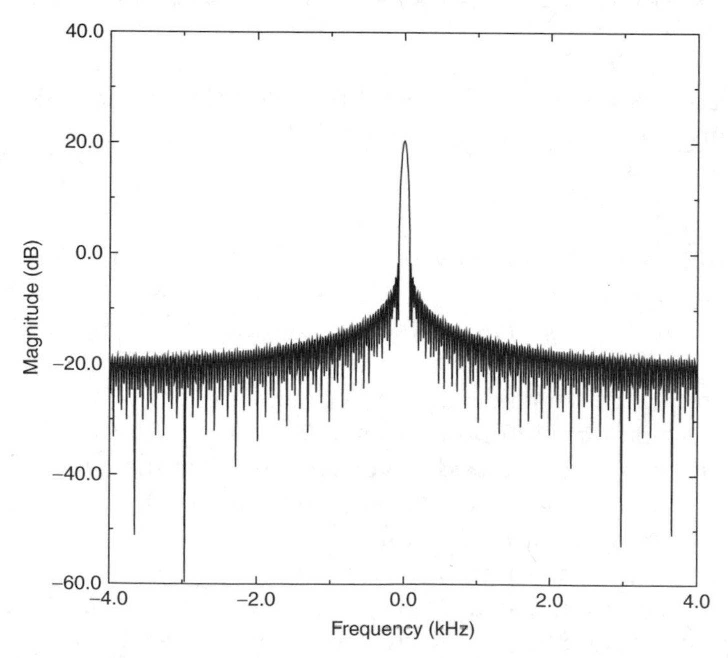

Figure 6.31 Frequency response of the Hamming window

the lth harmonic amplitude which is computed using:

$$A_l(\omega_0) = \frac{\sum_{m=\lceil a_l \rceil}^{\lceil b_l \rceil} S(m)W(m)}{\sum_{m=\lceil a_l \rceil}^{\lceil b_l \rceil} |W(m)|^2} \tag{6.61}$$

When creating the synthetic spectrum, it is very important to adjust the position of $W(m)$ and the size of the transform used, to make sure that the peak of the window is centred on the harmonic and dies down to a very small value at $\pm 0.5\,\omega_0$ around each harmonic. As can be seen from the formulation above, the synthetic spectrum is assumed to be all voiced. However, the speech spectrum is not all voiced and, although the synthetic spectrum will be very similar to the original spectrum in the voiced regions, it will have larger differences in the unvoiced regions. Therefore, this similarity (or dissimilarity) measure can be used to make a reasonably correct voicing decision by comparing it against a pre-determined threshold. The value of the threshold is set to give the proper mix of voiced and unvoiced energy. Listening tests can be used to set the adaptive threshold function to values where the ratio of voiced and unvoiced energy is perceptually optimum. To determine the voicing decisions, the normalized error, D_k, for each frequency band is compared with this adaptive threshold, $\Delta_k(\omega_0)$ given by [28]

$$\Delta_k(\omega_0) = (\alpha + \beta\omega_0)\,[1.0 - \varepsilon(k-1)\omega_0]\,M(E_0, E_{av}, E_{min}, E_{max}) \tag{6.62}$$

where $\alpha = 0.35$, $\beta = 0.557$, and $\varepsilon = 0.4775$ are the factors that give good subjective quality and,

$$M(E_0, E_{av}, E_{min}, E_{max}) = \begin{cases} 0.5 & ;\ E_{av} < 200 \\ \dfrac{(E_0 + E_{min})(2E_0 + E_{max})}{(E_0 + \mu E_{max})(E_0 + E_{max})} & ;\ \begin{array}{l} E_{av} \geq 200 \textit{ and} \\ E_{min} < \mu E_{max} \end{array} \\ 1.0 & ;\ \textit{otherwise} \end{cases} \tag{6.63}$$

is the adaptation factor that controls the decision threshold for voicing decisions, and $\mu = 0.0075$. The parameters E_{av}, E_{max}, and E_{min} roughly correspond to the local average energy, the local maximum energy and the local minimum energy respectively. These three parameters are updated

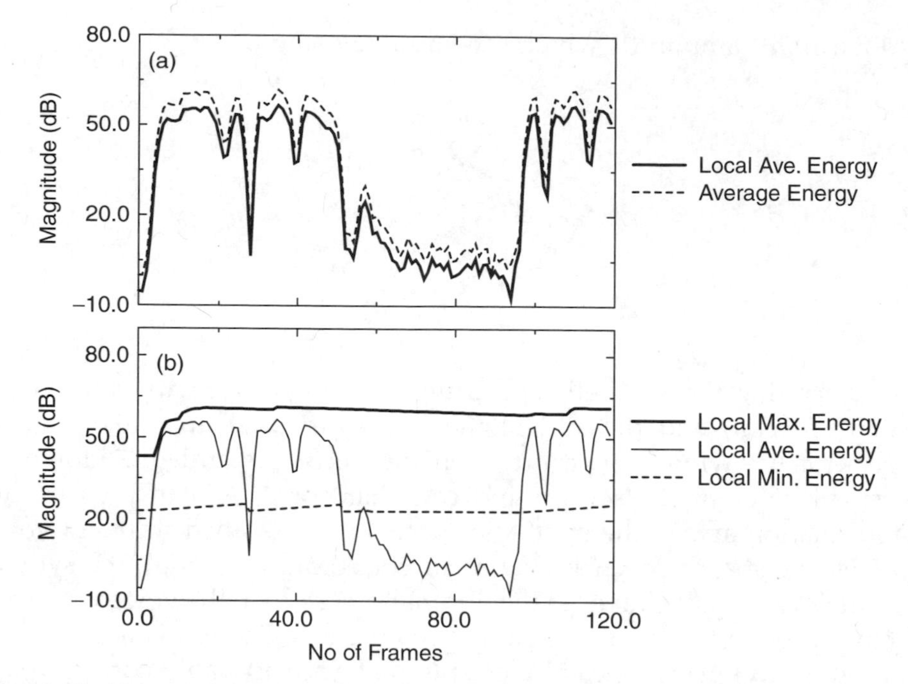

Figure 6.32 The relationship between the energy levels used in determining voiced and unvoiced speech

every speech frame according to [28],

$$E_{av}(n) = 0.7E_{av}(n-1) + 0.3E_0 \tag{6.64}$$

$$E_{max}(n) = \begin{cases} 0.5E_{max}(n-1) + 0.5E_0 & ; \quad \text{if } E_0 > E_{max}(n-1) \\ 0.99E_{max}(n-1) + 0.01E_0 & ; \quad \text{otherwise} \end{cases} \tag{6.65}$$

$$E_{min}(n) = \begin{cases} 0.5E_{min}(n-1) + 0.5E_0 & ; \quad \text{if } E_0 \leq E_{min}(n-1) \\ 0.975E_{min}(n-1) + 0.025E_0 & ; \quad \text{if } E_{min}(n-1) \leq E_0 < 2E_{min}(n-1) \\ 1.025E_{min}(n-1) & ; \quad \text{otherwise} \end{cases} \tag{6.66}$$

Relative variations of these energy levels are illustrated in Figure 6.32. The voicing decision for each band is made by comparing the normalized error for the band with the value of the threshold function which is computed using the above procedure. If the normalized error is less than the threshold function, the corresponding frequency band is declared voiced; otherwise, the frequency band is declared unvoiced. The variations of the threshold and the corresponding error function are shown in Figure 6.33.

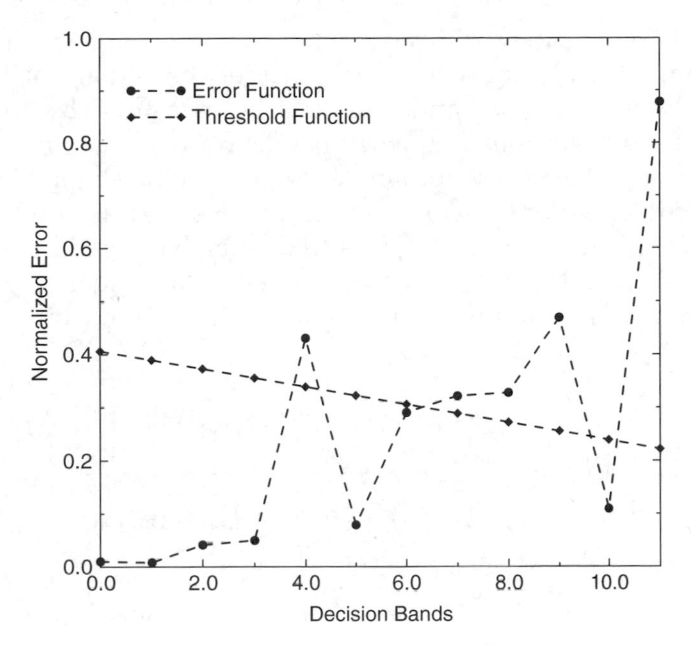

Figure 6.33 The error and threshold functions for one frame

Split-Band Mixed Voicing

Correct estimation of the threshold level for each band is the most important stage in MBE mixed-voicing estimation. The other important factor is that more than one bit will be needed during the coding of the mixed-voicing decision estimate. Since each band will require one bit, more bands will mean higher accuracy but an increased bit rate. When closely examined however, we see that if a spectrum contains an unvoiced band between two voiced bands, the unvoiced signal in the middle is usually relatively small and if it is declared as voiced, subjectively it would not make much difference. This is very important because it saves bits when coding the mixed voicing-decision. In this case a single point in the frequency spectrum can be used to identify the voiced (low frequency) and unvoiced (high frequency) regions. There may be several ways to obtain the single frequency marker or cut-off point which separates the voiced and unvoiced parts. For example, using MBE mixed-voicing, above, one can assume that the spectrum is voiced up to the highest frequency voiced band. Alternatively, the total number of voiced bands obtained in MBE mixed-voicing can be counted and used to set the same number of low frequency bands to voiced. In this case, some high-frequency voiced bands will be swapped with low-frequency unvoiced bands. Although these methods may give good quality in the majority of mixed-signal frames, they still rely on hard-decision voicing in individual bands. A more reliable approach is to consider the actual voicing value or voicing likelihood in each band. This can be measured by the degree of harmonic structure in each frequency band. If a harmonic band is voiced, then its content will have a shape similar to the spectral shape of the window used prior to the Fourier transform, while unvoiced bands will be random in nature. Hence the level of voicing in a frequency band can be measured by the normalized correlation between the content of a frequency band and the spectral shape of the window positioned on each harmonic bin.

$$V(l) = \frac{\left[\sum_{m=l}^{\omega_0} S(m - l\omega_0 + 0.5\,\omega_0)W(m)\right]^2}{\sum_{m=l}^{\omega_0} W^2(m) \sum_{m=l}^{\omega_0} S^2(m - l\omega_0 + 0.5\,\omega_0)} \tag{6.67}$$

where $S(m)$ is speech spectrum and $W(m)$ is the Fourier transform of the analysis window. $W(m)$ is usually up-sampled by computing its Fourier transform using a larger transform size (compared to $S(m)$) and then down-sampled with respect to the fundamental frequency, so as to have the same number of points within each harmonic region of $S(m)$. The voicing $V(l)$ has a value between 0.0 and 1.0, which indicate fully unvoiced and voiced respectively. Similar to MBE mixed-voicing, $V(l)$ is compared against a threshold in each band. Since the voicing probability varies in each band as well as in each frame, the threshold value needs to be adaptive. This threshold can be computed by combining the voicing indicators, such as pitch similarity, zero crossing, peakiness, low-band to high-band energy ratio, E_0/E_{max}, etc. Having computed $V(l)$ and the threshold $T(l)$ for each band, we need to make a decision to choose the best cut-off frequency. Since this cut-off point will be quantized before transmission, it is more convenient to test each quantizer level against a measure so that the selected cut-off frequency is also quantized. For each quantizer value i, a matching measure $M(i)$ can be computed as given below:

$$M(i) = \sum_{l=1}^{L} (V(l) - T(l)v_i(l)E(l)B(l) \tag{6.68}$$

This takes into account the energy of each harmonic, $E(l)$, and a biasing, $B(l)$, which represents the perceptual weighting. For a given quantizer level i, individual voicings $v_i(l)$ will have values of $+1$ up to the cut-off i and -1 for the higher harmonics. The weighting, $B(l)$, is usually set to 1.0 when unvoiced ($T(l) > V(l)$) and higher for voiced. The above voicing process takes

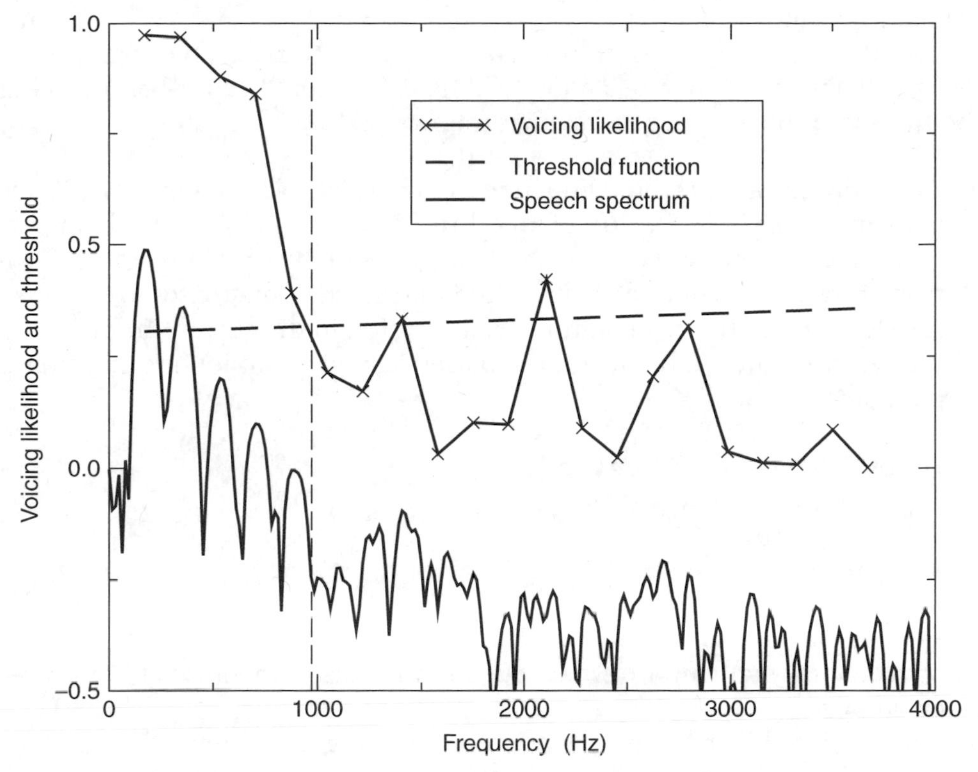

Figure 6.34 Original speech spectrum with voicing likelihood and threshold function; the voicing cut-off frequency is indicated by the vertical dashed line

into account the difference between $V(l)$ (the voicing likelihood) and the threshold $T(l)$, which replaces the hard decision used in MBE mixed-voicing with a soft decision in each band. An example of a voicing likelihood and threshold function is shown in Figure 6.34.

It is also possible to use this weighted-sum approach on the voicing measure used in MBE. However, the MBE approach requires the computation and generation of a synthetic spectrum, as described above. This is not required for the voicing likelihood method discussed here. However, as for the MBE and MELP voicing-decision algorithms, the most important stage during split-band voicing estimation is the calculation of the threshold function. Using a limited number of speech characteristics for the threshold computation does not lead to good voicing determination. For example, the energy alone is not a reliable enough voicing indicator, since there can be high-energy unvoiced speech segments and low-level voiced speech. The peakiness factor is not entirely reliable either: single spikes can lead to high peakiness, but they should be declared as unvoiced for optimal speech quality. Likewise, the periodic similarity measure has its limits: when the pitch varies, the normalized autocorrelation may be quite low whereas the speech is clearly voiced. It is therefore necessary to make full use of the speech characteristics described above to generate a good threshold function. In split-band voicing, the threshold function is generated as follows [29]:

1. An initial linear threshold function is generated which starts at 0.4 and goes up to 0.55. The value of the threshold is increased for harmonics which correspond to the unvoiced harmonics in the previous frame. If the previous frame is completely unvoiced the threshold increases to 0.55–0.65 (increasing the chance of an unvoiced decision in the current frame).
2. The voicing-threshold function is biased using the following individual parameters:

   - Low- to full-band energy ratio
   - Pre-emphasis energy ratio
   - Zero-crossing rate
   - Frame energy

   These parameters have their high and low thresholds set and, if they are triggered, the voicing threshold function is biased towards either voiced or unvoiced.
3. The voicing-threshold function is biased using the pitch value. A high number of harmonics present in the speech implies that the harmonic bands are narrow and contain a small number of frequency bins. As a result, the voicing likelihood tends to increase, as the matching is performed on fewer points. The voicing threshold function needs to be biased to compensate for this effect.
4. Finally, very specific cases detected in individual speech characteristics are used to bias the threshold. For example, very high periodic similarity is used to increase the voiced likelihood and very high zero-crossing rate (in clean conditions) is used to increase the unvoiced likelihood.

This voicing determination method provides very robust detection accuracy, even under significant background noise conditions.

6.4 Summary

Developments in the field of fast DSP technology have allowed the use of more and more sophisticated algorithms required for accurate pitch estimation and voiced–unvoiced classification. With the new multi-domain (frequency and time) pitch estimation, it is possible to get good performance even under noisy conditions. However, even the latest and most complex pitch estimation algorithms are not perfect. In some speech segments, the pitch is not well-defined and some errors are inevitable. Overall performance of the pitch-estimation algorithms, however, can be considered to be pretty good. Voiced–unvoiced classification, on the other hand, has moved from a single (binary) indicator, where each block of speech was classified either as voiced or unvoiced, to more elaborate frequency-domain mixed decisions. This has increased the quality of synthetic speech dramatically. The performance of voicing estimation under noisy conditions has also been improved with developments in mixed-voicing classification.

Bibliography

[1] L. R. Rabiner, M. J. Cheng, A. E. Rosenberg, and C. A. McGonegal (1976) 'A comparative performance study of several pitch detection algorithms', in *IEEE Trans. on Acoust., Speech and Signal Processing*, 24(5):399–418.

[2] W. J. Hess (1992) 'Pitch and voicing determination', in *Advances in Speech Signal Processing* by S. Furui and M. M. Sondhi (Eds), pp. 3–48. New York: Marcel Dekker Inc.

[3] L. Rabiner (1977) 'On the use of autocorrelation analysis for pitch detection', in *IEEE Trans. on Acoust., Speech and Signal Processing*, 25(1):24–33.

[4] M J. Ross, H. L. Shaffer, A. Cohen, R. Freudberg, and H. J. Manley (1974) 'Average magnitude difference function pitch extractor', in *IEEE Trans. on Acoust., Speech and Signal Processing*, 22(5):353–62.

[5] C. K. Un and S.-H. Yang (1977) 'A pitch extraction algorithm based on LPC inverse filtering and AMDF', in *IEEE Trans. on Acoust., Speech and Signal Processing*, 25(6):565–72.

[6] ITU-T (1996) *Dual rate speech coder for multimedia communications transmitting at 5.3 and 6.3 kbit/s*, ITU-T Rec. G.723.1.

[7] ITU-T (1996) *Coding of speech at 8 kbit/s using conjugate-structure algebraic-code-excited linear prediction (CS-ACELP)*, ITU-T Rec. G.729.

[8] ETSI (1997) *Digital cellular telecommunications system (phase 2+); Half rate speech; Half rate speech transcoding*, GSM 06.20 v5.1.0 (draft ETSI ETS 300 969).

[9] ETSI (1998) *Digital cellular telecommunications system (phase 2); Enhanced full rate (EFR) speech transcoding*, GSM 06.60 v4.1.0 (ETS 301 245), June.

[10] ETSI (1998) *Digital cellular telecommunications system (phase 2+); Adaptive multi-rate (AMR) speech transcoding*, GSM 06.90 v7.2.0 (draft ETSI EN 301 704).

[11] FIPS (1997) *Analog to digital conversion of voice by 2,400 bit/second mixed excitation linear prediction (MELP)*, Draft. Federal Information Processing Standards

[12] A. M. Noll (1967) 'Cepstrum pitch determination', in *Journal of the Acoustic Soc. of America*, 41:293–309.

[13] D. Griffin and J. S. Lim (1988) 'Multiband excitation vocoder', in *IEEE Trans. on Acoust., Speech and Signal Processing*, 36(8):1223–35.

[14] R. J. McAulay and T. F. Quatieri (1995) 'Sinusoidal coding', in *Speech coding and synthesis* by W. B. Kleijn and K. K. Paliwal (Eds), pp. 121–74. Amsterdam: Elsevier Science

[15] L. P. Nguyen and S. Imai (1977) 'Vocal pitch detection using generalized distance function associated with a voiced-unvoiced decision logic', in *Bull, P.M.E. (T.I.T)*, 39:11–12.

[16] W. Hess (1983) *Pitch Determination of Speech Signals: algorithms and devices.* Berlin: Springer-Verlag

[17] I. Atkinson (1997) 'Advanced linear predictive speech compression at 3.0 kbit/s and below', Ph.D. thesis, CCSR, University of Surrey, UK.

[18] D. B. Paul (1981) 'The spectral envelope estimation vocoder', in *IEEE Trans. on Acoust., Speech and Signal Processing*, 29(4):786–94.

[19] Y. D. Cho (2001) 'Speech detection enhancement and compression for voice communications', Ph.D. thesis, CCSR, University of Surrey, UK.

[20] I. Atkinson, S. Yeldener, and A. Kondoz (1997) 'High quality split-band LPC vocoder operating at low bit rates', in *Proc. of Int. Conf. on Acoust., Speech and Signal Processing*, pp. 1559–62. May 1997. Munich

[21] ITU-T (1996) *Software tool library.*

[22] M. M. Sondhi (1968) 'New methods of pitch extraction', in *IEEE Trans. on Audio and Electroacoustics*, 16:262–6, June.

[23] J. D. Markel (1972) 'The SIFT algorithm for fundamental frequency estimation', in *IEEE Trans. Audio and Electroacoustics*, 20:367–77.

[24] M. Schroeder and B. Atal (1979) 'Predictive coding of speech signals and subjective error criteria', in *IEEE Trans. on Acoust., Speech and Signal Processing*, 27:247–54.

[25] J. J. Dubnowski, R. W. Schafer, and L. R. Rabiner (1976) 'Real-time digital hardware pitch detector', in *IEEE Trans. on Acoust., Speech and Signal Processing*, 24:2–8.

[26] J. Sundberg (1979) 'Maximum speed of pitch changes in singers and untrained subjects', in *Journal of Phonetics*, 7:71–9.

[27] W. Ma, A. M. Kondoz, and B. G. Evans (1992) *The Real-Time Implementation of INMARSAT Standard-M Codec.*

[28] DVSI (1991) *INMARSAT M Voice Codec*, Version 1.3. Digital Voice Systems Inc.

[29] S. Villette (2001) 'Sinusoidal speech coding for low and very low bit rate applications', Ph.D. thesis, University of Surrey, UK.

7

Analysis by Synthesis LPC Coding

7.1 Introduction

The broad classification of speech coding techniques that attempt to reproduce the original speech waveform as best as possible can be split into two basic groups, namely analysis-and-synthesis (AaS) schemes and analysis-by-synthesis (AbS) schemes. Although AaS schemes, such as APC [1, 2], ATC [3] and SBC [4], have been successful at rates around 16 kb/s and above, below 16 kb/s they can no longer reproduce good quality speech. In addition, AaS coders that have been used at bit rates of around 9.6–16 kb/s can not achieve true toll quality performance (MOS≥4). There are two main reasons for their shortcomings: first, the coded speech is not analysed to see if the coding procedure is operating efficiently, i.e. there is no check on or control over the distortions of the reconstructed speech; and secondly, in adaptive schemes, the errors accumulated from previous frames are not usually considered in the current frame of analysis, hence the errors propagate into the following frames without any form of resetting. In AbS schemes, particularly AbS-LPC schemes [5, 6], these two factors are incorporated in the coding process. In AbS-LPC coding systems, a *closed-loop* optimization procedure is used to determine the excitation signal, which produces a perceptually optimum synthesized speech signal when used to excite the *model filter*. It is this closed-loop approach which enables AbS-LPC coding schemes to be far more successful at 4.8 to 16 kb/s than conventional AaS schemes such as APC and SBC.

The method of AbS is not unique to speech coding, but is a general technique used in other areas of estimation and identification. The basic idea behind AbS is as follows. First it is assumed that the signal can be observed

Digital Speech. A. Kondoz
© 2004 John Wiley & Sons, Ltd ISBN 0-470-87007-9 (HB)

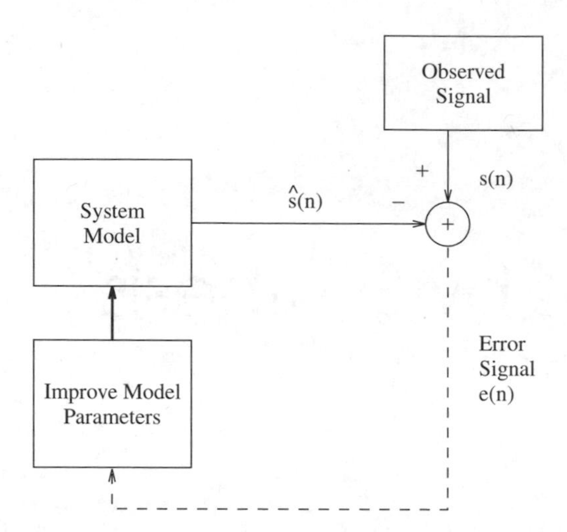

Figure 7.1 General block diagram of analysis-by-synthesis closed-loop analysis

and represented in some form, e.g. the time or frequency domain. Then a theoretical form of the signal production model is assumed, as depicted in Figure 7.1. The model has a number of parameters which can be varied to produce different variations of the observable signal. In order to derive a representation of the model that is of the same form as the true signal model, a trial and error procedure can be applied. By varying the parameters of the model in a systematic way, it is possible to find a set of parameters that can produce a synthetic signal which matches the real signal with minimum possible error (assuming the model is valid to begin with). Therefore, when such a match is calculated, the parameters of the model are assumed to be the parameters of the true signal.

The AbS procedure outlined above was applied to speech processing in the earlier days of formant estimation [7] but, because of its obvious complexity, it was not re-applied until Atal outlined the basis of Multi-pulse LPC (MPLPC) in [8] for low bit-rate coding. In Atal's work, the time-domain representation of speech was used and a model very similar to the conventional *source-filter* model was selected. However, AbS with other domains and models are equally applicable [9]. In the following sections a unified presentation of the various AbS-LPC schemes using Atal's modelling is described.

7.2 Generalized AbS Coding

The basic structure of an AbS-LPC coding system is illustrated in Figure 7.2. There are basically three blocks in the model that can be varied to match our true model and, hence, obtain a good synthesized speech signal: *time-varying filter*, *excitation signal* and *perceptually-based error minimization procedure*. As our

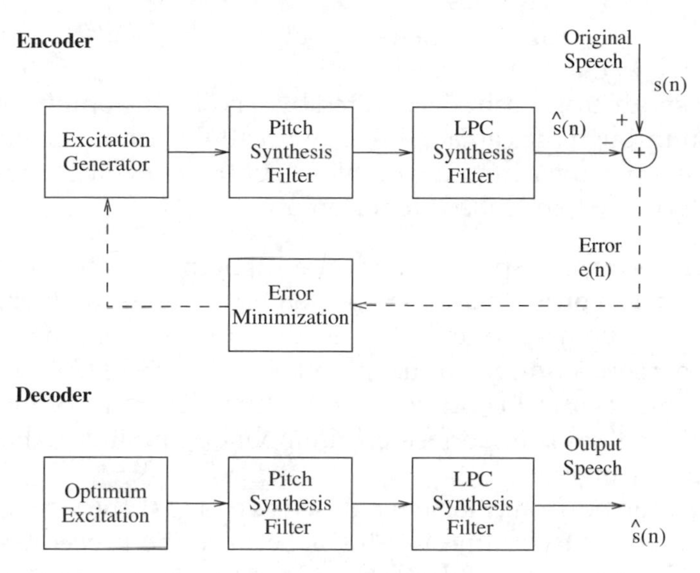

Figure 7.2 Block diagram of AbS-LPC coding scheme

model requires frequent updating of the parameters to yield a good match with the original signal, the analysis procedure of the system is carried out in blocks, i.e. the input speech is partitioned into suitable blocks of samples. The update rate of the analysis block or frame determines the bit rate or capacity of the coding schemes. The basic operation of an AbS-LPC scheme is as follows:

1. Initialize the contents of the time-varying filter (LPC and pitch) to predetermined values (usually zero or low level random noise).
2. A frame of speech samples is buffered and a set of LPC coefficients are computed, using LPC analysis on the frame.
3. As the LPC analysis frame is usually too large for efficient analysis to determine the excitation, the frame is subdivided into a subframes.
4. For each subframe:
   (a) Using the computed LPC coefficients (usually interpolated for each subframe) in the LPC filter, its memory effect (filter-ringing response) is computed and subtracted from the original signal, which is usually perceptually weighted.
   (b) The pitch filter delay (the pitch or its integer multiples) and its associated scaling factor (pitch gain) are then calculated. This calculation is performed in such a way that the difference between the synthetically-generated speech and the remaining original signal is minimized.
   (c) Once the pitch filter parameters are found, the pitch and LPC synthesis filter can be grouped together to form a cascaded filter. Using this cascaded filter, the best secondary excitation is determined in such a

way as to minimize the difference between the synthetically-generated speech and the original speech.

5. The final synthetic speech is generated by passing the optimum secondary excitation through the cascaded filter with all the initial memory contents of the filters (left over from the previous subframe synthesis) restored.
6. Repeat steps 2 to 5 for subsequent frames.

Note that the synthetic speech is generated at both the encoder and decoder. This is necessary in order to update the memory contents of the time-varying filters such that both encoder and decoder possess replica conditions in their filter memories. In fact, a major concern of AbS-LPC schemes is how to preserve this identical condition at both encoder and decoder when the transmitting medium is imperfect, e.g. in mobile radio links where the error rates can be very high.

It can be observed from the above descriptions that the AbS-LPC scheme is not truly analysis-by-synthesis. This is because the procedure is actually sequential in nature, i.e. the LPC filter parameters are calculated and fixed, then the pitch filter parameters are calculated, followed by the computation of the secondary excitation. Consequently, although the secondary excitation is obtained optimally with respect to the original reference signal, its optimality is limited by the optimality of the filters it uses. The best combination of the excitation and the filters is desired, which means optimizing all the parameters in parallel. Obviously, this joint procedure is very complicated as well as being very computationally intensive, thus it is split into the sequential stages described above.

It is interesting to note that this model is very similar to that of the classical source-filter vocoders [7]. However, there is one major difference between basic vocoders and AbS-LPC coders. In classical vocoders, the source excitation is classified into voiced (pulse excitation) and unvoiced (random noise excitation), which is a major source of model inaccuracy. However, in AbS-LPC, this categorization is not explicit and therefore the excitation signal can be anything from pulse-like to noise-like in characteristic, thus enabling much better quality speech to be synthesized.

7.2.1 Time-Varying Filters

The block representing the time-varying filter in our model is usually made of two linear predictors, namely the LPC or short-term predictor (STP) and the pitch or long-term predictor (LTP). The LPC models the short-term correlation in the speech signal (the spectral envelope) and is given by,

$$\frac{1}{A(z)} = \frac{1}{1 - \sum_{i=1}^{p} a_i z^{-i}} \tag{7.1}$$

where a_i are the LPC coefficients and p is the filter order. It is made time-varying to reflect the change in the speech spectrum with adaptation rates of typically around 20–30 ms. The order of the filter, p, is usually chosen to be around 8 to 12.

The pitch filter models the long-term correlation in speech (the fine spectral structure) and is given by,

$$\frac{1}{P(z)} = \frac{1}{1 - \sum_{i=-I}^{I} b_i z^{-(D+i)}} \tag{7.2}$$

where D is a pointer to long-term correlation which usually corresponds to the pitch period or its multiples and b_i are the pitch (or LTP) gain coefficients. Again, this is a time-varying filter but it usually has higher adaptation rates than the LPC, e.g. 5–10 ms. The number of filter taps typically takes the form $I = 0$, i.e. 1 tap, and $I = 1$, i.e. 3 taps. Note that because of the recursive nature of the two filters, both contain *memory* in their working buffers carried over from the previous frame of analysis. The preservation and inclusion of this filter memory in the AbS analysis is very important as it reflects the past history of the analysis, and includes any errors incurred in the previous frames. Also, it provides a smoothing effect to the distortions caused by the block-oriented analysis, such as edge effects.

7.2.2 Perceptually-based Minimization Procedure

The AbS-LPC coder of Figure 7.2 minimizes the error between the original $s(n)$ and the synthesized signal $\hat{s}(n)$ according to a suitable error criterion, by varying the excitation signal and the LPC and pitch filters. As described earlier, this is achieved via a sequential procedure. First the time-varying filter parameters are determined, then the excitation is optimized.

The optimization criterion used for both procedures is the commonly used mean squared error, which offers simplicity and adequate performance. However, at low bit-rates there is one or fewer bit per sample coding capacity, thus it is more difficult to match the waveform closely than in, say, higher than 16 kb/s schemes, where more than 1 bit/sample is available. Consequently, the mean squared error between the original and reconstructed signal is less meaningful and less than adequate. What is required is an error criterion which is more in sympathy with human perception. Although much work on auditory perception is in progress, no satisfactory error criterion has yet emerged. In the meantime, however, a popular but not totally satisfactory method is the use of a weighting filter in AbS-LPC schemes. This weighting filter is given by,

$$W(z) = \frac{A(z)}{A(z/\gamma)} \tag{7.3}$$

$$= \frac{1 - \sum_{i=1}^{p} a_i z^{-i}}{1 - \sum_{i=1}^{p} a_i \gamma^i z^{-i}}, \quad 0 \le \gamma \le 1$$

This weighting filter is the same as that proposed by Atal [2] for APC schemes and a typical plot of its frequency response is shown in Figure 7.3. The effect of the factor γ does not alter the centre formant frequencies but just broadens the bandwidth of the formants by Δf given by,

$$\Delta f = -\frac{f_s}{\pi} \ln \gamma \quad (Hz) \tag{7.4}$$

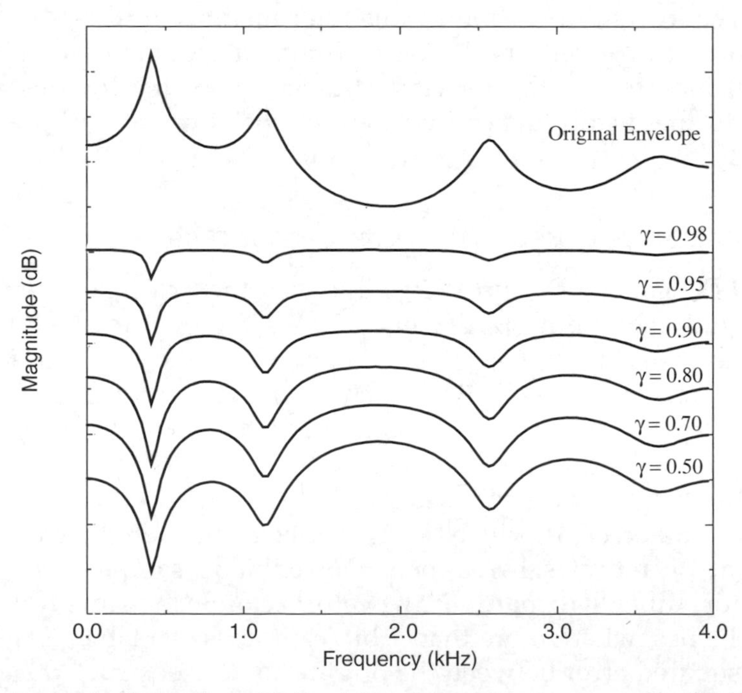

Figure 7.3 Typical plots of weighting filter spectra compared with the original speech envelope

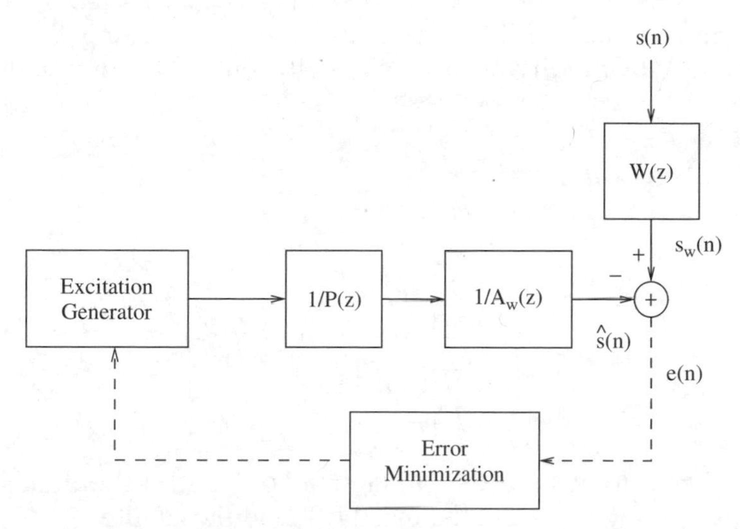

Figure 7.4 Modified AbS-LPC encoder with the weighting filter moved to the two branches of the error minimization procedure

where f_s is the sampling frequency in Hz. As can be observed from Figure 7.3, the weighting filter de-emphasizes the frequency regions corresponding to the formants as determined by the LPC analysis. By allocating larger distortion in the formant regions, noise that is more subjectively disturbing in the formant nulls can be reduced. The amount of de-emphasis is controlled by γ which introduces a broadening effect and must lie between 0 and 1. The most suitable value of γ is selected subjectively by listening tests; for 8 kHz sampling, γ is usually around 0.8 to 0.9.

Although the weighting filter can be used as it is in its normal position (after subtraction of $\hat{s}(n)$ from $s(n)$), it can also be modified in a computationally-advantageous way by moving it to the two branches contributing to the subtraction operation, as illustrated in Figure 7.4. This results in a block of the input samples being weighted only once prior to the AbS search. At the same time $W(z)$ is combined with the LPC filter to form a modified all-pole synthesis filter.

$$\frac{1}{A_w(z)} = \frac{1}{A(z)}.W(z) \tag{7.5}$$

$$= \frac{1}{1 - \sum_{i=1}^{p} a_i \gamma^i z^{-i}}$$

Note that, in the latest CELP coders, the above weighting filter has been slightly changed by modifying its zeros (coefficients in the numerator) as well as its poles.

$$W(z) = \frac{A(z/\beta)}{A(z/\gamma)} = \frac{1 - \sum_{i=1}^{p} a_i \beta^i z^{-i}}{1 - \sum_{i=1}^{p} a_i \gamma^i z^{-i}} \quad 0 \le \gamma \le \beta \le 1 \tag{7.6}$$

If this structure is to be used then in Figure 7.4, $1/A_w(z)$ should contain both the above weighting filter and the usual LPC synthesis filter.

7.2.3 Excitation Signal

The excitation signal represents the input to the AbS-LPC model and is therefore an important block of the model shown in Figure 7.2. It provides any residual structures that are not represented by the spectral model of the time-varying filters, including pitch or long-term dependent structures that exhibit significant correlation which is not covered by the pitch filter and random structures that cannot be modelled efficiently by deterministic methods. A proper excitation model is vital to the pitch-filtering efficiency as the pitch filter memory is built up of its scaled versions. Therefore, the excitation is usually represented by a shape vector with its associated gain or scale factor. The various shapes that have been reported include multi-pulse, regular-pulse, codebook, etc. Some mixtures of the above excitation schemes have also been used.

Codebook Excitation

In codebook excitation (CELP) [10], the excitation vector is chosen from a stored collection of C possible unity variance stochastic sequences with an associated scaling or gain factor. In the AbS procedure, the C possible sequences are systematically passed through the combined synthesis filter (pitch, LPC and perceptual filters); the vector that produces the lowest error is the desired sequence and is then scaled by its gain. Since the set of sequences are present at both the encoder and the decoder, only an index to the codebook and the gain level are required to be transmitted. Therefore, less than 1 bit/sample coding is possible.

As the codebook is of finite dimension, it must be populated with representative vectors of the excitation. In Atal's original proposal, unit-variance white

Gaussian random numbers were used. This choice of population was reported to give very good results, partly due to the fact that the probability-density function of the prediction error samples, produced by inverse filtering the speech through both the LPC and pitch filters, is very close to having a Gaussian shape. Another popular choice of codebook entries are centre-clipped Gaussian vectors, which reduce complexity and improve performance. However later developments in the codebook design such as VSELP and, more significantly, ACELP structures reduce the codebook storage and search complexity as well as improving the resultant speech quality.

Self-Excitation

Self-excitation (SELP) [11] can use the excitation signals derived from the past history of the coded excitation function itself, using a structure similar to a pitch filter in the form more than one long-term predictor (LTP). The self-excitation LTP is 'started' by initially filling its memory with some random contents. Then at each analysis-by-synthesis procedure, a sequence equal to the block length L, indexed in time by k, is selected and passed through a combined synthesis filter. The best vector with index k_{opt} is the vector which minimizes the difference error. When the best vector is found and used to synthesize the current block, it is fed back into the LTP with the oldest L samples discarded. The self-excitation LTP is effectively a CELP coder with an adaptive codebook. However, in SELP, the C possible sequences are not codebook entries but a windowed version of optimum excitations.

Multi-Pulse and Regular Pulse Excitation

Multi-pulse LPC (MPLPC) [8, 12] was the first of the AbS-LPC coding schemes. In MPLPC, the rigorous division of the excitation into voiced and unvoiced classes is avoided by making no prior assumptions about the nature of the excitation signal. In MPLPC, rather than selecting an optimum sequence from a codebook as in CELP, the excitation is specified by a small set of pulses with different amplitudes located at nonuniformly-spaced intervals. The encoding involves the determination of the pulse positions and the amplitudes of the excitation which produce the minimum error. The only prior information required is the number of pulses per analysis block (or an error threshold). A typical pulse rate for good quality speech synthesis is around one pulse per 4–8 samples.

The MPLPC can be viewed as a CELP system with a very large codebook, the size of which is determined by the number of pulses and the number of bits used to quantize the pulse amplitudes and locations.

The MPLPC makes no restriction on the spacing or spread of the pulses except that the number of pulses must be fixed in advance. This arrangement

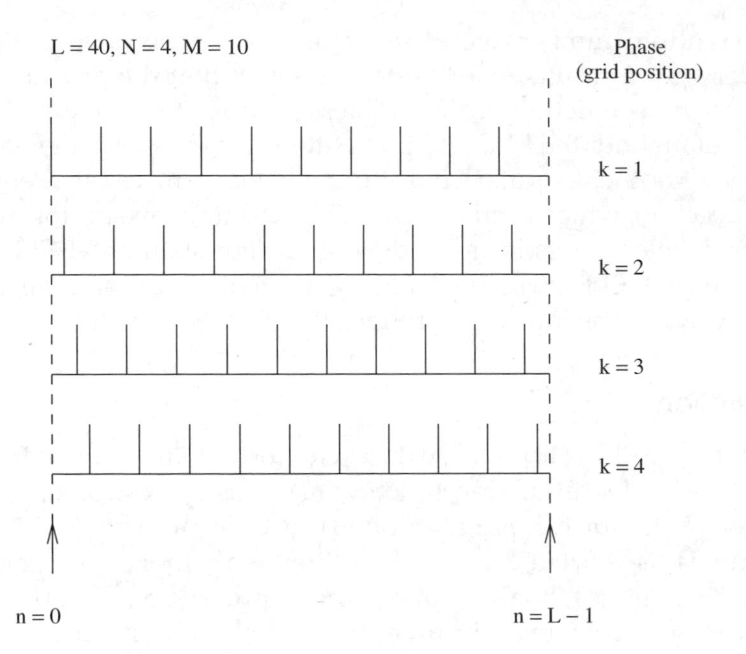

Figure 7.5 Typical make-up of an RPELPC pulse-positioning structure

obviously requires a large number of bits to encode the pulse positions. Therefore, a more structured allocation of the pulses would be more desirable both in terms of bit savings and complexity in determining the optimum positions. A sparse codebook CELP is effectively an MPLPC with severe restrictions on both the positions and amplitudes of the pulses. This is obviously a very drastic compromise, but if structure is only imposed on the pulse-positioning then the amplitudes can vary. This is the structure of the regular pulse excited LPC (RPELPC) [13] shown in Figure 7.5. In an RPELPC excitation frame, the pulses are equally spaced, with spacing N, and their positions are specified completely by the position of the first pulse. In Figure 7.5, the excitation frame size L is 40 and N is 4. The total number of pulses per frame is $M = L/N$. Thus, in RPELPC, the N-sequences are passed through combined synthesis filters and the sequence which minimizes the error is chosen as the best sequence.

7.2.4 Determination of Optimum Excitation Sequence

In the previous section, the different forms of excitation signals were described without any detailed description of their determination. While each of the excitation techniques best models different types of structure that may exist in the residual, the formulation for determining the optimum excitation sequence for each is the same [6, 14]. The only difference is the function space from which an optimum excitation can be chosen. In CELP, a sample space of random functions corresponding to the L-point Gaussian random sequences contained in the codebook is searched. In MPLPC and RPELPC, the search is in time, but through a set of delayed impulse response functions. For a given technique, the criterion for finding the optimum excitation function is the same. The objective is to determine the shape matrix $\mathbf{X}$ and the associated gain $\mathbf{g}$ (assuming MPLPC and RPELPC have normalized shape vectors) so that $\mathbf{gX}$ produces a synthetic signal that minimizes the weighted error $e(n)$ shown in Figure 7.4, i.e.

$$\mathbf{e}_k = \mathbf{s}_w - \hat{\mathbf{s}}_k \tag{7.7}$$

where $\mathbf{s}_w$ is the weighted original reference signal, $\hat{\mathbf{s}}_k$ is the synthesized signal (with pitch, LPC and perceptual-weighting filter contributions), and k denotes the particular excitation.

Let $\mathbf{H}$ be an $L \times L$ matrix whose j^{th} row contains the (truncated) combined impulse response $h(n)$ of the pitch, LPC and perceptual weighting filters caused by a unit impulse $\delta(n-j)$, i.e.

$$\mathbf{H} = \begin{bmatrix} h(0) & h(1) & \cdots & h(L-1) \\ 0 & h(0) & \cdots & h(L-2) \\ \vdots & \vdots & \vdots & \vdots \\ 0 & 0 & \cdots & h(0) \end{bmatrix} \tag{7.8}$$

If $\mathbf{s}_m$ denotes the output of the cascaded filters with zero input, i.e. the memory hangover from previously synthesized frames, then the reference signal $\tilde{\mathbf{s}}$ to be matched can be described as,

$$\tilde{\mathbf{s}} = \mathbf{s}_w - \mathbf{s}_m \tag{7.9}$$

$$\Rightarrow \mathbf{e}_k = \tilde{\mathbf{s}} - \mathbf{g}_k\mathbf{X}_k\mathbf{H} \tag{7.10}$$

$$= \tilde{\mathbf{s}} - \mathbf{g}_k\hat{\mathbf{s}}_k \tag{7.11}$$

where,

$$\hat{\mathbf{s}}_k = \mathbf{X}_k\mathbf{H} \tag{7.12}$$

and $\mathbf{X}_k$ and $\mathbf{g}_k$ are the kth excitation shape and gain vectors. The criterion is minimum-squared error, thus our objective is to minimize E_k where,

$$E_k = \mathbf{e}_k\mathbf{e}_k^T \tag{7.13}$$

and T denotes transpose. The optimum amplitude vector $\mathbf{g}_k$ for the kth candidate excitation can be computed from equations (7.11) and (7.13) by requiring the error $\mathbf{e}_k$ to be orthogonal to our estimation $\hat{\mathbf{s}}_k$, i.e.

$$\mathbf{e}_k\hat{\mathbf{s}}_k^T = 0 \tag{7.14}$$

Therefore,

$$(\tilde{\mathbf{s}} - \mathbf{g}_k\hat{\mathbf{s}}_k).\hat{\mathbf{s}}_k^T = 0 \tag{7.15}$$

$$\Rightarrow \mathbf{g}_k = \tilde{\mathbf{s}}\hat{\mathbf{s}}_k^T[\hat{\mathbf{s}}_k\hat{\mathbf{s}}_k^T]^{-1} \tag{7.16}$$

By substituting equation (7.16) into equation (7.11), equation (7.13) can be rewritten as

$$E_k = \tilde{\mathbf{s}}[\mathbf{I} - \hat{\mathbf{s}}_k^T[\hat{\mathbf{s}}_k\hat{\mathbf{s}}_k^T]^{-1}\hat{\mathbf{s}}_k]\tilde{\mathbf{s}}^T \tag{7.17}$$

where $\mathbf{I}$ is the identity matrix. The vector $\mathbf{g}_k$ and matrix $\mathbf{X}_k$ that yield the minimum value of E_k over all k are then selected as the optimum excitation.

The above expression for E_k is generalized for all the possible forms of excitations and is, therefore, rather more complicated than required in practical cases. The $[\hat{\mathbf{s}}_k\hat{\mathbf{s}}_k^T]^{-1}$ inversion, for instance, is unnecessary in most cases, as illustrated below using codebook excitation.

$$\hat{\mathbf{s}}_k\hat{\mathbf{s}}_k^T = \sigma \quad \text{(scalar)} \tag{7.18}$$

$$\mathbf{g}_k = \frac{\tilde{\mathbf{s}}\hat{\mathbf{s}}_k^T}{\sigma} = g_k \quad \text{(scalar)} \tag{7.19}$$

$$\Rightarrow E_k = \tilde{\mathbf{s}}\tilde{\mathbf{s}}^T - \mathbf{g}_k\hat{\mathbf{s}}_k\tilde{\mathbf{s}}^T \tag{7.20}$$

$$= \tilde{\mathbf{s}}\tilde{\mathbf{s}}^T - \mathbf{Q}_k \tag{7.21}$$

Rewriting equations (7.19) and (7.20) in time-domain samples form,

$$g_k = \frac{\sum_{i=0}^{L-1} \tilde{s}(i)\hat{s}_k(i)}{\sum_{i=0}^{L-1} \hat{s}_k^2(i)} \tag{7.22}$$

$$E_k = \sum_{i=0}^{L-1} \tilde{s}^2(i) - g_k \sum_{i=0}^{L-1} \hat{s}_k(i)\tilde{s}(i) \tag{7.23}$$

and, substituting g_k into equation (7.23), we can rewrite equation (7.21) as,

$$E_k = \sum_{i=0}^{L-1} \tilde{s}^2(i) - \frac{\left[\sum_{i=0}^{L-1} \hat{s}_k(i)\tilde{s}(i)\right]^2}{\sum_{i=0}^{L-1} \hat{s}_k^2(i)} \tag{7.24}$$

and $\mathbf{Q}_k$ is given by,

$$\mathbf{Q}_k = \frac{\left[\sum_{i=0}^{L-1} \hat{s}_k(i)\tilde{s}(i)\right]^2}{\sum_{i=0}^{L-1} \hat{s}_k^2(i)} \tag{7.25}$$

The scalar factor g_k is simply the cross-correlation of the weighted speech with the synthesized excitation response divided by the squared sum of the synthesized excitation response. The squared error E_k is the difference between the energy of the weighted speech and $\mathbf{Q}_k$. In practice, we find the maximum of $\mathbf{Q}_k$ to select the best excitation.

In MPLPC, the above procedure to determine the excitation shape is not practical as it would involve searching for all possible combinations of pulse location, e.g. for $L = 40$ and $M = 4$, the number of pulse position vectors is 91 390. Therefore to simplify computation, suboptimal strategies are usually used. A simple and popular method involves locating one pulse at a time. The optimum location for any of these pulses is found by computing the error for all possible pulse locations in a given interval L and locating the minimum error location. Once that location is known, the contribution of the pulse at position k is subtracted from the reference signal (similarly to equation (7.9)), and the procedure is repeated for the next pulse until all pulses are found. This is summarized below:

Let $\tilde{\mathbf{s}}_0 = \mathbf{s}_w - \mathbf{s}_m$, where $\mathbf{s}_m$ is the combined filter memory response, and $\mathbf{H}_k$ be the combined impulse response scaled by the kth pulse.
For $i = 1, \ldots, M$,

1. Find $E_i^k = min\{\tilde{\mathbf{s}}_i \tilde{\mathbf{s}}_i^T - \mathbf{g}_k \mathbf{H}_k \tilde{\mathbf{s}}_i^T\}$, for $k = 1, \ldots, L$
2. $\tilde{\mathbf{s}}_{i+1} = \tilde{\mathbf{s}}_i - \mathbf{g}_{i,opt}\mathbf{H}_i$

Finally, $\hat{\mathbf{s}}_{mp} = \sum_{i=1}^{M} \mathbf{g}_{i,opt}\mathbf{H}_i$

This pulse-at-a-time procedure is obviously suboptimal and procedures which try to add more optimality have been extensively reported. For example, a popular post-processing method is to re-optimize the pulse amplitudes after the pulse positions are found by performing an M by M matrix inversion. Most of these involve substantially more computations with some reported improvements especially if M is large compared with the size of the analysis block L. In RPELPC, the positions of the pulses are fixed, therefore there is no requirement to locate one pulse at a time. Thus in RPELPC, the amplitude vector is jointly optimized in one step, repeated N times (for the number of possible amplitude sequences), and the vector that minimizes the mean squared error is chosen, also identifying the grid position (phase position $0 \leq k < N$).

7.2.5 Characteristics of AbS-LPC Schemes

Before investigating AbS-LPC schemes in more detail in the form of code-excited linear prediction (CELP), it is worthwhile to highlight some of the similarities and differences in the way the different AbS-LPC schemes operate. These mainly lie in the characteristics of the excitation behaviour as described below.

CELP

In the CELP system, the objective is to select from the codebook the vector which best matches the original reference vector. Typical plots of consecutive vector searches are shown in Figures 7.6 and 7.7, where the output error (in fact $\mathbf{Q}_k$ for each codebook vector) is plotted. In Figure 7.6, the best matching vector (v1) from the codebook is fairly distinct from the remaining vectors. However, in Figure 7.7, we find that the error is less distinctive, as illustrated by the similarity in the first, second, and third best matching vectors. The mean squared error is clearly inadequate as a selection criterion in AbS-LPC coding schemes (indeed, in speech coding in general): the selection of an optimum candidate is by no means readily controllable. Although the selection of v1 in Figure 7.6 is probably correct, even subjectively, the selection of v1 in Figure 7.7 is not as clear-cut. What subjective difference would result if, say, the second-best vector was selected instead? This test was performed and, not surprisingly, the quality of the processed speech was not noticeably degraded. This prompts the question as to whether or not the codebook vectors can be better optimized such that there is a clearer distinction, both objectively and subjectively, between the best and the second-best vectors. Unfortunately, trained codebooks (whether multi-pulse characteristic codebooks, glottal-pulse codebooks, or other types) have been largely unsuccessful in this respect.

SELP

In SELP, the best excitation is generated from previous excitations. Here, we assumed that the number of secondary long-term predictors was fixed at one. As the secondary long-term prediction (LTP2) tries to model long-term correlations not modelled by the primary (and previously much shorter) excitation memory pitch filter (LTP1), it can be expected that some form of

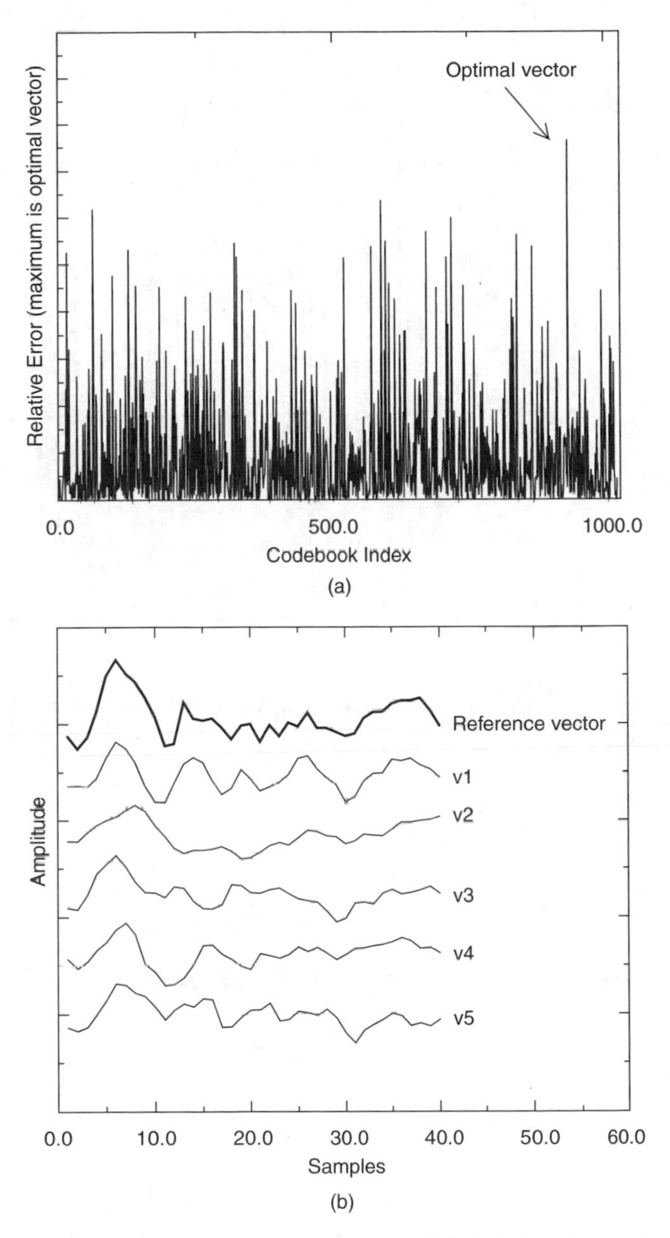

Figure 7.6 Typical example of a distinctive codebook vector selection for CELP (a) Plot of error versus codebook index for CELP, and (b) Synthesized best codebook entries compared with original

structure may exist in the selected time index for LTP2. Figure 7.8 shows a plot of the distribution of the time-delay index for both LTP1 and LTP2 for a female speaker. From the histogram of LTP1, it is clear that the speaker used for the test had a pitch period peaking at around 42 samples. However,

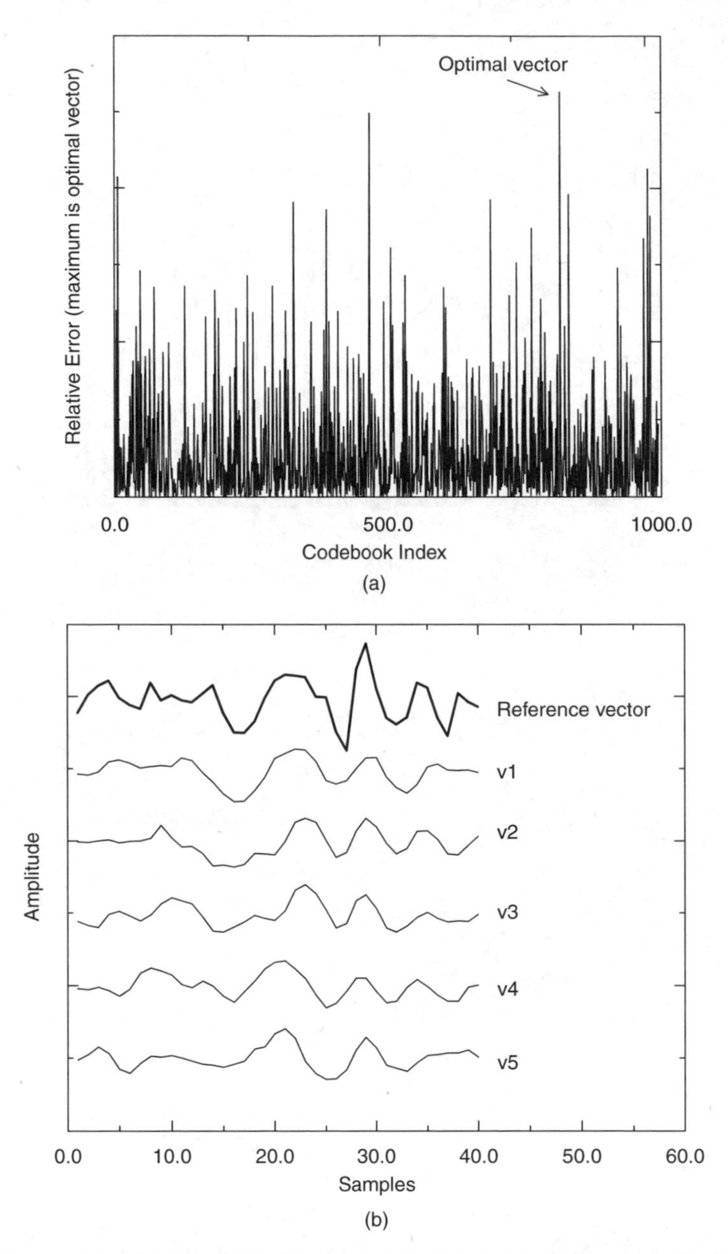

Figure 7.7 Example of a less distinctive codebook vector selection for CELP (a) Plot of error versus codebook index for CELP, and (b) Synthesized best codebook entries compared with original

from the similar plot for LTP2 very little structure can be deduced. It would appear that if there were any extra long-term structure in the test speech then the SELP did not model it properly. Alternatively, the assumption of the existence of extra long-term structures as used in SELP could be at fault.

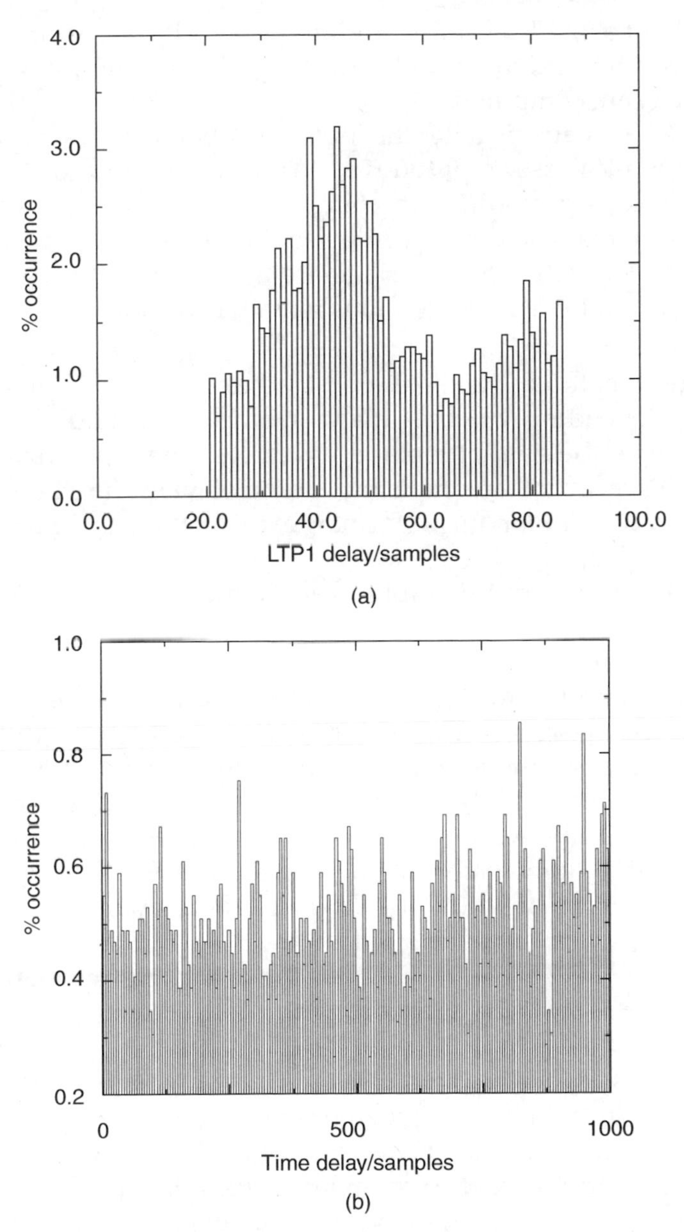

Figure 7.8 Distribution of delays for LTP1 and LTP2 in SELP (a) Plot of the distribution of delays for LTP1 for SELP, and (b) Plot of the distribution of delays for LTP2 for SELP

MPLPC

In MPLPC the best set of randomly-placed pulses that minimizes the output error is selected. In order to obtain the pulses, three search strategies are generally applied:

- **Method 1:** At stage j, all pulse amplitudes and locations up to stage $j-1$ are assumed to be known and only the pulse location n_j and the pulse amplitude g_j are computed.
- **Method 2:** At stage j, only the pulse locations $n_1, n_2, \ldots, n_{j-1}$ remain constant and the pulse amplitudes up to g_{j-1} are optimized. As all the pulse amplitudes can be modified to compensate for inaccuracies of previous pulses, they remain accurate even when they are closely spaced.
- **Method 3:** Only after the last stage are all the amplitudes $g_1, g_2, \ldots, g_M$ re-optimized and the pulse locations remain constant.

As expected, method 2 gives the best performance both objectively and subjectively depending on the analysis block size and the number of pulses. However, the gain of method 2 reduces when quantization is introduced. Therefore, from a complexity point of view, method 3 is preferable as it is very similar in performance to method 2, but has only one re-optimization loop.

The variation in the pulse amplitudes for method 2, normalized to unit variance in each frame, are shown in Figure 7.9. As can be observed from the plot without the pitch prediction (LTP) the histogram is bimodal with little content around zero, and with most amplitudes lying within $\pm 3\sigma$. This is expected as small pulses contribute little energy to the error minimization process. For the amplitude plot with the LTP, the range is even shorter. This is also expected as the majority of the large energy pulses would have been removed by the LTP. The more confined spread of values indicates that the quantization of the pulses in MPLPC with LTP can be much more efficient than without LTP. This efficiency in pulse quantization is very noticeable because, for similar bit-rate MPLPC schemes with and without LTP, the MPLPC with LTP is generally preferred both objectively and subjectively as the output speech with the LTP becomes smoother and more natural.

The histograms of the pulse locations are also interesting (see Figure 7.10). Note that pulse locations at the beginning of the frame are favoured more than the other locations since these locations allow large errors inside the frame to be reduced. In order to compensate for this uneven spread of pulse positions due to the autocorrelation type of analysis used in the derivations, the covariance form of analysis has been suggested. The covariance method attempts to account for the block edge effects of the autocorrelation analysis by taking into account the part of the impulse response of the cascaded filter that spills outside the analysis block. The positions of the pulses also have an impact on the choice of subframe sizes. Ideally, large subframe sizes are better suited to MPLPC because the limited number of pulses can be put in the most useful locations, e.g. pitch pulse locations. With small subframes, pulses are assigned even for relatively unimportant details of the speech signal which

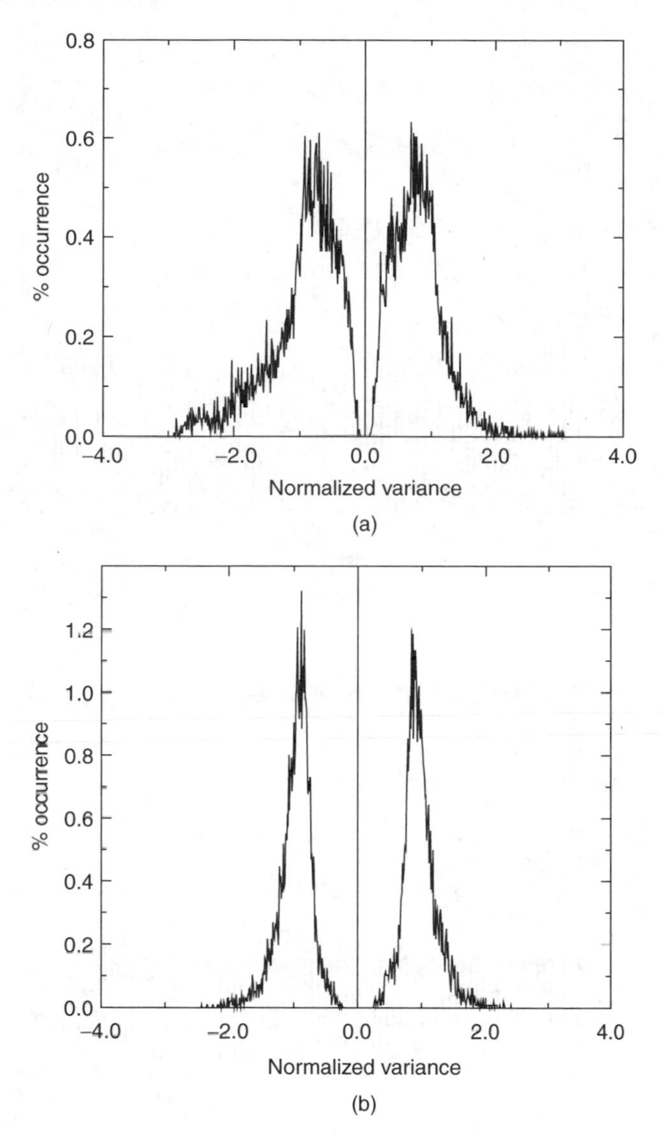

Figure 7.9 Distribution plot of MPLPC pulse amplitudes (a) The distribution of MPLPC pulses with no LTP, and (b) The distribution of MPLPC pulses with LTP

lowers the coding efficiency. The disadvantage with large subframe sizes is that the complexity is increased.

RPELPC

As described earlier, RPELPC is similar to MPLPC except that the pulse locations are pre-structured. The amount of structuring obviously determines the amount of freedom that the pulses have in estimating the reference signal.

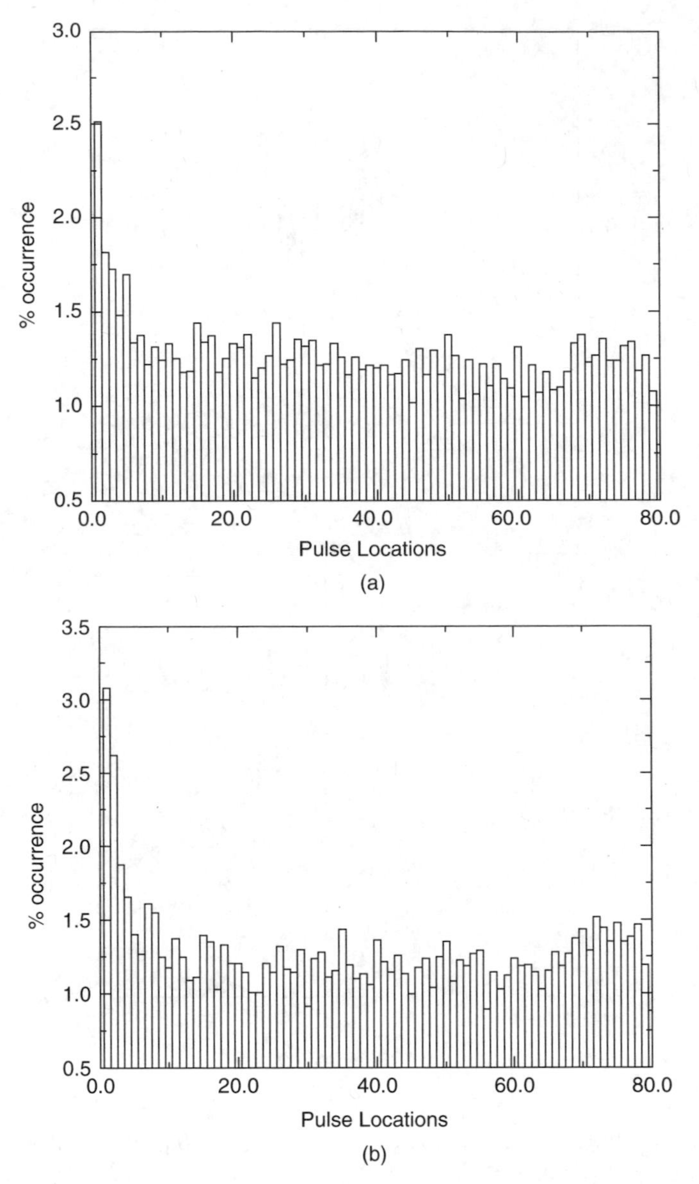

Figure 7.10 Distribution plots of MPLPC pulse locations with subframe size of 80 samples (a) The distribution of MPLPC pulse locations with no LTP, and (b) The distribution of MPLPC pulse locations with LTP

The effect of the decimation ratio on the performance of RPELPC is shown in Figure 7.11 for analysis frames of 40 and 80 samples with and without LTP. The relative SNR differences are partly dependent on the speech material but, generally, the inclusion of the LTP improves the performance of the RPELPC especially at higher decimation ratios. As can be observed from the plot, the

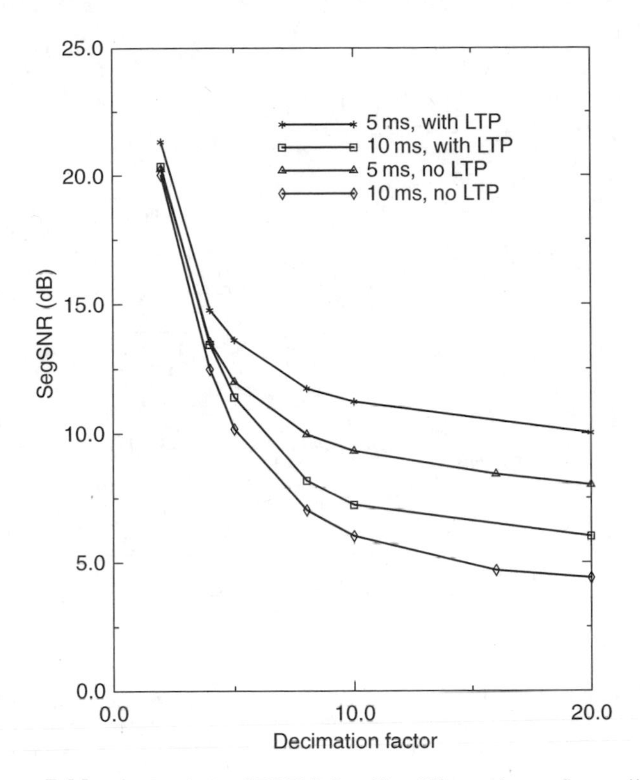

Figure 7.11 SNR plot of RPELPC with different configurations

performance at a lower analysis frame size is better. This is because it has more degrees of freedom to vary the excitation to match the reference vector.

7.3 Code-Excited Linear Predictive Coding

Amongst the variations of AbS-LPC schemes, the most widely-reported scheme used at 8 kb/s and below is code-excited linear predictive coding (CELP). As the name suggests, the excitation of the time-varying filters is provided by a codebook. CELP operates as follows (see the simplified block diagram in Figure 7.12):

1. The original speech, $s(n)$, is partitioned into analysis frames of around 20–30 ms. LPC analysis is performed on the frame of $s(n)$ to give a set of LPC coefficients which are used in the short-term LPC predictor to model the spectral envelope of the speech.
2. Having computed the LPC parameter, the frame is usually split into a number of subframes (usually 40 or 60 samples long). The following processing is carried out for each subframe.

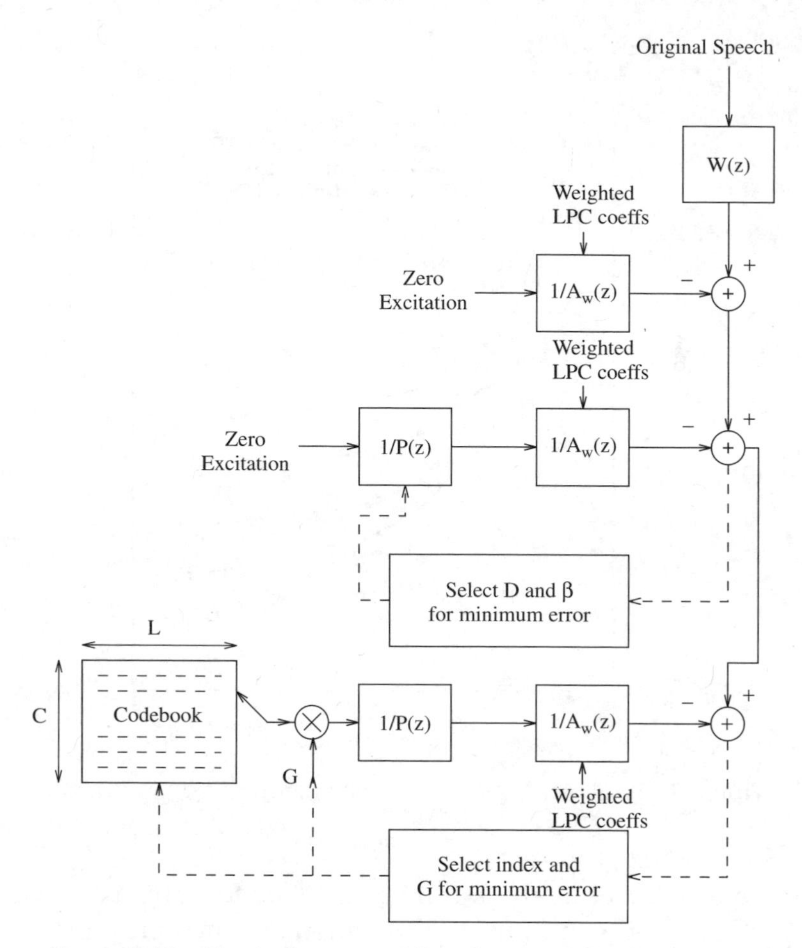

Figure 7.12 Block diagram of the standard CELP algorithm

(a) The memory of the combined LPC and perceptual weighting filters (the initial conditions) is removed from the reference (the perceptually-weighted original speech) to give a memoryless filter for subsequent analysis.

(b) Pitch prediction proceeds to deal with the long-term redundancies. The pitch analysis is performed by testing all possible pitch lags and selecting the lag D that minimizes the difference between the reference (the original speech remaining after the above step) and the speech produced by passing the pitch excitation at each possible pitch delay through memoryless LPC and perceptual weighting filters. Having selected the best delay, D, its associated gain β, is then computed.

This process is computationally very complex. In order to simplify it, an open-loop pitch may be computed first and only a limited range around this open-loop pitch is searched. The open-loop pitch computation is usually carried out on the perceptually-weighted original speech to obtain a good idea of the likely pitch period before a closed-loop search is applied around this value. The pitch contribution is then subtracted from the reference signal to update it for the next stage (the codebook search). Since the pitch can change up to 1%/ms, the pitch delay is updated more frequently than the LPC for accurate voice periodicity generation in the synthesized speech.

(c) Once the parameters of the two synthesis filters are found, the excitation is determined. Each codebook vector is passed through the memoryless LPC and perceptual weighting filters and the codebook vector which gives the minimum squared difference between the output it produces and the reference signal is selected and its corresponding scaling factor is computed. Note that if the delay D in the pitch filter is greater than the subframe size, it will not affect the synthesized codebook vector. In addition, the pitch filter is usually implemented as an adaptive codebook operating in parallel with the stochastic codebook and, hence, its response is eliminated from the stochastic codebook search loop.

(d) Finally the initial conditions (i.e. the memory) of the filters are restored, and the synthetic speech is generated by filtering the scaled optimum codebook sequence through the filters so as to update the filters for processing the next subframe.

3. In the synthesizer (decoder), the initial conditions (i.e. the memory) of the filters are restored and the synthetic speech is generated by filtering the scaled optimum codebook sequence through the filters without any perceptual weighting.

From the above description it is clear that the computation can be broken down into three blocks: LPC analysis to compute the LPC parameters; pitch analysis to compute the long-term predictor parameters; and a codebook search to determine the shape and gain of the excitation vector.

7.3.1 LPC Prediction

The role of LPC prediction is to represent the general shape of the speech spectrum. Therefore, in the CELP synthesizer, the (ideally flat) excitation is shaped by the spectral envelope of the LPC filter. The LPC parameters can be computed by a number of methods as discussed in Chapter 4. However, most CELP coders use a 10th-order LPC filter based on autocorrelation estimation. The speech signal, which is usually 20 ms long, is passed through a Hamming window which is usually placed half a frame ahead so as to enable accurate parameter interpolation for each subframe. However, many delay-sensitive applications and standards use an asymmetric window to give more weighting to the latest samples contained in the analysis frame. The delay problem can also be solved by employing backward forms of LPC analysis, i.e. using quantized (or past) samples only, to estimate the LPC coefficients as in the 16 kb/s LD-CELP (G.728) proposed for the ITU standard [15]. However, such backward techniques can only operate successfully at around 16 kb/s, because the prediction accuracy reduces rapidly with the increase in the quantization noise of the encoded speech.

7.3.2 Pitch Prediction

Pitch prediction is an essential part of all CELP coders. Since the early versions of CELP had Gaussian-noise-populated excitation codebooks, pitch filtering was required to introduce the necessary pitch of the voiced speech parts. The order of the pitch filter is usually less than the order of the LPC filter and is given in its general form as,

$$P(z) = 1 - \sum_{i=-I}^{I} \beta_i z^{(-D-i)} \tag{7.26}$$

The pitch predictor in CELP generates long-term correlation, either due to the actual pitch excitation or other long-term similarities. Thus the term 'long-term predictor' (LTP) is usually preferred to 'pitch predictor', which is somewhat misleading in describing the action of this filter for unvoiced speech and even, to some extent, for voiced speech when D is equal to pitch multiples. In CELP and other AbS-LPC schemes, the LTP analysis is usually performed in a closed loop [16] with single or multiple taps. In CELP, one is interested in minimizing the error between the weighted original and the synthesized output speech. By this definition, analysis of the signal to derive the desired LTP parameters must minimize the error between the weighted original and the synthesized speech, and not minimize the LTP prediction error (or second residual) as is the case in older analysis and synthesis systems. Assuming that the LPC parameters have already been calculated, the remaining undetermined parameters are $Gx(n)$, D, and β_k. Although these parameters can be obtained by exhaustively searching for all $Gx(n)$ as well as the LTP parameters, the procedure becomes very computationally-intensive and thus suboptimal solutions have to be used. One way of reducing the complexity is by obtaining the LTP and $Gx(n)$ in two sequential steps. First we assume $Gx(n)$ is zero, and calculate the LTP parameters such that $e(n)$ is minimized. Next the LTP is held constant and $Gx(n)$ is computed. Thus, let the codebook excitation be zero, i.e.

$$x(n) = 0, \quad 0 \le n \le L - 1 \tag{7.27}$$

The synthetic speech is produced only by the LTP excitation passing through the combined LPC and perceptual weighting filters.

$$\Rightarrow \hat{s}(n) = \sum_{i=-I}^{I} \beta_i \sum_{k=0}^{n} \hat{r}(n-k-D-i)h(k) \tag{7.28}$$

Then the weighted squared error E for the delay D is given by,

$$E(D) = \sum_{n=0}^{L-1} e^2(n) = \sum_{n=0}^{L-1} (\tilde{s}(n) - \hat{s}(n))^2 \tag{7.29}$$

where,

$$\tilde{s}(n) = s_w(n) - s_m(n) \tag{7.30}$$

and $s_m(n)$ is the memory response contribution of the combined LPC and perceptual weighting filters, and $s_w(n)$ is perceptually weighted original speech. Therefore,

$$\frac{\partial E}{\partial \beta_i} = 2\left[\sum_{n=0}^{L-1} \tilde{s}(n) - \sum_{i=-I}^{I} \beta_i \sum_{k=0}^{n} \hat{r}(n-k-D-i)h(k)\right] \tag{7.31}$$

$$\times \left[-\sum_{k=0}^{n} \hat{r}(n-k-D-j)h(k)\right] = 0 \tag{7.32}$$

$$\text{Let } Z_i(n) = \sum_{k=0}^{n} \hat{r}(n-k-D-i)h(k) \tag{7.33}$$

$$\Rightarrow \sum_{n=0}^{L-1} \tilde{s}(n)Z_j(n) - \left[\sum_{i=-I}^{I} \beta_i \sum_{n=0}^{L-1} Z_i(n)Z_j(n)\right] = 0, \quad -I \leq j \leq I \tag{7.34}$$

Hence, in matrix form, assuming a 3-tap filter,

$$\begin{bmatrix} \beta_{-1} \\ \beta_0 \\ \beta_1 \end{bmatrix} = \begin{bmatrix} \phi(-1,-1) & \phi(0,-1) & \phi(1,-1) \\ \phi(-1,0) & \phi(0,0) & \phi(1,0) \\ \phi(-1,1) & \phi(0,1) & \phi(1,1) \end{bmatrix}^{-1} \begin{bmatrix} B(-1) \\ B(0) \\ B(1) \end{bmatrix} \tag{7.35}$$

and for a single-tap filter where $I = 0$,

$$\beta_0 = \frac{B(0)}{\phi(0,0)} \tag{7.36}$$

where,

$$\phi(i,j) = \sum_{n=0}^{L-1} Z_i(n)Z_j(n) \tag{7.37}$$

$$B(i) = \sum_{n=0}^{L-1} \tilde{s}(n)Z_i(n) \tag{7.38}$$

Once the LTP gain coefficients are found, they are substituted back into equation (7.29) and the delay D for which $E(D)$ is a minimum gives the optimum delay D_{opt} and the corresponding gains β_i. The excitation $Gx(n)$ can then be found with D_{opt} and β_{opt} fixed. In practice however, the optimum delay D_{opt} is usually found before computing the gain coefficients.

There may be problems with the LTP when the delay D is less than the subframe L, i.e. when the LTP recurses within the same analysis subframe [16]. The basic problem in solving for the gain and delay coefficients for lags less than the subframe size is that the weighted mean squared error equation becomes nonlinear in the coefficients for $D < L$. Consider the case in which a single LTP coefficient is being determined and the LTP lag lies in the interval $L/2 \leq D \leq L-1$. The signal takes one of two forms:

$$\hat{r}(n) = \begin{cases} \beta\hat{r}(n-D) & 0 \leq n \leq D-1 \\ \beta^2\hat{r}(n-2D) & D \leq n \leq L-1 \end{cases} \tag{7.39}$$

The weighted squared error, E, can then be expressed as,

$$E = \sum_{n=0}^{L-1} e^2(n) = \sum_{n=0}^{L-1} [\tilde{s}(n) - \hat{s}(n)]^2 \tag{7.40}$$

Defining,

$$Z_D(n) = \sum_{k=0}^{n} \hat{r}(n-k-D)h(k) \tag{7.41}$$

we can expand the error equation as

$$E = \sum_{n=0}^{D-1} \left[\tilde{s}^2(n) + \beta^2 Z_D^2(n) - 2\beta\tilde{s}(n)Z_D(n)\right] + \sum_{n=D}^{L-1} \left[\beta^4 Z_D^2(n) - 2\beta^2\tilde{s}(n)Z_D(n)\right] \tag{7.42}$$

where the first part of the right hand side is for the first D samples and the second is for the recursive part (which may cause some difficulties). As can be seen from the above, when $D = L$ the second term has no effect. To solve the equation with respect to β we get:

$$\frac{\partial E}{\partial \beta} = 2\beta^3 \sum_{n=D}^{L-1} Z_D^2(n) + \beta \left[\sum_{n=0}^{D-1} Z_D^2(n) - 2 \sum_{n=D}^{L-1} \tilde{s}(n) Z_D(n) \right] - \sum_{n=0}^{D-1} \tilde{s}(n) Z_D(n) = 0 \tag{7.43}$$

From the above we can see that the solution to E involves solving a cubic in β. This is obviously very costly as it is required for every value of D less than L. One solution to the above is to adopt a trial-and-error method based on quantized values for β. In this method, the sum terms are precomputed, and each of the possible quantized values of β is substituted into the equation. The value of β which gives the smallest squared error is thus the desired value. Obviously the computation involved is still quite large, e.g. if $L = 50$, $D_{min} = 20$, and $\beta = 3$ bits, the number of searches is $2^3 \times (50 - 20) = 240$, with the addition for $D > L$.

A second method for $D < L$ is to use an adaptive codebook formulation of the LTP [17], to periodically repeat the past LTP output, i.e.

$$\hat{r}(n) = \begin{cases} \beta\hat{r}(n-D) & 0 \le n \le D-1 \\ \beta\hat{r}(n-2D) & D \le n \le 2D-1 \\ \vdots & \\ \beta\hat{r}(n-aD) & aD \le n \le L-1 \end{cases} \tag{7.44}$$

In other words, the previously undefined part of the LTP excitation in a subframe is constructed by repeating its defined part with periodicity D. Using this method only β terms needs to be solved. This scheme does not allow for pitch pulses in a subframe to change amplitude from one period to another. Using this adaptive method, the CELP synthesis procedure is as shown in Figure 7.13, where the LPC and perceptual weighting filters have been represented by a single short-term filter (STP).

Fractional-Delay LTP

In the above LTP computation, the matching of the reference signal with the LTP contribution is achieved via a cross-correlation procedure. A major restriction of this is the inherent sampling resolution of the signal, i.e. for our cross-correlation to be most effective we would ideally like a continuous signal such that the best instance of similarity between the reference and the synthetic signal can be obtained. However, as our delay, D, is restricted to

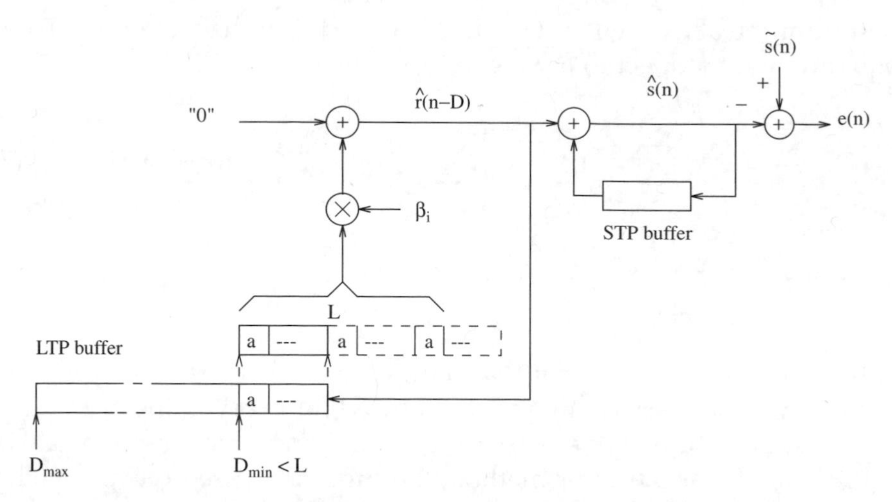

Figure 7.13 Block diagram illustrating pitch repetition for delays less than the optimization interval

integer values of the sampling rate, the LTP is not able to cope with arbitrary lag intervals without replacing, in some way, the optimum noninteger delay by an integer value which may degrade the performance of the LTP in terms of objective matching. As explained in the previous section, higher-order LTPs can be used where its multiple coefficients can provide interpolation between the adjacent samples around D even if the lag value does not correspond to an integer number of samples. However, the disadvantage of higher-order LTPs is the increased coding capacity required to code the additional gains. Based on this observation, in order to achieve a greater LTP delay resolution but to minimize coding capacity, an up-sampling procedure [18] can be used.

Increased Resolution by Up-sampling

The LTP delay, D, is expressed as an integer number of samples at sampling rate f_s. When trying to replace D by a real number, D_r, it is necessary to convert the discrete time signal $s(n)$, to a continuous time signal as noninteger values are not defined by $s(n)$. As our signal $s(n)$ is sampled according to the Nyquist rate, the continuous form, $s_c(t)$, can be recovered at any time instant by filtering through an ideal noncausal low-pass filter,

$$s_c(t) = \sum_{k=-\infty}^{\infty} \frac{s(k)\sin(\pi(t-k))}{\pi(t-k)} \tag{7.45}$$

As we are only interested in the submultiples of D, the $s_c(t)$ signal is not required, but a higher sampled signal, $s_{up}(m)$, is required. The ideal low-pass prototype filter is replaced by a finite length filter. The up-sampling of $s(n)$

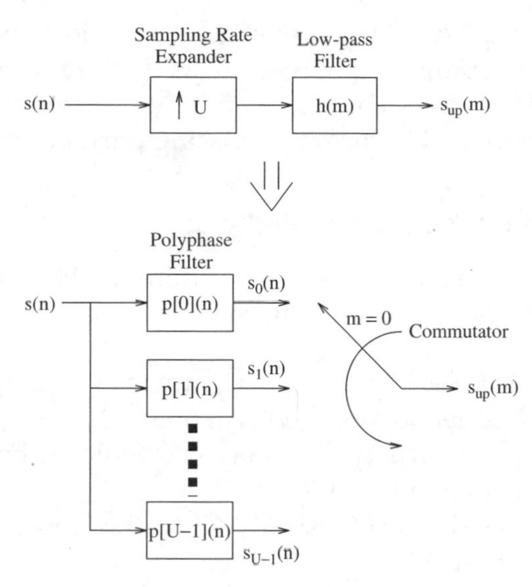

Figure 7.14 Polyphase structure for implementing interpolation

to $s_{up}(m)$ is illustrated in Figure 7.14 where U is the up-sampling factor and a polyphase structure is used.

From the up-sampling procedure a noninteger lag value of $(D + d/U)$ at sampling rate f_s now corresponds to an integer delay of $(UD + d)$ where $d = 0, 1, \ldots, U - 1$ at a rate of Uf_s. Therefore, to implement a delay of noninteger values, one simply takes the appropriate branch of the polyphase structure of the interpolation filter (see Figure 7.14). An important aspect of the interpolation process is the choice of the low-pass filter both in terms of performance and complexity. As suggested by Kroon in [18], a simple but effective filter design is to use a Hamming windowed $\sin(x)/x$ design, which has three advantages:

1. The resultant FIR filter has exactly linear phase and a fixed delay.
2. The characteristics of the filter are adequate with only a short filter length, i.e. the aliasing components are small.
3. The original signal can be obtained without any phase-shift, i.e. the top branch of the polyphase structure in Figure 7.14. This means that the number of filtering operations is decreased as the top branch is effectively just a delay operation.

In order to obtain the above advantages, the filter length N must be chosen such that the filter delay, $(N - 1)/2$, at sampling rate Uf_s is an integer multiple of U, i.e.

$$N = 2IU + 1 \tag{7.46}$$

where I is the delay of the low-pass filter at sampling rate f_s.

As for the integer delay LTP, there exists a problem when the candidate delay is less than the optimization interval, as the filter then recurses. Again, the technique of adaptive codebook structure, where the available part of the LTP buffer can be repeated to form the missing part, is applicable.

Performance Comparison of LTP Methods

In order to assess the performance of the different LTP analysis methods, an unquantized CELP has been used with the following LTPs:

- One-tap (CL1) and three-tap (CL3) using the closed-loop analysis method with the adaptive codebook method when $D < L$.
- Modified CL3 (MCL3), where the delay is calculated by CL1 but the gain coefficients are calculated by CL3.
- Fractional delay closed-loop (basically CL1 with up-sampling of the LTP) with $U = 2$ (F2CL1) and $U = 4$ (F4CL1).

The configuration of the CELP coder is shown in Table 7.1. In order to assess the performance of the LTP, the overall SNR of the CELP coder is split into three parts (see Figure 7.15): (i) the LPC filter memory contribution, shown as short-term prediction (STP), (ii) the LTP contribution, and (iii) the codebook contribution. In this test no perceptual weighting filter is used.

Table 7.2 shows the result of the comparison between the different LTP methods using the configuration in Table 7.1. Note that the SNRs given in this table have been computed using only a few short speech sentences and are intended to give a quick comparison for the LTP methods. The SNRs are dependent on the input signal and may vary significantly for other input samples, so we should only consider their relative variations. From the comparison test and segmental SNR values shown in Table 7.2 several interesting points can be gathered:

- As expected, the contribution from the memory of the LPC filter (the short-term predictor) is more or less constant. A plot of the SNR values of

Table 7.1 CELP coder configuration for the LTP comparison test

Sampling freq.	8 kHz
Parameters	Update Rate
LPC order = 10	160 samples
LTP, various	40 samples
Codebook, 10-bit	40 samples
Weighting	None

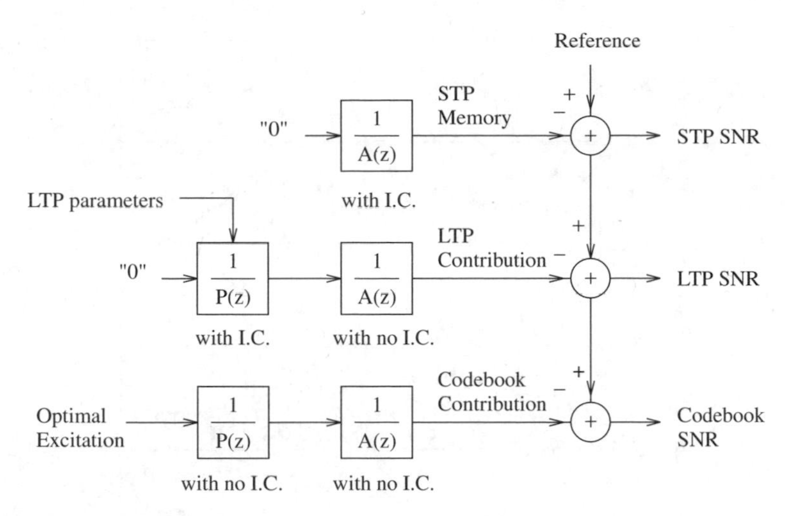

Figure 7.15 Breakdown of the CELP scheme subsystems which contribute to the overall SNR (I.C. means 'initial conditions')

Table 7.2 Breakdown of segmental SNR values for different LTPs

Scheme	Delay range	STP (dB)	LTP (dB)	Codebook (dB)	Overall (dB)
CL1	20–147	1.76	7.28	2.74	11.77
CL3	21–146	1.81	9.13	2.33	13.26
MCL3	21–146	1.77	8.98	2.42	13.16
F2CL1	20–147	1.79	7.37	2.79	11.95
F4CL1	20–147	1.80	7.68	2.76	12.24

the LPC *ringing* is shown in Figure 7.16. It can be seen from the fluctuation in the SNR that the LPC contribution does not always provide a positive SNR to the overall total, i.e. the memory is actually making the rest of the coding process work harder. This obviously reflects the past history of the encoding process, i.e. if the previous subframe was poorly matched, then the LPC memory will also be poor. This generally occurs during speech transitions.

- The fractional resolution closed-loop LTPs objectively perform slightly better than the integer resolution closed-loop LTPs. However, subjectively, the improvements are more substantial than the SNR suggests. The speech becomes cleaner, especially for female speakers. Only uniform spacing of the delays have been investigated. The performance can be improved if nonuniform spacing is used as reported in [18].
- From Table 7.2 we can see that the three-tap integer-delay closed-loop LTPs provide significantly higher SNR than the one-tap integer and fractional delay LTPs. The LTP contributions are significantly higher than the

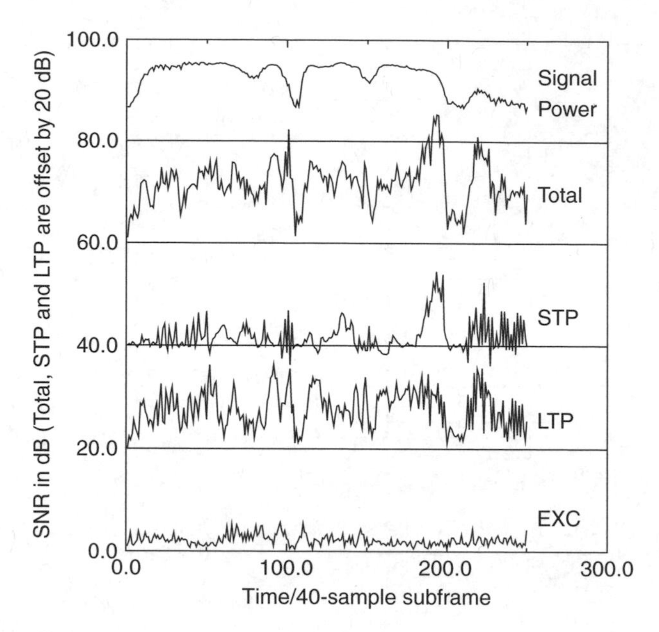

Figure 7.16 Breakdown of SNR values for the CL3 LTP structure

codebook contribution. This is due to the fact that the codebook provides contribution to match the remaining signal after the LPC and LTP contributions have been subtracted from the original target. So the higher the matching in LTP, the lower the contribution of the codebook will be. In order to fill in the remaining information, the 10-bit codebook can only provide up to a certain threshold in the overall SNR. In order to provide more contribution, the LTP and codebook can be jointly optimized [19], or a better codebook excitation source can be used.

Limited informal subjective listening tests have also shown that three-tap LTPs are generally better than one-tap LTPs, although the difference between the three-tap integer delay and the one-tap fractional delay (up-sampling four times) is not very noticeable.

7.3.3 Multi-Pulse Excitation

Early versions of CELP coding used multi-pulse excitation with and without LTP to match the original input signal. As was briefly discussed earlier, a low-complexity MPLPC coder sequentially determines the locations and amplitudes of the excitation pulses so as to minimize the error between the original and the synthesized speech. The optimum pulse locations are found by computing the error for all possible pulse locations with their optimum amplitudes in a given analysis block and selecting the allowable number of locations and their amplitudes that result in the minimum error. A block

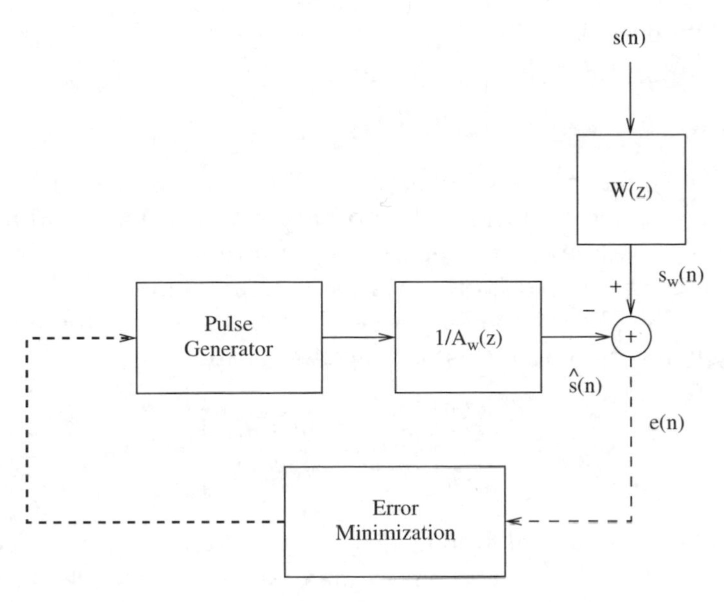

Figure 7.17 Block diagram of a simple MPLPC coder

diagram of a simple MPLPC is shown in Figure 7.17. Assuming that $h(n)$ is the impulse response of the combined LPC and perceptual weighting filters, the squared error for a single pulse excitation at location m_i with amplitude g_i can be written as,

$$E = \sum_{n=0}^{L-1} [\tilde{s}(n) - g_i h(n - m_i)]^2 \tag{7.47}$$

where $\tilde{s}(n)$ is the perceptually-weighted target signal with the combined LPC and perceptual-weighting filter memory effect subtracted. The optimum pulse location is obtained by differentiating equation (7.47) with respect to g_i and setting the derivative to zero,

$$\frac{\partial E}{\partial g_i} = -2 \sum_{n=0}^{L-1} [\tilde{s}(n) - g_i h(n - m_i)] \times h(n - m_i) = 0 \tag{7.48}$$

This yields,

$$g_i = \frac{\sum_{n=0}^{L-1} \tilde{s}(n) h(n - m_i)}{\sum_{n=0}^{L-1} h^2(n - m_i)} \tag{7.49}$$

which in general can be written as,

$$g_i = \frac{\Psi(m_i)}{\phi(m_i m_i)} \tag{7.50}$$

where $\Psi(m_i)$ is the cross-correlation between the perceptually-weighted target speech and the combined LP and perceptual-weighting filter impulse response; $\phi(m_i m_i)$ is the autocorrelation of the combined LPC and perceptual-weighting filter impulse response at positions m_i; and $0 \leq i \leq M-1$. Substituting equation (7.50) into (7.47) gives an expression for the perceptually-weighted squared error in terms of the pulse locations,

$$E = \sum_{n=0}^{L-1} \left[\tilde{s}^2(n) - \frac{\Psi^2(m_i)}{\phi(m_i m_i)} \right] \tag{7.51}$$

To minimize the error in equation (7.51), it can be seen that the best position for a single pulse is that value of m_i which maximizes the term, $\Psi^2(m_i)/\phi(m_i m_i)$. Once the search for the optimum pulse has finalized, the effect of this newly-found pulse is removed from the perceptually-weighted input speech to give a new reference signal to be used in determining the next pulse location. Hence the updated reference speech is,

$$\tilde{s}_{i+1}(n) = \tilde{s}_i(n) - g_i h(n - m_i) \tag{7.52}$$

The steps carried out from equations (7.49)–(7.52) are repeated to find the remaining pulse locations and amplitudes. Figure 7.18 shows a typical example of speech signal and the excitation signal produced in an AbS manner, as discussed above. It can clearly be seen from Figure 7.18 that the multi-pulse structure is very effective in producing a flexible excitation signal in modelling the glottal characteristics, especially the pitch pulses.

Optimum Amplitude Excitation MPLPC

The sequential AbS method described is simple and fast but it has several shortcomings. Successive optimization of individual pulses becomes inaccurate when the number of pulses per frame increases. In order to improve the performance one needs to consider the interactions amongst all the pulses during optimization. To consider the interaction between the pulses, let the weighted mean squared error after having placed M pulses at positions m_i, be given by,

$$E = \sum_{n=0}^{L-1} \left[\tilde{s}(n) - \sum_{i=0}^{M-1} g_i h(n - m_i) \right]^2 \tag{7.53}$$

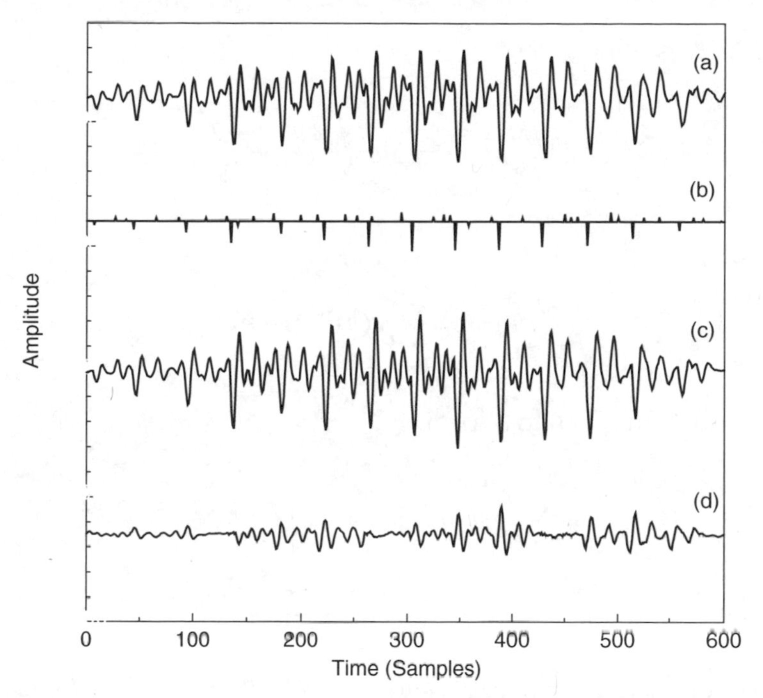

Figure 7.18 Waveform illustrations of the MPLPC coder: (a) original speech, (b) multi-pulse excitation, (c) synthesized speech, and (d) error signal

Differentiating the above equation with respect to the amplitudes g_i gives a solution for optimum amplitudes,

$$\frac{\partial E}{\partial g_i} = -2\sum_{n=0}^{L-1}\left[\tilde{s}(n) - \sum_{i=0}^{M-1} g_i h(n-m_i)\right] \times h(n-m_k) = 0 \quad k = 0, 1, \ldots, M-1 \tag{7.54}$$

or,

$$\sum_{n=0}^{L-1} \tilde{s}(n)h(n-m_k) = \sum_{n=0}^{L-1}\left[\sum_{i=0}^{M-1} g_i h(n-m_i)\right] \times h(n-m_k) \quad k = 0, 1, \ldots, M-1 \tag{7.55}$$

Rearranging the summation on the right hand side,

$$\sum_{n=0}^{L-1} \tilde{s}(n)h(n-m_k) = \sum_{i=0}^{M-1} g_i \left[\sum_{n=0}^{L-1} h(n-m_i)h(n-m_k)\right] \quad k = 0, 1, \ldots, M-1 \tag{7.56}$$

Now defining,

$$\phi(k,i) = \sum_{n=0}^{L-1} h(n-k)h(n-i) \tag{7.57}$$

and,

$$\Psi(k) = \sum_{n=0}^{L-1} \tilde{s}(n)h(n-k) \tag{7.58}$$

simplifies equation (7.56) to a form,

$$\Psi(m_k) = \sum_{i=0}^{M-1} g_i\phi(m_i, m_k) \qquad k = 0, 1, \ldots, M-1 \tag{7.59}$$

which can be written in the form of a correlation matrix as,

$$\begin{bmatrix} \Psi(m_0) \\ \Psi(m_1) \\ \vdots \\ \Psi(m_{M-1}) \end{bmatrix} = \begin{bmatrix} g_0 \\ g_1 \\ \vdots \\ g_{M-1} \end{bmatrix} \begin{bmatrix} \phi(m_0m_0) & \phi(m_0m_1) & \ldots & \phi(m_0m_{M-1}) \\ \phi(m_1m_0) & \phi(m_1m_1) & \ldots & \phi(m_1m_{M-1}) \\ \vdots & \vdots & \vdots & \vdots \\ \phi(m_{M-1}m_0) & \phi(m_{M-1}m_1) & \ldots & \phi(m_{M-1}m_{M-1}) \end{bmatrix} \tag{7.60}$$

The optimum amplitudes g_i can now be solved utilizing the Cholesky decomposition of the correlation matrix.

Using the above analysis, two forms of pulse amplitude re-optimization procedure can be used [12]. One can re-optimize the amplitudes after all of the M pulses have been located within a subframe, or after each new pulse is located. Of course the latter method has the greater computational burden of the matrix inversion, but the overall quality compared to the former method is superior. If amplitude re-optimization takes place once at the end of each subframe, the required matrix inversion size is $(M \times M)$. If it takes place after each new pulse is located, amplitude re-optimization occurs M times with matrix sizes of (1×1) up to $(M \times M)$. Figure 7.19 shows the variation of number of pulses versus *segSNR* for three different algorithms. Curve (*c*) is for a basic sequential MPLPC coder. It shows increasing *segSNR* as the number of pulses increases but, after 30 pulses per 160 samples, its performance tends to saturate. Curves (*b*) and (*a*) are for the improved MPLPC algorithms, i.e. amplitude re-optimization after all pulses have been located and amplitude re-optimization after location of each new pulse respectively. Objective results show curve (*b*) giving lower *segSNR* than (*a*), as expected.

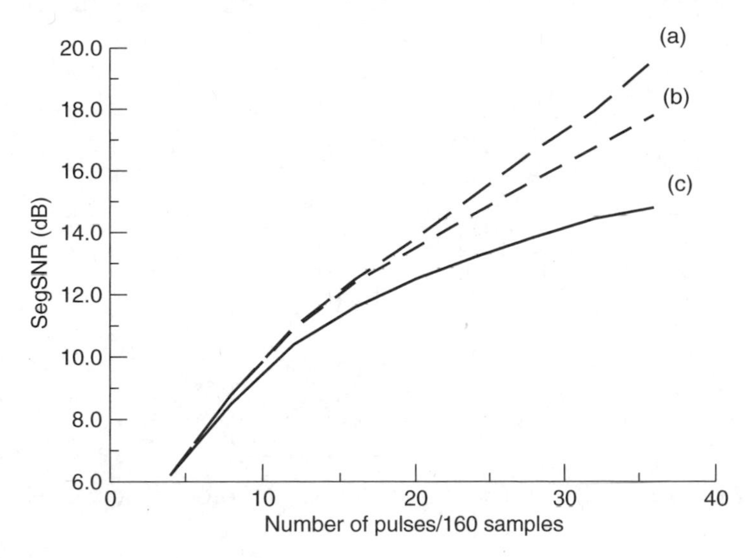

Figure 7.19 The number of pulses against the segmental SNR: (a) amplitude re-optimization after each pulse, (b) amplitude re-optimization after all pulses, and (c) sequential MPLPC with no amplitude re-optimization

Subjectively, curves (*a*) and (*b*) are superior only when a high number of pulses (eight or more every 20 ms) are employed in the process of amplitude re-optimization. This is expected since the re-optimization process improves the performance of closely placed pulses. On average, five pulses per 4–5 ms are adequate to achieve good speech quality. One major problem during the search for the pulses is pulse-doubling. Pulse-doubling usually occurs in voiced regions with greater than about eight pulses every 10 ms and involves the re-selection of already-selected pulse positions. In order to avoid this effect, the newly-found pulse amplitude is added to the existing amplitude or the already-found pulse locations are excluded from further pulse position selection. If joint pulse amplitude re-optimization is applied every time a new pulse is positioned, pulse-doubling is eliminated automatically.

MPLPC with Pitch Prediction

A basic multi-pulse coder produces satisfactory speech quality at medium bit rates. However as the bit rate is lowered, degradations in the speech quality become noticeable. This is especially true for the higher-pitched voiced regions which usually occur with female speakers. This is due to a limited number of pulses being available, the majority of which are used to model the fundamental pitch pulses and hence relatively few pulses remain for the modelling of the remaining excitation signal. With the introduction of a pitch predictor into the AbS loop as shown in Figure 7.20 such effects can be reduced. In 1989, Singhal and Atal [12] proposed a closed-loop solution

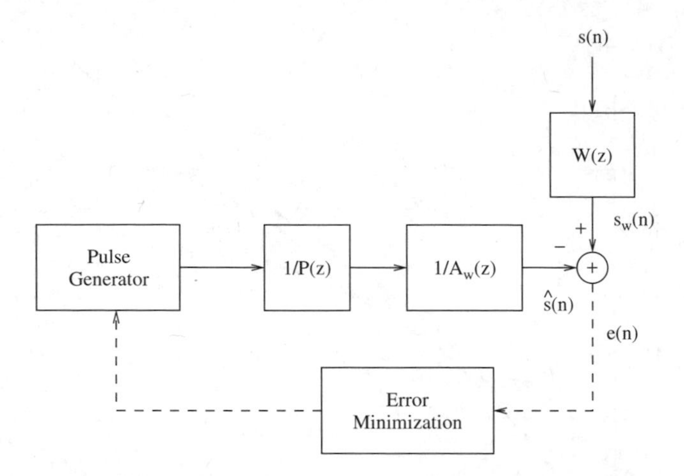

Figure 7.20 Block diagram of MPLPC AbS procedure with LTP

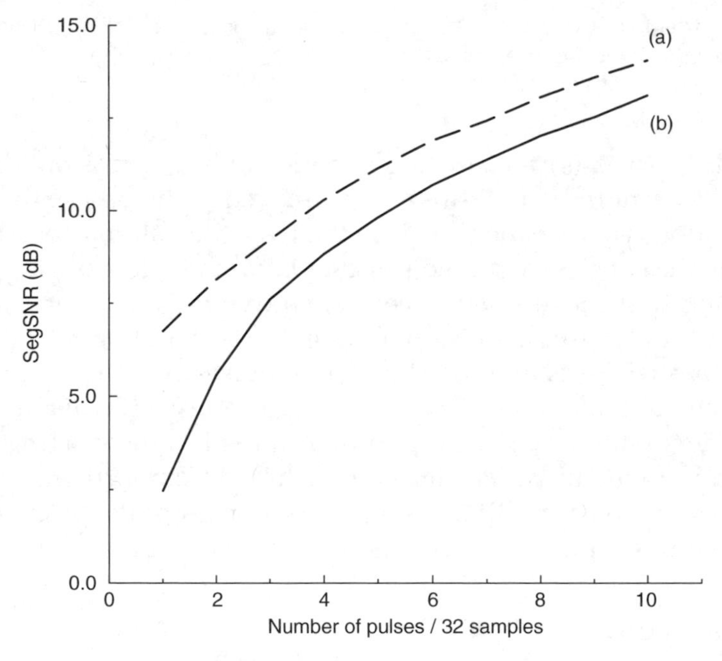

Figure 7.21 Performance of MPLPC (a) with and (b) without a pitch predictor

which gave optimum values of the pitch predictor where the delay in the predictor was integer multiples of the pitch. The aim of pitch prediction was to model the long-term similarities in speech and hence it was also called the long-term predictor (LTP). The most popular pitch prediction used in MPLPC is the one-tap predictor. Figure 7.21 shows the objective performance of the MPLPC with a (one-tap closed-loop) pitch predictor and without a pitch predictor. It is clear that pitch prediction provides higher *segSNR* at all pulse rates. At high pulse rates, the subjective difference between the MPLPC with and without LTP saturates, since at higher rates the pulses from the multi-pulse excitation can model the fundamental pitch pulses accurately.

Pulse Position Coding

The coding of pulse positions is usually performed by enumerative source coding techniques [6]. The number of possibilities for placing M pulses in a subframe of L samples is given by,

$$\Theta = \begin{pmatrix} L \\ M \end{pmatrix} = \frac{L!}{M!(L-M)!} \tag{7.61}$$

Hence the minimum number of bits required for coding these positions is

$$B_{min} = \lceil(\log_2 \Theta)\rceil \tag{7.62}$$

where $\lceil . \rceil$ is the nearest integer greater than or equal to Θ. Such methods of pulse-position coding can be considered as a vector quantization of the pulse positions. These techniques are not very favourable in bad channel conditions and alternative coding methods are pursued. Another method is the independent coding of the pulse positions. Although this method leads to a higher bit rate, it is more robust to channel errors. The number of bits required for independent pulse position coding is given by,

$$B_{min} = M \times \lceil(\log_2 L)\rceil \tag{7.63}$$

Figure 7.22 shows the comparison between combinational and independent coding of pulse positions for $L = 32$. Clearly, combinational coding is by far the more efficient at very high pulse rates, but at very low number of pulses the difference is small (i.e. 3 bits). At such coding rates, the disadvantage of an extra few bits per analysis frame for independent coding is reflected in the complexity and the coder robustness.

Pulse Amplitude Coding

Efficient normalization is necessary for coding of the pulse amplitudes, because the pulse amplitudes have a large dynamic range and direct quantization requires a large number of bits. Normalization can be carried out by the *rms* of the amplitudes. In such methods, the *rms* value must also be included in the transmission. This inevitably leads to higher bit rates. In most MPLPC designs, quantization of the first pulse is accomplished by incorporating a large number of nonuniform quantizer levels (usually five bits or more) and the rest of the pulses in the subframe are normalized with

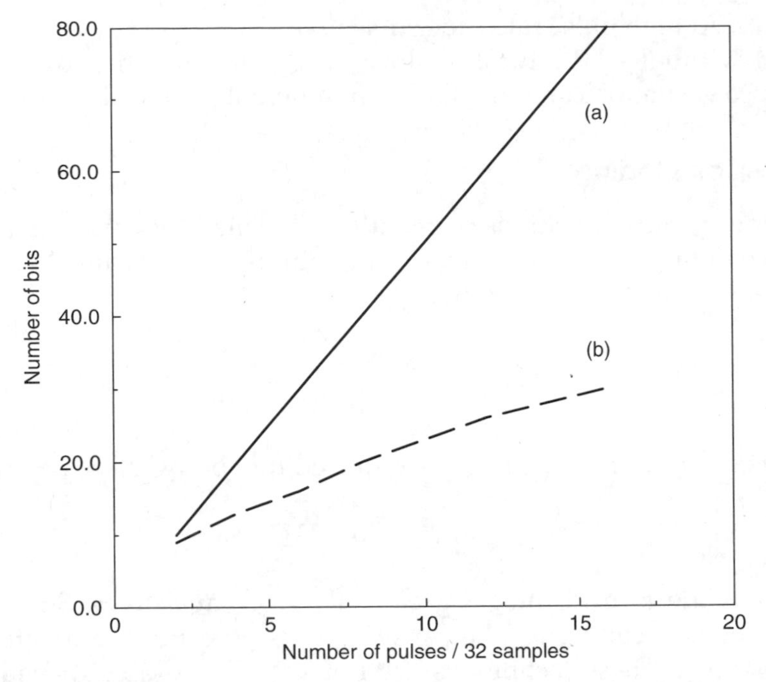

Figure 7.22 Comparison between (a) independent and (b) combinational pulse-position coding

the first pulse magnitude and coded using fewer quantization levels (typically three bits) [6]. This assumes that the first pulse usually has the largest magnitude and, if independent pulse position coding is used, this can be advantageous. Otherwise the largest magnitude pulse may be used to limit the pulses within ± 1.

7.3.4 Codebook Excitation

The vectors contained in the excitation codebook form a very important part of the CELP coding algorithm. They serve two main purposes:

- They provide the start-up information to the LTP memory, including any sudden changes in the speech not adequately tracked by the pitch prediction.
- They supply the 'filling-in' information that the pitch predictor has omitted. This is especially the case during unvoiced regions.

Thus, how the codebook of a CELP is populated and the method by which the optimum vector is computed are very important issues as indicated by the many publications on this subject [20–22]. Another related issue is the computational cost and storage of the codebook search procedure. The search process for the best vector in CELP can be broken down into four stages.

1. Synthesis of the codebook vector to obtain the output $\hat{s}_k(n)$.
2. Calculation of the cross-correlation between the reference $\tilde{s}(n)$ (LPC, perceptual weighting and pitch effects removed) and the synthetic estimate $\hat{s}_k(n)$.
3. Calculation of the autocorrelation of the synthetic estimate $\hat{s}_k(n)$.
4. Testing for the minimum error, or the maximum normalized correlation.

To reduce complexity and memory, and improve quality, many versions of the codebook excitations have been developed and used. Here, we will consider three secondary excitation types.

Gaussian Excitation

Almost all of the early versions of CELP used a form of Gaussian codebook as the source of secondary excitation. This is mainly because speech, after inverse filtering with the LPC followed by the pitch, has residual characteristics similar to Gaussian. One problem with this approach is the size of the memory required to store the Gaussian codebook vectors. For example, if a 10-bit codebook is used to match L reference samples, the number of storage locations will be $2^{10} \times L$. Assuming $L = 40$, this will correspond to 40 960 storage locations, which can be very large for real-time memory-restricted implementations. In order to overcome this problem, the Gaussian vectors are represented as a one-dimensional array, where most of the L samples of two consecutive vectors are common. The most popular versions of this *overlapping codebook* are those with one or two shifts. In other words, to generate a new vector, one or two samples at the end of the previously-used vector are dropped and one or two new samples are introduced at the beginning for one or two shifts respectively. An overlapping codebook with single shift can be represented as an LTP filter where the minimum and maximum delays are $L - 1$ and $C + L - 1$ respectively, assuming that the locations start from 0 and the size of the codebook is C.

In some coders, a centre-clipped version is used with a clipping threshold of 1.2 for a unit variance vector. This is found to produce sharper output speech. One reason for this is that, when matching the codebook vectors with the reference vector, a few higher-magnitude vector elements dominate the selection causing errors in the lower-magnitude vector elements. By making the smaller vector elements zero, as well as minimizing the error they cause, the matching of the larger magnitude samples is improved in the absence of the erroneous smaller samples in the vector.

Training Gaussian codebooks does not improve the quality significantly if the codebook size is eight bits or more. Therefore, they can simply be populated by using a Gaussian random number generator. Some applications [23] use ternary codebooks, where each Gaussian number amplitude is set to 1 if it is positive, left as zero if it is zero and set to -1 if it is negative. This is especially useful in fixed point implementations. In addition to the above versions, one other popular version of the standard codebooks is called the sparse codebook, where each nonzero vector element is followed by a fixed number of zeros. This is very similar to regular pulse excited LPC.

Overlapping codebooks are useful in reducing the computation of the codebook search as well as requiring less storage. The fact that the adjacent vectors have similarities can be used to reduce the convolution (filtering) process to generate the synthetic output. If $h(n)$ represents the combined impulse response of the LPC and perceptual-weighting filters, the synthesized vector $\hat{s}_k(n)$ due to the kth excitation vector $x_k(n)$ of a single shift codebook is,

$$\hat{s}_k(n) = \sum_{i=0}^{n} x_k(n-i)h(i) \tag{7.64}$$

(When the pitch filter is implemented as an adaptive codebook for $D < L$, it can be assumed to be parallel with the fixed codebook since it is not considered during the codebook search and hence will not affect the codebook search process.) In a single shift codebook, the difference between two consecutive vectors is only one sample at either end of the two vectors and the synthesized vector $\hat{s}_{k+1}$ can be written in terms of $\hat{s}_k$ as,

$$\hat{s}_{k+1}(n) = x_{k+1}(0)h(n) + \hat{s}_k(n-1) \tag{7.65}$$

where $\hat{s}_k(-1) = 0$. As can be seen from the above expression, by shifting the previous output by one sample and adding this to the impulse response $h(n)$ scaled by the new sample, most of the convolution computations can be simplified. As the number of shifts in the codebook increases, however, complexity increases, and when the shift equals the vector size, the overlapping codebook becomes a standard codebook containing independent vectors. Note that if centre-clipping is used, the zero values of $x_k(n)$ in equation (7.64) will not need multiplication with $h(i)$, and hence will reduce computational complexity. However, once the first vector is fully synthesized, more savings will be made using equation (7.65). Every time a zero-valued excitation sample helps to produce the new vector, the first term of equation (7.65) will be zero, which means that the new synthetic vector is simply the shifted version of the previous with its first sample set to zero. Using an unquantized CELP coder as defined by Table 7.3, four different versions of the standard Gaussian codebook were compared. The results are shown in Table 7.4 (the overlapping codebooks have a two-sample shift). Subjective listening shows that the speech quality is generally improved with centre-clipping compared with standard Gaussian codebooks. The difference between the overlapping and nonoverlapping codebooks of the same type is negligible. Coupled

Table 7.3 CELP coder parameter definition for the comparison test

Parameter	Update rate
Sampling	8 kHz
LPC analysis	160 samples
LTP 1- & 3-tap	40 samples
10-bit codebook	40 samples
Weighting	$\gamma = 0.9$

Table 7.4 Performance of four standard codebooks

Scheme	1-tap (dB)	3-tap (dB)	Storage (words)
Gaussian	11.11	12.52	1024×40
Centre-clipped Gaussian	11.20	12.53	1024×40
Overlapping Gaussian	11.16	12.49	$(2 \times 1023) + 40$
Overlapping centre-clipped Gaussian	11.18	12.55	$(2 \times 1023) + 40$

with the objective results in Table 7.4, the overlapping centre-clipped Gaussian codebook is very attractive for its reduced memory and computational requirements.

Vector Sum Excitation

In the normal filtering approach of CELP, $\tilde{s}(n)$ is matched by exhaustively searching a finite number of sequences $\hat{s}_k(n)$ and the best match, $\hat{s}_{opt}(n)$, is the sequence which gives the minimum mean square error between $\tilde{s}(n)$ and $\hat{s}_k(n)$. How good a match between $\tilde{s}(n)$ and $\hat{s}_k(n)$ is determined by the degree of freedom in $\hat{s}_k(n)$, i.e. the size and characteristics of the codebook. In the method previously described, the freedom in $\hat{s}_k(n)$ was obtained by synthesizing many versions of $\hat{s}_k(n)$, i.e. the degree of freedom in $\hat{s}_k(n)$ is limited by $x(n)$ at the residual side of the analysis. However, it can be noted that if the same degree of freedom can be achieved at the synthetic signal side of the analysis whilst retaining the fact that all candidate $\hat{s}_k(n)$ are spectrally-shaped by the LPC and perceptual-weighting filters, then less complexity and equal performance could be obtained. Therefore, the aim is to limit the amount of synthesis operations and perform the vector combinations to give the necessary freedom in $\hat{s}_k(n)$ at the output side of the analysis. One such method is vector sum excitation (VSE) [24].

As for the majority of speech-coding analysis, in VSE, the mean squared error approximation is used. The formulation of VSE to derive the optimum excitation $x_{opt}(n)$ and hence $\hat{s}_{opt}(n)$ is as follows. Let each candidate synthesized signal be given by,

$$\hat{s}_k(n) = \sum_{i=1}^{M} a_i \hat{s}_i(n), \quad n = 0, 1, \ldots, L-1 \tag{7.66}$$

where the sequences $\hat{s}_i(n)$ are derived from exciting the combined LPC and perceptual-weighting filter with M different excitation sequences, $x_i(n)$; a_i are variable scaling factors; and L is the excitation subframe size. An optimum synthesized signal $\hat{s}_{opt}(n)$ can be derived by choosing the set of coefficients a_i, $i = 1, 2, \ldots, M$ which minimizes the weighted mean squared error between $\hat{s}(n)$ and $\tilde{s}(n)$ given by,

$$E = \sum_{n=0}^{L-1} \left[\tilde{s}(n) - \hat{s}_k(n)\right]^2 = \sum_{n=0}^{L-1} \left[\tilde{s}(n) - \sum_{i=1}^{M} a_i \hat{s}_i(n)\right]^2 \tag{7.67}$$

This minimization is achieved by solving a set of equations produced by the partial derivatives of equation (7.67), with respect to each of the variables a_i, to be zero,

$$\frac{\partial}{\partial a_j}\left[\sum_{n=0}^{L-1}[\tilde{s}(n) - \sum_{i=1}^{M} a_i \hat{s}_i(n)]^2\right] = 0 \tag{7.68}$$

which can be simplified into,

$$\sum_{i=1}^{M} a_i R(i,j) = \phi(j), \quad j = 1, 2, \ldots, M \tag{7.69}$$

where,

$$R(i,j) = \sum_{i=0}^{L-1} \hat{s}_i(n) \hat{s}_j(n) \tag{7.70}$$

and,

$$\phi(j) = \sum_{n=0}^{L-1} \tilde{s}(n) \hat{s}_j(n) \tag{7.71}$$

Assuming that the vectors $\hat{s}_i(n)$ are made independent and orthogonal, $\bar{s}_i(n)$, then,

$$R(i,j) = \sum_{n=0}^{L-1} \bar{s}_i(n) \bar{s}_j(n) \quad \begin{cases} = 0 & \text{for } i \neq j \\ \neq 0 & \text{for } i = j \end{cases} \tag{7.72}$$

Thus, substituting equation (7.72) into equation (7.69) and rearranging, the a_i can be calculated as,

$$a_i = \phi(i)/R(i,i), \quad i = 1,2,\ldots,M \tag{7.73}$$

The best estimate for $\tilde{s}(n)$ is then given by,

$$\hat{s}_{opt}(n) = \sum_{i=1}^{M} a_i \bar{s}_i(n), \quad n = 0,1,\ldots,L-1 \tag{7.74}$$

The above procedure comprises a minimum mean square error approximation in which the optimum solution can be derived only if the set of sequences $\hat{s}_i(n)$ are linearly independent (i.e. form a basis) and are orthogonal to each other. A popular method for achieving this orthogonalization is the Gram–Schmidt procedure, summarized below.

Consider a set of $m+1$ vectors $p_i(n)$ each of length L which form a basis. The objective is to construct orthogonal vectors $q_i(n)$ so that

$$\sum_{n=0}^{L-1} q_i(n)q_j(n) \quad \begin{cases} = 0 & \text{for } i \neq j \\ \neq 0 & \text{for } i = j, \end{cases} \quad i,j = 0,1,\ldots,m \tag{7.75}$$

Let $q_0(n) = p_0(n)$ and define $q_1(n)$ as a linear combination of $q_0(n)$ and $p_1(n)$, then

$$q_1(n) = p_1(n) - \alpha_{01} q_0(n) \tag{7.76}$$

Then for $q_1(n)$ to be orthogonal to $q_0(n)$,

$$\sum_{n=0}^{L-1} q_1(n)q_0(n) = 0 \tag{7.77}$$

i.e.

$$\sum_{n=0}^{L-1} p_1(n)q_0(n) - \sum_{n=0}^{L-1} \alpha_{01} q_0^2(n) = 0 \tag{7.78}$$

and the correlation α_{01} is given by,

$$\alpha_{01} = \frac{\sum_{n=0}^{L-1} p_1(n)q_0(n)}{\sum_{n=0}^{L-1} q_0^2(n)} \tag{7.79}$$

With this α_{01}, the two functions $q_0(n)$ and $q_1(n)$ are now orthogonal. In order to build the other functions $q_i(n)$, linearly independent $p_i(n)$ are added one at a time until all are constructed. This can be formulated in general as

$$q_k(n) = p_k(n) - \sum_{j=0}^{k-1} \alpha_{jk} q_j(n) \tag{7.80}$$

where,

$$\alpha_{jk} = \frac{\sum_{n=0}^{L-1} p_k(n) q_j(n)}{\sum_{n=0}^{L-1} q_j^2(n)}, \quad j = 0, 1, \ldots, k-1 \tag{7.81}$$

Therefore, the basic task in VSE is to obtain $\hat{s}_{opt}(n)$ by deriving the set of orthogonal vectors $\bar{s}_i(n)$ and their optimum scaling factors a_i. The optimum excitation $x_{opt}(n)$ can then be derived by passing $\hat{s}_{opt}(n)$ through the combined LPC and perceptual-weighting inverse filter.

In VSE, it is very important to construct the basis vectors in a perceptually advantageous way and training is required. As in CELP with Gaussian excitation, in VSE linear prediction (VSELP), the pitch filter is treated as an adaptive codebook for lag values less than the subframe size and, hence, it is not included in the codebook search process. Finally, the total excitation is obtained by adding the gain-scaled secondary excitation $Gx_{opt}(n)$ to the pitch predictor excitation. A block diagram of vector processing of a VSELP coder where the scaling factors a_i are assumed to be ± 1 is shown in Figure 7.23.

The M basis vectors, $[x_i(n)]_{i=1}^{M}$, are first synthesized to give M synthetic basis vectors $[\hat{s}_i(n)]_{i=1}^{M}$. These are then made orthogonal to the pitch predictor contribution signal via the Gram–Schmidt orthogonalization process to get $[\bar{s}_i(n)]_{i=1}^{M}$. This ensures that the secondary excitation does not cover the vector space that has already been covered by the pitch predictor contribution. After orthogonalization, the individual basis vector scaling factors a_i are computed to form a Mth element vector, $\mathbf{a}$. The vector $\mathbf{a}$ is then quantized as ± 1. The final synthetic signal from the excitation, $\hat{s}_{opt}(n)$, is obtained by first summing up the properly signed (± 1) orthogonal basis vectors and then gain scaling the summed vector. In order to obtain the quantized secondary excitation, $\hat{s}_{opt}(n)$ is then inverse-filtered with the combined LPC and perceptual-weighting filters at the encoder. At the decoder, as there is no need for the perceptual-weighting filtering, only the LPC inverse filtering is used. The overall scaling factor is obtained as in standard CELP. The final

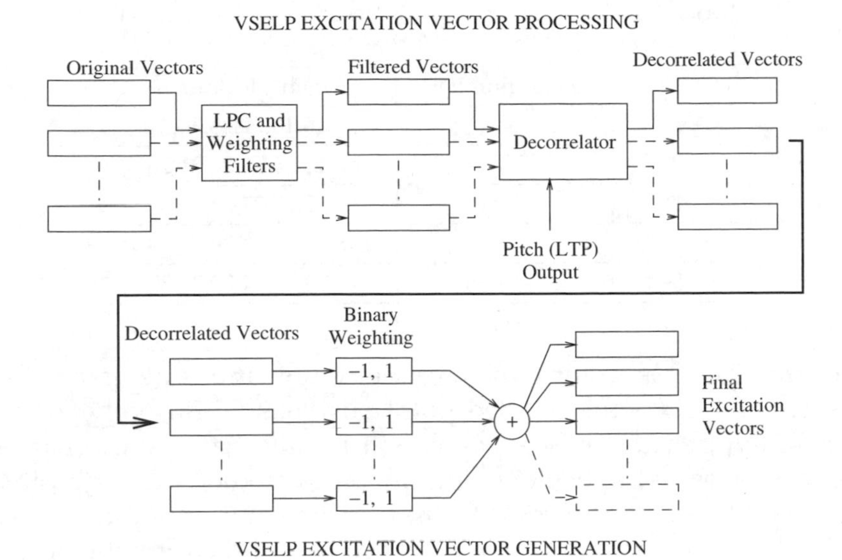

Figure 7.23 Block diagram of VSELP vector processing

synthetic speech is obtained by adding the scaled secondary excitation and the pitch predictor contribution and synthesizing this through the combined LPC and perceptual-weighting filters with initial memories restored. Note that the orthogonalization procedure and inverse filtering is required both at the encoder and the decoder.

Algebraic Codebook Excitation

As discussed earlier multi-pulse excitation is very useful in tracking the changes in speech accurately. Its main disadvantage is the number of bits required to encode its pulse positions as well amplitudes. Regular pulse excitation, on the other hand, is a very restricted version of multi-pulse excitation in terms of pulse positions which may degrade speech quality if decimation factors of five or more are used. The codebook types discussed above are restrictive both in terms of pulse positions and amplitudes, i.e. the codebook contains preset vectors. The VSE makes an attempt to modify the vectors by the pitch predictor contribution but still has fixed basis vectors. Although they are very efficient in coding capacity, they may suffer from quality degradations especially in speech transitions where the pitch predictor fails to perform its usual function adequately. Algebraic codebooks have overcome these problems by cleverly representing excitation pulses where they have some freedom in position [25, 26]. In algebraic codebooks, only a small number of pulses are used and they are positioned in interleaved tracts (for efficient coding); hence, although each pulse position is severely restricted,

Table 7.5 Typical 5-pulse algebraic codebook tracks for a 40-sample subframe

Track	Pulse number	Possible locations
1	i_0	0,5,10,15,20,25,30,35
2	i_1	1,6,11,16,21,26,31,36
3	i_2	2,7,12,17,22,27,32,37
4	i_3	3,8,13,18,23,28,33,38
5	i_4	4,9,14,19,24,29,34,39

together they are able to form most of the combinations necessary for adequate excitation. Since the selected pulse positions will usually correspond to the remaining major pulses, which will usually have somewhat similar magnitudes (expected after removing the pitch predictor contribution), the pulse amplitudes are also restricted to having the same amplitude, usually set to ± 1. However, in order to have efficient coding of the formations (indices) of the excitation vectors and enable fast search, the overall combination of the nonzero samples is usually restricted to four or five interleaved tracks. Only one or two nonzero pulses with either positive or negative signs are placed in each track. Table 7.5 shows typical five-pulse interleaved track positions in a 40-sample excitation subframe. Using the possibilities shown in Table 7.5, the codebook vector $x(n)$ is formed by setting only five unity pulses in a possible 40-sample vector with all other locations being set to zero.

$$x(n) = s_0\delta(n - m_0) + s_1\delta(n - m_1) + s_2\delta(n - m_2) + s_3\delta(n - m_3) + s_4\delta(n - m_4),$$
$$n = 0, \ldots, 39 \tag{7.82}$$

where s_i and m_i are the sign and position of the ith pulse and $\delta(0)$ represents unity pulse amplitude.

The total possible number of excitation vector combinations that an algebraic codebook can produce is quite large. Therefore full searching of all possible excitations becomes prohibitive for real-time implementations. However, algebraic codebooks are designed to reduce this complexity significantly. Having got the synthetic output for each excitation vector, the cross-correlation of the synthesized signal with the target signal (LPC and perceptual-weighting filters' memory response and the pitch predictor contribution removed from weighted input speech) and the synthesized signal energy need to be computed. The best excitation sequence is then selected by maximizing:

$$A_k = \frac{(\mathbf{d}^t\mathbf{x}_k)^2}{\mathbf{x}_k^t\mathbf{\Phi}\mathbf{x}_k} \tag{7.83}$$

where $\mathbf{d} = \mathbf{H}^t\tilde{\mathbf{s}}$ is the correlation matrix between the target signal $\tilde{s}(n)$ and the combined LPC and perceptual-weighting filter impulse response $h(n)$, $\mathbf{x}_k$ is the kth excitation vector, $\mathbf{H}$ is the lower triangular Toeplitz convolution matrix with diagonal $h(0), h(1), \ldots, h(39)$, and $\boldsymbol{\Phi} = \mathbf{H}^t\mathbf{H}$ is the matrix correlation of $h(n)$. Before the fixed codebook search starts, both $\mathbf{d}$ and $\boldsymbol{\Phi}$ are computed,

$$d(n) = \sum_{i=n}^{39} \tilde{s}(i)h(i-n) \quad ; \quad n = 0, \ldots, 39 \tag{7.84}$$

$$\phi(i,j) = \sum_{n=j}^{39} h(n-i)h(n-j) \quad ; \quad j \geq i \tag{7.85}$$

Once the correlation of the impulse response, $h(n)$, and the target signal, $\tilde{s}(n)$, is computed for every possible pulse position, the overall correlation (i.e. the numerator of equation (7.83)) becomes a simple summation of the correlations at only nonzero excitation pulse positions.

$$C = \sum_{i=0}^{N_p-1} p_i d(m_i) \tag{7.86}$$

where N_p is the number of pulses, $p_i = s_i\delta(0)$, in the excitation. The denominator of equation (7.83) is given by:

$$D = \sum_{i=0}^{N_p-1} \phi(m_i, m_i) + 2\sum_{i=0}^{N_p-2}\sum_{j=i+1}^{N_p-1} p_i p_j \phi(m_i, m_j) \tag{7.87}$$

However the above equation can be simplified significantly if we assume that the pulses p_i have unity amplitudes. Before the codebook search, $d(n)$ is decomposed into its absolute value $|d(n)|$ and sign $sign[d(n)]$. Using the sign information, ϕ is modified,

$$\phi'(i,j) = sign[d(i)]sign[d(j)]\phi(i,j), \quad i = 0, \ldots, 39, \; j = i+1, \ldots, 39. \tag{7.88}$$

The main-diagonal elements of ϕ are scaled to remove the factor of two in equation (7.87).

$$\phi'(i,i) = 0.5\phi'(i,i) \quad i = 0, \ldots, 39 \tag{7.89}$$

The correlation in equation (7.86) can now be computed over the nonzero pulses as (five nonzero pulses are assumed, to follow the example),

$$C = |d(m_0)| + |d(m_1)| + |d(m_2)| + |d(m_3)| + |d(m_4)| \tag{7.90}$$

since the sign is separately coded, the absolute value of $d(n)$ is used in the above equation. The denominator of equation (7.83) expressed in equation (7.87) can be computed by;

$$\begin{aligned}E/2 = {} & \phi'(m_0, m_0) \\ & +\phi'(m_1, m_1) + \phi'(m_0, m_1) \\ & +\phi'(m_2, m_2) + \phi'(m_0, m_2) + \phi'(m_1, m_2) \\ & +\phi'(m_3, m_3) + \phi'(m_0, m_3) + \phi'(m_1, m_3) + \phi'(m_2, m_3) \\ & +\phi'(m_4, m_4) + \phi'(m_0, m_4) + \phi'(m_1, m_4) + \phi'(m_2, m_4) + \phi'(m_3, m_4)\end{aligned} \tag{7.91}$$

Having simplified the search process shown above, further reductions in search complexity are achieved by a focused search approach. A precomputed threshold is tested before entering the last loop (in a nested search to locate the pulses in five tracks). The threshold determines if the first four pulses have already produced a good combination and whether it is worthwhile to continue into the last loop. The threshold is based on the correlation C in equation (7.86). The maximum absolute correlation and the average correlation produced by the first four pulses, max_4 and av_4, are used to compute the threshold,

$$TH_4 = av_4 + K_4(max_4 - av_4)\ ;\ 0 \le K_4 < 1. \tag{7.92}$$

The last loop is entered (to search for the fifth pulse) only if the absolute correlation due to first four pulses exceeds TH_4. K_4 controls the percentage of combinations searched (set to 0.4 in G.729 for the fourth loop where four nonzero pulses are used). Since this will result in variable search computation, the last loop is entered only a predetermined fixed number of times.

As explained above, an algebraic codebook is an excellent compromise between the very restricted regular pulse excitation and multi-pulse excitation. Algebraic codebooks are computationally very efficient as well as producing good performance. However the combinations of the interleaved track positions still require a large number of bits. For example in a 40-sample subframe using five tracks with eight locations in each, the pulse positions would require three bits each plus a sign bit, giving a total of 20 bits per subframe. This will result in $20 \times \frac{8000}{40} = 4\,\text{kb/s}$ which is only applicable to bit rates of around 6 kb/s and above. Algebraic codebooks have been extensively used in many CELP coders such as G.729, EFR, G.723.1, etc.

Pitch Adaptive Mixed Excitation

In the case of Gaussian codebooks, it is assumed that the size of the codebook is large enough to cater for both voiced and unvoiced speech excitations. In the VSE case, orthogonalization with respect to the pitch predictor output enables the secondary excitation vectors to cover the space that is not covered by the pitch prediction, resulting in a better system. However at low bit-rates (increased vector sizes), during voiced onsets and transitions where the pitch cannot build up fast enough to track the changes, the speech quality deteriorates significantly. The advantage of algebraic codebooks also reduces at low bit-rates (i.e. at around 4.8 kb/s) as the number of pulse combinations need to be severely restricted in order to allocate fewer bits for the secondary excitation which results in distorted speech. Other important issue at low bit-rates is the amount of noise added to speech from the secondary excitation during steady state voiced regions. A constrained gain approach [27] helps to produce cleaner voiced speech by limiting the power of secondary excitation during steady state voiced regions. This section describes an adaptive codebook excitation where the excitation pulse-positioning is made adaptive with the pitch lag computed for the same subframe. This can be seen as a subset of the algebraic codebook approach where the pulse positions are severely restricted but made adaptive with respect to the pitch so as to increase their chances of positioning them to locations where they are needed most.

In pitch adaptive mixed excitation (PAME), the static codebook is split into two parts. The first part is made adaptive with respect to the pitch lag as follows. The excitation buffer is filled with a unit sample amplitude every D samples starting from the first location. The rest of the vector elements are set to zero. During the search of the codebook, this vector is synthesized and its phase position is determined by shifting its synthetic response one sample at a time for $D-1$ times. Each phase position is then treated as a new excitation vector. In order to guard against pitch-doubling errors in the LTP search, if the lag D is greater than $2D_{min}$ the same process is applied again by placing the excitation pulses every $D/2$ samples. The total number of excitation vectors searched is then found by adding the total phase positions considered. This is similar to regular pulse excitation with the decimation factor of D and $D/2$. After selecting the best excitation vector from the pitch-adaptive section of the codebook using C_a phase positions, the search continues in the second part of the codebook which is fixed and contains centre-clipped overlapping excitation. Here, a further $C_f = C - C_a$ vectors are searched and the best performing vector index from the overall search process is transmitted to the receiver. At the receiver, after decoding the pitch lag, the corresponding excitation vector is decoded.

By forcing the secondary excitation to have pitch structure, it is possible to match voiced onsets more accurately. This is because the pitch predictor memory builds more quickly to track the incoming periodicity more accurately and the secondary excitation provides the required periodicity where the pitch predictor fails. This, of course, depends on the accurate computation of the periodicity by the pitch predictor in the first place. Many other adaptation schemes may be used to accurately place the secondary excitation pulses every pitch period. The pitch predictor lag adaptation is useful because it does not require extra computation or bits. Encoding and decoding processes of the codebook index in this algorithm have three possibilities: $D \geq L$; $L/2 \leq D < L$; and $D < L/2$, assuming $D_{min} < L/2$. The total phase positions considered in each possibility can be calculated as follows:

1. In the case of $D \geq L$, there will be a single excitation pulse located in the first position of the secondary excitation vector, hence, a possible L phase positions will be considered. If the submultiple is also greater than L, then the process stops. However, if $D/2 < L$, then $L - D/2$ more phase positions will be considered where the excitation vector will have an extra pulse located at position $D/2$. Therefore, the total phase positions will be $2L - D/2$.
2. In the case of $L/2 \leq D < L$ the secondary excitation vector will have two pulses, placed at the first and D^{th} positions. Therefore, the total phase positions to be considered is D. If $D/2 \geq D_{min}$ then, a further $D/2$ phase positions will be searched giving a total phase positions of $3D/2$.
3. Finally, when $D < L/2$, the secondary excitation will have pulses at every D samples starting from the first position, resulting in a possible D phase positions. If, however, $D/2 \geq D_{min}$ then a further $D/2$ phase positions are considered giving a total of $3D/2$.

The above possibilities are indicated to the receiver by the fixed subframe size and the decoded pitch predictor lag D.

In informal listening comparisons of VSE, centre-clipped Gaussian and PAME, PAME produced the best result by making the overall speech sharper. This was the result of the periodic excitation part of the secondary excitation matching voiced speech faster and hence more accurately. This is illustrated in Figure 7.24 where, the pitch of a voiced onset is better reproduced by the pitch adaptive excitation. In this figure we can also see that PAME tracks voice changes much faster. It must be noted, however, that the performance of PAME can be affected if the pitch predictor lag is chosen wrongly in the first place. Therefore, it is important that during the LTP search, the correct lag or its integer multiples are selected. The dependency of the PAME performance on the pitch predictor lag can be removed if all the possible lags (in a subframe) and phase positions are exhaustively searched. This, however, requires more index values to be coded in the adaptive part of the codebook. In this case, a set of primary excitation vectors are formed by placing a unit amplitude pulse at the start of the excitation buffer $\mathbf{x}$, and then after every P samples. P is varied from D_{min} (the smallest possible pitch) to L (the subframe size) to get all primary vectors. Whilst D_{min} is related to the minimum pitch, it may also be varied to enhance fidelity. Therefore, for each P, the primary

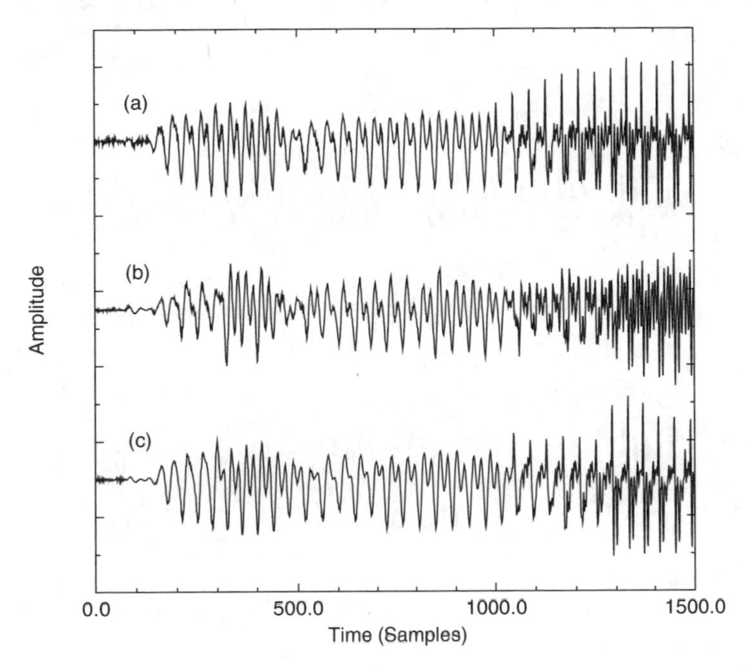

Figure 7.24 Speech waveforms of (a) original, (b) overlapping centre-clipped Gaussian excited CELP output and, (c) PAME excited CELP output

candidate excitation is derived as follows:

$$x_j(n) = \begin{cases} 1 & n = iP < L, \quad i = 0, 1, 2, \ldots \\ 0 & otherwise \end{cases} \tag{7.93}$$

In order to form all possible phase positions, for each primary vector x_j, $P-1$ further vectors x_{j+k}, $k = 1, 2, \ldots, P-1$, are derived by shifting as,

$$x_{j+k}(n) = \begin{cases} 0 & n = 0, 1, 2, \ldots, k-1 \\ x_j(n-k) & n = k, k+1, \ldots, L-1 \end{cases} \tag{7.94}$$

It should be noted that the number of candidate excitation vectors C_a depends on L and D_{min} such that,

$$C_a = L + \sum_{I=D_{min}}^{L/2} I + \sum_{I=1}^{L/2\,-1} I \tag{7.95}$$

Thus the number of bits required by the adaptive excitation index range is $I = \lceil \log_2 C_a \rceil$. If C_a does not correspond to an integer power of 2, a further $2^I - C_a$ vectors are then searched in the fixed codebook. As with all algebraic

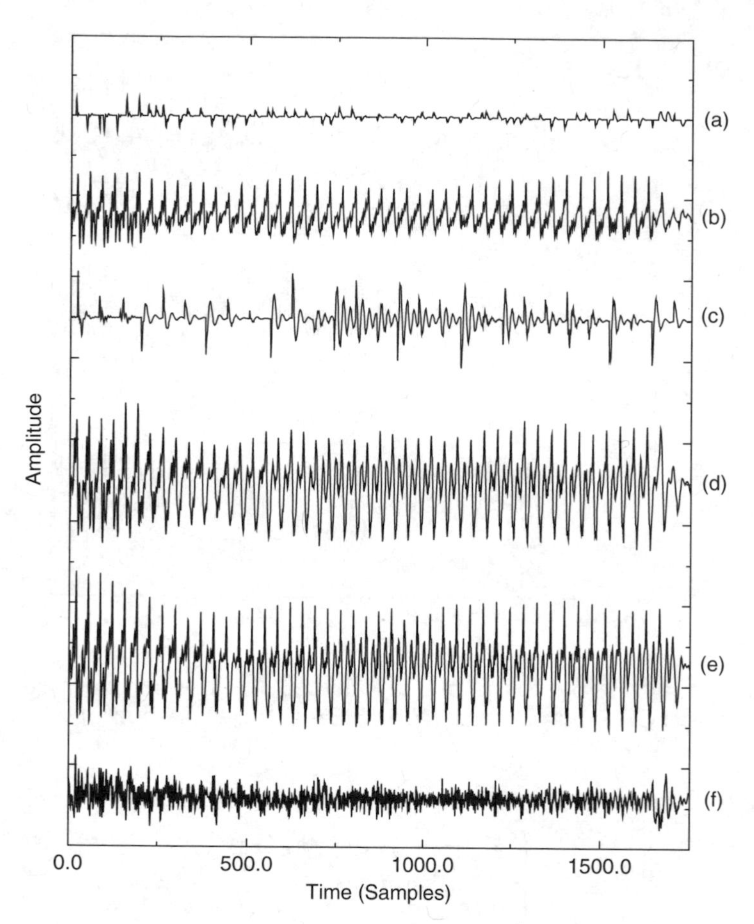

Figure 7.25 Typical CELP waveform plots: (a) codebook contribution, (b) LTP (pitch) contribution, (c) LPC memory contribution, (d) total output, (e) original speech, and (f) final error

codebooks, the above codebook type does not require codebook storage and all codebook search simplifications are applicable. Typical CELP waveforms which use the above excitation are shown in Figure 7.25. The overall rate of this coder was around 4.8 kb/s.

7.3.5 Joint LTP and Codebook Excitation Computation

In the above analysis, where the pitch predictor contribution and secondary excitations are computed sequentially, the final LPC excitation which is formed by adding the pitch predictor contribution and secondary excitations is suboptimum. When the pitch predictor is optimized, the effect of the secondary excitation is assumed to be zero. Thus, when the secondary excitation is computed the pitch prediction excitation becomes suboptimum. Since the whole idea of AbS is to compute an optimum excitation, the pitch prediction and secondary excitations should ideally be computed jointly. As this process would require a huge number of combination (or computation) possibilities, it has not been applied in practice. However assuming that the pitch delay D has been selected correctly during the pitch search (which should be a reasonable assumption for voiced speech), an approximation to joint optimization of pitch prediction contribution and codebook excitations can be made by jointly computing their gains. Assuming a single-tap pitch predictor, the weighted mean squared error for a subframe can be written as,

$$E(k,D) = \sum_{n=0}^{L-1}\left[s_w(n) - s_m(n) - \sum_{i=0}^{n} g_k x_k(n-i)h(i) - \sum_{i=0}^{n} \beta_D \hat{r}(n-i-D)h(i)\right]^2 \tag{7.96}$$

Since the zero input memory response of the LPC and perceptual-weighting filter, $s_m(n)$, cannot be changed, and does not affect the selection of k and D, we can substitute $\tilde{s}(n)$ in the place of $s_w(n) - s_m(n)$. Thus,

$$E(k,D) = \sum_{n=0}^{L-1}\left[\tilde{s}(n) - g_k \sum_{i=0}^{n} x_k(n-i)h(i) - \beta_D \sum_{i=0}^{n} \hat{r}(n-i-D)h(i)\right]^2 \tag{7.97}$$

The above equation is searched for $D_{min} \leq D \leq D_{max}$ and $0 \leq k \leq C-1$. To further simplify the above equation, substitute,

$$Z_D(n) = \sum_{i=0}^{n} \hat{r}(n-i-D)h(i) \tag{7.98}$$

and,

$$V_k(n) = \sum_{i=0}^{n} x_k(n-i)h(i) \tag{7.99}$$

Thus,

$$E(k,D) = \sum_{n=0}^{L-1}[\tilde{s}(n) - g_k V_k(n) - \beta_D Z_D(n)]^2 \tag{7.100}$$

In equation (7.100), the variables k and D should be jointly selected such that with their optimum gains g_k and β_D, the overall error $E(k,D)$ is minimized.

To minimize $E(k, D)$, we differentiate equation (7.100) with respect to g_k and β_D which are functions of k and D respectively. Thus,

$$\frac{\partial E(k,D)}{\partial g_k} = \sum_{n=0}^{L-1} [\tilde{s}(n) - g_k V_k(n) - \beta_D Z_D(n)] \times (-2) V_k(n) = 0 \tag{7.101}$$

and similarly,

$$\frac{\partial E(k,D)}{\partial \beta_D} = \sum_{n=0}^{L-1} [\tilde{s}(n) - g_k V_k(n) - \beta_D Z_D(n)] \times (-2) Z_D(n) = 0 \tag{7.102}$$

Rearranging the above equations we get

$$\sum_{n=0}^{L-1} \tilde{s}(n) V_k(n) = g_k \sum_{n=0}^{L-1} V_k^2(n) + \beta_D \sum_{n=0}^{L-1} Z_D(n) V_k(n) \tag{7.103}$$

$$\sum_{n=0}^{L-1} \tilde{s}(n) Z_D(n) = g_k \sum_{n=0}^{L-1} V_k(n) Z_D(n) + \beta_D \sum_{n=0}^{L-1} Z_D^2(n) \tag{7.104}$$

In matrix form,

$$\begin{bmatrix} g_k \\ \beta_D \end{bmatrix} = \begin{bmatrix} G_1 & G_2 \\ G_2 & G_3 \end{bmatrix}^{-1} \begin{bmatrix} K_1 \\ K_2 \end{bmatrix} \tag{7.105}$$

where,

$$G_1 = \sum_{n=0}^{L-1} V_k^2(n) \tag{7.106}$$

$$G_2 = \sum_{n=0}^{L-1} Z_D(n) V_k(n) \tag{7.107}$$

$$G_3 = \sum_{n=0}^{L-1} Z_D^2(n) \tag{7.108}$$

$$K_1 = \sum_{n=0}^{L-1} \tilde{s}(n) V_k(n) \tag{7.109}$$

$$K_2 = \sum_{n=0}^{L-1} \tilde{s}(n) Z_D(n) \tag{7.110}$$

The gains g_k and β_D can then be found by solving the above equation as,

$$g_k = \frac{K_1 G_3 - K_2 G_2}{G_1 G_3 - G_2^2} \tag{7.111}$$

$$\beta_D = \frac{K_2 G_1 - K_1 G_2}{G_1 G_3 - G_2^2} \tag{7.112}$$

The variables k and D are searched through all combinations using the above analysis and the combination giving the minimum error is selected for transmission. Even though, the speech quality increases significantly with this joint computation of the pitch predictor and secondary excitations, the search computation required is extremely high. Therefore the pitch lag D is usually limited to a narrow range (i.e. ± 2 samples) around the selected value during the LTP search (and its submultiples) before the codebook vector is selected in this joint optimization process.

7.3.6 CELP with Post-Filtering

As discussed earlier, the function of the perceptual weighting filter is to shape the noise spectrum so as to hide it under the speech spectrum [2]. However, at low rates such as 4.8 kb/s, where the average noise level is relatively large, it is very difficult to suppress the noise below the masking threshold at all frequencies. Therefore, in order to improve CELP speech quality at lower bit rates (or, indeed, all rates), further subjective noise reduction techniques are required.

As CELP is essentially a waveform type speech coder, the coded speech can be considered to be the original speech corrupted by additive Gaussian-type noise. Therefore, any speech enhancement technique that deals with this problem can be used to reduce the noise. One such method is that of post-filtering [28]. Adaptive post-filtering (APF) has been used successfully in enhancing ADPCM-coded speech and APC-type schemes [29]. For AbS-LPC coders, the APF as reported by Chen [30] and given by equation (7.113), has been widely accepted.

$$H_{apf}(z) = \frac{(1 - \mu z^{-1})\left(1 - \sum_{i=1}^{p} a_i \beta^i z^{-i}\right)}{\left(1 - \sum_{i=1}^{p} a_i \alpha^i z^{-i}\right)} \tag{7.113}$$

The function of the APF is to attenuate the components in the spectral valleys. However, to achieve this successfully, the simple all-pole APF used in earlier schemes is not adequate. If an all-pole APF is used alone, then,

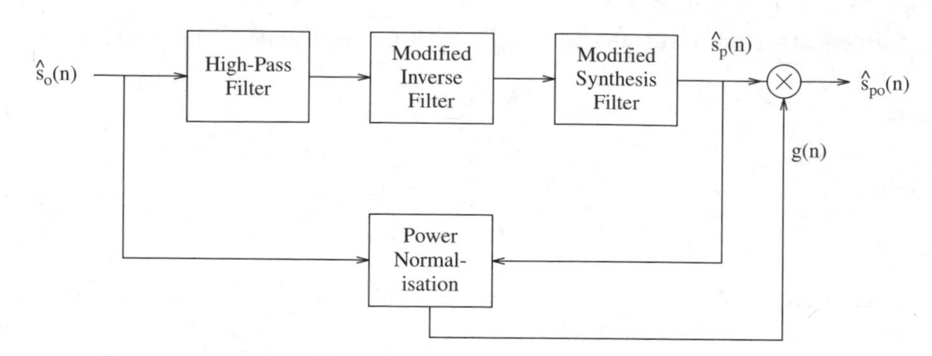

Figure 7.26 Block diagram of the adaptive post-filter

although the perceived noise level is lowered, the output speech is severely low-pass, giving a muffling effect. In order to compensate for this low-pass effect the spectral tilt of the all-pole APF can be modified such that its response is somewhere between an all-pass response and the signal spectrum. The best APF combination was found to be that shown in Figure 7.26.

The simple high-pass filter in the first stage provides a slightly high-pass spectral tilt and thus helps to reduce muffling. The pole-zero second-stage filter provides 'shaping' of the spectral envelope. Finally, a gain control is added to scale the post-filtered speech such that it has roughly the same power as the unfiltered noisy speech. This is necessary as the cascaded filters are not unity gain filters. One technique used to normalize the output signal power is to estimate the power of the un-filtered and filtered speech separately, then determine an appropriate scaling factor based on the ratio of the two estimated power values. The speech power is estimated by an exponential-average gain estimator, i.e. the two estimated power values δ_o^2 and δ_p^2 are given by,

$$\delta_o^2(n) = \zeta\delta_o^2(n-1) + (1-\zeta)\hat{s}_o^2(n) \tag{7.114}$$

$$\delta_p^2(n) = \zeta\delta_p^2(n-1) + (1-\zeta)\hat{s}_p^2(n) \tag{7.115}$$

where $\hat{s}_o(n)$ is the original synthetic speech and $\hat{s}_p(n)$ is the post-filtered speech. A suitable leakage factor ζ is 0.96. At each sampling instant, $\delta_o^2(n)$ and $\delta_p^2(n)$ are computed as above, then the ratio and the square root are computed in order to obtain the gain factor $g(n) = \sqrt{\delta_o^2(n)/\delta_p^2(n)}$. Therefore, the final post-filtered speech is given by

$$\hat{s}_{po}(n) = g(n)\hat{s}_p(n) \tag{7.116}$$

The above procedure is quite computationally-intensive as it requires a divide and square root operation per sample. Therefore, instead of the

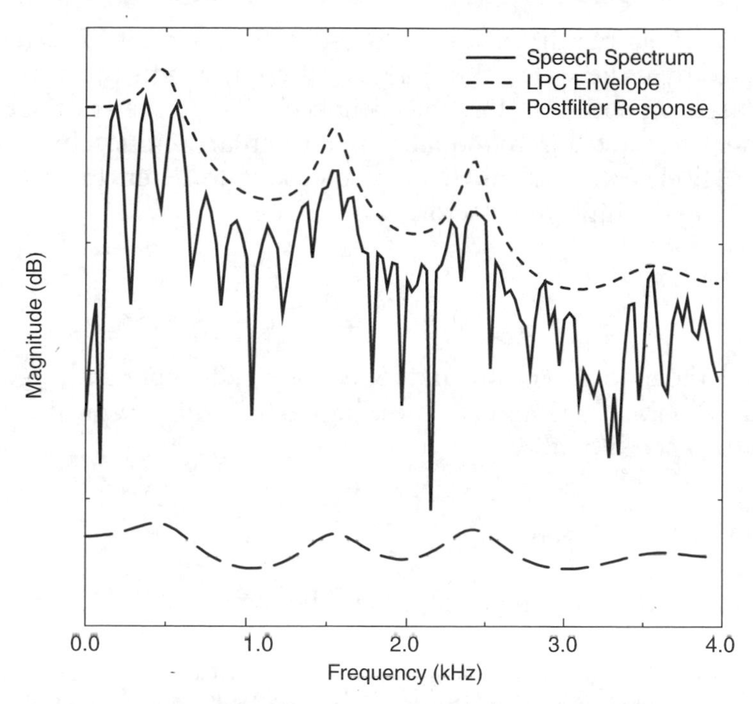

Figure 7.27 Typical response of APF together with the original LPC envelope and speech spectrum ($\mu = 0.3, \beta = 0.6$, and $\alpha = 0.9$)

sample-by-sample normalization, block-wise normalization can be used, i.e. sum the values of $\delta_o^2(n)$ and $\delta_p^2(n)$ for a block and use the average. Small block sizes (e.g. 10 samples) generally produce indistinguishable results. The effect of the APF of Figure 7.26 can be seen in Figure 7.27 alongside a typical example of the original LPC envelope.

Generally, the following parameter ranges have been found to give reasonable subjective results:

$$0.2 \leq \mu \leq 0.4 \tag{7.117}$$

$$0.5 \leq \beta \leq 0.7 \tag{7.118}$$

$$0.8 \leq \alpha \leq 0.9 \tag{7.119}$$

The factor μ controls the 'brightness' of the speech, and hence larger values tend to bring in more high-frequency background noise. The factors β and α control the degree of spectral filtering, and the difference between the parameters determines the filtering effect. Subjectively, large differences give quieter speech, but this is usually accompanied by an unnatural 'deep' voice effect. Applying APF with the correct subjectively-selected control parameters to the coders produces significant subjective noise reduction with almost negligible distortion in the speech. The post-filtered speech is characterized by its lack of background noise components (quiet room effect) and increased smoothness for voiced speech. For lower-rate CELPs, this enhancement to the subjective quality is particularly noticeable: the speech sounds much cleaner and much more pleasant to listen to. As suggested in [17], making the high-pass factor, μ, adaptive as

$$\mu = \varepsilon |k_1| \tag{7.120}$$

where $|k_1|$ is the modulus of the first reflection coefficient computed from the quantized LP parameters and ε is a tuning factor with a typical value of 0.3, improves the speech quality.

7.4 Summary

Analysis by synthesis coding of speech in the form of MPLPC and CELP has been very popular for the past couple of decades. At bit rates of 6 kb/s and above they produce good performance and the various versions reported in the literature differ mainly on the way the secondary (codebook) excitation is generated or represented. In early days, random Gaussian numbers were used to populate the codebooks, but they were complex to store and search and did not produce the best quality. The use of vector sum excitation improved the situation both in terms of the cost of implementation and the overall speech quality. However the most successful CELP coders have been produced after the invention of algebraic codebooks. ACELPs are currently being used in many international standards.

Although ACELPs have been very dominant at bit rates of 6 kb/s and above, they rely heavily on objective measures (although perceptual weighting is used) and as the bit rate is lowered their quality deteriorates rapidly. It would therefore be very difficult to produce a toll-quality 4 kb/s CELP coder unless significant modifications are made to the basic structure described above.

Bibliography

[1] B. Atal and M. Schroeder (1970) 'Adaptive predictive coding of speech signals', in *Bell Sys. Technical Journal*, pp. 1973–87. October 1970.

[2] M. Schroeder and B. Atal (1979) 'Predictive coding of speech signals and subjective error criteria', in *IEEE Trans. on Acoust., Speech and Signal Processing*, 27:247–54.

[3] R. Zelinski and P. Noll (1977) 'Adaptive transform coding of speech signals', in *IEEE Trans. on Acoust., Speech and Signal Processing*, 25(4):299–309.

[4] J. Tribolet and R. Crochiere (1979) 'Frequency domain coding of speech', in *IEEE Trans. On Acoust., Speech and Signal Processing*, 27:512–30, October.

[5] N. Gouvianakis and C. Xydeas (1987) 'Advances in analysis-by-synthesis LPC speech coders', in *Journal of IERE*, 57:272–86.

[6] P. Kroon and E. Deprettere (1988) 'A class of analysis-by-synthesis predictive coders for high quality speech coding at rates between 4.8 and 16 kbit/s', in *IEEE Journal on Selected Areas in Communications*, 6(2):353–363.

[7] L. Rabiner and R. Schafer (1978) *Digital Processing of Speech Signals*. Englewood Cliffs, NJ: Prentice-Hall

[8] B. Atal and J. Remde (1982) 'A new model of LPC excitation for producing natural-sounding speech at low bit rates', in *Proc. of Int. Conf. on Acoust., Speech and Signal Processing*, pp. 614–17.

[9] J. Chung and R. Schafer (1989) 'A 4.8 Kbit/s homomorphic vocoder using analysis by synthesis excitation analysis', in *Proc. of Int. Conf. on Acoust., Speech and Signal Processing*, pp. 144–7.

[10] M. Schroeder and B. Atal (1985) 'Code excited linear prediction (CELP): high quality speech at very low bit rates', in *Proc. of Int. Conf. on Acoust., Speech and Signal Processing*, pp. 937–40. Tampa, FL

[11] R. Rose and T. Barnwell (1986) 'The self-excited vocoder-alternative approach to toll quality at 4800 bits/s', in *Proc. of Int. Conf. on Acoust., Speech and Signal Processing*, pp. 453–6.

[12] S. Singhal and B. Atal (1989) 'Amplitude optimization and pitch prediction in multipulse coders', in *IEEE Trans. On Acoust., Speech and Signal Processing*, 37(3):317–27.

[13] P. Kroon, E. Deprettere, and R. Sluyter (1986) 'Regular-pulse excitation: A novel approach to effective and efficient multipulse coding of speech', in *IEEE Trans. on Acoust., Speech and Signal Processing*, 34:1054–63.

[14] Offer, D. Malah, and A. Dembo (1989) 'A unified framework for LPC excitation representation in residual speech coders', in *Proc. of Int. Conf. on Acoust., Speech and Signal Processing*, pp. 41–4.

[15] J. H. Chen (1990) 'High quality 16 kbit/s speech coding with a one-way delay less than 2 ms', in *Proc. of Int. Conf. on Acoust., Speech and Signal Processing*, pp. 453–6.

[16] R. P. Ramachandran and P. Kabal (1989) 'Pitch prediction filters in speech coding', in *IEEE Trans. On Acoust., Speech and Signal Processing*, 37:467–78.

[17] W. B. Kleijn, D. J. Krasinski, and R. H. Ketchum (1988) 'Improved speech quality and efficient vector quantisation in SELP', in *Proc. of Int. Conf. on Acoust., Speech and Signal Processing*, pp. 155–8.

[18] P. Kroon and B. Atal (1990) 'Pitch predictors with high temporal resolution', in *Proc. of Int. Conf. on Acoust., Speech and Signal Processing*, pp. 661–4. Albuquerque, NM, USA
[19] P. Kabal and R. P. Ramachandran (1988) 'Joint solutions for formant and pitch predictors in speech processing', in *Proc. of Int. Conf. on Acoust., Speech and Signal Processing*, pp. 315–18.
[20] P. Dymarski, N. Moreau, and A. Vigier (1990) 'Optimal and sub-optimal algorithms for selecting the excitation in linear predictive coders', in *Proc. of Int. Conf. on Acoust., Speech and Signal Processing*, pp. 485–8.
[21] N. Ireton and C. Xydeas (1989) 'On improving VEC through the use of spherical lattice codebooks', in *Proc. of Int. Conf. on Acoust., Speech and Signal Processing*, pp. 57–60.
[22] R. Salami (1989) 'Binary CELP: New approach to CELP coding of speech without codebooks', in *IEE Electronics Letters*, 26(6):401–403.
[23] J. Campbell *et al.* (1990) 'The proposed Federal Standard 1016 4800 bit/s voice coder: CELP', in *Speech Technology*, pp. 58–64. April 1990.
[24] I. Gerson and M. Jasiuk (1990) 'Vector sum excited linear prediction (VSELP) speech coding at 4.8 kbps', in *Proc. of Int. Mobile Satellite Conf*, pp. 678–83. Ottawa
[25] J.-P. Adoul, P. Mabilleau, M. Delprat, and S. Morissette (1987) 'Fast CELP coding based on algebraic codes', in *Int. Conf. on Acoust., Speech and Signal Processing*, pp. 1957–60.
[26] C. Laflamme, J.-P. Adoul, H. Su, and S. Morissette (1990) 'On reducing computational complexity of codebook search through the use of algebraic codes', in *Int. Conf. on Acoust., Speech and Signal Processing*, pp. 177–80.
[27] Y. Shoham (1990) 'Constraint-stochastic excitation coding of speech at 4.8 kb/s', in *Proc. of Int. Conf. on Spoken Language Processing*, pp. 645–8. Kobe
[28] N. Jayant and V. Ramamoorthy (1986) 'Adaptive postfiltering of 16 kbit/s ADPCM speech', in *Proc. of Int. Conf. on Acoust., Speech and Signal Processing*, pp. 829–32.
[29] J. Chen and A. Gersho (1987) 'Real-time vector APC speech coding at 4800 b/s with adaptive postfiltering', in *Proc. of Int. Conf. on Acoust., Speech and Signal Processing*, pp. 2185–8. Dallas, USA
[30] J. H. Chen and A. Gersho (1987) 'Real-time VAPC Speech Coding at 4800 Bit/s with Adaptive Postfiltering', in *icassp*, pp. 2185–8.

8

Harmonic Speech Coding

8.1 Introduction

A general sinusoidal analysis and synthesis concept was introduced by McAulay [1] when he developed the Sinusoidal Transform Coder (STC) [2] to demonstrate the applicability of the technique in low bit-rate speech coding. Sinusoidal coding does not restrict the component sinusoids of the synthesized speech to be harmonics of the fundamental frequency. The frequency tracks of the sinusoids may vary independently of each other. However in harmonic coding the higher frequency sinusoids are restricted to be integer multiples of the fundamental frequency [3]. Therefore harmonic coding can be seen as a subset of a generalized sinusoidal transform coding. At low bit-rates, STC also restricts the frequency tracks to be harmonics of the fundamental frequency, and deduces the harmonic phases at the decoder, simply because the available bits are not sufficient to encode the large number of parameters of the general sinusoidal representation.

The STC was introduced as an alternative to the source filter model, and its analysis and synthesis was directly applied to the original speech signal. The binary voicing decision of the source filter model is one of its major limitations. The STC employs a more general mixed-voicing scheme by separating the speech spectrum into voiced and unvoiced components, using a voicing transition frequency above which the spectrum is declared unvoiced. However, one of the most recent harmonic coders operates in the LPC residual domain, i.e. Split Band LPC (SB-LPC) [4]. SB-LPC replaces the binary excitation of the source-filter model with a more general mixed excitation, and filters the excitation signal using an LPC filter. The LPC residual has a simpler phase spectrum than the original speech. The residual harmonic phases

Digital Speech. A. Kondoz
 ISBN 0-470-87007-9 (HB)

can be approximated by using the integrals of the component frequencies. Moreover, LPC models the large variation in the speech magnitude spectrum and simplifies the harmonic amplitude quantization.

8.2 Sinusoidal Analysis and Synthesis

Figure 8.1 depicts block diagrams of the sinusoidal analysis and synthesis processes introduced by McAulay. The speech spectrum is estimated by windowing the input speech signal using a Hamming window and then computing the Discrete Fourier Transform (DFT). The frequencies, amplitudes, and phases corresponding to the peaks of the magnitude spectrum become the model parameters of the sinusoidal representation. Employing a pitch-adaptive analysis window length of two and a half times the average pitch improves the accuracy of peak estimation. The synthesizer generates the sine waves corresponding to the estimated frequencies and phases, and modulates them using the amplitudes. Then all the sinusoids are summed to produce the synthesized speech. The block edge effects are smoothed out by applying overlap and add, using a triangular window. Overlap and add is effectively a simple interpolation technique and, in sinusoidal synthesis, it requires parameter update rates of at least every 10–15 ms for good quality speech synthesis. At lower frame rates the spectral peaks need to be properly aligned between the analysis frames to form frequency tracks. The amplitudes of the frequency tracks are linearly interpolated, and the instantaneous phases are generated using a cubic polynomial [1] as shown in Figure 8.2.

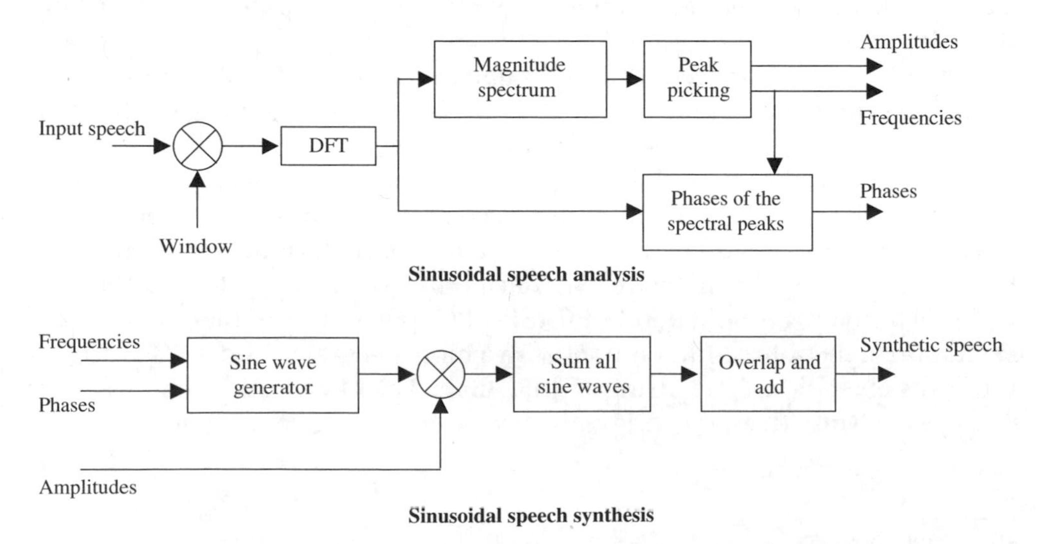

Figure 8.1 General sinusoidal analysis and synthesis

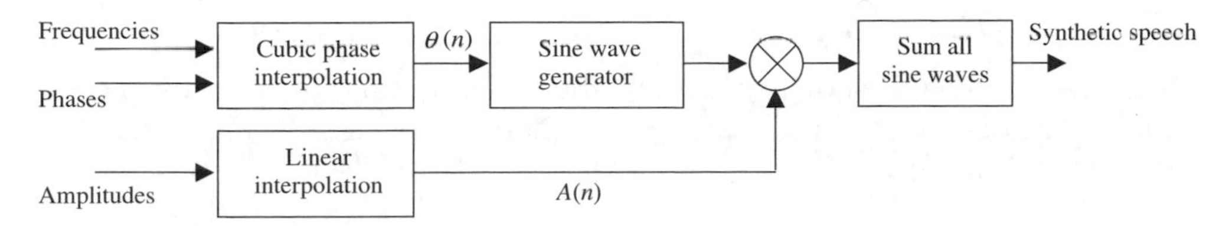

Figure 8.2 Sinusoidal synthesis with matched frequency tracks

8.3 Parameter Estimation

Low bit-rate sinusoidal coders estimate the amplitudes at the harmonics of the fundamental frequency. At low bit-rates, the harmonic phases are not transmitted. Instead the phases are deduced from the spectral envelope on the assumption that it is the gain response of a minimum phase transfer function and added to the integrals of the component frequencies. STC implements the harmonic phases explicitly and LPC-based coders implement the phases implicitly through the time-domain LPC synthesis filter. Improved multi-band excitation (IMBE) coders do not use any kind of phase information and the phases are evolved as the integrals of the component harmonic frequencies. Restricting the component frequencies to the harmonics and modelling the phases at the decoder is well suited for stationary voiced segments of speech. However, in general, the speech signal is not stationary voiced and consists of a mixture of voiced and unvoiced segments. When those segments are synthesized with the phase models described above, the synthesized speech sounds buzzy. In order to remove this 'buzzyness' the concept of frequency-domain voicing was introduced into low bit-rate harmonic coders [5]. Frequency-domain voicing allows the synthesis of mixed voiced signals, by separating the speech spectrum into frequency bands marked as either voiced or unvoiced.

Frequency-domain voicing decisions are usually made for each harmonic of the speech spectrum. Therefore, an accurate pitch estimate is a prerequisite of harmonic amplitude and voicing determination. The frequency-domain voicing determination techniques based on spectral matching need a high precision pitch estimate for good performance. A small error in the pitch will cause large deviations at the high frequency harmonics, and subsequent declaration of them as unvoiced. Furthermore, female voices with short pitch periods are more sensitive to small pitch error. In order to reduce the complexity of a high-precision pitch estimation, an initial pitch estimate is usually further refined by performing a limited search around the initial estimate. Having determined an accurate pitch the harmonic coding usually proceeds with voicing and spectral amplitude estimation processes.

8.3.1 Voicing Determination

There are many ways of performing the voicing classification of speech, which was discussed in Chapter 6, but here we briefly summarize two common techniques.

Multi-Band Approach

Harmonic voicing is estimated by computing the normalized mean squared error of a synthetic voiced spectrum, $\hat{S}_w(\omega, \omega_0)$, with respect to the speech spectrum, $S_w(\omega)$, and comparing it against a threshold function for each harmonic band [6]. The normalized mean squared error, D_k, of the kth harmonic band is given by,

$$D_k = \frac{\displaystyle\int_{(k-0.5)\omega_0}^{(k+0.5)\omega_0} \left[S_w(\omega) - \hat{S}_w(\omega, \omega_0) \right]^2 d\omega}{\displaystyle\int_{(k-0.5)\omega_0}^{(k+0.5)\omega_0} S_w^2(\omega)\, d\omega} \quad \text{for} \quad k = 1, 2, \ldots, K \tag{8.1}$$

where $K = \lfloor \pi / \omega_0 \rfloor$ and ω_0 is the normalized fundamental frequency. Figure 8.3 illustrates D_k values of two speech spectra with the corresponding synthetic spectra. If D_k is below the threshold function, i.e. a small error and a good spectral match, the kth band is declared voiced. The initial multi-

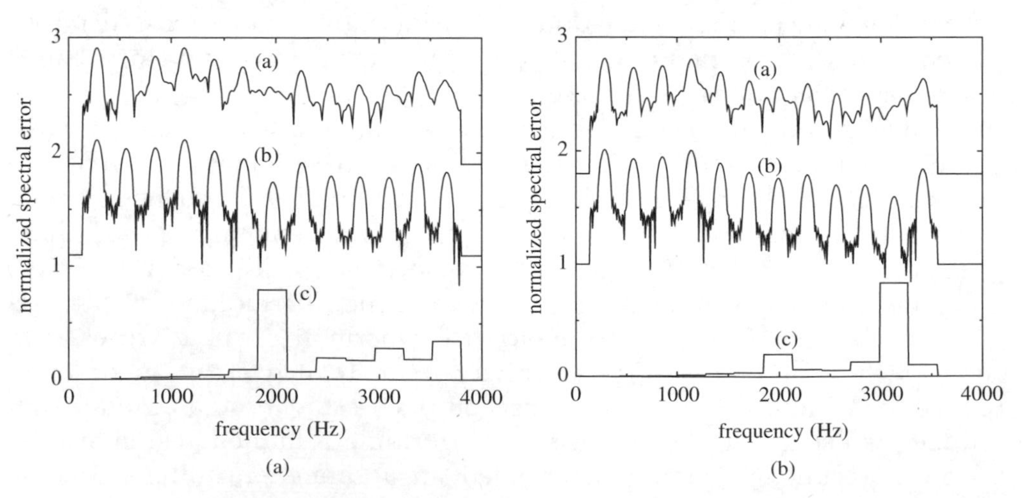

Figure 8.3 Two speech spectra: (a) original spectrum $S_w(\omega)$, (b) synthetic spectrum $\hat{S}_w(\omega, \omega_0)$, and (c) normalized D_k

band excitation (MBE) coders used a constant threshold for all the bands. However the most recent versions use several heuristic rules to obtain a better performance [7], e.g. as the frequency increases the threshold function is decreased, if the same band of the previous frame was unvoiced, if the high-frequency energy exceeds the low-frequency energy, and if the speech energy approaches the energy of the background noise.

Sinusoidal Model Approach

McAulay *et al.* proposed a different voicing determination technique for his sinusoidal transform coder (STC) [2]. The speech spectrum is divided into two bands, determined by a voicing transition frequency above which the spectrum is declared unvoiced. This method estimates the similarity between the harmonically-synthesized signal, $\hat{s}(n, \omega_0)$, and the original speech signal $s(n)$. The signal to noise ratio (SNR), δ, between $s(n)$ and $\hat{s}(n, \omega_0)$ is given by,

$$\delta = \frac{\sum_{n=0}^{N-1} s^2(n)}{\sum_{n=0}^{N-1} \left[s(n) - \hat{s}(n, \omega_0)\right]^2} \tag{8.2}$$

where N is the analysis frame length and $\hat{s}(n, \omega_0)$ is given by

$$\hat{s}(n, \omega_0) = \sum_{l=1}^{K(\omega_0)} \overline{A}_l \exp\left(jnl\omega_0 + j\theta_l\right) \tag{8.3}$$

where the harmonic amplitudes, $\overline{A}_l$, are obtained from the spectral envelope and θ_l are the harmonic phases. McAulay simplified equation (8.2) for reduced computational complexity, and the simplified δ is given by,

$$\delta = \frac{\sum_{l=1}^{L} A_l^2}{\sum_{l=1}^{L} A_l^2 - 2N\rho(\omega_0)} \tag{8.4}$$

where A_l are the harmonic-frequency spectral amplitudes of the original signal as shown below,

$$s(n) = \sum_{l=1}^{L} A_l \exp\left(jn\omega_l + j\phi_l\right) \tag{8.5}$$

and $\rho(\omega_0)$ is given by,

$$\rho(\omega_0) = \sum_{l=1}^{K(\omega_0)} \overline{A}_l \left\{ \max_l \left[A_l D(\omega_l - k\omega_0) \right] - \frac{1}{2} \overline{A}_l \right\} \tag{8.6}$$

where $K(\omega_0) = \lfloor \pi / \omega_0 \rfloor$,

$$D(\omega_l - k\omega_0) = \frac{\sin\left(2\pi \frac{\omega_l - k\omega_0}{\omega_0}\right)}{2\pi \frac{\omega_l - k\omega_0}{\omega_0}} \quad \text{for} \quad \left|\omega_l - k\omega_0\right| \leq \frac{\omega_0}{2} \tag{8.7}$$

and $D(\omega_l - k\omega_0) = 0$ otherwise.

The voicing level (probability), $L_v(\delta)$ (i.e. the ratio of the voiced bandwidth to the speech bandwidth, $0 \leq L_v(\delta) \leq 1$), is defined as,

$$L_v(\delta) = \begin{cases} 1 & \delta > 13\,dB \\ \frac{1}{9}(\delta - 4) & 4\,dB \leq \delta \leq 13\,dB \\ 0 & \delta < 4\,dB \end{cases} \tag{8.8}$$

The advantage of estimating the voicing for independent bands is that it essentially removes the spectral tilt, i.e. all the components are equally weighted. When the voicing is based on a single metric, i.e. δ, the large amplitudes contribute more to the overall decision. If they have been corrupted by background noise, it may result in a large voicing error [2]. Therefore, the voicing estimates based on independent bands are more robust against background noise.

8.3.2 Harmonic Amplitude Estimation

The harmonic coding algorithms require the spectral amplitudes of the harmonics, which can be estimated in a number of ways.

Peak-picking of the Magnitude Spectrum

Harmonic amplitudes may be estimated by simple peak-picking of the magnitude spectrum and searching for the largest peak in each harmonic band. The peak amplitude value, $S_w(m_k)$ should be normalized by a factor depending on the window function used, as follows:

$$A_k = \frac{|S_w(m_k)|}{\kappa} \quad \text{for} \quad -\frac{\omega_0}{2} < \frac{2\pi}{N} m_k - k\omega_0 < \frac{\omega_0}{2} \quad \text{and} \quad k = 1, 2, \ldots, K \tag{8.9}$$

where ω_0 is the normalized fundamental frequency, $K = \lfloor \pi / \omega_0 \rfloor$, $\kappa = \sum_{n=0}^{N-1} w(n)$, $w(n)$ is the window function, N is the length of the window, and $S_w(m)$, the windowed speech spectrum, is given by,

$$S_w(m) = \sum_{n=0}^{N-1} s(n)\, w(n)\, e^{-j\frac{2\pi}{N}mn} \quad \text{for } m = 0, 1, 2, \ldots, N \tag{8.10}$$

Spectral Correlation

Harmonic amplitudes may be estimated by computing the normalized cross-correlation between the harmonic lobes of the speech spectrum and the main lobe of the window spectrum. This method is based on the fact that the spectrum of the windowed speech is equivalent to the convolution between the speech spectrum and the window spectrum. It is also assumed that the speech signal is stationary during the windowed segment and the spectral leakage due to the side lobes of the window spectrum is negligible.

$$A_k = \frac{\sum_{m=a_k}^{b_k-1} S_w(m)\, W^*(2\pi m/N - k\omega_0)}{\sum_{m=a_k}^{b_k-1} W^2(2\pi m/N - k\omega_0)} \quad \text{for } k = 1, 2, \ldots, K \tag{8.11}$$

where $a_k = \max\left[\left\lceil \frac{N}{2\pi}\left(k - \frac{1}{2}\right)\omega_0 \right\rceil, 0\right]$ and $b_k = \min\left[a_{k+1}, N/2\right]$, and $W(\omega)$ is the spectrum of the window function, given by,

$$W(\omega) = \sum_{n=0}^{N-1} w(n)\, e^{-j\omega n} \tag{8.12}$$

In practice, $W(\omega)$ is computed with a high-resolution FFT, e.g. 2^{14} samples, by zero-padding the window function, and stored in a lookup table. The high-resolution FFT is required because, in general, the spectral samples m of $S_w(m)$ do not coincide with the harmonic locations, $k\omega_0$, of the fundamental frequency. Hence $W(\omega)$ is shifted to the harmonic frequency and down-sampled to coincide with the corresponding spectral samples of $S_w(m)$, as shown in equation (8.11). $W(\omega)$ is pre-computed and stored in order to reduce the computational complexity.

The spectral cross-correlation-based amplitude estimation gives the optimum gain of the harmonic lobes with respect to the main lobe of the window spectrum, hence it is a more accurate estimate than the simple peak-picking.

However the cross-correlation-based method has a higher complexity and requires a high-precision pitch estimate.

The unvoiced amplitudes are calculated as the *rms* spectral energy over the unvoiced spectral bandwidth, given by,

$$A_{k\_uv} = \frac{1}{\kappa}\sqrt{\frac{\sum_{m=a_k}^{b_k} S_w^2(m)}{b_k - a_k}} \tag{8.13}$$

The harmonic amplitude estimation techniques described may be applied to either the speech spectrum or the LPC residual spectrum.

8.4 Common Harmonic Coders

This section describes three examples of low bit-rate harmonic coders: sinusoidal transform coding (STC) [2], improved multi-band excitation (IMBE) [8], and split-band linear predictive coding (SB-LPC) [4]. The STC and IMBE apply sinusoidal analysis and synthesis techniques to the original speech signal and SB-LPC uses the LPC residual signal. All three examples restrict the synthesis of sinusoidal components to be harmonics of the fundamental frequency.

8.4.1 Sinusoidal Transform Coding

The sinusoidal transform coding (STC) operating at 4.8 kb/s divides the speech spectrum into two voicing bands using the sinusoidal model approach described in Section 8.3.1. The lower part of the spectrum, which is declared as voiced, is synthesized as follows:

$$\hat{s}_v(n) = \sum_{l=1}^{L_v} \overline{A}(l\omega_0^k)\exp\left(jl\phi_0(n) + j\phi_s\left(l\omega_0^k\right)\right) \quad \text{for } -N/2 \le n \le N/2 \tag{8.14}$$

where

$$\phi_0(n) = n\omega_0^k + \phi_0^k \tag{8.15}$$

and

$$\phi_0^k = \phi_0^{k-1} + \left(\omega_0^{k-1} + \omega_0^k\right)N'/2 \tag{8.16}$$

where $N+1$ is the frame length, ω_0^k is the normalized fundamental frequency of the kth frame, N' is the duration between the analysis points, $\overline{A}(\omega)$ is the spectral envelope obtained by interpolating the selected peaks of the magnitude spectrum, $\phi_s(\omega)$ is the phase spectrum derived from the spectral envelope on the assumption that it is the gain response of a minimum phase transfer function, and L_v is the harmonic just below the voicing transition frequency.

The upper part of the spectrum, which is declared as unvoiced, is synthesized as follows:

$$\hat{s}_{uv}(n) = \sum_{l=L_v+1}^{K\left(\omega_0^k\right)} \overline{A}\left(l\omega_0^k\right) \exp\left(jl\phi_0(n) + j\phi_s\left(l\omega_0^k\right) + jU[-\pi,\pi]\right) \tag{8.17}$$

where $K(\omega_0^k) = \left\lfloor \pi \middle/ \omega_0^k \right\rfloor$ and $U[-\pi,\pi]$ denotes a uniformly distributed random variable in the range $-\pi$ and π. When a frame is fully unvoiced the pitch estimate is meaningless and pitch frequencies greater than 150 Hz may degrade the perceptual quality of unvoiced speech. In order to synthesize the noise-like unvoiced speech with adequate quality, the number of sinusoids with random phases should be sufficiently large. Therefore, the pitch frequency is set to 100 Hz for unvoiced speech. The synthesized speech of the kth frame is then given by,

$$\hat{s}(n) = \hat{s}_v(n) + \hat{s}_{uv}(n) \tag{8.18}$$

The overlap and add method is used with a triangular window to produce the final speech output. Therefore, the frame length is equal to twice the duration between the analysis points, i.e. $N = 2N'$. The frequency response of the spectral envelope is given by,

$$H(\omega) = \overline{A}(\omega) \exp\left(j\phi_s(\omega)\right) \tag{8.19}$$

which is approximated by an all-pole model,

$$H(\omega) \cong \frac{g}{1 - \sum_{i=1}^{p} a_i z^{-i}} \quad \text{for} \quad |z| = 1 \tag{8.20}$$

where g is the gain and a_i are the predictor coefficients. The conventional time-domain all-pole LPC analysis is performed on the original speech signal and the maximum filter order is usually limited to half the smallest pitch period. The limitation is imposed so that the LPC models the formant spectral envelope, since LPC filters with a large number of taps tend to resolve the harmonic structure. However in the case of STC, all-pole modelling is applied to the estimated spectral envelope. Hence, the filter order is not restricted and can be increased depending only on the desired accuracy of the spectral envelope and the bit rate. The 4.8 kb/s STC uses a 14th-order all-pole model and quantizes the predictor coefficients in the LSF domain. In addition to the LSFs, the STC transmits gain, pitch, and voicing.

8.4.2 Improved Multi-Band Excitation, INMARSAT-M Version

Improved multi-band excitation (IMBE) operating at 4.15 kb/s for INMARSAT-M divides the speech spectrum into several voiced and unvoiced frequency bands, using the multi-band approach described in Section 8.3.1. However, IMBE makes the voicing decisions for groups of three harmonics and a single bit is allocated for each group. The total number of voicing bits B_v is limited to a maximum of 12 and the harmonics beyond the coverage of voicing are declared unvoiced. The refined pitch is transmitted using eight bits. The frame length is 20 ms giving 83 bits per frame at 4.15 kb/s and the remaining bits, i.e. $83-8-B_v$, are allocated for spectral amplitudes. The voiced amplitudes are estimated using equation (8.11) and the unvoiced amplitudes are estimated using equation (8.13). The voiced bands are synthesized as follows:

$$\hat{s}_v(n) = \sum_{k=voiced} A_k \cos\left(k\phi_0(n)\right) \quad \text{for} \quad n = 0,1,2,\ldots,N-1 \tag{8.21}$$

where N is the frame length and the fundamental phase evolution, $\phi_0(n)$, is defined by the following equations:

$$\phi_0(n) = \phi_0(n-1) + \omega_0(n) \tag{8.22}$$

$$\omega_0(n) = \frac{1}{N}(N-n)\,\omega_0^{l-1} + n\omega_0^{l} \tag{8.23}$$

where $\phi_0(-1)$ is $\phi_0(N-1)$ of the previous frame and ω_0^l is the normalized fundamental frequency estimated at the end of the lth frame. The amplitudes of the voiced harmonics are linearly interpolated between the analysis points. If the corresponding harmonic of one analysis point does not exist or is declared unvoiced then its amplitude is set to zero and the harmonic frequency stays constant (set to the frequency of the existing voiced harmonic). However if the pitch estimate is not steady, neither the pitch nor the amplitudes are interpolated for any harmonics; instead overlap and add method is used.

The unvoiced component is synthesized using filtered white Gaussian noise. White noise is generated in the time domain and transformed into the frequency domain; the bands corresponding to the voiced components are set to zero; and the unvoiced bands are scaled according to the unvoiced gain factors. The inverse Fourier transform of the modified spectrum gives the unvoiced component, $\hat{s}_{uv}(n)$, which is produced using the overlap and

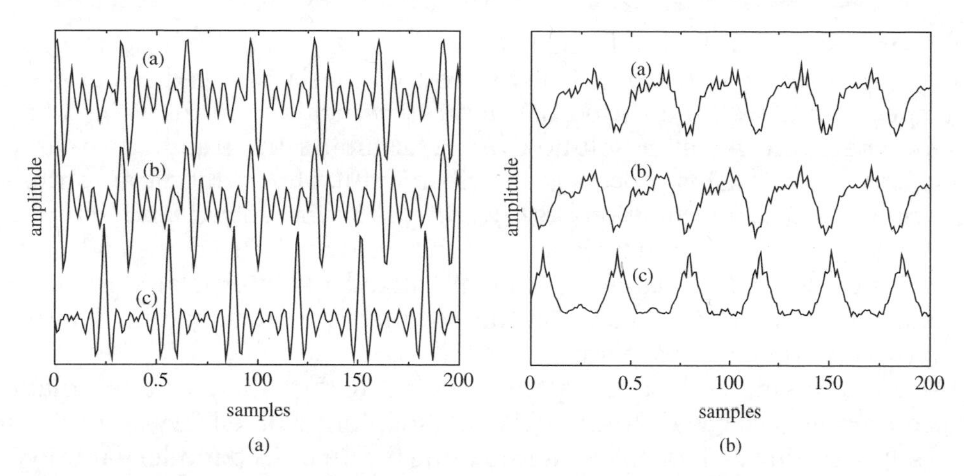

Figure 8.4 Harmonic speech synthesis: (a) original speech, (b) original harmonic phases, and (c) IMBE

add method with the unvoiced part of the preceding frame. The synthesized speech $\hat{s}(n)$ is then given by,

$$\hat{s}(n) = \hat{s}_v(n) + \hat{s}_{uv}(n) \tag{8.24}$$

An interesting feature of the IMBE coder is its simple phase model. The fundamental phase is computed as the integral of the linearly-interpolated pitch frequency, and the multiples of the fundamental phase are used as the harmonic phases. The effect of this phase model is illustrated in Figure 8.4. The coherent phase model used in IMBE concentrates the speech energy at the phase locations corresponding to the multiples of 2π of the fundamental phase. For reference, the speech waveforms synthesized using the original harmonic phases are also shown and they are very similar to the original speech waveforms.

8.4.3 Split-Band Linear Predictive Coding

The split-band linear predictive coding (SB-LPC) coder operating at 4 kb/s employs time-domain LPC filtering and uses a multi-band type of excitation signal. However the excitation signal of SB-LPC consists of only two bands, separated by a frequency marker, below which the spectrum is declared voiced and above which it is declared unvoiced. The estimation of the frequency marker of SB-LPC is different from the technique used in STC. The SB-LPC estimates a voicing decision for each harmonic band using a similar multi-band approach described in Section 8.3.1. The estimated voicing decisions are used to determine the voicing frequency marker, which has eight possible equally-spaced locations in the spectrum, the first being fully unvoiced and the last being fully voiced. One method of deciding the frequency marker is placing it at the end of the last voiced harmonic of the spectrum, i.e. all the voiced harmonics are included in the voiced band of the spectrum. A better solution for determining the frequency marker, based on a soft decision process is described in [9]. The harmonic amplitudes are estimated using equations (8.11) and (8.13) for voiced and unvoiced harmonics respectively, however the LPC residual is used instead of the speech signal. The LPC parameters are quantized and interpolated in the LSF domain. The shape of the harmonic amplitudes is vector-quantized and the gain is scalar-quantized separately.

At the receiving end, speech is synthesized with parameter interpolation based on pitch cycle waveform (PCW). First, intermediate PCWs for the current subframe are generated by interpolating the quantized model parameters of the last and current subframes. The excitation signal $e_i(n)$, $0 \leq n < T_{0,i}$, for the ith PCW is produced as

$$e_i(n+n_i) = \sum_{l=1}^{V_c} A_{e,i}(l)\cos\{l\omega_{0,i}(n-n_i)\} + \sum_{l=V_c+1}^{H} A_{e,i}(l)\cos\{l\omega_{0,i}(n-n_i) + U[-\pi,\pi]\} \quad (8.25)$$

where H is the total number of harmonics, $\omega_{0,i} = 2\pi/T_{0,i}$ and $U[-\pi,\pi]$ denotes a random number with uniform distribution between $-\pi$ and π. The start position n_i for the ith PCW is given by

$$n_i = n_0 + \sum_{j=0}^{i-1} T_{0,j} \quad (8.26)$$

where n_0 is the start position corresponding to the last position of the previous subframe. The interpolated pitch $T_{0,i}$ for the ith PCW is calculated as

$$T_{0,i} = \alpha_i T_0^{(t-1)} + (1-\alpha_i)T_0^{(t)} \quad (8.27)$$

where $T_0^{(t)}$ is the received pitch of the tth subframe. The interpolation factor α_i is defined as

$$\alpha_i = \frac{G^{(t)}N_i}{G^{(t-1)}(N-N_i) + G^{(t)}N_i} \quad (8.28)$$

where N is the subframe size, $G^{(\cdot)}$ is the received gain, and N_i is the PCW position defined by,

$$N_i = n_i + 0.25(T_0^{(t-1)} + T_0^{(t)}) \quad (8.29)$$

The starting position $n_0^{(t+1)}$ for the next subframe is updated as

$$n_0^{(t+1)} = (n_I^{(t)} + T_{0,I}^{(t)})\%N \tag{8.30}$$

where % is the modulo operator and I is the total number of PCWs. The voicing cut-off index, V_c, is given by

$$V_c = \max\{V_c^{(t-1)}, V_c^{(t)}\} \tag{8.31}$$

The interpolated amplitude, $A_{e,i}(l)$, for the lth harmonic is computed as

$$A_{e,i}(l) = \begin{cases} \alpha_i A_e^{(t-1)}(l) + (1-\alpha_i)A_e^{(t)}(l), & \text{if } V^{(t-1)}(l) = V^{(t)}(l), \\ A_e^{(t-1)}(l), & \text{if } V^{(t-1)}(l) = 1 \;\&\; V^{(t)}(l) = 0 \\ A_e^{(t)}(l), & \text{if } V^{(t-1)}(l) = 0 \;\&\; V^{(t)}(l) = 1 \end{cases} \tag{8.32}$$

where $V^{(\cdot)}(l)$ is the voicing information for the lth harmonic and 1 and 0 in the voicing comparison denote voiced and unvoiced, respectively. The LPC coefficient for the ith PCW is interpolated in the same way as, obtaining the interpolated pitch. Finally, the normalized speech signal $\tilde{s}_i(n)$ is reconstructed by exciting the LPC synthesis filter $h_i(n)$ with the signal $e_i(n)$ in equation (8.25), as

$$\tilde{s}_l(n) = e_i(n) * h_i(n) \tag{8.33}$$

where $*$ is the convolution operator. In calculation of $\tilde{s}_i(n)$, the required memory for $e_i(n)$, $n < 0$, can be obtained from $e_{i-1}(n)$ or the excitation signal of the last subframe. The synthesized speech signal $s_i(n)$ for the ith PCW is produced by compensating for the gain as

$$s_i(n) = \sqrt{\frac{T_{0,i}}{\sum_{n=0}^{T_{0,i}-1} \tilde{s}_i^2(n)}} G_i \tilde{s}_i(n) \tag{8.34}$$

where G_i is the interpolated gain based on the relative position of the PCW in the subframe. Concatenation of each PCW in equation (8.34) forms the final speech signal.

The above description of excitation generation is based on the sinusoidal synthesis of voiced and random noise generation of unvoiced parts of the excitation. However, in practice, a DFT-based method (with the DFT size equal to the pitch period), where the unvoiced frequencies would have random phases, can be used to generate both voiced and unvoiced parts jointly [10, 11].

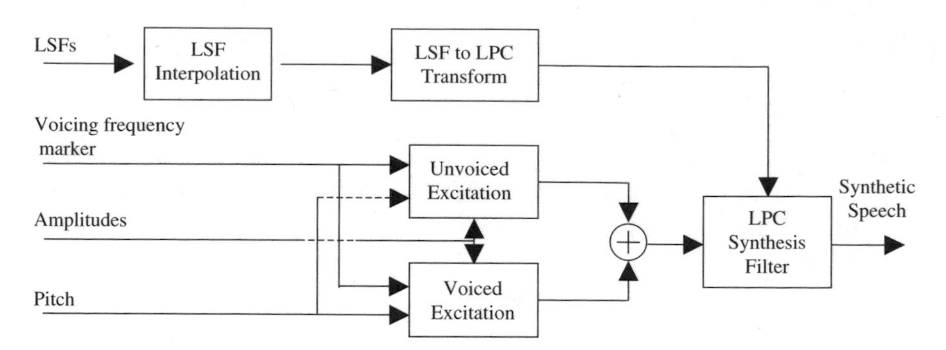

Figure 8.5 SB-LPC decoder

Table 8.1 Bit allocation of 4 kb/s SB-LPC coder for a 20 ms frame

Parameter	Bits
LSFs	23
Pitch	5 + 7
Parity bit	1
Voicing	3 + 3
Gain	5 + 5
Harmonic amplitudes	14 + 14

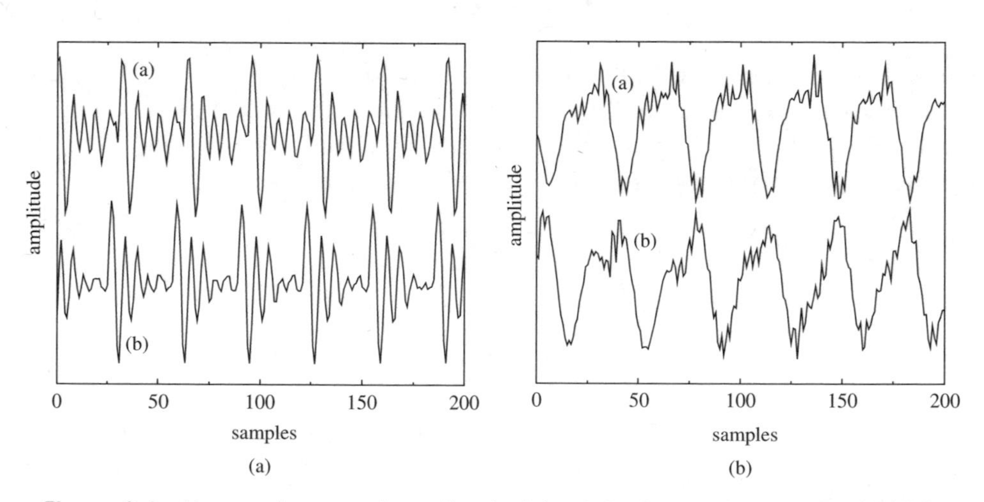

Figure 8.6 Harmonic speech synthesis: (a) original speech, and (b) SB-LPC

A block diagram of the SB-LPC decoder is shown in Figure 8.5 and the bit allocation is shown in Table 8.1. Figure 8.6 illustrates the same waveforms shown in Figure 8.4, but synthesized using the SB-LPC coder. The time-domain LPC filter adds its phase response to the coherent excitation signal of SB-LPC and disperses the energy of the excitation pulses. However the waveform shape of the synthesized speech is different from the original speech.

8.5 Summary

The fundamental sinusoidal speech analysis and synthesis techniques have been briefly discussed in this chapter. The basic sinusoidal model has been modified to reduce the number of parameters in order to adapt it for low bit-rates. At low bit-rates the frequencies of the sinusoids are restricted to be harmonics of the pitch frequency and the harmonic phases are modelled at the decoder. The concept of frequency-domain voicing is introduced to achieve a compromise between the hoarseness and buzzyness of harmonically-synthesized speech.

Three examples of low bit-rate harmonic coders have been presented: sinusoidal transform coding (STC), improved multi-band excitation (IMBE), and split-band linear predictive coding (SB-LPC). One of the main limitations of low bit-rate harmonic coders is their inability to produce adequate quality at the speech transitions.

Bibliography

[1] R. J. McAulay and T. F. Quatieri (1986) 'Speech analysis/synthesis based on a sinusoidal representation', in *IEEE Trans. on Acoust., Speech and Signal Processing*, 34(4):744–54.

[2] R. J. McAulay and T. F. Quatieri (1995) 'Sinusoidal coding', in *Speech coding and synthesis* by W. B. Kleijn and K. K. Paliwal (Eds), pp. 121–74. Amsterdam: Elsevier Science

[3] L. B. Almeida and F. M. Silva (1984) 'Variable frequency synthesis: an improved harmonic coding scheme', in *Proc. of Int. Conf. on Acoust., Speech and Signal Processing*, pp. 27.5.1–4.

[4] I. Atkinson, S. Yeldener, and A. Kondoz (1997) 'High quality split-band LPC vocoder operating at low bit rates', in *Proc. of Int. Conf. on Acoust., Speech and Signal Processing*, pp. 1559–62. May 1997. Munich

[5] J. Makhoul, R. Viswanathan, R. Schwartz, and A. W. F. Huggins (1978) 'A mixed source excitation model for speech compression and synthesis', in *Proc. of Int. Conf. on Acoust., Speech and Signal Processing*, pp. 163–6.

[6] D. Griffin and J. S. Lim (1988) 'Multiband excitation vocoder', in *IEEE Trans. on Acoust., Speech and Signal Processing*, 36(8):1223–35.

[7] A. Kondoz, (1994) *Digital Speech: coding for low bit rate communication systems*. New York: John Wiley & Sons Ltd

[8] DVSI (1991) *INMARSAT-M Voice Codec*, Version 1.7. September 1991. Digital Voice Systems Inc.

[9] I. Atkinson (1997) 'Advanced linear predictive speech compression at 3.0 kbit/s and below', Ph.D. thesis, CCSR, University of Surrey, UK.

[10] T. Wang, K. Koishida, V. Cuperman, A. Gersho, and J. S. Collura (2002) 'A 1200/24000 bps coding suite based on MELP', in *Proc. of IEEE Workshop on Speech Coding*, pp. 90–2.

[11] S. Villette (2001) 'Sinusoidal speech coding for low and very low bit rate applications', Ph.D. thesis, University of Surrey, UK.

9

Multimode Speech Coding

9.1 Introduction

Harmonic coders extract the frequency-domain speech parameters and speech is generated as a sum of sinusoids with varying amplitudes, frequencies and phases. They produce highly intelligible speech down to about 2.4 kb/s [1]. By using the unquantized phases and amplitudes, and by frequent updating of the parameters, i.e. at least every 10 ms, they can even achieve near transparent quality [2]. However this requires a prohibitive bit-rate, unsuitable for low bit-rate applications. For example, the earlier versions of multi-band excitation (MBE) coders (a typical harmonic coder) operated at 8 kb/s with harmonic phase information [3]. However, harmonic coders operating at 4 kb/s and below do not transmit phase information. The spectral magnitudes are transmitted typically every 20 ms and interpolated during the synthesis. The simplified versions used for low bit-rate applications are well suited for stationary voiced segment coding. However at the speech transitions such as onsets, where the speech waveform changes rapidly, the simplified assumptions do not hold and degrade the perceptual speech quality.

Figure 9.1 demonstrates two examples of harmonically-synthesized speech, Figure 9.1a shows a stationary voiced segment and Figure 9.1b shows a transitory speech segment. In both cases, (i) represents the original speech, i.e. 128 kb/s linear pulse code modulation, and (ii) represents the synthesized speech. The synthesized speech is generated using the split-band linear predictive coding (SB-LPC) harmonic coder operating at 4 kb/s [4]. The synthesized waveforms are shifted in the figures in order to compensate for the delay due to look-ahead and the linear phase deviation due to loss of phase information in the synthesis. The SB-LPC decoder predicts the evolution of harmonic phases using the linearly interpolated fundamental frequency, i.e. a quadratic phase evolution function. Low bit-rate harmonic

Digital Speech. A. Kondoz
 ISBN 0-470-87007-9 (HB)

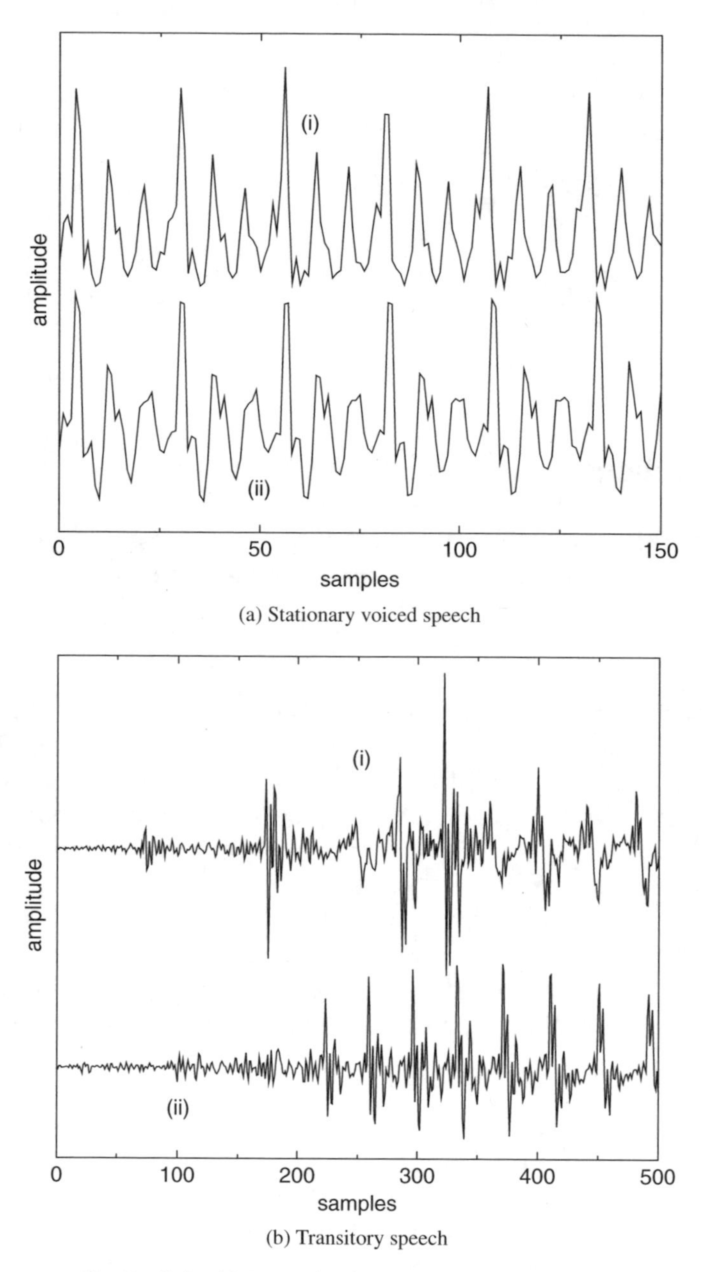

(a) Stationary voiced speech

(b) Transitory speech

Figure 9.1 Harmonically-synthesized speech

coders cannot preserve waveform similarity as illustrated in the figures, since the phase information is not transmitted. However, in the stationary voiced segments, phase information has little importance in terms of the perceptual quality of the synthesized speech. Stationary voiced speech has a strong, slowly-evolving harmonic content. Therefore extracting frequency domain speech parameters at regular intervals and interpolating them in the harmonic synthesis is well suited for stationary voiced segments. However at the transitions, where the speech waveform evolves rapidly, this low bit-rate simplified harmonic model fails. As depicted in Figure 9.1b, the highly nonstationary character of the transition has been smeared by the low bit-rate harmonic model causing reduction in the intelligibility of the synthesized speech.

CELP-type coders, such as ACELP [5, 6], encode the target speech waveform directly and perform relatively better at the transitions. However, at low bit-rates, analysis-by-synthesis (AbS) coders fail to synthesize stationary segments with adequate quality. As the bit rate is reduced, they cannot maintain clear periodicity of the stationary voiced segments [7]. CELP-type AbS coders perform waveform-matching for each frame or subframe and select the best possible excitation vector. This process does not consider the pitch cycles of the target waveform, and consecutive synthesized pitch cycles show subtle differences in the waveform shape. This artifact introduces granular noise into the voiced speech, perceptible up to about 6 kb/s. Preserving the periodicity of voiced speech is essential for high quality speech reproduction. Figure 9.2a shows a stationary voiced segment and 9.2b shows a transitory segment synthesized using ACELP at 4 kb/s. In Figure 9.2a, the consecutive pitch cycles have different shapes, which degrades the slowly-evolving periodicity of voiced speech, compared to Figure 9.1a. Therefore despite the fact that waveform similarity is less in Figure 9.1a, harmonically-synthesized voiced speech is perceptually superior to waveform-coded speech at low bit-rates. Figure 9.2b shows that ACELP can synthesize the highly nonstationary speech transitions better than harmonic coders (see Figure 9.1b). ACELP may also introduce granular noise at the transitions. However, the speech waveform changes rapidly at the transitions, masking the granular noise of ACELP, which is not perceptible down to about 4 kb/s. The above observations suggest a hybrid coding approach, which selects the optimum coding algorithm for a given segment of speech: coding stationary voiced segments using harmonic coding and transitions using ACELP. Unvoiced and silence segments can be encoded with CELP [8] or white-noise excitation.

Harmonic coders suffer from other potential problems such as voicing and pitch errors that may occur at the transitions. The pitch estimates at the transitions, especially at the onsets may be unreliable due to the rapidly-changing speech waveform. Furthermore, pitch-tracking algorithms do not have history at the onsets and should be turned off. Inaccurate pitch estimates also account for inaccurate voicing decisions, in addition to the spectral mismatches due to the nonstationary speech waveform at the transitions. These voicing decision errors declare the voiced bands as unvoiced and increase the hoarseness of synthetic speech. Encoding the transitions using ACELP eliminates those potential problems of harmonic coding.

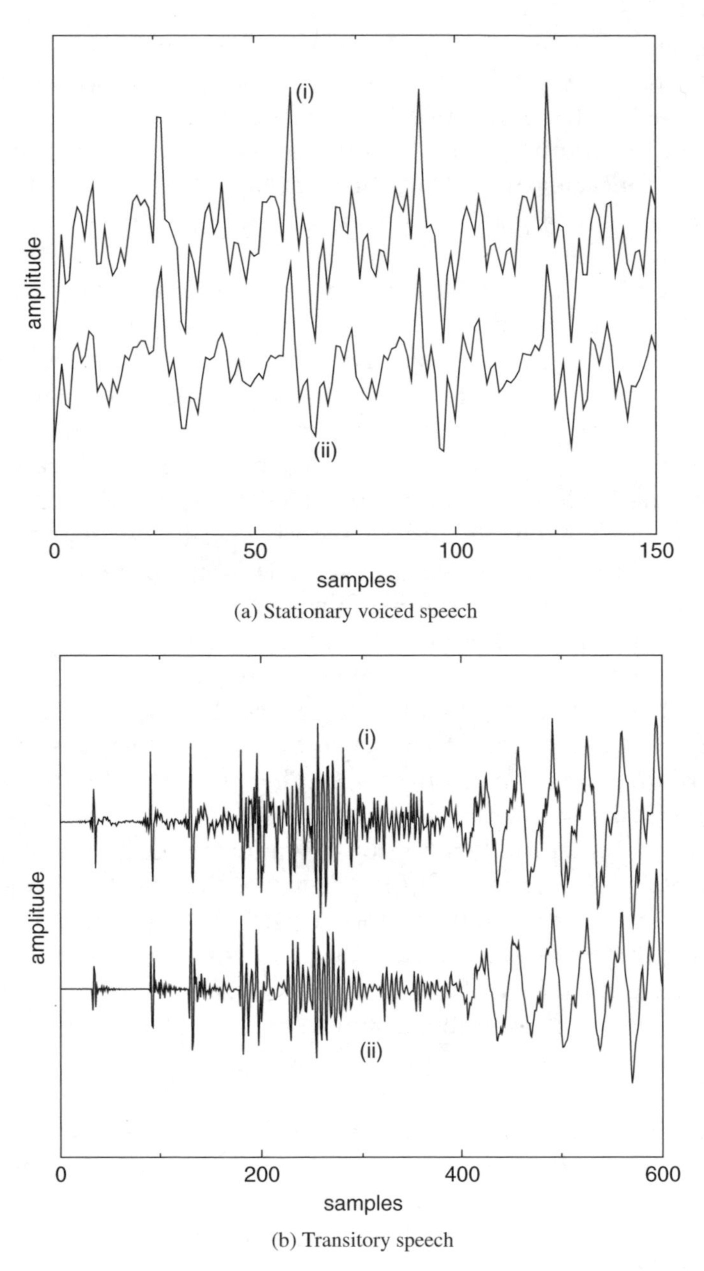

(a) Stationary voiced speech

(b) Transitory speech

Figure 9.2 Speech synthesized using ACELP

9.2 Design Challenges of a Hybrid Coder

The main challenges in designing a hybrid coder are reliable speech classification and phase synchronization when switching between the coding modes. Furthermore, most of the speech-coding techniques make use of a look-ahead and parameter interpolation. Interpolation requires the parameters of the previous frame; when switched from a different mode, those parameters may not be directly available. Predictive quantization schemes also require the previous memory. Techniques which eliminate these initialization/memory problems are required.

9.2.1 Reliable Speech Classification

A voice activity detector (VAD) can be used to identify speech and silence segments [9], while classification of speech into voiced and unvoiced segments can be seen as the most basic speech classification technique. However, there are coders in the literature which use up to six phonetic classes [10]. The design of such a phonetic classification algorithm can be complicated and computationally complex, and a simple classification with two or three modes is sufficient to exploit the relative merits of waveform and harmonic coding methods. The accuracy of the speech classification is critical for the performance of a hybrid coder. For example, using noise excitation for a stationary voiced segment (which should operate in harmonic coding mode) can severely degrade the speech quality, by converting the high-voiced energy of the original speech into noise in the synthesized speech; use of harmonic excitation for unvoiced segments gives a tonal artifact. ACELP can generally maintain acceptable quality for all the types of speech since it has waveform-matching capability. During the speech classification process, it is essential that the above cases are taken into account to generate a fail-safe mode selection.

9.2.2 Phase Synchronization

Harmonic coders operating at 4 kb/s and below do not transmit phase information, in order to allocate the available bits for accurate quantization of the more important spectral magnitude information. They exploit the fact that the human ear is partially phase-insensitive and the waveform shape of the synthesized speech can be very different from the original speech, often yielding negative SNRs. On the other hand, AbS coders preserve the waveform similarity. Direct switching between those two modes without any precautions will severely degrade the speech quality due to phase discontinuities.

9.3 Summary of Hybrid Coders

The hybrid coding concept has been introduced in the LPC vocoder [11], which classifies speech frames into voiced or unvoiced, and synthesizes the excitation using periodic pulses or white noise, respectively. Analysis-by-synthesis CELP coders with dynamic bit allocation (DBA), which adaptively distribute the bits among coder parameters in a given frame while maintaining a constant bit rate, by classifying each frame into a certain mode, have also been reported [12]. However, we particularly focus here on hybrid coders, which combine AbS coding and harmonic coding. The advantages and disadvantages of harmonic coding and CELP, and the potential benefits of combining the two methods have been discussed by Trancoso *et al.* [13]. Improving the speech quality of the LPC vocoder by using a form of multi-pulse excitation [14] as a third excitation model at the transitions has also been reported [15].

9.3.1 Prototype Waveform Interpolation Coder

Kleijn introduced prototype waveform interpolation (PWI) in order to improve the quality of voiced speech [7]. The PWI technique extracts prototype pitch cycle waveforms from the voiced speech at regular intervals of 20–30 ms. Speech is reconstructed by interpolating the pitch cycles between the update points. The PWI technique can be applied either directly to the speech signal or to the LPC residual. Since the PWI technique is not suitable for encoding unvoiced speech segments, unvoiced speech is synthesized using CELP. Even though the motivation behind using two coding techniques is different in the PWI coder (i.e. waveform coding is not used for transitions), it combines harmonic coding and AbS coding. The speech classification of the PWI coder is relatively easier, since it only needs to classify speech into either voiced or unvoiced.

At the onset of a voiced section, the previously estimated prototype waveform is not present at the decoder for the interpolation process. Kleijn suggests three methods to solve this problem:

- Extract the prototype waveform from the reconstructed CELP waveform of the previous frame.
- Set to a single pulse waveform (filtered through LPC) with its amplitude determined from the transmitted information.
- Use a replica of the prototype transmitted at the end of the current synthesis frame.

The starting phase of the pitch cycles at the onsets can be determined at the decoder from the CELP encoded signal. At the offsets, the linear phase deviation between the harmonically synthesized and original speech is measured and the original speech buffer is displaced, such that the AbS coder begins exactly where the harmonic coder ended.

9.3.2 Combined Harmonic and Waveform Coding at Low Bit-Rates

This coder, proposed by Shlomot *et al.*, consists of three modes: harmonic, transition, and unvoiced [16, 17]. All the modes are based on the source filter model. The harmonic mode consists of two components: the lower part of the spectrum or the harmonic bandwidth, which is synthesized as a sum of coherent sinusoids, and the upper part of the spectrum, which is synthesized using sinusoids of random phases. The transitions are synthesized using pulse excitation, similar to ACELP, and the unvoiced segments are synthesized using white-noise excitation.

Speech classification is performed by a neural network, which takes into account the speech parameters of the previous, current, and future frames, and the previous mode decision. The classification parameters include the speech energy, spectral tilt, zero-crossing rate, residual peakiness, residual harmonic matching SNRs, and pitch deviation measures. At the onsets, when switching from the waveform-coding mode, the harmonic excitation is synchronized by shifting and maximizing the cross-correlation with the waveform-coded excitation. At the offsets, the waveform-coding target is shifted to maximize the cross-correlation with the harmonically-synthesized speech, similar to the PWI coder.

9.3.3 A 4 kb/s Hybrid MELP/CELP Coder

The 4 kb/s hybrid MELP/CELP coder with alignment phase encoding and zero phase equalization proposed by Stachurski *et al.* consists of three modes: strongly-voiced, weakly-voiced, and unvoiced [18, 19]. The weakly-voiced mode includes transitions and plosives, which is used when neither strongly-voiced nor unvoiced speech segments are clearly identified. In the strongly-voiced mode, a mixed excitation linear prediction (MELP) [20, 21] coder is used. Weakly-voiced and unvoiced modes are synthesized using CELP. In unvoiced frames, the LPC excitation is generated from a fixed stochastic codebook. In weakly-voiced frames, the LPC excitation consists of the sum of a long-term prediction filter output and a fixed innovation sequence containing a limited number of pulses, similar to ACELP.

The speech classification is based on the estimated voicing strength and pitch. The signal continuity at the mode transitions is preserved by transmitting an 'alignment phase' for MELP-encoded frames, and by using 'zero phase equalization' for transitional frames. The alignment phase preserves the time-synchrony between the original and synthesized speech. The alignment phase is estimated as the linear phase required in the MELP-encoded excitation generation to maximize the cross-correlation between the MELP excitation and the corresponding LPC residual. Zero-phase equalization modifies the CELP target signal, in order to reduce the phase discontinuities, by removing the phase component, which is not coded in MELP. Zero phase equalization is implemented in the LPC residual domain, with a Finite Impulse Response (FIR) filter similar to [22]. The FIR filter coefficients are derived from the smoothed pitch pulse waveforms of the LPC residual signal. For unvoiced frames the filter coefficients are set to an impulse so that the filtering has no effect. The AbS target is generated by filtering the zero-phase-equalized residual signal through the LPC synthesis filter.

9.3.4 Limitations of Existing Hybrid Coders

PWI coders and low bit-rate coders that combine harmonic and waveform coding use similar techniques to ensure signal continuity. At the onsets, the initial phases of the harmonic excitation are extracted from the previous excitation vector of the waveform-coding mode. This can be difficult at rapidly-varying onsets, especially if the bit-rate of the waveform coder is low. Moreover, inaccuracies in the onset synchronization will propagate through the harmonic excitation and make the offset synchronization more difficult. At the offsets, the linear phase deviation between the harmonically-synthesized and original speech is measured and the original speech buffer is displaced, such that the AbS coder begins exactly where the harmonic coder has ended. This method needs the accumulated displacement to be reset during unvoiced or silent segments, and may fail to meet the specifications of a system with strict delay requirements.

Another problem arises when a transition occurs within a voiced speech segment as shown in Figure 9.3, where there are no unvoiced or silent segments after the transition to reset the accumulated displacement. Even though the accumulated displacement can be minimized by inserting or eliminating exactly complete pitch cycles, the remainder will propagate into the next harmonic section. Furthermore, a displacement of a fraction

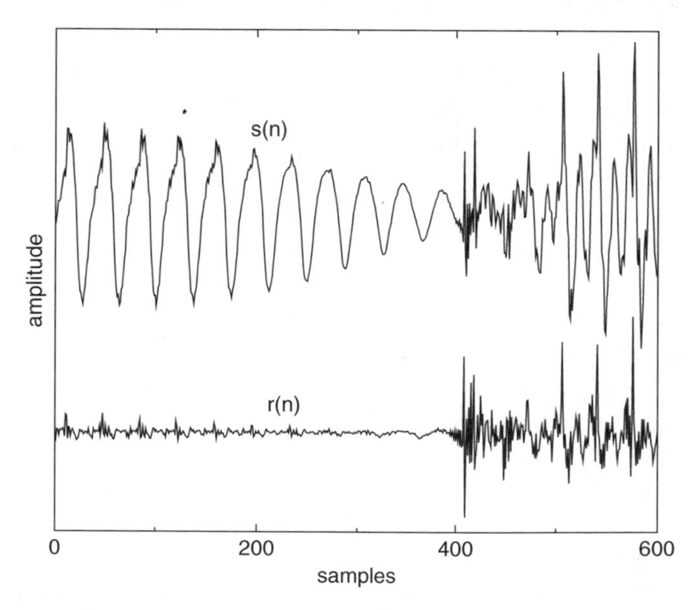

Figure 9.3 A transition within voiced speech

of a sample can introduce audible high frequency distortion, especially in segments with short pitch periods. Consequently, the displacements should be performed with a high resolution. The MELP/CELP coder preserves signal continuity by transmitting an alignment phase for MELP-encoded frames and using zero phase equalization for transitional frames. Zero phase equalization may reduce the benefits of AbS coding by modifying the phase spectrum, and it has been reported that the phase spectrum is perceptually important [23–25]. Furthermore, zero phase equalization relies on accurate pitch pulse position detection at the transitions, which can be difficult.

Harmonic excitation can be synchronized with the LPC residual by transmitting the phases, which eliminates the above difficulties. However this requires a prohibitive capacity making it unsuitable for low bit-rate applications. As a compromise, Katugampala [26] proposed a new phase model for the harmonic excitation called synchronized waveform-matched phase model (SWPM). SWPM facilitates the integration of harmonic and AbS coders, by synchronizing the harmonic excitation with the LPC residual. SWPM requires only two parameters and does not alter the perceptual quality of the harmonically-synthesized speech. It also allows the ACELP mode to target the speech waveform without modifying the perceptually-important phase components or the frame boundaries.

9.4 Synchronized Waveform-Matched Phase Model

The SWPM maintains the time-synchrony between the original and the harmonically-synthesized speech by transmitting the pitch pulse location (PPL) closest to each synthesis frame boundary [27, 28, 26]. The SWPM also preserves sufficient waveform similarity, such that switching between the coding modes is transparent, by transmitting a phase value that indicates the pitch pulse shape (PPS) of the corresponding pitch pulse. PPL and PPS are estimated in every frame of 20 ms. SWPM needs to detect the pitch pulses only in the stationary voiced segments, which is somewhat easier than detecting the pitch pulses in the transitions as in [18]. The SWPM has the disadvantage of transmitting two extra parameters (PPL and PPS) but the bottleneck of the bit allocation of hybrid coders is usually in the waveform-coding mode. Furthermore, in stationary voiced segments the location of the pitch pulses can be predicted with high accuracy, and only an error needs to be transmitted. The same argument applies to the shape of the pitch pulses.

In the harmonic synthesis, cubic phase interpolation [2] is applied between the pitch pulse locations, setting the phases of all the harmonics equal to PPS. This makes the waveform similarity between the original and the synthesized speech highest in the vicinity of the selected pitch pulse locations. However this does not cause difficulties, since switching is restricted to frame boundaries and the pitch pulse locations closest to the frame boundaries are selected. Furthermore, SWPM can synchronize the synthesized excitation and the LPC residual with fractional sample resolutions, even without up-sampling either of the waveforms.

9.4.1 Extraction of the Pitch Pulse Location

The TIA Enhanced Variable Rate Coder (EVRC) [29], which employs relaxed CELP (RCELP) [30], uses a simple method based on the energy of the LPC residual to detect the pitch pulses. EVRC determines the pitch pulse locations by searching for a maximum in a five-sample sliding energy window within a region larger than the pitch period, and then finding the rest of the pitch pulses by searching recursively at a separation of one pitch period. It is possible to improve the performance of the residual-energy-based pitch pulse location detection by using the Hilbert envelope of windowed LP residual (HEWLPR) [31, 32]. A robust pitch pulse detection algorithm based on the group delay of the phase spectrum has also been reported [33], however this method has a very high computational complexity.

The SWPM requires a pitch pulse detection algorithm that can detect the pulses at stationary voiced segments with a high accuracy and has a low computational complexity. However the ability to detect the pitch pulses at the onsets and offsets is beneficial, since this will increase the flexibility of transition detection. Therefore an improved pitch pulse detection algorithm, based on the algorithm used in EVRC, is developed for SWPM. Figure 9.4 depicts a block diagram of the pitch pulse location detection algorithm. Initially, all the possible pitch pulse locations are determined by considering the localized energy of the LPC residual and an adaptive threshold function,

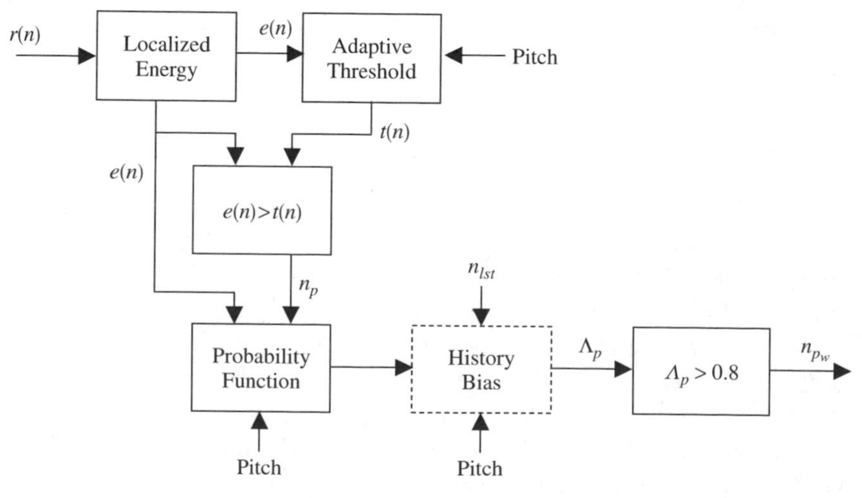

Figure 9.4 Block diagram of the pitch pulse detection algorithm

$t(n)$. The localized energy, $e(n)$, of the LPC residual, $r(n)$, is given by,

$$e(n) = \frac{1}{5} \sum_{j=-2}^{2} |r(n+j)| \quad \text{for } 2 \le n < N-2 \tag{9.1}$$

where $N = 240$ is the length of the residual buffer.

The adaptive threshold function, $t(n)$, is updated for each half pitch period, by taking 0.7 of the maximum of $e(n)$ in the pitch period symmetrically-centred around the half pitch period chosen to calculate $t(n)$, and $t(n)$ is given by,

$$t\left(n_k - \tau_{1/4} + n_{\tau/2}\right) = 0.7 \max\left[e\left(n_k - \tau_{1/2} + n_\tau\right)\right]$$
$$\text{for } 0 \le n_\tau < \tau_1 \quad \text{and} \quad 0 \le n_{\tau/2} < \tau_{1/2} \tag{9.2}$$

where $\tau_1 = \left\lfloor \tau + \frac{1}{2} \right\rfloor$, $\tau_{1/2} = \left\lfloor \frac{\tau}{2} + \frac{1}{2} \right\rfloor$, $\tau_{1/4} = \left\lfloor \frac{\tau}{4} + \frac{1}{2} \right\rfloor$, $n_k = k\tau_{1/2}$ for $1 \le k < \left\lfloor \frac{2N}{\tau} \right\rfloor$, and τ is the pitch period.

The exceptions corresponding to the analysis frame boundaries are given in,

$$e(0) = e(1) = e(N-2) = e(N-1) = 0 \tag{9.3}$$

$$t(m) = 0.7 \max\left[e\left(\tau_{1/2}\right)\right] \quad \text{for } 0 \le m < \tau_{1/4} \tag{9.4}$$

$$n_{\left\lfloor \frac{2N}{\tau} \right\rfloor} = N - \tau_1 \tag{9.5}$$

The sample locations, for which $e(n) > t(n)$, are considered as the regions which may contain pitch pulses. If $e(n) > t(n)$ for more than eight consecutive samples, those regions are ignored, since in those regions the residual energy is smeared, which is not a feature of pitch pulses. The centre of the each remaining region is taken as a possible pitch pulse location, n_p. If any of the two candidate locations are closer than eight samples (i.e. half of the minimum pitch), the one which has the higher $e(n_p)$ is taken.

Applying an adaptive threshold to estimate the pitch pulse locations from the localized energy $e(n)$ is advantageous, especially for segments where the energy of the LPC residual varies rapidly, giving rise to spurious pulses. Figure 9.5 demonstrates this for a male offset and a female onset. The male speech segment has a pitch period of about 80 samples and the two high-energy irregular pulses which do not belong to the pitch contour are clearly visible. The female speech segment has a pitch of about 45 samples, which also contains two high-energy irregular pulses. The energy function $e(n)$ and the threshold function $t(n)$ are also depicted in Figure 9.5, shifted upwards for clarity. The figures also show that $e(n)$ at the irregular pulses may be higher than $e(n)$ at the correct pitch pulses. Therefore selecting the highest $e(n)$ to detect a pitch pulse location as in [34] may lead to errors. Since $e(n) > t(n)$, for some of the irregular pulses as well as for correct pitch pulse locations,

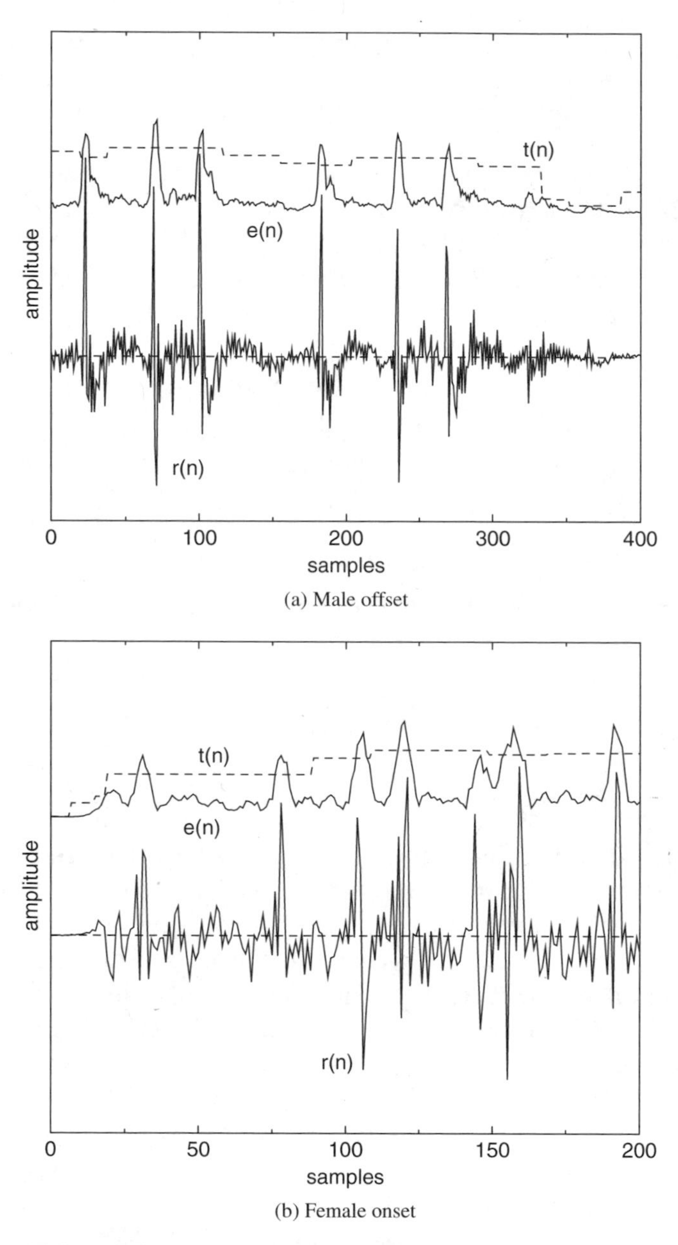

(a) Male offset

(b) Female onset

Figure 9.5 Irregular pulses at the onsets and offsets

further refinements are required. Moreover, the regions where $e(n) > t(n)$, gives only a crude estimation of the pitch pulse location. The algorithm relies on the accuracy of the estimated pitch used for the computation of $t(n)$ and in the refinement process described below. However SWPM needs only the pitch pulses in the stationary voiced segments, for which the pitch estimate is reliable.

For each selected location n_p, the probability of it being a pitch pulse is estimated, using the pitch and the energy of the neighbouring locations. First, a total energy metric, E_{p0} for the candidate pulse at n_{p0} is computed recursively as follows,

$$E_{p0} = \sum_l e(n_l) \tag{9.6}$$

where $l = p_0$ and any q which satisfies the condition,

$$|n_l \pm \tau - n_q| < 0.15\tau \tag{9.7}$$

For each term, $+\tau$ and $-\tau$, if more than one q satisfies equation (9.7), only the one which minimizes $|n_l \pm \tau - n_q|$ is chosen. Then further locations n_q that satisfy equation (9.7) are searched recursively, with any n_q which have already satisfied equation (9.7) taken as n_l in the next iteration. Therefore, E_{p0} can be defined as the sum of $e(n_p)$ of the pitch contour corresponding to the location n_{p0}. This process eliminates the high-energy irregular pulses, since they do not form a proper pitch contour and equation (9.7) detects them as isolated pulses. The probability of the candidate location, n_{p0}, containing a pitch pulse, Λ_{p0}, is given by,

$$\Lambda_{p0} = \frac{E_{p0}}{\max\left[E_p\right]} \tag{9.8}$$

If pitch pulse locations were detected in the previous frame and any of the current candidate pitch pulse locations form a pitch contour which is a continuation of the previous pitch contour, a history bias term is added. Adding the history bias term enhances the performance at the offsets, especially at the resonating tails. Furthermore, the history bias helps to maintain the continuity of the pitch contour between the frames, at the segments, where the pitch pulses become less significant, as shown in Figure 9.6. A discontinuity in the pitch contour adds a reverberant character into voiced speech segments. The biased term Λ'_l for any location n_l which satisfies equations (9.10) or (9.11) is given by,

$$\Lambda'_l = \Lambda_l + 0.2 \tag{9.9}$$

The initial value for l is given by equation (9.10), with ϵ being the minimum possible integer value which satisfies equation (9.10). If more than one l satisfies equation (9.10) with the same minimum ϵ, the one which maximizes $e(n_l)$ is taken.

$$|n_{lst} + \epsilon\tau - n_l| < 0.1\tau \tag{9.10}$$

where n_{lst} is the pitch pulse location selected in the last analysis frame. Then any location n_q which satisfies equation (9.11) is searched and further n_l are

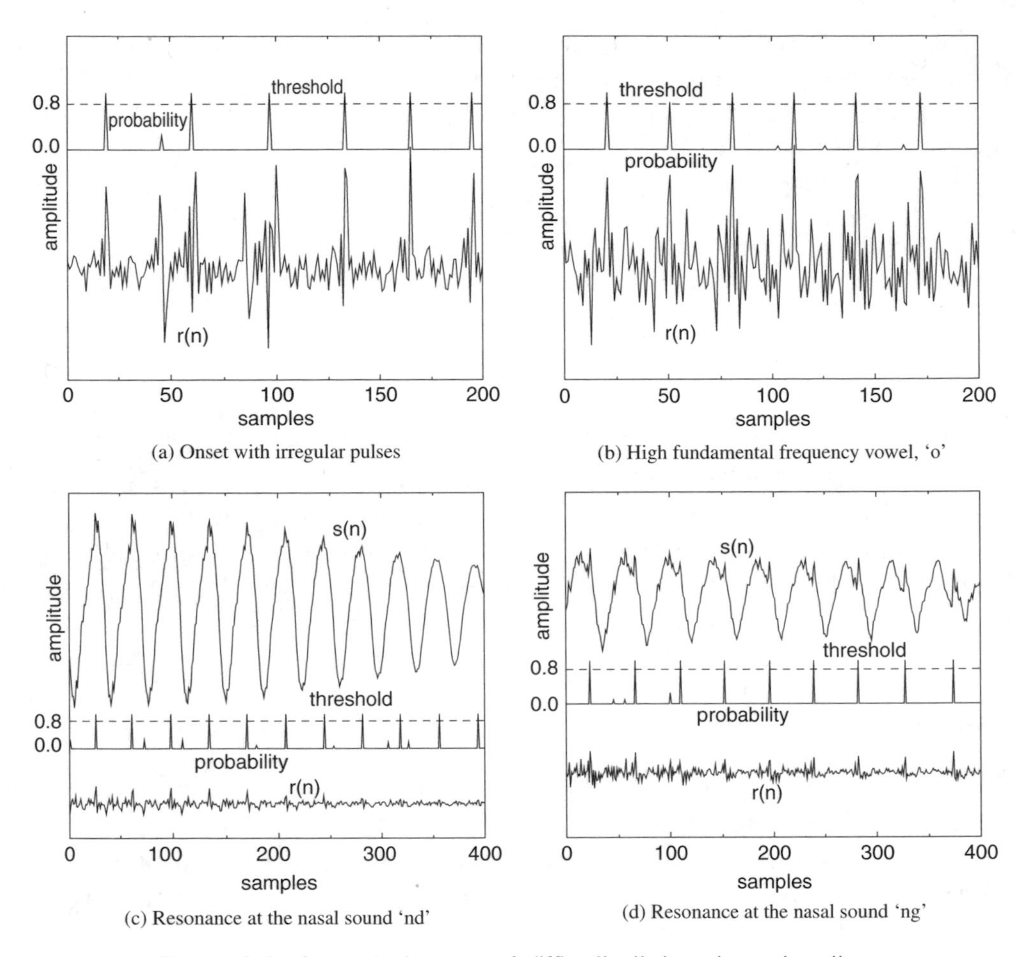

(a) Onset with irregular pulses

(b) High fundamental frequency vowel, 'o'

(c) Resonance at the nasal sound 'nd'

(d) Resonance at the nasal sound 'ng'

Figure 9.6 Some instances of difficult pitch pulse extraction

found recursively, with any n_q which have already satisfied equation (9.11) taken as n_l in the next iteration. If more than one n_q satisfies equation (9.11), the one which minimizes $|n_l + \tau + n_q|$ is chosen.

$$|n_l + \tau + n_q| < 0.15\tau \tag{9.11}$$

The final probability of the candidate location n_{p0} containing a pitch pulse Λ_{p0} is recalculated,

$$\Lambda_{p0} = \frac{\Lambda'_{p0}}{\max\left[\Lambda'_p\right]} \tag{9.12}$$

A set of positions, n_{p_w} which have probabilities, $\Lambda_p > 0.8$, are selected as the pitch pulse locations, and they are further refined in order to select the pitch pulse closest to the synthesis frame boundary. Figure 9.6 shows some instances of difficult pitch pulse detection along with the estimated probabilities, Λ_p, and the threshold value. In Figures 9.6c and 9.6d, the resonating speech waveforms are also shown.

The problem illustrated in Figure 9.6b can be explained in both the time and frequency domains. In speech segments with a short pitch period, the short-term LPC prediction tends to remove some of the pitch correlation as well, leaving an LPC residual without any clearly distinguishable peaks. Shorter pitch periods in the time domain correspond to fewer harmonics in the frequency domain. Hence the inter-harmonic spacing becomes wider and the formants of the short-term predictor tend to coincide with some of the harmonics (see Figure 9.7). The speech spectrum in Figure 9.7 is lowered by 80 dB in order to emphasize the coinciding points of the spectra. The excessive removal of some of the harmonic components by the LPC filter disperses the energy of the residual pitch pulses. It has been reported that large errors in the linear prediction coefficients occur in the analysis of sounds with high pitch frequencies [35]. In the case of nasal sounds, the speech waveform has a very high low-frequency content (see Figure 9.6c). In such cases, the LPC filter simply places a pole at the fundamental frequency. A pole in the LPC synthesis filter translates to a zero in the inverse filter, giving rise to a fairly random-looking LPC residual signal. The figures demonstrate that the estimated probabilities, Λ_p exceed the threshold value only at the required pitch pulse locations, despite those difficulties.

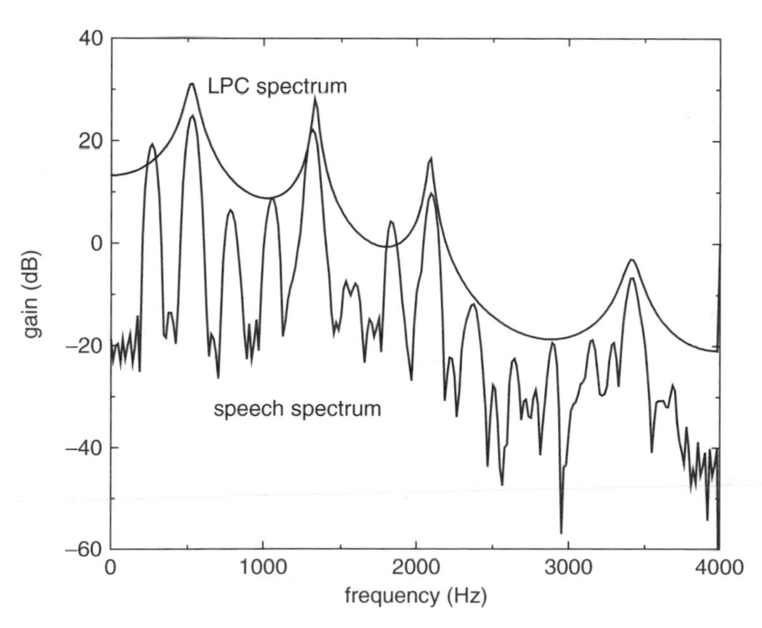

Figure 9.7 Speech and LPC spectra of a female vowel segment

9.4.2 Estimation of the Pitch Pulse Shape

Figure 9.8 depicts a complete pitch cycle of the LPC residual, which includes a selected pitch pulse, and the positive half of the wrapped phase spectrum obtained from its DFT. The integer pitch pulse position is taken as the time origin of the DFT, and the phase spectrum indicates that most of the harmonic phases are close to an average value. This average phase value varies with the shape of the pitch pulse, hence it is called pitch pulse shape (PPS). In the absence of a strong pitch pulse, the phase spectrum becomes random and varies between $-\pi$ and π.

Figure 9.9 depicts a block diagram of the pitch pulse shape estimation algorithm. This algorithm employs an AbS technique in the time domain to estimate PPS. A prototype pulse, $P(n_s)$, is synthesized as follows:

$$p\left(n_s\right) = \sum_{k=1}^{K} a_k \cos\left(k\omega n_s + \alpha_q\right) \quad \text{for} \quad -4 \leq n_s \leq 4 \tag{9.13}$$

where $\omega = 2\pi/\tau$, τ is the pitch period, K is the number of harmonics, a_k are the harmonic amplitudes, and the candidate pitch pulse shapes, α_q, are given by,

$$\alpha_q = 2\pi q/8 \quad \text{for} \quad 0 \leq q < 8 \tag{9.14}$$

Figure 9.10 depicts the synthesized pulses, $p\left(n_s\right)$, for two different candidate pitch pulse shapes, i.e. values of α_q. A simpler solution to avoid estimating the spectral amplitudes, a_k for equation (9.13) is to assume a flat spectrum. However, the use of spectral amplitudes, a_k, gives the relative weight for

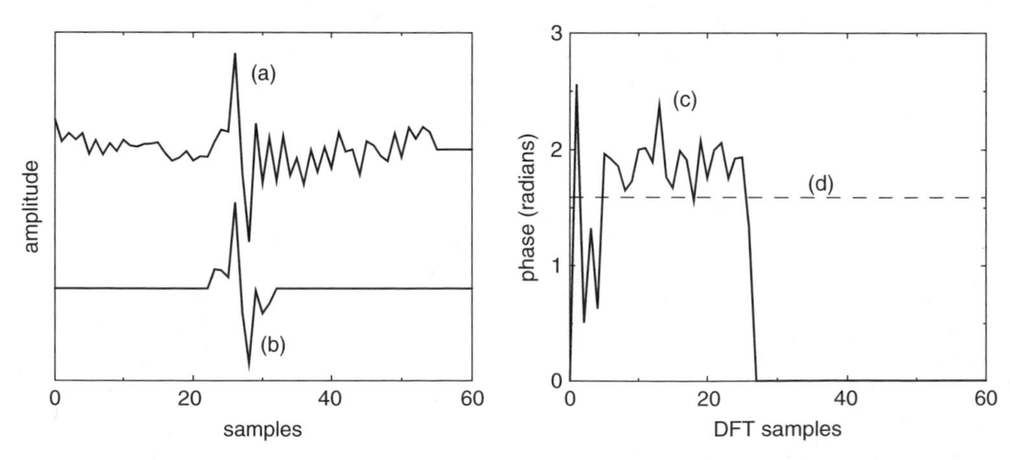

Figure 9.8 (a) a complete pitch cycle of the LPC residual, (b) the pitch pulse synthesized using PPS, (c) the positive half of the phase spectrum obtained from the DFT, and (d) the estimated PPS

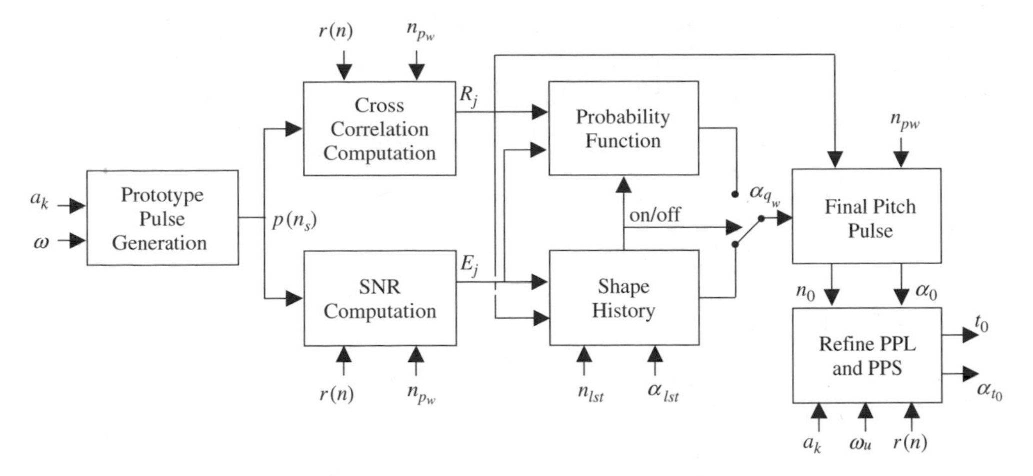

Figure 9.9 Block diagram of the pitch pulse shape estimation

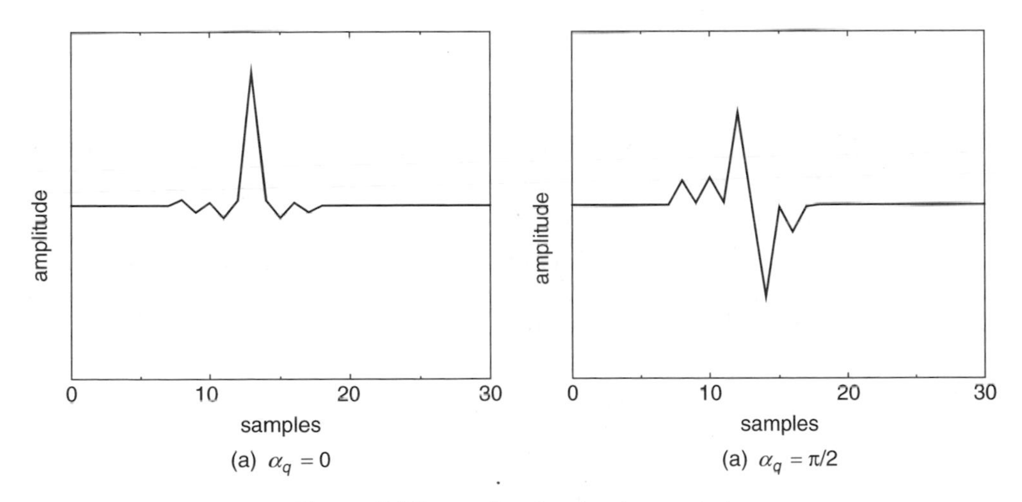

Figure 9.10 synthesized pulses, $p(n_s)$

each harmonic, which is beneficial in estimating the pitch pulse shape. For example, if a harmonic component which is relatively small in the LPC residual signal is given equal weight in the prototype pulse, $p(n_s)$, this may lead to inaccurate estimates in the subsequent AbS refinement process. Considering the frequency domain, those relatively small amplitudes may be affected by spectral leakage from the larger amplitudes, giving large errors in the phase spectrum. However, since computing the spectral amplitudes for each pitch pulse is a very intensive process, as a compromise, the same spectral amplitudes are used for the whole analysis frame, and are also transmitted to the decoder as the harmonic amplitudes of the LPC residual. Then the normalized cross-correlation, R_j, and SNR, E_j, are estimated between the synthesized prototype pitch pulse $p(n_s)$ and each of the detected LPC residual pitch pulses, at the locations n_p, where $n_p \in n_{p_w}$. R_j and E_j are estimated for each candidate pitch pulse shape, α_q,

$$R_j = \frac{\sum_{n_s=-4}^{4} r\left(n_p + n_s + j\right) p\left(n_s\right)}{\sqrt{\sum_{n_s=-4}^{4} p^2\left(n_s\right) \sum_{n_s=-4}^{4} r^2\left(n_p + n_s + j\right)}} \quad \text{for } -3 \leq j \leq 3 \tag{9.15}$$

$$E_j = \frac{\sum_{n_s=-4}^{4} r^2\left(n_p + n_s + j\right)}{\sum_{n_s=-4}^{4} \left[p\left(n_s\right) - r\left(n_p + n_s + j\right)\right]^2} \quad \text{for } -3 \leq j \leq 3 \tag{9.16}$$

The term j is introduced in R_j and E_j in order to shift the relative positioning of the LPC residual pulse and the synthesized pulse. This compensates for the approximate pitch pulse locations, n_p, estimated by the algorithm described in Section 9.4.1, by allowing the initial estimates to shift around, with a resolution of one sample. All the combinations of n_p, α_q, and j for which $E_j \leq 1.0$ are excluded from any further processing. $E_j \leq 1.0$ corresponds to an SNR of less than or equal to 0 dB. Then probability of the candidate shape, α_{q_0}, being the pitch pulse shape is estimated,

$$\Lambda_{q_0} = \frac{N_{q_0}}{\max\left[N_q\right]} \tag{9.17}$$

where N_q is the total number of residual pulses for a given q, for which $R_j > 0.5$. If more than one j satisfies the condition $R_j > 0.5$, for a particular set of q and n_p, N_q is incremented only once. The set of pitch pulse shape values, α_{q_w}, which have probabilities, $\Lambda_q > 0.7$ are chosen for further refinement. If $\max\left[N_q\right]$ is zero, then all the Λ_q are set to zero, i.e. no pitch pulses are detected. Figure 9.11 shows the LPC residual of an analysis frame and the estimated probability density function (PDF) of α_q in the range $-\pi \leq \alpha_q < \pi$. The pitch pulses of the LPC residual in Figure 9.11a have similar shapes to the shape of the synthesized pulse shown in Figure 9.10a. Consequently the PDF is maximum around $\alpha_q = 0$, for the pitch pulse shape used to synthesize the pulse shown in Figure 9.10a. If a history bias is used in pitch pulse location detection, then the probability term, Λ_q is not estimated. Instead the pitch pulse shape search is limited to three candidates, α_L, around the pitch pulse shape of the previous frame. During the voiced segments, the

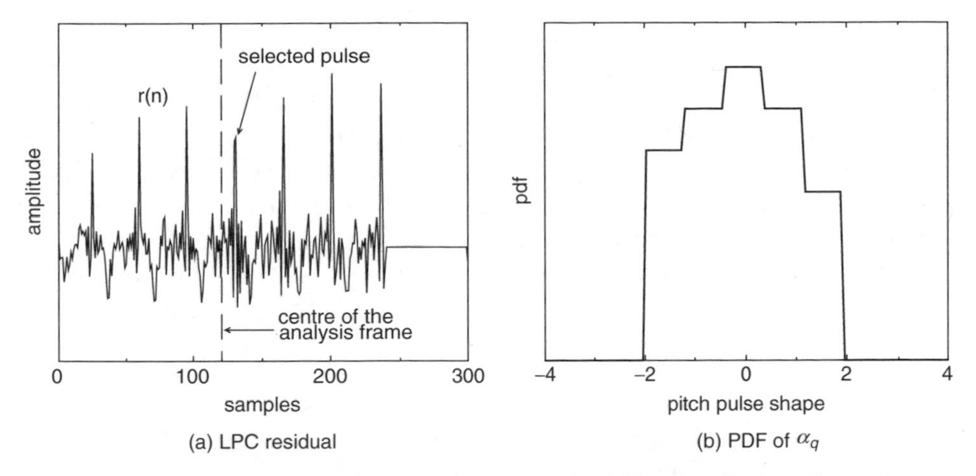

Figure 9.11 An analysis frame and the probability density function of α_q

pitch pulse shape is fairly stationary and restricting the search range around the previous value does not reduce the performance. Restricting the search range has advantages such as reduced computational complexity and efficient differential quantization of the pitch pulse shape. Furthermore, restricting the search range avoids large variations in the pitch pulse shape. Large variations in the pitch pulse shape introduce a reverberant character into the synthesized speech.

$$\alpha_L = 2\pi\left(q_L + \delta\right)/8 \quad \text{for} \quad -1 \le \delta \le 1 \tag{9.18}$$

where,

$$q_L = \left\lfloor \frac{\alpha_{lst}}{2\pi} 8 + \frac{1}{2} \right\rfloor \quad \text{for } 0 \le \alpha_{lst} < 2\pi \tag{9.19}$$

Then R_j and E_j are estimated as before, substituting α_q with α_L, and all the combinations of n_p, α_L, and j for which $E_j \le 1.0$ or $R_j \le 0.5$ are excluded from any further refinements. If no combination of n_p, α_L, and j are left, the search is extended to all the α_q, and Λ_q is estimated as before, otherwise the remaining α_L are chosen for further refinement, i.e. the remaining α_L form the set α_{q_w}. The pitch pulse closest to the centre of the analysis frame, i.e. closest to the synthesis frame boundary for which $R_j > \xi$ is selected as the final pitch pulse. The threshold value, ξ, is given by,

$$\xi = 0.7 \max\left[R_j\right] \quad \text{for} \quad \alpha_q \in \alpha_{q_w} \tag{9.20}$$

If more than one set of j and α_q satisfy the condition $R_j > \xi$ for the same pitch pulse closest to the synthesis frame boundary, the set of values which maximizes R_j is chosen. The pitch pulse shape and the integer pitch pulse location are given by the chosen, α_q and $n_p + j$ respectively. Figure 9.11a shows the centre of the analysis frame and the selected pitch pulse. It is also possible to select the pitch pulse closest to the centre of the analysis frame from the set n_{p_w} and estimate the shape of the selected pulse. However estimating the PDF of α_q for the whole analysis frame and including it in the selection process improves the reliability of the estimates, which enables the selection of the most probable α_q. Then the integer pitch pulse location is refined to a 0.125 sample accuracy, and the initial pitch pulse shape is refined to a $2\pi/64$ accuracy. In the refinement process, a synthetic pulse $p_u(n_u)$ is generated in an eight times up-sampled domain, i.e. at 64 kHz. If the selected integer pitch pulse location and shape are n_0 and α_0, respectively, then,

$$p_u\left(n_u\right) = \sum_{k=1}^{K} a_k \cos\left(k\omega_u n_u + \alpha_i\right) \quad \text{for} \quad -40 \leq n_u < 40 \tag{9.21}$$

where $\omega_u = 2\pi/8\tau$, and α_i is given by,

$$\alpha_i = \alpha_0 + 2\pi i/64 \quad \text{for} \quad -4 \leq i \leq 4 \tag{9.22}$$

Then equation (9.23) is used to compute the normalized cross-correlation $R_{i,j}$ for all i and j, and the indices corresponding to the maximum $R_{i,j}$ are used to evaluate the refined PPS and PPL, as shown in equations (9.22) and (9.25) respectively.

$$R_{i,j} = \frac{\sum_{n_r=-4}^{4} r\left(n_0 + n_r\right) p_j\left(n_r\right)}{\sqrt{\sum_{n_r=-4}^{4} p_j^2\left(n_r\right) \sum_{n_r=-4}^{4} r^2\left(n_0 + n_r\right)}} \tag{9.23}$$

where $p_j(n_r)$, is the shifted and down-sampled version of $p_u(n_u)$ given by,

$$p_j\left(n_r\right) = p_u\left(8n_r + j\right) \quad \text{for} \quad -8 \leq j < 8 \quad \text{and} \quad -4 \leq n_r \leq 4 \tag{9.24}$$

The final PPL, t_0, refined to a 0.125 sample resolution is given by,

$$t_0 = n_0 - j/8 \tag{9.25}$$

Fractional PPL is important for segments with short pitch periods and when the pitch pulse is close to or at the synthesis frame boundary. When the pitch period is short, a small variation in the pitch pulse location can induce a large percentage pitch error. The pitch pulses closest to the synthesis frame boundaries are chosen in SWPM in order to maximize the waveform similarity at the frame boundaries, since the mode changes are limited to synthesis frame boundaries. However if the selected pitch pulse is on the frame boundary or within a few samples of it, the pulse must be synthesized smoothly across this boundary, in order to avoid audible artifacts. In such cases, high resolution PPL and PPS are essential to maintain the phase continuity across the frame boundaries. It is also possible to compute the cross-correlation between $p_u(n_u)$ and the eight times up-sampled residual signal, in order to evaluate the best indices i and j. However this requires more computations and an equally good result is obtained by shifting $p_u(n_u)$ in the up-sampled domain and then computing the cross-correlation in the down-sampled domain, as shown in equations (9.23) and (9.24).

At the offsets, if no pitch pulses are detected, PPL is predicted from the PPL of the previous frame using the pitch, and PPS is set to equal to the PPS of the previous frame. This does not introduce any deteriorating artifacts, since the encoder checks the suitability of the harmonic excitation in the mode selection process. The prediction of PPL and PPS is particularly useful at offsets with a resonant tail, where pitch pulse detection is difficult.

9.4.3 Synthesis using Generalized Cubic Phase Interpolation

In the synthesis, the phases are interpolated cubically, i.e. by quadratic interpolation of the frequencies. In [2], phases are interpolated for the frequencies and phases available at the frame boundaries. But in the case of SWPM the frequencies are available at the frame boundaries and the phases at the pitch pulse locations. Therefore a generalized cubic phase interpolation formula is used, to incorporate PPL and PPS.

The phase $\theta_k(n)$ of the kth harmonic of the $i+1^{th}$ synthesis frame is given by,

$$\theta_k(n) = \theta_{k_i} + k\omega_i n + \alpha_k n^2 + \beta_k n^3 \quad \text{for } 0 \leq n < N \tag{9.26}$$

where N is the number of samples per frame and θ_{k_i} and ω_i are the phase of the kth harmonic and the fundamental frequency, respectively, at the end of synthesis frame i, and α_k and β_k are given by,

$$\begin{pmatrix} t_0^2 & t_0^3 \\ 2N & 3N^2 \end{pmatrix} \begin{pmatrix} \alpha_k \\ \beta_k \end{pmatrix} = \begin{pmatrix} \theta_{t_0} - \theta_{k_i} - k\omega_i t_0 + 2\pi M_k \\ k\omega_{i+1} - k\omega_i \end{pmatrix} \tag{9.27}$$

where t_0 is the fractional pitch pulse location (PPL), θ_{t_0} is the PPS estimated at t_0, and M_k represents the phase unwrapping and is chosen according to the 'maximally smooth' criterion used by McAulay [2]. McAulay chose M_k such that $f(M_k)$ is a minimum,

$$f(M_k) = \int_0^T \left[\ddot{\theta}_k(t, M_k)\right]^2 dt \tag{9.28}$$

where $\theta_k(t, M_k)$ represents the continuous analogue form of $\theta_k(n)$, and $\ddot{\theta}_k(t, M_k)$ is the second derivative of $\theta_k(t, M_k)$ with respect to t. Although M_k is integer-valued, since $f(M_k)$ is quadratic in M_k, the problem is most easily solved by minimizing $f(x_k)$ with respect to the continuous variable x_k and then choosing M_k to be an integer closest to x_k. For the generalized case of SWPM, $f(x_k)$ is minimized with respect to x_k and $x_{k_{\min}}$ is given by,

$$x_{k_{\min}} = \frac{1}{2\pi}\left(\theta_{k_i} - \theta_{t_0} + k\omega_i t_0 + \frac{k\left(\omega_{i+1} - \omega_i\right) t_0^2}{2N}\right) \tag{9.29}$$

$M_{k_{\min}} = \lfloor x_{k_{\min}} + 0.5 \rfloor$ is substituted in equation 9.27 for M_k to solve for α_k and β_k and in turn to unwrap the cubic phase interpolation function $\theta_k(n)$.

The initial phase θ_{k_i} for the next frame is $\theta_k(N)$, and the above computations should be repeated for each harmonic, i.e. k. It should be noted that there is no need to synthesize the phases, $\theta_k(n)$ in the up-sampled domain, in order to use the fractional pitch pulse location, t_0. It is sufficient to use t_0 in solving the coefficients of $\theta_k(n)$, i.e. α_k and β_k.

9.5 Hybrid Encoder

A simplified block diagram of a typical hybrid encoder that operates on a fixed frame size of 160 samples is shown in Figure 9.12. For each frame, the mode that gives the optimum performance is selected. There are three possible modes: scaled white noise coloured by LPC for unvoiced segments; ACELP for transitions; and harmonic excitation for stationary voiced segments.

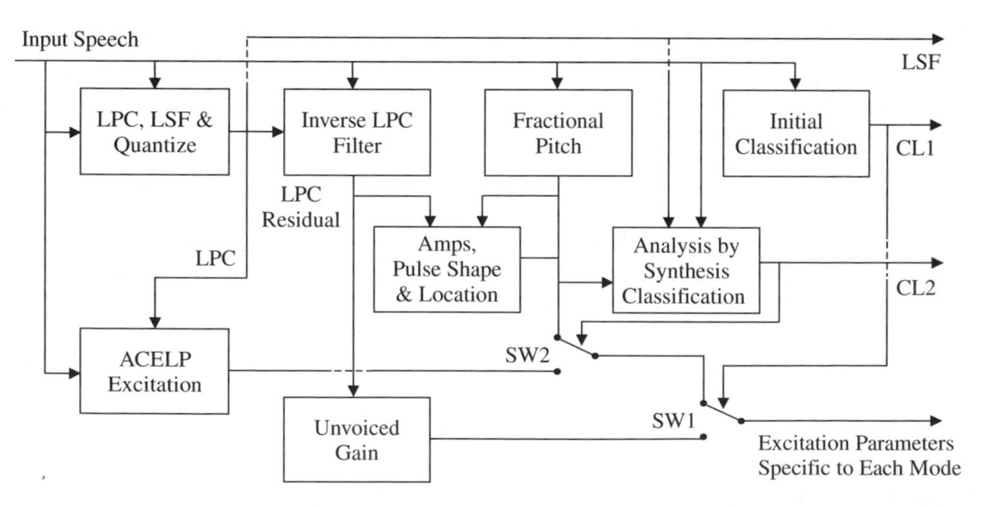

Figure 9.12 Block diagram of the hybrid encoder

Any waveform-coding technique can be used instead of ACELP. In fact this hybrid model [27] does not restrict the choice of coding technique for speech transitions, it merely makes the mode decision and defines the target waveform. In white noise excited mode, the gain estimated from the LPC residual energy is transmitted for every 20 ms. The LPC parameters are common for all the modes and estimated every 20 ms (with a 25 ms window length), which are usually interpolated in the LSF domain for every subframe in the synthesis process. In order to interpolate the LSFs, the LPC analysis window is usually centred at the synthesis frame boundary which requires a look-ahead.

A two-stage speech classification algorithm is used in the above coder. An initial classification is made based on the tracked energy, low-band to high-band energy ratio, and zero-crossing rate, and determines whether to use the noise excitation or one of the other modes. The secondary classification, which is based on an AbS process, makes a choice between the harmonic excitation or ACELP. Segments of plosives with high-energy spikes are synthesized using ACELP. When the noise excitation mode is selected, there is no need to estimate the excitation parameters of the other modes. If noise excitation is not selected, the harmonic parameters are always estimated and the harmonic excitation is generated at the encoder for the AbS transition detection. The speech classification is described in detail in Section 9.6.

For simplicity, details of LPC and adaptive codebook memory update are excluded from the block diagram. The encoder maintains an LPC synthesis filter synchronized with the decoder, and uses the final memory locations for ACELP and AbS transition detection in the next frame. Adaptive codebook memory is always updated with the previous LPC excitation vector regardless of the mode. In order to maintain the LPC and the adaptive codebook memories, the LPC excitation is generated at the encoder, regardless of the mode.

9.5.1 Synchronized Harmonic Excitation

In the harmonic mode, the pitch and harmonic amplitudes of the LPC residual are estimated for every 20 ms frame. The estimation windows are placed at the end of the synthesis frames, and a look-ahead is used to facilitate the harmonic parameter interpolation. The pitch estimation algorithm is based on the sinusoidal speech-model matching proposed by McAulay [36] and improved by Atkinson [4] and Villette [37, 38]. The initial pitch is refined to 0.2 sample accuracy using synthetic spectral matching proposed by Griffin [3]. The harmonic amplitudes are estimated by simple peak-picking of the magnitude spectrum of the LPC residual.

The harmonic excitation $e_h(n)$ is generated at the encoder for the AbS transition detection and to maintain the LPC and adaptive codebook memories, which is given by,

$$e_h(n) = \sum_{k=1}^{K} a_k(n) \cos(\theta_k(n)) \quad \text{for } 0 \le n < N \tag{9.30}$$

where K is the number of harmonics. Since two analysis frames are interpolated to produce a synthesis frame, K is taken as the higher number of harmonics out of the two analysis frames and the missing amplitudes of the other analysis frame are set to zero. N is the number of samples in a synthesis frame and $\theta_k(n)$ is given in equation (9.26) for continuing harmonic tracks, i.e. each harmonic of an analysis frame is matched with the corresponding harmonic of the next frame. For terminating harmonics, i.e. when the number of harmonics in the next frame is smaller, $\theta_k(n)$ is given by,

$$\theta_k(n) = \theta_{k_i} + 2\pi kn/\tau_i \tag{9.31}$$

where θ_{k_i} is the phase of the harmonic k and τ_i is the pitch at the end of synthesis frame i. For emerging harmonics, $\theta_k(n)$ is given by,

$$\theta_k(n) = \theta_{t_0} + 2\pi k(n - t_0)/\tau_{i+1} \tag{9.32}$$

where t_0 is the PPL, θ_{t_0} is the corresponding PPS, and τ_{i+1} is the pitch, all at the end of synthesis frame $i+1$. Continuing harmonic amplitudes are linearly interpolated,

$$a_k(n) = a_{k_i} + \frac{\left(a_{k_{i+1}} - a_{k_i}\right) n}{N} \quad \text{for } 0 \le n < N \tag{9.33}$$

where a_{k_i} is the amplitude estimate of the kth harmonic at the end of synthesis frame i. For terminating harmonic amplitudes a trapezoidal window, unity for 55 samples and linearly decaying for 50 samples, is applied from the beginning of the synthesis frame,

$$a_k(n) = a_{k_i} \quad \text{for } 0 \le n < 55, \quad a_k(n) = a_{k_i}\frac{105 - n}{50} \quad \text{for } 55 \le n < 105 \tag{9.34}$$

For emerging harmonic amplitudes a trapezoidal window, linearly rising for 50 samples and unity for 55 samples, is applied starting from the 56th sample of the synthesis frame,

$$a_k(n) = a_{k_{i+1}}\frac{n - 55}{50} \quad \text{for } 55 \le n < 105, \quad a_k(n) = a_{k_{i+1}} \quad \text{for } 105 \le n < 160 \tag{9.35}$$

9.5.2 Advantages and Disadvantages of SWPM

Figure 9.13 shows some examples of waveforms synthesized using the harmonic excitation technique described in Section 9.5.1. In each example, (i) represents the LPC residual or the original speech signal and (ii) represents the LPC excitation or the synthesized speech signal. Figure 9.13a shows the LPC residual and the harmonic excitation of a segment which has strong pitch pulses and Figure 9.13b shows the corresponding speech waveforms. It can be seen that the synthesized speech waveform is very similar to the original. Figure 9.13c shows the LPC residual and the harmonic excitation of a segment which has weak or dispersed pitch pulses and Figure 9.13d shows the corresponding speech waveforms. The synthesized speech is time-synchronized with the original, however the waveform shapes are slightly

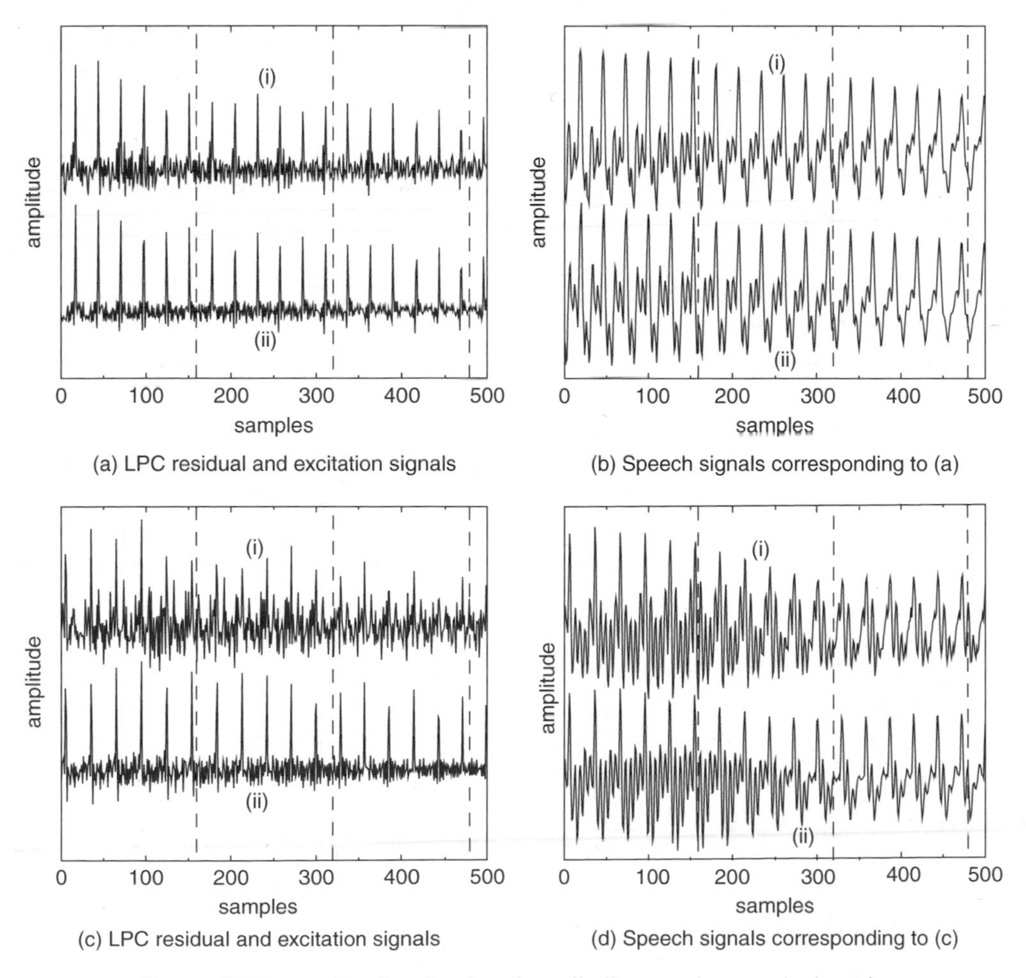

Figure 9.13 synthesized voiced excitation and speech signals

different, especially between the major pitch pulses. The waveform similarity is highest at the major excitation pulse locations and decreases along the pitch cycles. This is due to the fact that SWPM models only the major pitch pulses and it cannot model the minor pulses present in the residual signal when the LPC residual energy is dispersed. Furthermore, the dispersed energy of the LPC residual, becomes concentrated around the major pitch pulses in the excitation signal. The synthesized speech also exhibits larger variations in the amplitude around the pitch pulse locations, compared with the original speech.

In order to understand the effects on subjective quality due to the above observations, an informal listening test was conducted by switching between the harmonically-synthesized speech and the original speech waveforms at desired synthesis frame boundaries. The informal listening tests showed occasional audible artifacts at the mode transitions, when switching from the harmonic mode to the waveform-coding mode. However there were no audible switching artifacts when switching from waveform-coding to harmonic-coding mode, i.e. at the onsets. It was found that this is due to two reasons: difficulties in reliable pitch pulse detection and limitations in representing the harmonic phases using the pitch pulse shape at some segments. At some highly resonant segments, the LPC residual looks like random noise and it is not possible even to define the pitch pulses. The predicted pitch pulse location, assuming a continuing pitch contour, may be incorrect at resonant tails. At such segments, the pitch pulse locations are determined by applying AbS techniques in the speech domain, such that the synthesized speech signal is synchronized with the original, as described in the next subsection. In the speech segments illustrated using Figure 9.13c, it is possible to detect dominant pitch pulses. However the LPC residual energy is dispersed throughout the pitch periods, making the pitch pulses less significant, as described in Section 9.4.1. This effect reduces the coherence of the LPC residual harmonic phases at the pulse locations and the DFT phase spectrum estimated at the pulse locations look random. Female vowels with short pitch periods show these characteristics. A dispersed phase spectrum reduces the effectiveness of the pitch pulse shape, since the concept of pitch pulse shape is based on the assumption that a pitch pulse is the result of the superimposition of coherent phases, which have the same value at the pitch pulse location. This effect is illustrated in Figure 9.14. The synthesized pitch pulse models the major pulse in the LPC residual pitch period and concentrates the energy at the pulse location. This is due to the single phase value used to synthesize the pulse, as opposed to the more random-looking phase spectrum of the original pitch cycle. This phenomenon introduces phase discontinuities, which accounts for the audible switching artifacts. However the click and pop sounds present at the mode transitions in speech synthesized with SWPM are less annoying than those in a conventional

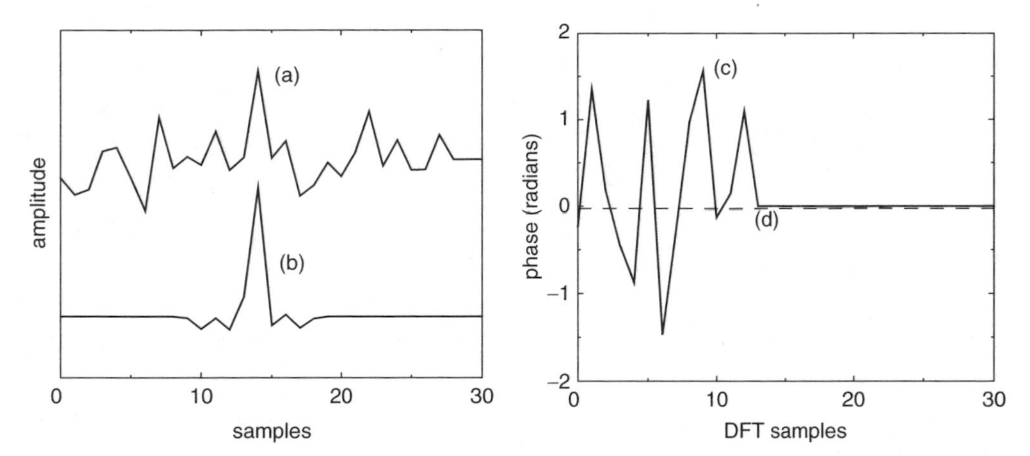

Figure 9.14 PPS at a dispersed pitch period: (a) a complete pitch cycle of the LPC residual, (b) the pitch pulse synthesized using PPS, (c) the positive half of the phase spectrum obtained from the DFT, and (d) estimated PPS

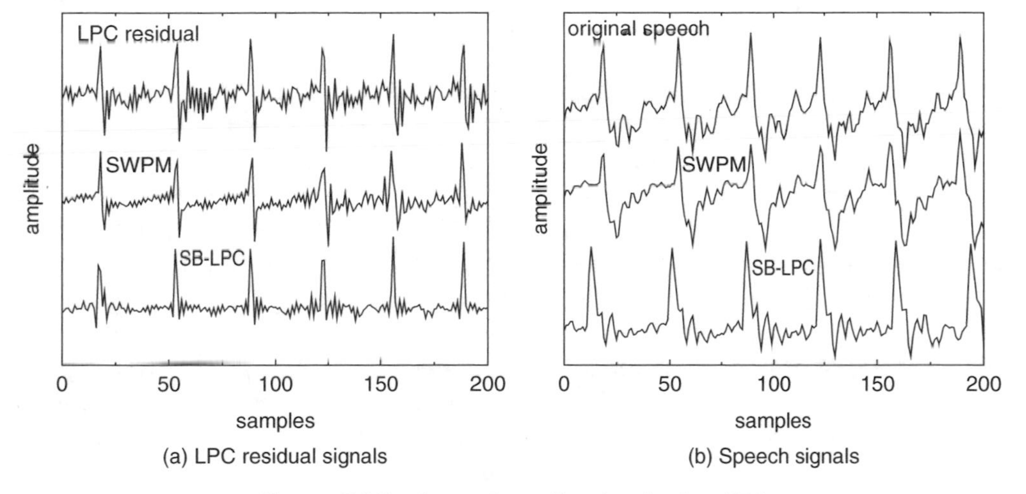

Figure 9.15 Speech synthesized using PPS

zero-phase excitation, even if the pitch pulse locations are synchronized. This is because SWPM has the additional flexibility of choosing the most suitable phase value (PPS) for pitch pulses, such that the phase discontinuities are minimized. Figure 9.15 illustrates the effect of PPS on the LPC excitation and the synthesized speech signals. For comparison, it includes the original signals and the signals synthesized using the SB-LPC coder [4] which assumes a zero-phase excitation.

The absence of audible switching artifacts at the onsets is an interesting issue. There are two basic reasons for the differences between switching artifacts at the onsets and at the offsets: the nature of the excitation signal and the LPC memory. At the onsets, even though the pitch pulses may be irregular due to the unsettled pitch of the vocal cords, they are quite strong and the residual energy is concentrated around them. Resonating segments and dispersed pulses do not occur at the onsets. Therefore the only difficulty at the onsets is in identifying the correct pulses and, as long as the pulse identification process is successful, SWPM can maintain the continuity of the harmonic phases at the onsets. The pitch pulse detection algorithm described is capable of accurate detection of the pitch pulses at the onsets as described in Section 9.4.1. Furthermore at the onsets, waveform coding preserves the waveform similarity, which also ensures the correct LPC memory, since LPC memory contains the past synthesized speech samples. Therefore the mode transition at the onsets is relatively easier and SWPM guarantees a smooth mode transition at the onsets. However at the offsets, the presence of weak pitch pulses is a common feature and the highly resonant impulse response LPC filter carries on the phase changes caused by the past excitation signal, especially when the LPC filter gain is high. Therefore, the audible switching artifacts remain at some of the offset mode transitions. These need to be treated as special cases.

At the resonant tails the LPC residual looks like random noise, and the pitch pulses are not clearly identifiable. In those cases AbS techniques can be applied directly on the speech signal to synchronize the synthesized speech. This process is applied only for the frames, which follow a harmonic frame and have been classified as transitions.

Synthesized speech is generated by shifting the pitch pulse location (PPL) at the end of the synthesis frame, $\pm\tau/2$ around the synthesis frame boundary with a resolution of one sample, where τ is the pitch period. The location which gives the best cross-correlation between the synthesized speech and the original speech is selected as the refined PPL. The pitch pulse shape is set equal to the pitch pulse shape of the previous frame. The excitation and the synthesized speech corresponding to the refined PPL are input to the closed-loop transition detection algorithm, and form the harmonic signal if the transition detection algorithm classifies the corresponding frame as harmonic, otherwise waveform coding is used.

9.5.3 Offset Target Modification

The SWPM minimizes the phase discontinuities at the mode transitions, as described in Section 9.5.2. However at some mode transitions such as the offsets after female vowels, which have dispersed pulses, audible phase discontinuities still remain. These discontinuities may be eliminated by transmitting more phase information. This section describes a more economical solution to remove those remaining phase discontinuities at the offsets, which does not need the transmission of additional information. The proposed method modifies some of the harmonic phases of the first frame of the waveform-coding target, which follows the harmonic mode. The remaining phase discontinuities can be corrected within the first waveform-coding frame, since SWPM keeps the phase discontinuities at a minimum and the pitch periods are synchronized.

As a first approach the harmonic excitation is extended into the next frame and the synthesized speech is linearly interpolated with the original speech at the beginning of the frame in order to produce the waveform-coding target. Listening tests were carried out with different interpolation lengths. The waveform-coding target was not quantized, in order to isolate the distortions due to switching. The tests were extended in order to understand the audibility of the phase discontinuities with the frequency of the harmonics, by manually shifting one phase at a time and synthesizing the rest of the harmonics using the original phases. Phase shifts of $\pi/2$ and π are used. Listening tests show that for various interpolation lengths the phase discontinuities below 1 kHz are audible, and an interpolation length as small as 10 samples is sufficient to mask distortions in the higher frequencies. Furthermore, male speech segments with long pitch periods, around 80 samples and above, do not cause audible switching artifacts. Male speech segments with long pitch periods have well-resolved short-term and long-term correlations, and produce clear and sharp pitch pulses, which can be easily modeled by SWPM. Therefore only the harmonics below 1 kHz of the segments with pitch periods shorter than 80 samples are considered in the offset target modification process.

The harmonic excitation is extended beyond the mode transition frame boundary, and the synthesized speech is generated in order to estimate the harmonic phases at the mode transition frame boundary. The phase of the kth harmonic of the excitation is computed as follows:

$$\theta_{k_{i+1}}(n) = \theta_{k_i} + 2\pi kn/\tau_i \quad \text{for } 0 \leq n < N \tag{9.36}$$

where θ_{k_i} is the phase of the kth harmonic and τ_i is the pitch at the end of synthesis frame i. The excitation signal is given by,

$$e_{h_{i+1}}(n) = \sum_{k=1}^{K} a_{k_i} \cos\left(\theta_{k_{i+1}}(n)\right) \tag{9.37}$$

where K is the number of harmonics and a_{k_i} is the amplitude of the kth harmonic estimated at the end of the synthesis frame i. The excitation signal is filtered through the LPC synthesis filter to produce the synthesized speech signal, with the coefficients estimated at the end of the synthesis frame i. The LPC memories after synthesizing the ith frame are used as the initial memories. The speech samples synthesized for the ith and $i+1^{th}$ frames are concatenated and windowed with a Kaiser window of 200 samples ($\beta = 6.0$) centred at the frame boundary. The harmonic phases, φ_{k_i}, are estimated using a 512 point FFT.

Having analysed the synthesized speech, the original speech is windowed at three points: at the end of the synthesis frame i, at the centre of the synthesis frame $i+1$, and at the end of the synthesis frame $i+1$, using the same window function as before. The corresponding harmonic amplitudes, $A_{k_i}, A_{k_{i+1/2}}, A_{k_{i+1}}$ and the phases $\phi_{k_i}, \phi_{k_{i+1/2}}, \phi_{k_{i+1}}$ are estimated using 512 point FFTs. Then the signal component $s_l(n)$, which consists of the harmonics below 1 kHz, is synthesized by,

$$s_l(n) = \sum_{k=1}^{L} A_k(n) \cos(\Theta_k(n)) \quad \text{for } 0 \le n < N \tag{9.38}$$

where L is the number of harmonics below 1 kHz at the end of the ith synthesis frame, $A_k(n)$ is obtained by linear interpolation between $A_{k_i}, A_{k_{i+1/2}}$, and $A_{k_{i+1}}$, and $\Theta_k(n)$ is obtained by cubic phase interpolation [2] between $\phi_{k_i}, \phi_{k_{i+1/2}}$, and $\phi_{k_{i+1}}$. Then the signal $s_m(n)$, which has modified phases is synthesized.

$$s_m(n) = \sum_{k=1}^{L} A_k(n) \cos(\Phi_k(n)) \quad \text{for } 0 \le n < N \tag{9.39}$$

and, finally, the modified waveform-coding target of the $i+1^{th}$ synthesis frame is computed by,

$$s_t(n) = s(n) - s_l(n) + s_m(n) \tag{9.40}$$

where $\Phi_k(n)$ is obtained by cubic phase interpolation between φ_{k_i} and $\phi_{k_{i+1}}$. Thus the modified signal, $s_m(n)$ has the phases of the harmonically-synthesized speech at the beginning of the frame and the phases of the original speech at the end of the frame. In other words, $\dot{\Theta}_k(n)$ (the rate of change of each harmonic phase) is modified such that the phase discontinuities are eliminated, by keeping $\dot{\Theta}_k(n)$ equal to the harmonic frequencies at the frame boundaries. There is a possibility that such phase modification operations induce a reverberant character in the synthesized signals. However, large phase mismatches close to π are rare, because SWPM minimizes the phase discontinuities. Furthermore, the modifications are applied only for the speech segments, which have pitch periods shorter than 80 samples, thus a phase mismatch is smoothed out in a few pitch cycles. The listening tests confirm that the synthesized speech does not possess a reverberant character. Limiting the phase modification process for the segments with pitch periods shorter than 80 samples also improves the accuracy of the spectral estimations, which use a window length of 200 samples. Figure 9.16

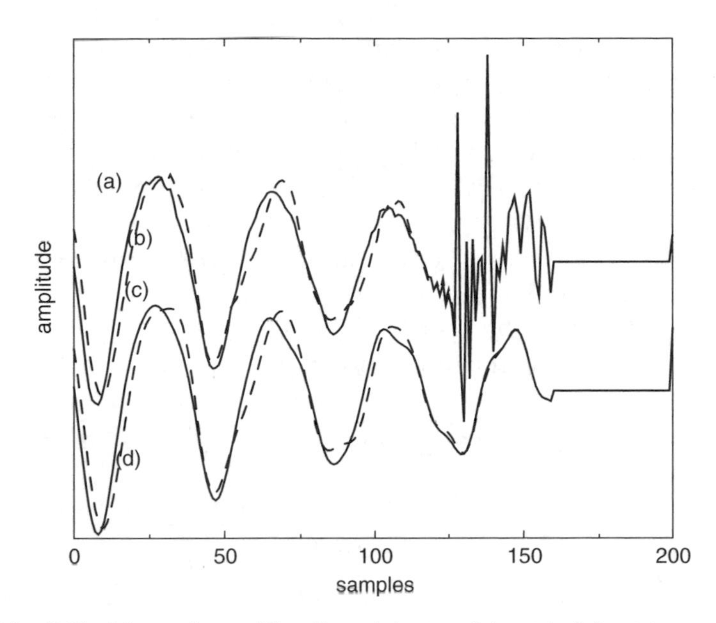

Figure 9.16 Offset target modification: (a) $s(n)$, (b) $s_t(n)$, (c) $s_l(n)$, and (d) $s_m(n)$

illustrates the waveforms of equation (9.40). It can be seen that the phases of the low frequency components of the original speech waveform, $s(n)$, are modified in order to obtain $s_t(n)$. The waveforms in Figures 9.16c and 9.16d depict $s_l(n)$ and $s_m(n)$, respectively, the low frequency components, which have been modified. The phase relationships between the high-frequency components account more for the perceptual quality of speech [25], and the high-frequency phase components are unchanged in the process.

Some speech signals show rapid variations in the harmonic structure at the offsets, which may reduce the efficiency of the phase modification process. In order to limit those effects the spectral amplitude and phase estimation process is not strictly confined to the harmonics of the fundamental frequency. Instead the amplitude and phase corresponding to the spectral peak closest to each harmonic frequency are estimated. The frequency of the selected spectral peak is taken as the frequency of the estimated amplitude and phase. When finding the spectral peaks closest to the harmonic frequencies, the harmonic frequencies are determined by the fundamental frequency at the end of the ith synthesis frame, since the pitch estimates at the transition frame are less reliable. In fact the purpose of the offset target modification process is to find the frequency components corresponding to the harmonics of the harmonically-synthesized frame in the ith frame and change the phase evolution of those components such that the discontinuities are eliminated. Moreover, the same set of spectral peak frequencies and amplitudes are used when synthesizing the terms $s_l(n)$ and $s_m(n)$, hence there is no need to restrict the synthesis process to the pitch harmonics.

Another important issue at the offsets is the energy contour of the synthesized speech. The harmonic coder does not directly control the energy of the synthesized speech, since it transmits the residual energy. However the waveform coders directly control the energy of the synthesized speech, by estimating the excitation gain using the synthesized speech waveform. This may cause discontinuities at the offset mode transition frame boundaries, especially when the LPC filter gain is high. The final target for the waveform coder is produced by linear interpolation between the extended harmonically synthesized speech and the modified target, $s_t(n)$ at the beginning of the frame for 10 samples. The linear interpolation ensures that the discontinuities due to variations of the energy contour are eliminated as well as the phase discontinuities, which are not accounted for in the phase modification process described above.

9.5.4 Onset Harmonic Memory Initialization

The harmonic phase evolution described in Section 9.4.3 and the harmonic excitation described in section 9.5.1 interpolate the harmonic parameters in the synthesis process, and assume that the model parameters are available at the synthesis frame boundaries. However, at the onset mode transitions, when switching from the waveform-coding mode, the harmonic model parameters are not directly available. The initial phases θ_{k_i}, the fundamental frequency ω_i in the phase evolution equation (9.26), and the initial harmonic amplitudes a_{k_i} in equation (9.33) are not available at the onsets. Therefore, they should be estimated at the decoder from the available information. The signal reconstructed by the waveform coder prior to the frame boundary and the harmonic parameters estimated at the end of the synthesis frame boundary are available at the decoder. The use of a waveform-coded signal in estimating the harmonic parameters at the onsets may be unreliable due to two reasons: the speech signal shows large variations at the onsets and, at low bit-rates, the ACELP excitation at the onsets reduces to a few dominant pulses, lowering the reliability of spectral estimates. Therefore the use of waveform-coded signal in estimating the harmonic parameters should be minimized. The waveform-coded signal is used only in initializing the amplitude quantization memories.

Since preserving the waveform similarity at the frame boundaries is important, the pitch is recomputed such that the previous pitch pulse location can be estimated at the decoder. Therefore the transmitted pitch represents the average over the synthesis frame. The other transmitted harmonic model parameters are unchanged, and are estimated at the end of the synthesis frame boundary. Let's define the pitch, τ_{i+1} and pitch pulse location, $t_{0_{i+1}}$, at the end of the $i+1^{th}$ synthesis frame, and the pitch pulse location at the end of the ith synthesis frame, t_{0_i}. The number of pitch cycles n_c between t_{0_i} and

$t_{0_{i+1}}$ is given by,

$$n_c = \left\lfloor \frac{t_{0_{i+1}} - t_{0_i}}{\tau_{i+1}} + \frac{1}{2} \right\rfloor \tag{9.41}$$

The recomputed pitch, τ_r, is given by,

$$\tau_r = \frac{t_{0_{i+1}} - t_{0_i}}{n_c} \tag{9.42}$$

Then τ_r and $t_{0_{i+1}}$ are transmitted, and t_{0_i} is computed at the decoder, as follows,

$$t_{0_i} = t_{0_{i+1}} - \tau_r \left\lfloor \frac{t_{0_{i+1}} - t'}{\tau_r} + \frac{1}{2} \right\rfloor \tag{9.43}$$

where t' is the starting frame boundary and t_{0_i} is the pitch pulse location closest to t'. The pitch pulse shape, θ_{0_i}, at the end of the ith synthesis frame is set equal to the pitch pulse shape, $\theta_{0_{i+1}}$, at the end of the $i+1^{th}$ synthesis frame. The initial phases θ_{k_i} in equation (9.26) are estimated as follows,

$$\theta_{k_i} = \theta_{0_i} - \frac{2\pi k t_{0_i}}{\tau_r} \tag{9.44}$$

Both fundamental frequency terms, ω_i and ω_{i+1}, in equation (9.27) are computed using τ_r, i.e. $\omega_i = \omega_{i+1} = 2\pi/\tau_r$. The harmonic amplitudes a_{k_i} in equation (9.33) are set equal to a_{ki+1}. Therefore, the phase evolution of the first harmonic frame of a stationary voiced segment becomes effectively linear and the harmonic amplitudes are kept constant, i.e. not interpolated.

9.5.5 White Noise Excitation

Unvoiced speech has a very complicated waveform structure. ACELP can be used to synthesize unvoiced speech and it essentially matches the waveform shape. However, a large number of excitation pulses are required to synthesize the noise-like unvoiced speech. Reducing the number of ACELP excitation pulses introduces sparse excitation artifacts in noise-like segments [39]. The synthesized speech also shows the sparse nature, and the pulse locations are clearly identifiable even in the LPC-synthesized speech. In fact, during unvoiced speech the short term correlation is small and the LPC filter gain has little effect.

Sinusoidal excitation can also be used to synthesize unvoiced segments, despite the fact that there is no harmonic structure. Speech synthesized by generating the magnitude spectrum every 80 samples (100 Hz) and uniformly-distributed random phases for unvoiced segments can achieve good quality

[40]. This method suits sinusoidal coders using frequency domain voicing without an explicit time-domain mode decision, since it facilitates the use of the same general analysis and synthesis structure for both voiced and unvoiced speech. However, this hybrid model classifies the unvoiced and silence segments as a separate mode, and, hence, uses a simpler unvoiced excitation generation model, which does not require any frequency-domain transforms. It has been shown that scaled white noise coloured by LPC can produce unvoiced speech with quality equivalent to μ-law logarithmic PCM [41, 42], implying that the complicated waveform structure of unvoiced speech has no perceptual importance. Therefore in terms of the perceptual quality, the phase information transmitted by ACELP is redundant and higher synthesis quality can be achieved at lower bit-rates using scaled white-noise excitation. Figure 9.17 shows a block diagram of the unvoiced gain estimation process and Figure 9.18 shows a block diagram of the unvoiced synthesis process. The band pass filters used are identical and have cut-off frequencies of 140 Hz and 3800 Hz. The transfer function of the fourth-order infinite impulse response (IIR) band pass filters is given by,

$$H_{bp}(z) = \frac{0.8278 - 1.6556z^{-2} + 0.8278z^{-4}}{1 - 0.0662z^{-1} - 1.6239z^{-2} + 0.0451z^{-3} + 0.6855z^{-4}} \tag{9.45}$$

and the unvoiced gain, g_{uv}, is given by,

$$g_{uv} = \sqrt{\frac{\sum_{n=0}^{N-1} r_{bp}^2(n)}{N}} \tag{9.46}$$

where $r_{bp}(n)$ is the band-pass-filtered LPC residual signal and N is the length of the residual vector, which is 160 samples including a look-ahead of 80 samples to facilitate overlap and add synthesis at the decoder.

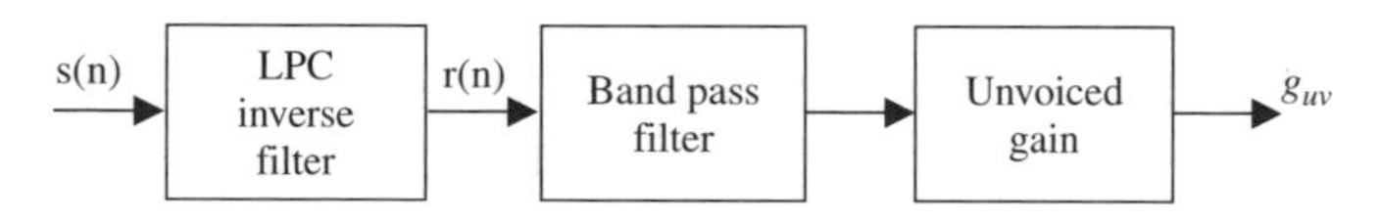

Figure 9.17 Unvoiced gain estimation

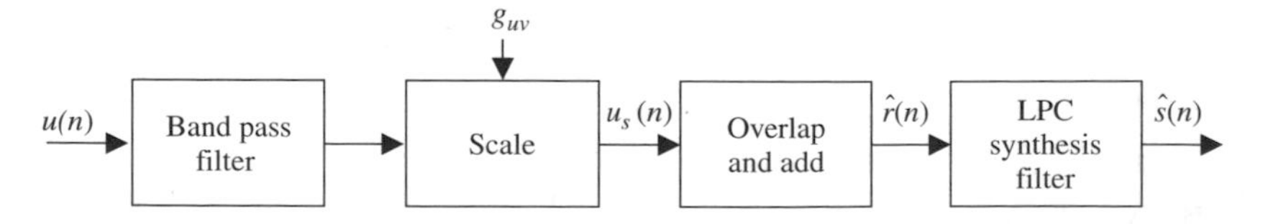

Figure 9.18 Unvoiced synthesis

White noise, $u(n)$, is generated by a random number generator with a Gaussian distribution (a Gaussian noise source has been found to be subjectively superior to a simple uniform noise source). The scaled white-noise excitation, $u_s(n)$, is obtained by,

$$u_s(n) = u_{bp}(n) \frac{g_{uv}}{\sqrt{\frac{\sum_{n=0}^{Z} u_{bp}(n)}{Z}}} \tag{9.47}$$

where $u_{bp}(n)$ is the band-pass-filtered white noise and Z is the length of the noise vector, 240 samples. For overlap and add, a trapezoidal window is used with an overlap of 80 samples. For each synthesis frame the filtered noise buffer, u_{bp}, is shifted by 80 samples and a new 160 samples are appended, this eliminates the need for energy compensation functions to remove the windowing effects [43]. In fact the overlapped segments are correlated, and the trapezoidal windows do not distort the *rms* energy.

No attempt is made to preserve the phase continuity when switching to or from the noise excitation. When switching from a different mode, the unvoiced gain, g_{uv}, of the previous frame is set equal to the current value. The validity of these assumptions are tested through listening tests and the results confirm that these assumptions are reasonable and do not introduce any audible artifacts. The average bit rate can be further reduced by the introduction of voice activity detection (VAD) and comfort noise generation at the decoder for silence segments [9, 44].

9.6 Speech Classification

The speech classification or mode selection techniques can be divided into three categories [45].

- **Open-loop mode selection:** Each frame is classified based on the observations of parameters extracted from the input speech frame without assessing how the selected mode will perform during synthesis for the frame concerned.
- **Closed-loop mode selection:** Each frame is synthesized using all the modes and the mode that gives the best performance is selected.
- **Hybrid mode selection:** The mode selection procedure combines both open-loop and closed-loop approaches. Typically, a subset of modes is first selected by an open-loop procedure, followed by further refinements using closed-loop techniques.

Closed-loop mode selection has two major difficulties: high complexity and difficulty in finding an objective measure which reflects the subjective quality of synthesized speech [46]. The existing closed-loop mode selection coders are based on CELP, and select the best configuration such that the weighted MSE is minimized [47, 48]. Open-loop mode selection is based on techniques such as: voice activity detection, voicing decision, spectral envelope variation, speech energy, and phonetic classification [10]. See [49] for a detailed description on acoustic phonetics.

In the following discussion, a hybrid mode selection technique is used, with an open-loop initial classification and a closed-loop secondary classification. The open loop initial classification decides to use either the noise excitation or one of the other modes. The secondary classification synthesizes the harmonic excitation and makes a closed loop decision to use either the harmonic excitation or ACELP. A special feature of this classifier is the application of closed-loop mode selection to harmonic coding. The SWPM [26] preserves the waveform similarity of the harmonically-synthesized speech, making it possible to apply closed-loop techniques in harmonic coding.

9.6.1 Open-Loop Initial Classification

The initial classification extracts the fully unvoiced and silence segments of speech, which are synthesized using white-noise excitation. It is based on tracked energy, the low-band to high-band energy ratio, and the zero-crossing rate of the speech signal. The three voicing metrics are logically combined to enhance the reliability, since a single metric alone is not sufficient to make a decision with high confidence. The metric combinations and thresholds are determined empirically, by plotting the metrics with the corresponding speech waveforms. A statistical approach is not suitable for deciding the thresholds, because the design of the classification algorithm should consider that a misclassification of a voiced segment as unvoiced will severely degrade the speech quality, but a misclassification of an unvoiced segment as voiced can be tolerated. A misclassified unvoiced segment will be synthesized using ACELP, however a misclassified voiced segment will be synthesized using noise excitation.

The tracked energy of speech, t_e is estimated as follows:

$$t_e = \frac{0.00025e_h + e}{0.01e_h + e} \tag{9.48}$$

where e is the mean squared speech energy, given by,

$$e = \frac{\sum_{n=0}^{N-1} s^2(n)}{N} \tag{9.49}$$

where N, the length of the analysis frames, is 160 and e_h is an autoregressive energy term given by,

$$e_h = 0.9e_h + 0.1e \quad \text{if } 8e > e_h \tag{9.50}$$

The condition $8e > e_h$ ensures that e_h is updated only when the speech energy is sufficiently high and e_h should be initialized to approximately the mean squared energy of voiced speech. Figure 9.19a illustrates the tracked energy over a segment of speech. The low-band to high-band energy ratio, γ_ω, is estimated as follows:

$$\gamma_\omega = \frac{\int_0^{1/4} S^2\left(\frac{\omega}{\omega_s}\right) d\left(\frac{\omega}{\omega_s}\right)}{\int_{1/4}^{1/2} S^2\left(\frac{\omega}{\omega_s}\right) d\left(\frac{\omega}{\omega_s}\right)} \tag{9.51}$$

where ω_s is the sampling frequency and $S(\omega)$ is the speech spectrum. The speech spectrum is estimated using a 512-point FFT, after windowing 240 speech samples with a Kaiser window of $\beta = 6.0$. Figure 9.19b illustrates the low-band to high-band energy ratio over a segment of speech, where the speech signal is shifted down for clarity.

The zero-crossing rate is defined as the number of times the signal changes sign, divided by the number of samples used in the observation. Figure 9.20a illustrates the zero-crossing rate over a segment of speech, where the speech

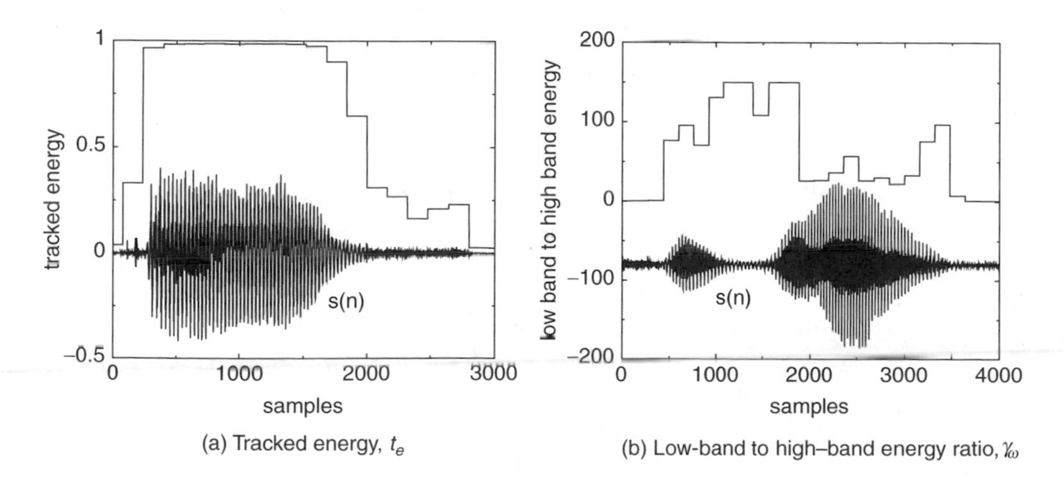

(a) Tracked energy, t_e

(b) Low-band to high–band energy ratio, γ_ω

Figure 9.19 Voicing metrics of the initial classification

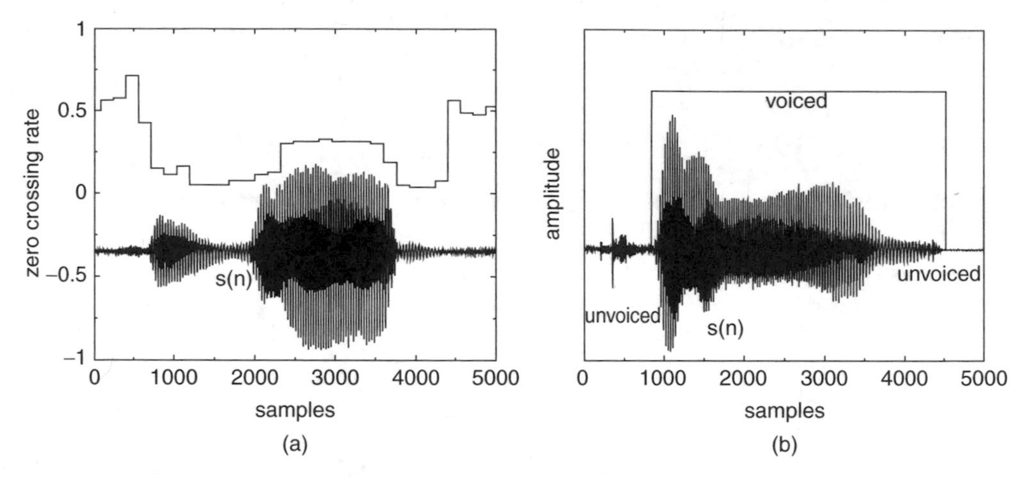

Figure 9.20 (a) Zero-crossing rate and (b) Voicing decision of the initial classification

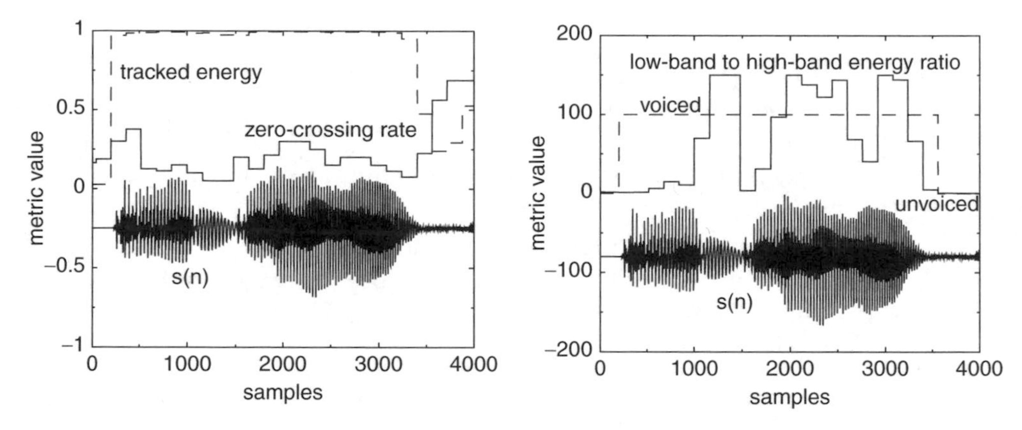

Figure 9.21 Voicing metrics of the initial classification

signal is shifted down for clarity. Figure 9.20b depicts the voicing decision made by the initial classification. Figure 9.21 depicts the three metrics used and the final voicing decision over the same speech segment.

Even though the plosives have a significant amount of energy at high frequencies and a high zero-crossing rate, synthesizing the high energy spikes of the plosives using ACELP instead of noise excitation improves speech quality. Therefore we need to detect the plosives, which are classified as unvoiced by the initial classification, and switch them to ACELP mode. A typical plosive is depicted at the beginning of the speech segment in Figure 9.20b.

9.6.2 Closed-Loop Transition Detection

AbS transition detection is performed on the speech segments [26, 27] that are declared voiced by the open-loop initial classification. A block diagram of the AbS classification process is shown in Figure 9.22. The AbS classification module synthesizes speech using SWPM and checks the suitability of the harmonic model for a given frame. The normalized cross-correlation and squared error are computed in both the speech domain and the residual domain for each of the selected pitch cycles within a synthesis frame. The pitch cycles are selected such that they cover the complete synthesis frame. The mode decision between harmonic and ACELP modes is then based on the estimated cross-correlation and squared error values. The squared error of the ith pitch cycle, E_i, is given by,

$$E_i = \frac{\sum_{j=0}^{T-1} \left[s\left(iT+j\right) - \hat{s}\left(iT+j\right)\right]^2}{\sum_{j=0}^{T-1} s^2\left(iT+j\right)} \quad \text{for } 0 \le i < I \qquad (9.52)$$

The normalized cross-correlation of the ith pitch cycle, R_i, is given by,

$$R_i = \frac{\sum_{j=0}^{T-1} s\left(iT+j\right)\hat{s}\left(iT+j\right)}{\sqrt{\sum_{j=0}^{T-1} s^2\left(iT+j\right) \sum_{j=0}^{T-1} \hat{s}^2\left(iT+j\right)}} \quad \text{for } 0 \le i < I \qquad (9.53)$$

where $T = \lfloor \tau + 0.5 \rfloor$, τ is the pitch period, $I = \lfloor N/\tau + 1 \rfloor$, and N is the synthesis frame length of 160 samples.

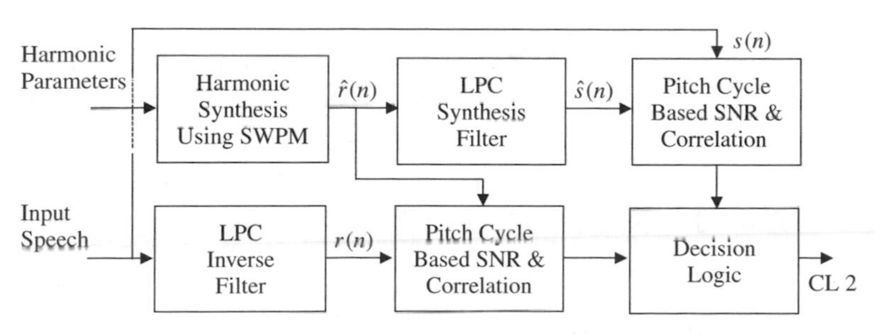

Figure 9.22 Analysis by synthesis classification

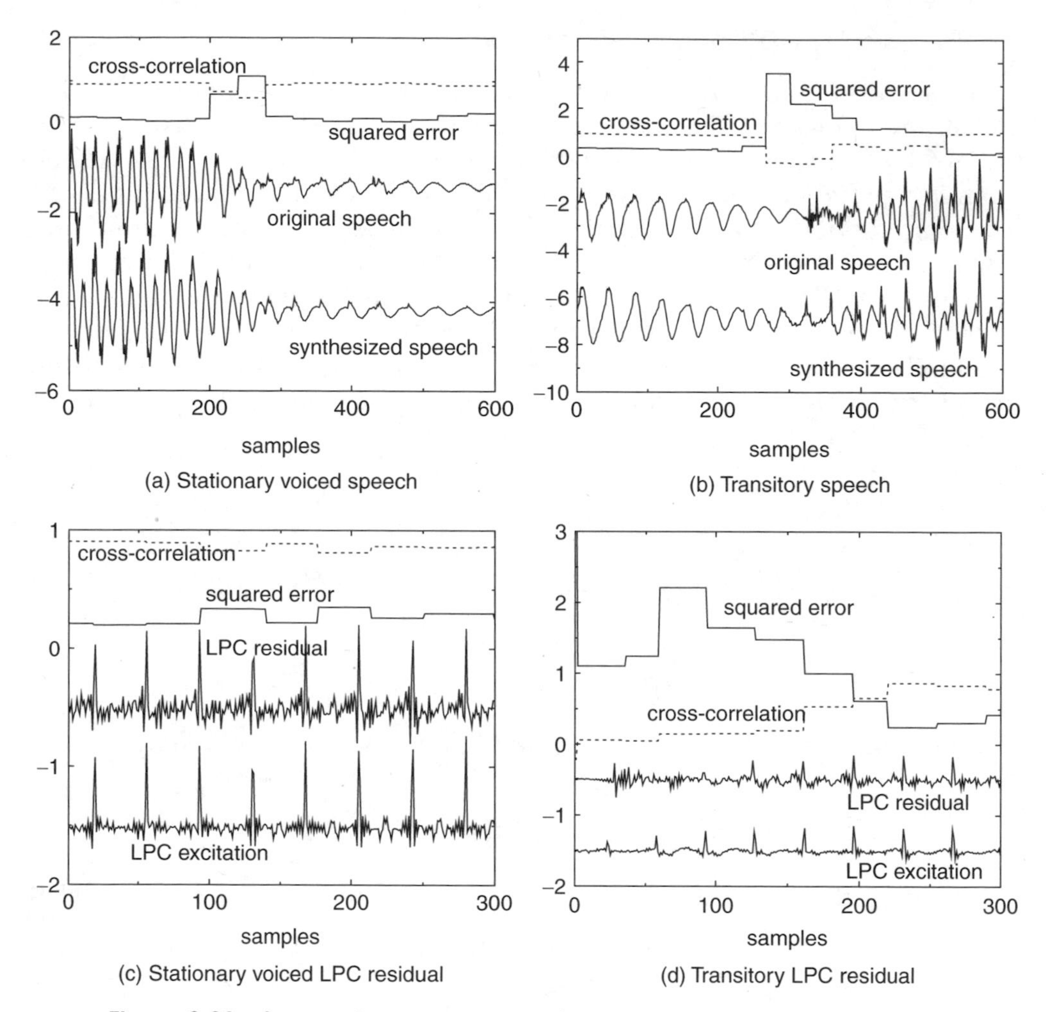

Figure 9.23 Squared error, E_i, E_{i_r}, and cross-correlation, R_i, R_{i_r}, values

In order to estimate the normalized residual cross-correlation, R_{i_r}, and residual squared error, E_{i_r}, equations (9.52) and (9.53) are repeated with $s(n)$ and $\hat{s}(n)$ replaced by $r(n)$ and $\hat{r}(n)$ respectively. Figure 9.23 depicts E_i, R_i, original speech $s(n)$, and synthesized speech $\hat{s}(n)$. E_i and R_i are aligned with the corresponding pitch cycles of the speech waveforms, and the speech waveforms are shifted down for clarity. Examples of the residual domain signals, LPC residual $r(n)$, LPC excitation $\hat{r}(n)$, E_{i_r}, and R_{i_r} are also shown in the figure.

For stationary voiced speech, the squared error, E_i, is usually much lower than unity and the normalized cross-correlation, R_i, is close to unity. However, the harmonic model fails at the transitions, which results in larger errors and lower correlation values. The estimated normalized cross-correlation and squared error values are logically combined to increase the reliability of the AbS transition detection. The combinations and thresholds are determined empirically by plotting the parameters with the corresponding speech waveforms. This heuristic approach is superior to a statistical approach, because it allows inclusion of the most important transitions, while the less important ones can be given a lower priority. AbS transition detection compares the harmonically synthesized speech with the original speech, verifies the accuracy of the harmonic model parameters, and decides to use ACELP when the harmonic model fails.

The cross-correlation and squared error values are estimated on the pitch cycle basis in order to determine the suitability of the harmonic excitation for each pitch cycle. Estimating the parameters over the complete synthesis frame may average out a large error caused by a sudden transition. In Figure 9.23a, the speech waveform has a minor transition. The estimated parameters also indicate the presence of such a transition. These minor transitions are synthesized using the harmonic excitation, and the mode is not changed to waveform coding. Changing the mode for these small variations leads to excessive switching, which may degrade the speech quality, when the bit-rate of the waveform coder is relatively low, due to the quantization noise of the waveform coding. Moreover, the harmonic excitation is capable of producing good quality speech despite those small variations in the waveform. In addition to maintaining the harmonic mode across those minor transitions, in order to limit excessive switching, the harmonic mode is not selected after ACELP when the speech energy is rapidly decreasing. Rapidly-decreasing speech energy indicates an offset and at some offsets the coding mode may fluctuate between ACELP and harmonic, if extra restrictions are not imposed. At such offsets, the accumulated error in the LPC memories through the harmonic mode is corrected by switching to the ACELP mode, which in turn causes a switch back to the harmonic mode. The additional measures taken to eliminate those fluctuations are described below.

In order to avoid mode fluctuations at the offsets, extra restrictions are imposed when switching to the harmonic mode after waveform coding. The *rms* energy of the speech and the LPC residual are computed for each frame, and a hysteresis loop is added using a control flag. The flag is set to zero when the speech or the LPC residual *rms* energy is less than 0.75 times the corresponding *rms* energy values of the previous frame. The flag is set to one when the speech or the LPC residual *rms* energy is more than 1.25 times the corresponding *rms* energy values of the previous frame. The flag is set to zero if the pitch is greater than 100 samples, regardless of the energy. When switching to harmonic mode after waveform coding, the control flag should be one, in addition to the mode decision of closed-loop transition detection. The flag is checked only at a mode transition, once the harmonic mode is initialized, the flag is ignored. This process avoids excessive switching at the offsets.

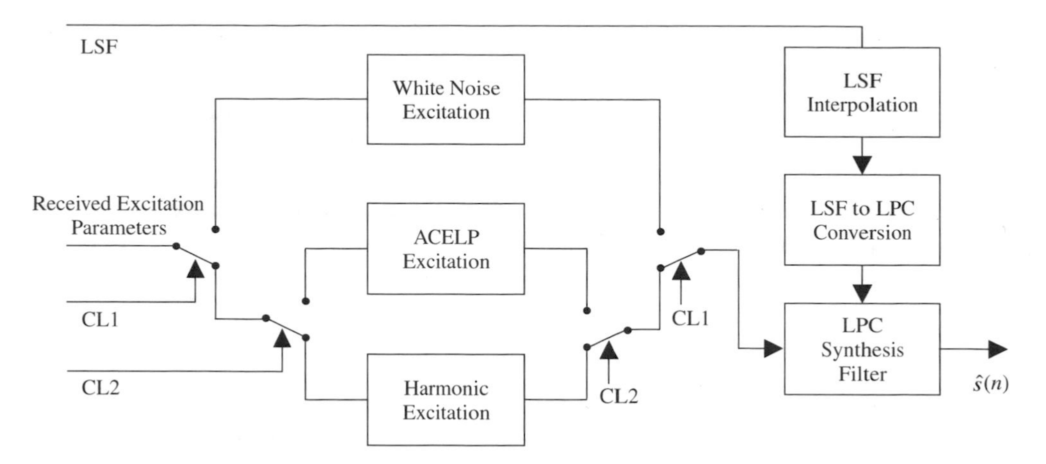

Figure 9.25 Block diagram of the hybrid decoder

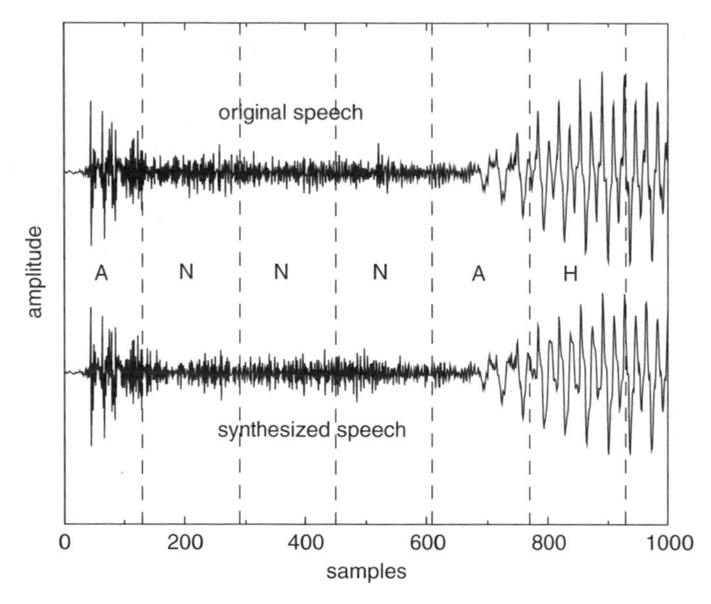

Figure 9.26 Synthesized speech and classification: A (ACELP), H (harmonic), and N (noise excitation)

9.8 Performance Evaluation

The hybrid coder [26] described above has been tested to evaluate its performance. The major tasks were developing a reliable classification technique and preserving the phase continuity when switching between the coding modes. The classification algorithm is tested using 64 seconds of modified

IRS-filtered speech, by comparing the mode decision against manually-classified waveforms. Eight English sentence pairs uttered by four male and four female speakers, taken from the Nippon Telegraph and Telephone [50] speech database are used as the test material. The silence segments are excluded from the analysis and synthesized using white-noise excitation. The initial classification detects all the voiced frames. Therefore the worst possible classification error, i.e. classifying a voiced frame as unvoiced, is eliminated. More than 90 % of the unvoiced frames are also detected and the rest of the unvoiced frames are misclassified as voiced. This bias towards voiced is preferable to misclassifying voiced frames as unvoiced, since the misclassified unvoiced frames will be classified as ACELP by the secondary classification, while a misclassified voiced frame will be synthesized using white-noise excitation. The plosive detection algorithm detects all the plosives in the unvoiced frames and does not misclassify other unvoiced frames as plosives.

The transition frames are manually marked by observing the waveforms, in order to test the closed-loop transition detection algorithm. Speech frames which have irregular pitch periods and show large variations in the energy are identified as transitions. The closed-loop transition detection classifies the frames already classified as voiced by the initial classification into transitory and harmonic. Consequently, all the frames classified as voiced by the initial classification are included in the test and the unvoiced frames that are classified as voiced are marked as transitions, since they are expected to be synthesized using ACELP. When testing the transition detection algorithm, the use of waveform coding for pitch periods longer than 100 samples is not activated. The transition detection algorithm detects more than 90 % of the transition frames and the rest of the transitions are classified as harmonic frames. It also detects more than 90 % of the harmonic frames and the rest of the stationary voiced frames are classified as transitions.

Misclassifications may restrict the maximization of the speech quality because of not choosing the best coding algorithm. However misclassifications of the secondary classification do not degrade the speech quality, due to its closed-loop nature. A misclassification of a stationary voiced segment as a transition indicates a harmonic parameter estimation error and such frames are synthesized using ACELP, perhaps a better solution than synthesizing with the inaccurate harmonic parameters. A misclassification of a transition as stationary voiced indicates that the harmonic mode is capable of synthesizing the particular transitory frame. This may be possible at some transitions, particularly offsets, which usually have a steady pitch contour and a smooth energy variation, where the harmonic interpolation model can fit in.

The phase continuity is tested by listening to the synthesized speech, without introducing quantization. The tests verify the validity of the hybrid model and there are no perceptible discontinuities. The speech synthesized

Table 9.1 Unquantized hybrid model vs 128 kb/s linear PCM

	Better	Slightly better	Same	Slightly worse	Worse
Male (%)	0.0	19.3	51.9	26.9	1.9
Female (%)	0.0	5.8	69.2	17.3	7.7
Average (%)	0.0	12.5	60.6	22.1	4.8

Table 9.2 Unquantized hybrid model vs 8 kb/s G.729

	Better	Slightly better	Same	Slightly worse	Worse
Male (%)	1.9	30.8	51.9	15.4	0.0
Female (%)	0.0	34.7	44.2	17.3	3.8
Average (%)	1.0	32.7	48.1	16.3	1.9

also indicates the upper bound of the quality achievable by the designed hybrid model. An informal listening test was conducted using 128 kb/s linear pulse code modulation (PCM), which is the best narrow-band speech quality, and 8 kb/s ITU G.729, a toll-quality speech coder, as the reference coders [26]. The speech material used for the test consists of eight sentences, four from male and four from female talkers, filtered by the modified IRS filter and a pair of headphones was used to conduct the test. Twelve listeners were asked to indicate their preferences for the randomized pairs of synthesized speech. Both experienced and inexperienced listeners participated in the test. The subjective test results are shown in Tables 9.1 and 9.2. As indicated by these results, the unquantized hybrid model performs better than G.729 and worse than 128 kb/s linear PCM. Therefore the quality of the unquantized hybrid model can be classified as being higher than toll quality and lower than transparent quality. In general, the speech encoded and decoded with unquantized hybrid coder model parameters does not sound too different from the original speech material. The perceived speech quality shows only a slight degradation, even after quantizing the harmonic mode parameters at 4 kb/s and white-noise excitation at 1.5 kb/s, with unquantized transitions (at 128 kb/s linear PCM). The hybrid coder achieves toll quality when transitions are quantized with 6 kb/s ACELP.

9.9 Quantization Issues of Hybrid Coder Parameters

9.9.1 Introduction

The above hybrid speech-coding model can be adopted for various applications with different quality requirements by quantizing the model parameters at different bit-rates. For applications which support variable bit rates, the model parameters of different modes may be quantized at different bit-rates, allocating the minimum number of bits required for each mode to maintain adequate quality.

In the example, here the LPC parameters are common for all the modes, and quantized using a fixed number of bits per frame. This is advantageous under noisy channel conditions, since the LPC parameters can be decoded correctly even when the mode bits are in error. The LPC parameters are quantized in the LSF domain using a multi-stage vector quantifier (MSVQ), with a first order moving average (MA) prediction [37]. Having quantized the LSFs, the excitation of the three modes are quantized differently.

9.9.2 Unvoiced Excitation Quantization

The hybrid coding algorithm synthesizes unvoiced speech using scaled white Gaussian noise as the LPC excitation. Therefore, only a gain term is required in addition to the LPC parameters to synthesize unvoiced speech. In order to synthesize the unvoiced plosives with adequate quality, the gain term should be updated at least every 5 ms. However listening tests show that synthesizing plosives using ACELP gives better perceptual quality. Therefore the plosives are synthesized using ACELP. The energy of the fricatives does not show rapid fluctuations and updating at the frame rate of every 20 ms is adequate to synthesize high-quality unvoiced fricatives.

The unvoiced gain g_{uv} is quantized using a logarithmic scalar quantizer. The quantized unvoiced gain g_{uv_i} is given by,

$$g_{uv_i} = k\left(\frac{g_{\max}+k}{k}\right)^{\frac{i}{N-1}} - k \quad \text{for} \ \ i = 0, 1, 2, \ldots, N-1 \tag{9.56}$$

where N is the number of quantizer levels, $g_{\max}$, defines the upper limit of g_{uv_i}, and k is a constant which controls the gradient of the exponential function. All the g_{uv} values larger than $g_{\max}$ are clipped at $g_{\max}$. The constant k is set as 16 and 32 quantizer levels were sufficient to produce high quality unvoiced speech. Hence five bits are required to transmit the quantized unvoiced gain, g_{uv_i}. Figure 9.27 depicts a typical plot of the unvoiced gain quantizer levels where the maximum $g_{\max} = 904$.

9.9.3 Harmonic Excitation Quantization

The stationary voiced speech segments are synthesized using the synchronized harmonic excitation model described earlier. The model parameters of the harmonic excitation with SWPM are pitch period, pitch pulse location (PPL), pitch pulse shape (PPS), harmonic amplitudes, and gain. The AbS transition detection algorithm synthesizes the harmonic excitation using SWPM at the encoder to evaluate the suitability of the harmonic mode.

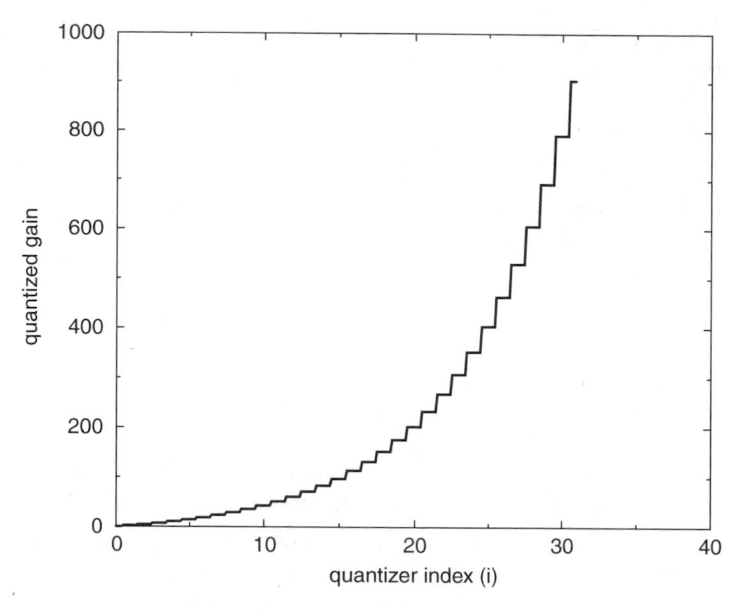

Figure 9.27 Unvoiced gain quantizer levels

Therefore, quantized or unquantized harmonic parameters may be used for the transition detection at the encoder. Generally, AbS algorithms include the quantization in the error minimization loop, so that the quantization noise is also accounted for in the parameter estimation process. However in this case, the solution is not straightforward, since the decision is between two modes, rather than the best set of parameters of a unimodal coder. One solution to this problem is to perform a full closed-loop mode decision with quantized parameters, i.e. synthesizing the speech frames with all the modes and selecting the best mode. A weighting factor may be required in the mode selection process, since the harmonic excitation with SWPM may give superior perceptual quality even with a slightly lower SNR compared to ACELP. However such a solution is computationally demanding, since ACELP excitation should be computed for all the frames, excluding the silence and unvoiced frames. Furthermore, defining a suitable weighting factor which reflects the perceptual quality is a difficult task.

A more practical solution is to decide the inclusion of the harmonic parameter quantization in the mode decision loop based on ACELP bit rate. The inclusion of the harmonic quantization in the closed-loop mode decision increases the number of ACELP mode frames. However, occasionally switching to ACELP between harmonic frames may degrade the perceptual quality, when the bit rate of the ACELP mode is below 8 kb/s, due to the sudden discontinuities introduced in the voiced harmonics. In general, ACELP operating at 8 kb/s or higher is capable of synthesizing perceptually-superior speech compared to harmonic coding (with no phase transmission), even at the stationary voiced segments. Therefore the harmonic quantization can be included in the closed-loop mode decision without worrying about the quality of ACELP coded frames between the harmonic frames (except, of course, the bit rate will be higher). However when the bit rate of ACELP mode is low, the quantization noise becomes audible; hence, trying to eliminate the quantization noise of the harmonic mode by switching to ACELP mode does not improve the perceptual quality. Therefore, in all the tests described here, harmonic parameter quantization is not included in the transition detection loop.

The sensitivity of AbS transition detection is different for each parameter. The sensitivity is high for the pitch period and PPL. Changes in these parameters dramatically reduce the cross-correlation of the original and the synthesized speech, due to the resulting time shifts. The spectral amplitudes and the LPC parameters are least sensitive. In fact, quantized and unquantized LPC parameters both produced the same classification decisions for the test speech material. The LPC memory locations of the transition detection algorithm are initialized for each frame with the memory locations of the LPC synthesis filter. This avoids drifting the LPC synthesis filter of the transition detection algorithm from the synthesized speech.

Pitch Quantization

The pitch period, τ, is quantized using a nonlinear scalar quantizer, reflecting the high sensitivity of the human ear to the pitch deviations at shorter pitch periods. A logarithmic scale is used for the pitch values from 16 to 60 samples and a linear scale is used for the pitch values from 60 to 160 samples (see Figure 9.28). The quantized pitch τ_i is given by,

$$\tau_i = \tau_{\min} \left(\frac{\tau_0}{\tau_{\min}} \right)^{\frac{i}{N_0-1}} \quad \text{for } i = 0, 1, 2, \ldots, N_0 - 1 \tag{9.57}$$

$$\tau_i = \tau_0 + \frac{\tau_{\max} - \tau_0}{N - N_0}(i - N_0 + 1) \quad \text{for } i = N_0, N_0 + 1, \ldots, N - 1 \tag{9.58}$$

where $\tau_{\min}$ is 16, $\tau_{\max}$ is 160, τ_0 is 60, N_0 is 156, and N is 256. Therefore eight bits are required to transmit the quantized pitch period.

Pitch Pulse Location Quantization

The pitch pulse location (PPL) is the location of the pitch pulse closest to the centre of the analysis frame. PPL may be defined as the distance to the pitch pulse concerned from the centre of the analysis frame, measured in samples. Assuming that the maximum possible pitch is 160 samples, PPL varies between −80 and 80. However the pitch pulse location may be normalized

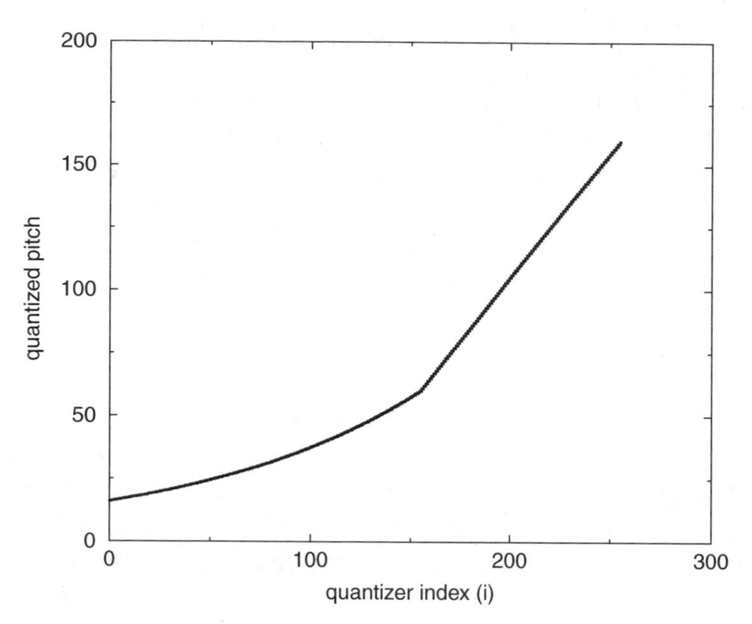

Figure 9.28 Pitch quantizer levels

with respect to the pitch so that the PPL varies between −0.5 and 0.5. Normalization of the PPL with respect to the pitch ensures the efficient use of quantizer dynamic range regardless of the pitch.

The accuracy of the PPL is more important when it is close to the centre of the analysis frames or the synthesis frame boundaries, i.e. PPL values close to zero. This is due to the fact that the mode changes between ACELP and harmonic excitation may take place at the synthesis frame boundaries. Preserving the continuity of the high-energy pitch pulses occurring at or close to the switching frame boundaries is essential to eliminate audible switching artifacts. Therefore the normalized PPL is quantized using a logarithmic scale, quantizing the PPL values close to zero more accurately. The quantized PPL, t_i is given by,

$$t_i = k\left(\frac{0.5+k}{k}\right)^{\frac{i-N/2-1}{N/2}} - k \quad \text{for } i = N/2-1, N/2, \ldots, N-1 \quad (9.59)$$

$$t_i = t_{N-2-i} \quad \text{for } i = 0, 1, \ldots, N/2-2 \quad (9.60)$$

where N is the number of quantizer levels and k is a constant that controls the gradient of the exponential function. The constant k is set to 0.125, and 128 quantizer levels are sufficient to eliminate audible switching artifacts. Hence seven bits are required to transmit the quantized normalized PPL. PPL is normalized using the quantized pitch so that the decoder can denormalize the

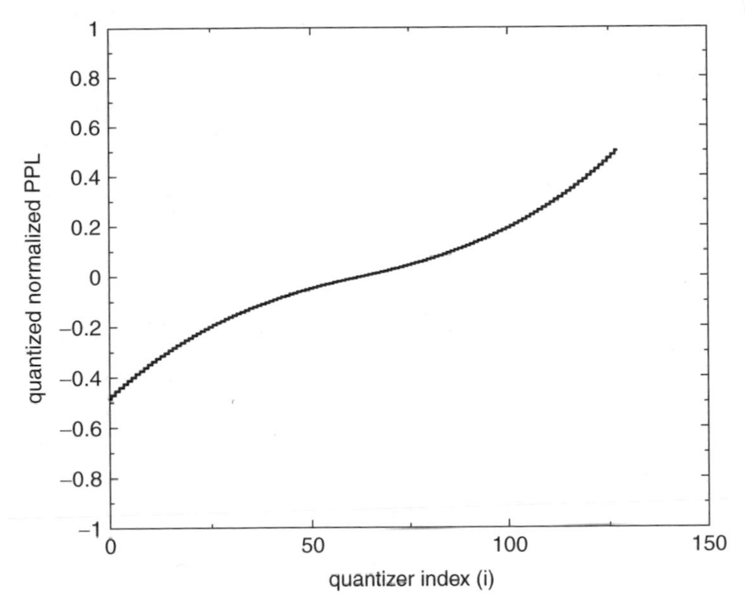

Figure 9.29 PPL quantizer levels

received PPL value accurately. Figure 9.29 depicts a plot of the normalized PPL quantizer levels.

Pitch Pulse Shape Quantization

Large variations in the PPS introduces a reverberant character into the synthesized speech, regardless of the PPS value. Therefore, in terms of the perceptual quality, all the PPS values are equally important and a linear quantizer is employed to quantize the PPS using 16 values. The quantized PPS, θ_i, is given by,

$$\theta_i = \frac{2\pi}{N}i - \pi \quad \text{for } i = 0, 1, \ldots, N-1 \text{ and } \quad -\pi \leq \theta_i < \pi \qquad (9.61)$$

where N, the number of quantizer levels, is 16 and four bits are required to quantize PPS.

Harmonic Amplitude Quantization

Harmonic amplitudes of the LPC residual are quantized using Switched Predictive Mel-scale-based Vector Quantization (SP-MVQ) [51]. SP-MVQ (see block diagram in Figure 9.30) converts the variable-dimension spectral-amplitude vectors into fixed-dimension vectors by warping the frequency axis using a logarithmic scale. The warping process emphasizes the low

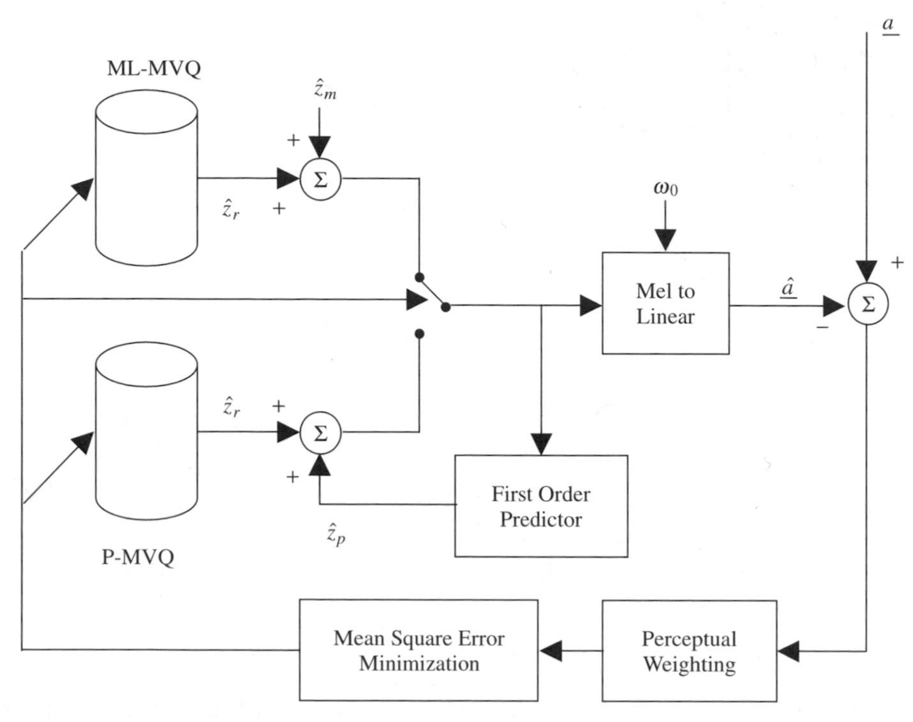

Figure 9.30 Block diagram of SP-MVQ

frequencies, taking into account the perceptual preferences of the human auditory system. The fixed dimension spectral vector, $\hat{z}$, is decomposed into a predicted vector, $\hat{z}_p$, and a prediction residual vector, $\hat{z}_r$, as follows:

$$\hat{z} = \hat{z}_p + \hat{z}_r \tag{9.62}$$

where the predicted vector, $\hat{z}_p$, is obtained using a first-order autoregressive method, given by,

$$\hat{z}_p = \Phi\left(\hat{z}_{-1} - \hat{z}_m\right) + \hat{z}_m \tag{9.63}$$

where $\hat{z}_{-1}$ is the most recently quantized $\hat{z}$, $\hat{z}_m$ is the mean vector, and Φ denotes a diagonal matrix of prediction coefficients. The prediction residual, $\hat{z}_r$ is quantized using a typical vector quantizer such as MSVQ [52]. The quantization becomes memoryless Mel-scale-based vector quantization (ML-MVQ) if all the prediction coefficients are zero, and autoregressive predictive MVQ (P-MVQ) otherwise. The predictive scheme is effective in stationary regions, and may increase spectral distortion at the transitions; therefore, a switching scheme is introduced to switch between P-MVQ and ML-MVQ. The decision between P-MVQ and ML-MVQ is made using AbS techniques and based on a weighted spectral-distortion measure. Therefore the quantization scheme is called switched predictive Mel-scale-based vector quantization (SP-MVQ). Moreover, the switching scheme restricts error propagation under noisy channel conditions.

SP-MVQ quantizes spectral amplitudes every 10 ms using 14 bits. The harmonic analysis/synthesis scheme described estimates the harmonic parameters every 20 ms. However there are sufficient bits for the allocation of 28 bits per 20 ms frame for spectral amplitudes at 4 kb/s (see Table 9.5). Therefore the harmonic analysis/synthesis scheme is modified to update the spectral amplitudes every 10 ms. However the pitch is transmitted only every 20 ms, and linearly interpolated to compute the number of harmonics corresponding to the centre of the synthesis frame or the first subframe, at the decoder. The spectral amplitude quantization uses the quantized (second subframe) or quantized and interpolated (first subframe) pitch to compute the number of harmonics, in order to ensure the correct dequantization of the spectral amplitude vectors. In the spectral amplitude quantization of the first subframe, if the actual number of harmonics is greater than the computed number of harmonics by interpolation, the higher harmonics are ignored. If the actual number of harmonics is less than the computed number of harmonics by interpolation, the amplitude vector is zero-padded. Usually the pitch values of the stationary voiced segments are fairly unchanged and linear interpolation of the number of harmonics gives a good approximation.

Harmonic Gain Quantization

The spectral amplitude vectors are normalized before the quantization, in order to improve the dynamic range. The shape components of the vectors are quantized using SP-MVQ, as described above, and the gain component is scalar quantized.

The normalized amplitude, a_{k_n}, of the kth harmonic is given by,

$$a_{k_n} = \frac{a_k}{g} \tag{9.64}$$

where a_k is the spectral amplitude estimated for the kth harmonic and g is the normalization factor, given by,

$$g = \sqrt{\frac{\sum_{k=1}^{K} a_k^2}{K}} \tag{9.65}$$

where K is the total number of harmonics. Normalization factor of the second subframe, g_2, is quantized using a logarithmic scale, given by,

$$g_{2_i} = k\left(\frac{g_{max} - g_{min} + k}{k}\right)^{\frac{i}{N-1}} - k + g_{min} \quad \text{for } i = 0, 1, \ldots, N-1 \qquad (9.66)$$

where k is eight (which controls the gradient of the exponential function), N (the number of quantizer levels) is 32, i.e. five bits are required to quantize the gain of the second subframe, and g_{max} and g_{min} are the maximum and minimum possible quantized normalization factors, respectively. The gain values beyond g_{max} and g_{min} are clipped by the quantizer. The term g_{min} is introduced in equation (9.66), because only the stationary voiced segments are synthesized using the harmonic excitation and the minimum gain is nonzero.

The normalization factor of the first subframe, g_1, is differentially quantized with respect to the mean of the adjacent two quantized g_2 values, as follows:

$$\delta = g_1 - \frac{g_2 + g_{2_{-1}}}{2} \qquad (9.67)$$

where $g_{2_{-1}}$ is the gain of the second subframe of the previous frame, i.e. the previous g_2, and δ is quantized using three bits. Finally the spectral amplitude vectors are denormalized by multiplying with the quantized normalization factors.

Onset Harmonic Parameter Quantization

The harmonic synthesis process interpolates the parameters between the synthesis frame boundaries. However, at the onsets, when switching from waveform-coding mode, the harmonic parameters of the initial synthesis frame boundary are not directly available. The pitch, PPL, and PPS are estimated, as described in Section 9.5.4, and quantized as described in the preceding sections.

The spectral amplitudes of the ACELP excitation signal used before the harmonic mode are estimated by windowing it using an asymmetric window function given by,

$$w(n) = 0.54 - 0.46\cos\left(\pi\frac{n}{n_1 - 1}\right) \quad \text{for } 0 \le n < n_1 \qquad (9.68)$$

$$w(n) = 0.08 + 0.92\cos\left(\frac{\pi}{2}\frac{n - n_1}{n_2 - 1}\right) \quad \text{for } n_1 \le n < n_1 + n_2 \qquad (9.69)$$

where n_1 is 140 and n_2 is 20. The asymmetric window function emphasizes the excitation signal close to the switching frame boundary.

The spectral amplitude vector of the windowed ACELP excitation signal is obtained by peak-picking of the magnitude spectrum, using the received pitch value for the harmonic frame. The *rms* normalization factor of the estimated spectral vector is used as $g_{2_{-1}}$ of the harmonic frame. The amplitude quantization memory, $\hat{z}_{-1}$ is initialized by quantizing the normalized shape vector, while forcing SP-MVQ to use memoryless quantization.

9.9.4 Quantization of ACELP Excitation at Transitions

The transitions are quantized using algebraic code excited linear prediction (ACELP). The pulse innovation of ACELP is capable of synthesizing highly nonstationary transitions. The long term prediction (LTP) is not very efficient at the onsets, since the LTP memory buffer has no information regarding the onsets. However LTP is employed, because it reduces the sparse excitation artifacts [39] and synthesizes a significant amount of the excitation at the offsets. Moreover, at the resonance offsets, where the gain of the excitation signal is small, the LTP gain acts as an adaptive gain term and compensates for an inadequate gain quantization dynamic range of the innovation pulses. Multi-tap and fractional delay LTP filters [53] are useful only for stationary voiced segments, consequently, only integer delays and single-tap filters are used to encode transitions.

The LTP gain is close to unity during the stationary voiced segments. However at the transitions, LTP gain shows large variations, due to the large variations in the speech energy. Therefore the LTP gain is quantized using a larger dynamic range. A drawback in allowing gain values larger than unity is that the LTP filter may become unstable under erroneous channel conditions. The high-energy pulses of plosives are synthesized using only the innovation sequence of ACELP. However the plosives are not classified as a separate mode; instead, when a plosive is detected, the LTP gain is forced to be zero.

9.10 Variable Bit Rate Coding

When using a 4 kb/s harmonic coder for steady state voiced segments and unvoiced segments quantized at 1.5 kb/s (as detailed in Table 9.5) with unquantized transitions, the synthesized speech quality shows only a slight degradation when compared with using the unquantized model parameters, which is nearly transparent. The quality versus the bit-rate limitation of this hybrid coder is therefore dependent on transition quantization by ACELP. Informal listening tests show that quantizing the transitions at 6 kb/s is sufficient to achieve toll quality. Three versions of the coder are tested and compared with standard coders by quantizing the transitions at 4, 6 and 8 kb/s.

9.10.1 Transition Quantization with 4 kb/s ACELP

The 4 kb/s version uses 10 ms subframes. For each subframe the LTP delay, LTP gain, locations, signs, and the gain of two innovation pulses are transmitted. The innovation gain terms of the two subframes are normalized with respect to the quantized *rms* energy of the speech signal and the normalization factor is transmitted for each 20 ms frame. The normalization reduces the dynamic range required to quantize the innovation sequence gain. Table 9.5 shows the bit allocation of the 4 kb/s ACELP parameters. The LTP delay range is from 20 to 147, and only integer delays are allowed, needing seven bits for the index. The LTP gain is quantized using four bits (see Table 9.3). The two innovation pulses cover only the first 64 locations of each 80-sample subframe. Each pulse is chosen from 32 possible locations, either even or odd, and five bits are required to transmit the location. The sign of each pulse is transmitted using one bit. The pulse gain and the common normalization factor of the frame are quantized using three bits each (see Table 9.4).

9.10.2 Transition Quantization with 6 kb/s ACELP

The 6 kb/s version uses 5 ms subframes. For each subframe the LTP delay, LTP gain, locations, signs, and the gain of two innovation pulses are transmitted. The pulse gain terms of the four subframes are normalized with respect to the quantized *rms* energy of the speech signal and the normalization factor is transmitted for each 20 ms frame. Table 9.5 shows the bit allocation of the 6 kb/s ACELP parameters. The LTP delay and gain are quantized in the same way to the 4 kb/s version, using seven bits and four bits respectively.

The two innovation pulses cover only the first 32 locations of each 40-sample subframe. Each pulse is chosen from 16 possible locations, either even

Table 9.3 LTP Gain quantizer table

Index	0	1	2	3	4	5	6	7
LTP Gain	0.00	0.15	0.30	0.40	0.50	0.60	0.70	0.80

Index	8	9	10	11	12	13	14	15
LTP Gain	0.90	1.05	1.20	2.00	3.50	5.50	8.00	10.00

Table 9.4 Innovation pulse gain quantizer table

Index	0	1	2	3	4	5	6	7
Pulse Gain	0.0	0.3	0.7	1.1	1.6	2.1	2.7	3.5
rms Gain	10	40	90	176	325	584	1030	1800

Table 9.5 Bit allocation for a 20 ms frame

Parameters	White noise	Harmonic	ACELP 4k	ACELP 6k
LPC	23	23	23	23
Pitch	–	8	–	–
PPL	–	7	–	–
PPS	–	4	–	–
Amplitudes	–	14 + 14	–	–
Gain	5	3 + 5	3	3
LTP Delay	–	–	7 + 7	7 + 7 + 7 + 7
LTP Gain	–	–	4 + 4	4 + 4 + 4 + 4
Pulse Locations	–	–	10 + 10	8 + 8 + 8 + 8
Pulse Signs	–	–	2 + 2	1 + 1 + 1 + 1
Pulse Gain	–	–	3 + 3	3 + 3 + 3 + 3
Mode	2	2	2	2
Total	30	80	80	120

or odd, and four bits are required to transmit the location. The signs of the two pulses are forced to be opposite in the error minimization process, hence only the sign of the first pulse is transmitted, using one bit. The pulse gain and the common normalization factor of the frame are quantized using three bits each (see Table 9.4).

9.10.3 Transition Quantization with 8 kb/s ACELP

The 8 kb/s version uses 5 ms sub frames. For each subframe the LTP delay, LTP gain, locations, signs, and the gain of four innovation pulses are transmitted. The pulse gain terms of the four subframes are normalized with respect to the quantized *rms* energy of the speech signal and the normalization factor is transmitted for each 20 ms frame. Table 9.8 shows the bit allocation of the 8 kb/s ACELP parameters. The LTP delay and gain are quantized in the same way as the 4 kb/s version, using seven bits and four bits, respectively.

The locations and the signs of the four pulses are shown in Table 9.6. The pulse gain of each subframe is quantized using four bits, as shown in Table 9.7. The common normalization factor, i.e. the *rms* energy of the original speech signal, in each frame is logarithmically quantized using seven bits, and the quantized value, g_{rms_i}, is given by,

$$g_{rms_i} = k\left(\frac{g_{\max} - g_{\min} + k}{k}\right)^{\frac{i}{N-1}} - k + g_{\min} \quad \text{for } i = 0, 1, \ldots, N-1 \qquad (9.70)$$

Table 9.6 Structure of the 17-bit algebraic codebook

Pulse	Amplitude	Position	Bits
0	±1	0, 5, 10, 15, 20, 25, 30, 35	1 + 3
1	±1	1, 6, 11, 16, 21, 26, 31, 36	1 + 3
2	±1	2, 7, 12, 17, 22, 27, 32, 37	1 + 3
3	±1	3, 8, 13, 18, 23, 28, 33, 38, 4, 9, 14, 19, 24, 29, 34, 39	1 + 4

Table 9.7 Innovation pulse gain quantizer table for 8 kb/s ACELP

Index	0	1	2	3	4	5	6	7
Pulse Gain	0.0	0.15	0.3	0.45	0.6	0.8	1.0	1.2

Index	8	9	10	11	12	13	14	15
Pulse Gain	1.5	1.8	2.1	2.4	2.8	3.2	3.7	4.3

Table 9.8 Bit allocation of 8 kb/s ACELP for a 20 ms frame

Parameters	ACELP 8k
LPC	23
Gain	7
LTP Delay	7 + 7 + 7 + 7
LTP Gain	4 + 4 + 4 + 4
Pulse Locations	13 + 13 + 13 + 13
Pulse signs	4 + 4 + 4 + 4
Pulse Gain	4 + 4 + 4 + 4
Mode	2
Total	160

Where k (a constant which controls the gradient of the exponential function) is 80, N (the number of quantizer levels) is 128, and $g_{\max}$ and $g_{\min}$ are 2720.5 and 0.5 respectively.

9.10.4 Comparison

Three informal listening tests were conducted to assess the speech quality of the hybrid coder, with transitions quantized at 4 kb/s, 6 kb/s, and 8 kb/s. The synthesized speech was compared against that from 5.3 kb/s ITU G.723.1, 6.3 kb/s ITU G.723.1, and 8 kb/s ITU G.729 coders. In all the tests, stationary voiced segments were quantized at 4 kb/s, and silence and unvoiced segments are quantized at 1.5 kb/s. The speech material used for each test consists of eight sentences, four from male and four from female talkers, filtered by modified IRS filter; a pair of headphones was used to conduct the test. Twelve listeners were asked to indicate their preferences for randomized pairs of synthesized speech. Both experienced and inexperienced listeners participated in the test. The subjective test results are shown in Tables 9.9, 9.10, and 9.11.

For the speech material used in the subjective tests, after discarding the silence frames, about 64 % of the frames used harmonic excitation, 22 % used ACELP, and 14 % used white-noise excitation. The 4 kb/s, 6 kb/s, and 8 kb/s ACELP mode hybrid coders give average bit-rates of 3.65 kb/s, 4.1 kb/s, and 4.53 kb/s, respectively. The 4 kb/s ACELP version performs slightly better than G.723.1 at 5.3 kb/s. The 6 kb/s ACELP version achieves similar quality to G.723.1 at 6.3 kb/s. The quality of the 8 kb/s ACELP version is also similar to G.729 at 8 kb/s, with an overall average bit rate of 4.53 kb/s.

Table 9.9 4 kb/s ACELP hybrid vs 5.3 kb/s G.723.1

	Better	Slightly better	Same	Slightly worse	Worse
Male (%)	6.2	34.4	28.2	31.2	0.0
Female (%)	9.4	31.2	37.5	18.8	3.1
Average (%)	7.8	32.8	32.8	25.0	1.6

Table 9.10 6 kb/s ACELP hybrid vs 6.3 kb/s G.723.1

	Better	Slightly better	Same	Slightly worse	Worse
Male (%)	0.0	31.3	43.7	18.8	6.2
Female (%)	6.3	28.1	37.5	21.9	6.2
Average (%)	3.2	29.7	40.6	20.3	6.2

Table 9.11 8 kb/s ACELP hybrid vs 8 kb/s G.729

	Better	Slightly better	Same	Slightly worse	Worse
Male (%)	0.0	9.6	65.4	23.1	1.9
Female (%)	1.9	11.5	55.8	30.8	0.0
Average (%)	1.0	10.6	60.5	26.9	1.0

9.11 Acoustic Noise and Channel Error Performance

Robustness to background noise and channel errors is an important factor for any practical speech-coding algorithm. The speech coders designed for mobile and military communication applications frequently encounter acoustic noise and channel errors. The background noise may be suppressed before the encoding process using a noise preprocessor [54]. However, this involves additional complexity and delay, which may not be desirable for mobile communication applications. Therefore the speech-coding algorithms are expected to produce intelligible synthetic speech even in the presence of background noise. Generally, AbS coders perform better than parametric coders under noisy background conditions. This inherent robustness of AbS coders is due to their waveform-matching process. The error minimization process attempts to synthesize the input waveform regardless of its contents. The model parameters estimated by the parametric coders may not be accurate when the input speech signal is corrupted with noise. Inaccurate model parameters may severely degrade the synthetic speech of a parametric coder.

Channel errors are usually divided into two classes: random errors and burst errors. A speech-coding algorithm should provide a reasonable output even if a small proportion of the received bit stream is incorrect due to random bit errors. Robustness against random channel errors can be increased by means of index assignment algorithms [55, 56], through proper quantizer design, and by adding redundancy into the transmitted information [57, 58, 59]. Unequal error protection techniques may be applied to provide a higher degree of protection to the most sensitive bits. For example, in CELP coders, the spectral envelope parameters are the most sensitive to errors, followed by the fixed codebook gain, the adaptive codebook index, the adaptive codebook gain, the sign of the fixed codebook gain, and the fixed codebook index [60]. In the case of sinusoidal coders, the gain is the most sensitive to errors, followed by the voicing, the pitch, the spectral envelope parameters, and the spectral amplitudes [61].

In the case of burst errors, error detection schemes are used to classify each frame of received bits as usable or unusable. A similar problem encountered in packet voice communication systems is lost packets due to transmission impairments and excessive delays. In order to reduce the annoying artifacts due to lost frames, concealment techniques based on waveform substitution can be used [62]. The burst errors may also be converted to occur in a more random fashion using interleaving techniques. The performance issues specific to a hybrid coding algorithm are the robustness of the classification algorithm under acoustic noise and the channel bit error performance of the coding mode; otherwise, the performance of hybrid coders will be similar to either ACELP or harmonic coding.

9.11.1 Performance Under Acoustic Noise

The classification algorithm was tested using 64 seconds of male and female speech corrupted with either babble or vehicular noise. The SNR of the corrupted speech is 10 dB.

Figure 9.31 depicts the classification of the female speech. The initial classification declares only the strongly-unvoiced segments as unvoiced and all

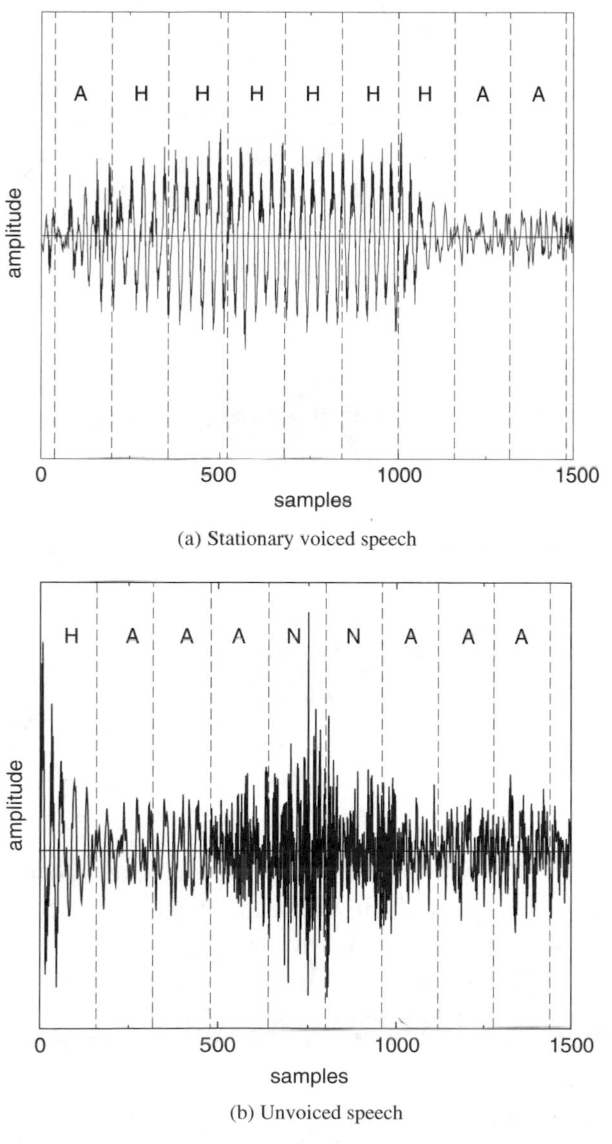

(a) Stationary voiced speech

(b) Unvoiced speech

Figure 9.31 Classification of female speech corrupted by babble noise (10 dB SNR): A (ACELP), H (harmonic), and N (noise excitation)

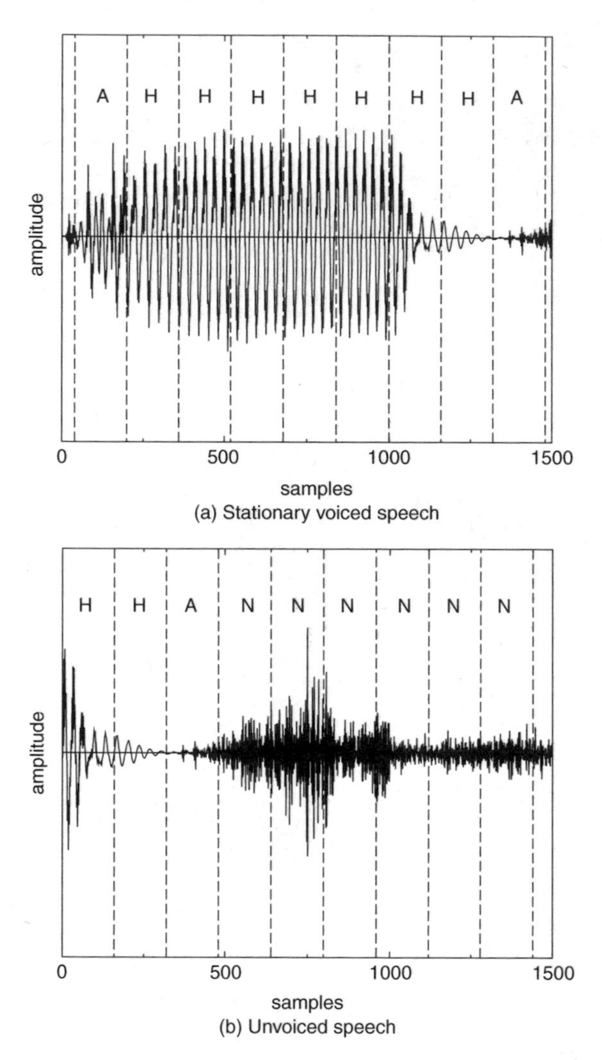

Figure 9.32 Classification of clean speech corresponding to Figure 9.31

the other frames are left to be encoded using either ACELP or harmonic excitation (compare Figures 9.31b and 9.32b). The weakly-unvoiced segments which have lower energy than the noise level are not detected as unvoiced. When corrupted with babble or vehicular noise, the silence and the low-energy unvoiced segments do not have the properties of unvoiced speech. It can be seen that the energy of the noise component is comparable with unvoiced speech and it has a significant low-frequency component (see Figure 9.35a). This is expected since babble noise is essentially attenuated and superimposed speech components. Figure 9.33 shows the classification of the male speech and Figure 9.34 shows the corresponding clean speech segments.

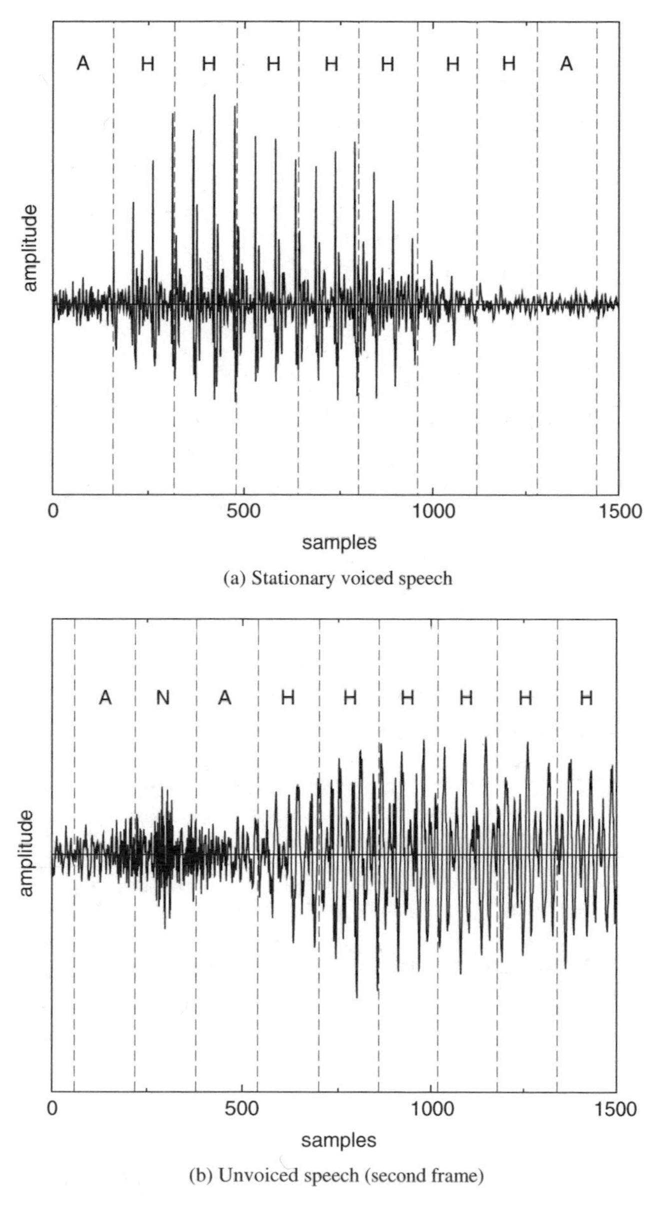

(a) Stationary voiced speech

(b) Unvoiced speech (second frame)

Figure 9.33 Classification of male speech corrupted by babble noise (10 dB SNR): A (ACELP), H (harmonic), and N (noise excitation)

The secondary classification performs very similarly under the clean speech conditions, except for the occasional classification of frames as ACELP, which were originally classified as harmonic under the clean speech conditions (compare Figures 9.31a and 9.32a). This is due to the inability of the harmonic model to adequately synthesize the corrupted signal and the model parameter

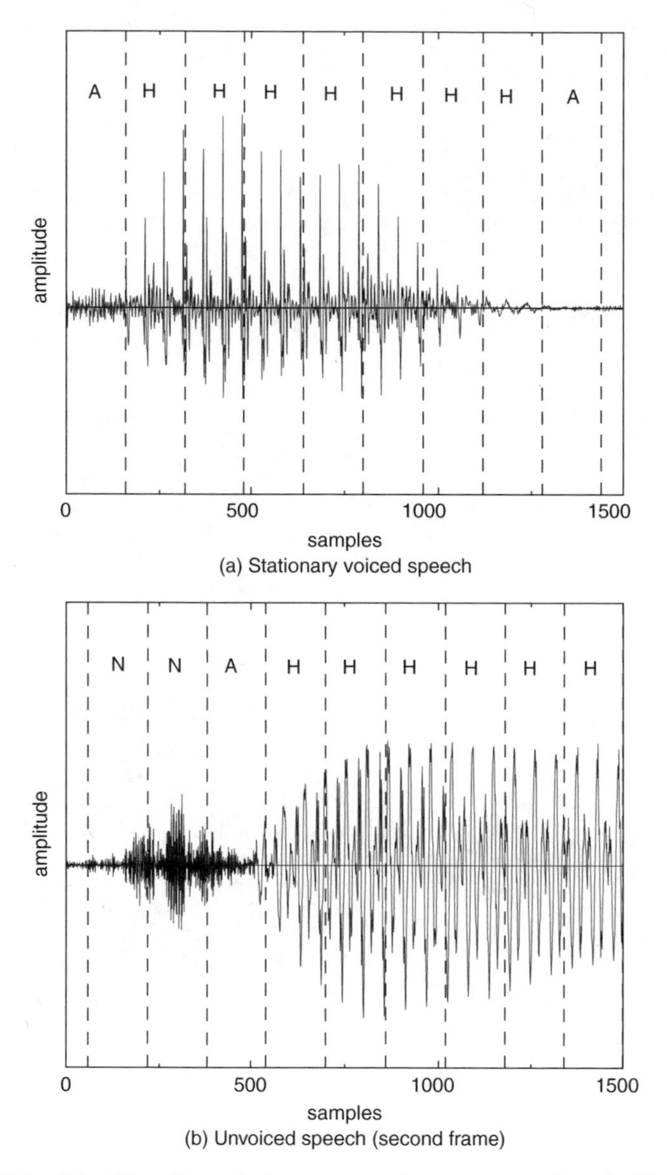

Figure 9.34 Classification of clean speech corresponding to Figure 9.33

estimation errors. Therefore, in general, in the presence of acoustic noise the speech classification algorithm declares more frames as ACELP. These include the silence frames of the original clean speech, unvoiced segments with lower energy than the noise level, and the stationary voiced frames with parameter estimation and harmonic modelling difficulties.

Neither white-noise excitation nor harmonic excitation is suitable for synthesizing the background noise. The spectra of babble and vehicular noise are not white, even after discarding the spectral envelope. synthesizing them

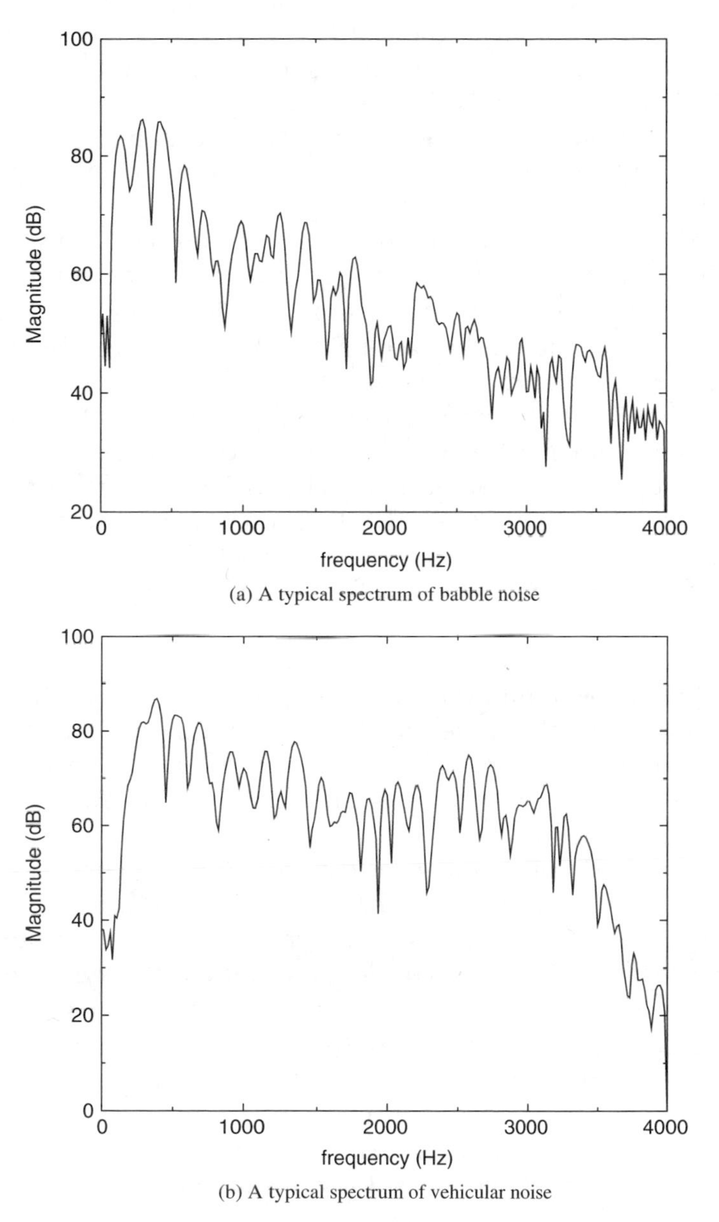

(a) A typical spectrum of babble noise

(b) A typical spectrum of vehicular noise

Figure 9.35 Typical acoustic noise spectra

using white-noise excitation will degrade the perceptual quality by introducing an unnaturally noisy background. Therefore, in fact, the classification algorithm detects the most suitable mode, i.e. ACELP, to synthesize background noise. However the drawback is a high average bit-rate, which may be reduced by using a robust voice activity detection (VAD) algorithm and comfort noise generation at the decoder end [9].

The correct classification of the stationary voiced segments as harmonic mode under noisy background conditions confirms the robustness of SWPM, since the AbS classification algorithm synthesizes speech using SWPM. Therefore, it can be concluded that the pitch pulse location (PPL) and the pitch pulse shape (PPS) detection algorithms described in Section 9.4 perform well under noisy background conditions.

An informal listening test was conducted to compare the speech quality of the hybrid coder under noisy background conditions with white noise, harmonic excitation, and ACELP quantized at 1.5 kb/s, 4 kb/s, and 6 kb/s, as discussed before. The synthesized speech was compared against the same noisy speech files synthesized using the 6.3 kb/s ITU G.723.1 coder. The speech material used for each test consists of eight sentences, four from male and four from female talkers, four corrupted with vehicular noise and four corrupted with babble noise (10 dB SNR); a pair of headphones was used to conduct the test. Twelve listeners were asked to indicate their preferences for the randomized pairs of synthesized speech. Both experienced and inexperienced listeners were participated in the test. The test results are shown in Table 9.12.

The informal listening test shows a clear preference for the 6.3 kb/s ITU G.723.1 coder. It was found that this is due to the metallic character of the stationary voiced speech synthesized by the harmonic excitation: it is cleaner, however, there is a pronounced metallic character. The test confirms that the listeners prefer more natural-sounding, noisy speech rather than metallic speech.

The metallic character is not so pronounced in noisy speech synthesized using a split-band LPC (SB-LPC) harmonic coder [4]. The SB-LPC coder divides the speech spectrum into two bands using a voicing frequency marker, where the upper band is declared unvoiced, and synthesized using a filtered noise excitation. For clean stationary voiced speech, most of the spectrum is declared voiced. However in the case of stationary voiced segments of noisy speech, some frequency bands are declared unvoiced. Therefore the voicing decision of SB-LPC reduces quality, synthesizing metallic sounds under noisy background conditions. The harmonic excitation model described in Section 9.5.1 was designed to synthesize stationary voiced segments and the complete spectrum is synthesized using harmonically related sinusoids.

Table 9.12 Hybrid vs 6.3 kb/s G.723.1 for noisy speech

	Better	Slightly better	Same	Slightly worse	Worse
Male(%)	0.0	9.6	38.5	40.4	11.5
Female(%)	0.0	21.2	21.2	42.3	15.3
Average(%)	0.0	15.4	29.9	41.3	13.4

Under noisy background conditions, there are strong spectral components which are not related to the fundamental frequency of the speech. These noise components change the harmonic amplitudes and are perceived as metallic sounds in harmonically synthesized speech (see Figure 9.36). Introducing a voicing frequency marker for the harmonic excitation, similar to SB-LPC, improves the speech quality of the hybrid coder, especially in noisy back-

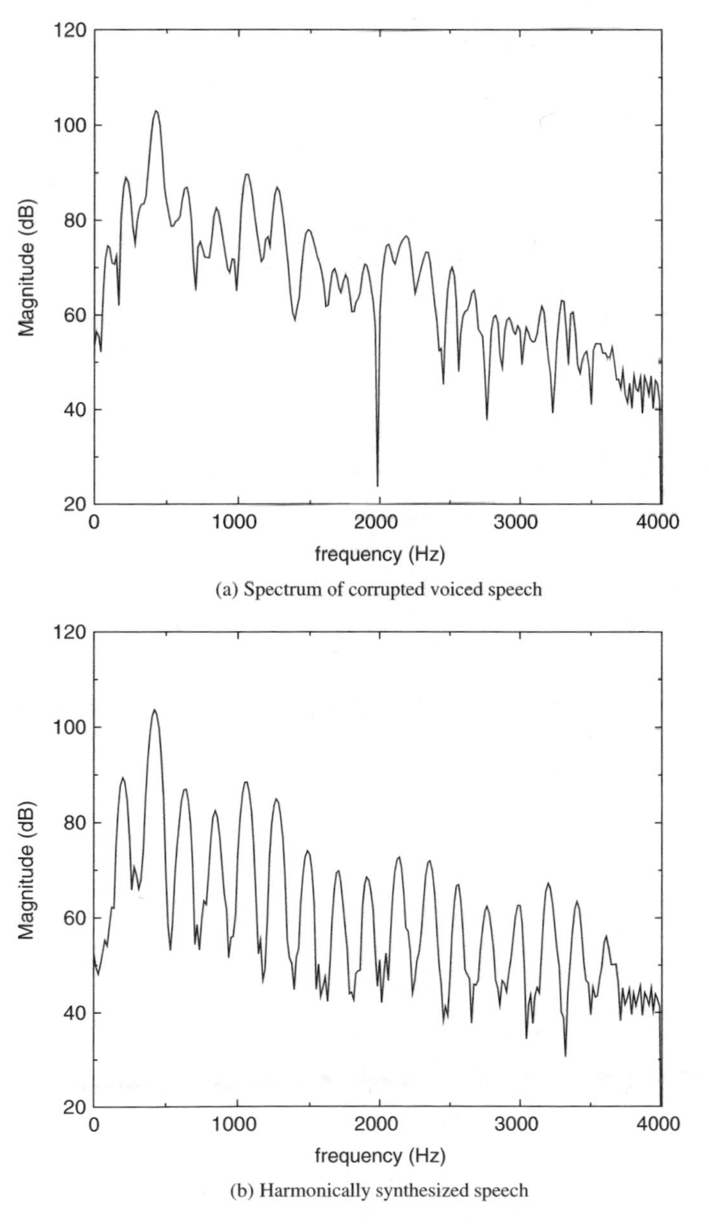

(a) Spectrum of corrupted voiced speech

(b) Harmonically synthesized speech

Figure 9.36 Speech corrupted with babble noise (10 dB SNR)

ground conditions. The hybrid coding algorithm described has three modes, and two bits are allocated to transmit the mode. Therefore an additional mode may be added to further improve the speech quality. The quality of speech corrupted by acoustic noise can be improved by using the additional mode as another harmonic mode with a constant voicing frequency marker, e.g. 80 % of the spectrum is voiced. Figure 9.37 depicts the spectrum of speech

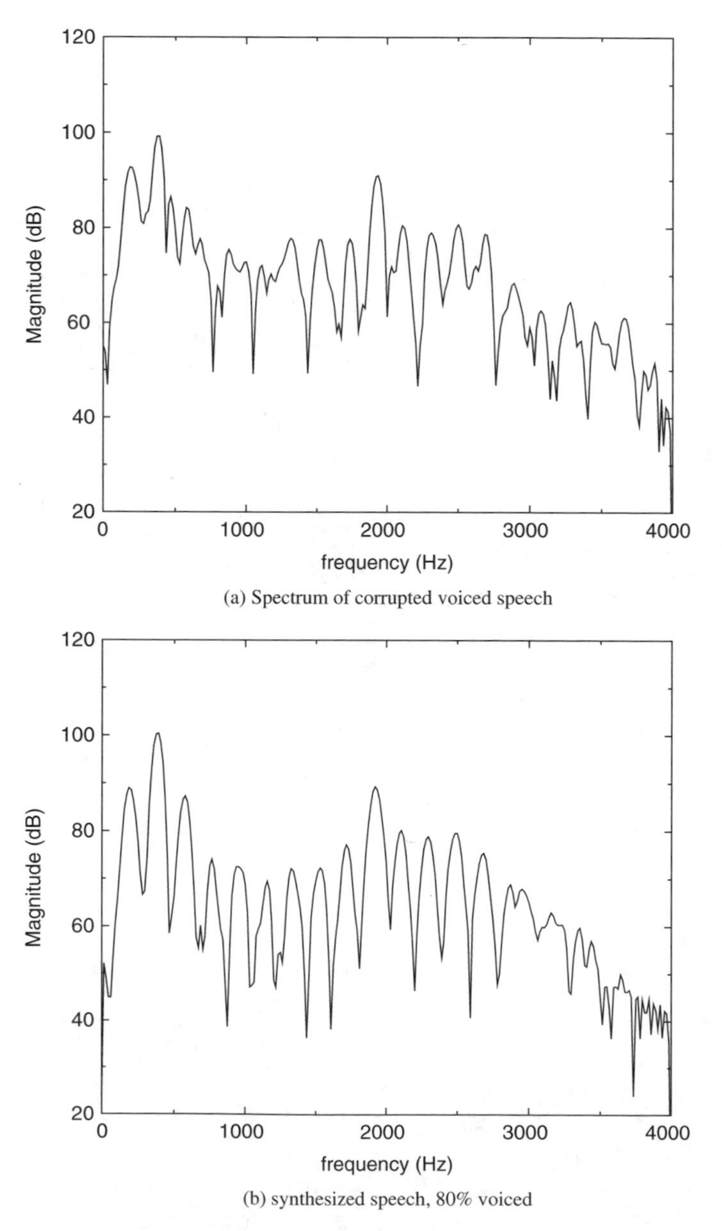

(a) Spectrum of corrupted voiced speech

(b) synthesized speech, 80% voiced

Figure 9.37 Speech corrupted with babble noise (10 dB SNR)

Table 9.13 Hybrid vs 6.3 kb/s G.723.1 for noisy speech

	Better	Slightly better	Same	Slightly worse	Worse
Male(%)	0.0	12.5	40.0	40.0	7.5
Female(%)	5.0	22.5	35.0	30.0	7.5
Average(%)	2.5	17.5	37.5	35.0	7.5

corrupted with babble noise (10 dB SNR) and the spectrum of the synthesized speech, with 80 % of the spectrum declared voiced and the remaining high frequency components synthesized using filtered and scaled Gaussian noise.

The same informal listening test was conducted to compare the speech quality. The informal test results are shown in Table 9.13. Comparing with the results shown in Table 9.12, the introduction of the harmonic voicing significantly improves the performance under background noise which indicates that there is still some room to retune the harmonic coder for the hybrid coding operation. The same is perhaps true for ACELP, and it should be designed specifically for hybrid operation.

9.11.2 Performance Under Channel Errors

The inherent robustness of the hybrid coder to mode bit errors was tested by simulating all the possible mode errors. The hybrid coder has three modes, hence there are six possible mode errors, i.e. each mode may be erroneously decoded with the other two modes. The bit stream of the hybrid coder is shown in Tables 9.14 and 9.15. For each parameter, the most significant bit (MSB) is transmitted first. When erroneously decoding a lower-rate mode as a higher-rate mode, e.g. decoding a white-noise excitation frame as harmonic, the remaining bits are set to 1. Simulations show that setting the remaining bits to 1 has the worst effect, since the higher indices are mapped to the higher-energy levels in the gain quantizers. Using the LTP gain quantizer shown in Table 9.3 results in blasts when the white noise or harmonic frames are erroneously decoded as ACELP. Therefore the maximum LTP gain is limited to 1.2.

All the modes quantize the LSFs using 23 bits, consequently they are transmitted using the same bits. Therefore the LSFs are independent of the mode and the mode bit errors can only affect the excitation parameters. This is particularly attractive for the LSF interpolation and quantization with first-order moving average prediction. The most significant bits of the gain parameters are also transmitted using the same bits. However the gain of each mode is estimated using different criteria. Hence the gain quantizers of each mode have different dynamic ranges, and mode errors affect the dequantization of the gain.

Table 9.14 Transmission bit stream of the hybrid coder

Parameters	White noise	Harmonic	ACELP 6k
Mode	1–2	1–2	1–2
LSF	3–25	3–25	3–25
Gain (2nd subframe)	26–30	26–30	26–28
Gain (1st subframe)	–	31–33	–
Pitch	–	34–41	–
PPL	–	42–48	–
PPS	–	49–52	–
Amplitudes 1st subframe	–	53–66	–
Amplitudes 2nd subframe	–	67–80	–

Table 9.15 Bit stream of 6 kb/s ACELP subframes

Parameters	Subframe 1	Subframe 2	Subframe 3	Subframe 4
LTP Delay	29–35	52–58	75–81	98–104
LTP Gain	36–39	59–62	82–85	105–108
Pulse Sign	40	63	86	109
Pulse track 1	41–44	64–67	87–90	110–113
Pulse track 2	45–48	68–71	91–94	114–117
Innovation Gain	49–51	72–74	95–97	118–120

White Noise Excitation Mode Errors

Figure 9.38 illustrates erroneous decoding of white-noise excitation frames as harmonic and ACELP. It shows that the errors are contained within the frames which have mode errors. This is because the decoder does not interpolate the unvoiced gain at switching. The present gain is used to synthesize the entire frame when switched from a different mode. However if the next frame after decoding a noise excitation frame as ACELP is also ACELP, the LTP memory propagates the errors, similar to the error propagation of CELP coders [60]. The hybrid coding algorithm has the advantage of limiting the error propagation, by switching to a different mode, which also refreshes the LTP memory.

Harmonic Mode Errors

Figure 9.39 illustrates erroneous decoding of harmonic excitation frames as unvoiced and ACELP. It shows that the errors are contained within the frames which have mode errors. This is because the decoder reinitializes the harmonic excitation memories when switched from a different mode, and

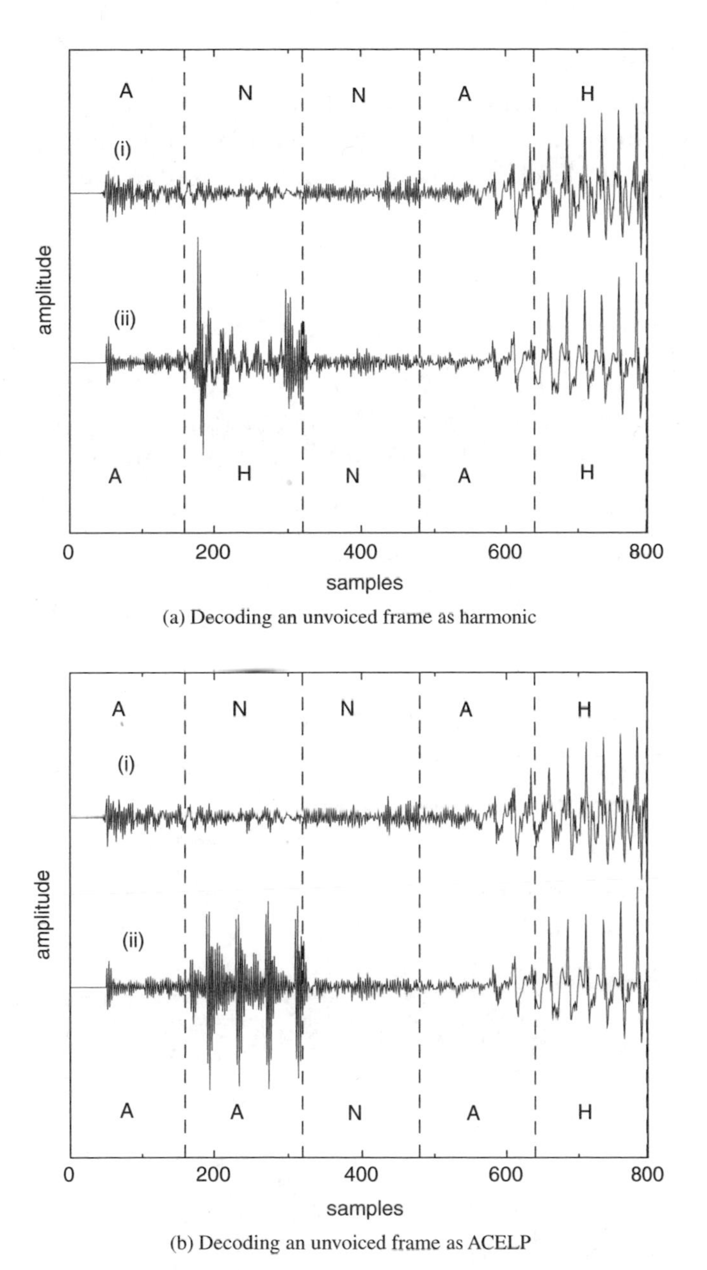

(a) Decoding an unvoiced frame as harmonic

(b) Decoding an unvoiced frame as ACELP

Figure 9.38 Erroneous decoding of white-noise excitation frames: (i) Original speech, (ii) synthesized speech: A (ACELP), H (harmonic), and N (noise excitation)

use of the previous excitation vector is minimized. However if the next frame after decoding a harmonic excitation frame as unvoiced is also unvoiced, the unvoiced overlap and add process spreads the incorrect gain into the next frame.

ACELP Mode Errors

Figure 9.40 illustrates erroneous decoding of ACELP frames as unvoiced and harmonic. In Figure 9.40a the error is contained within the frame which has the mode error. For the next frame the harmonic mode reinitializes the

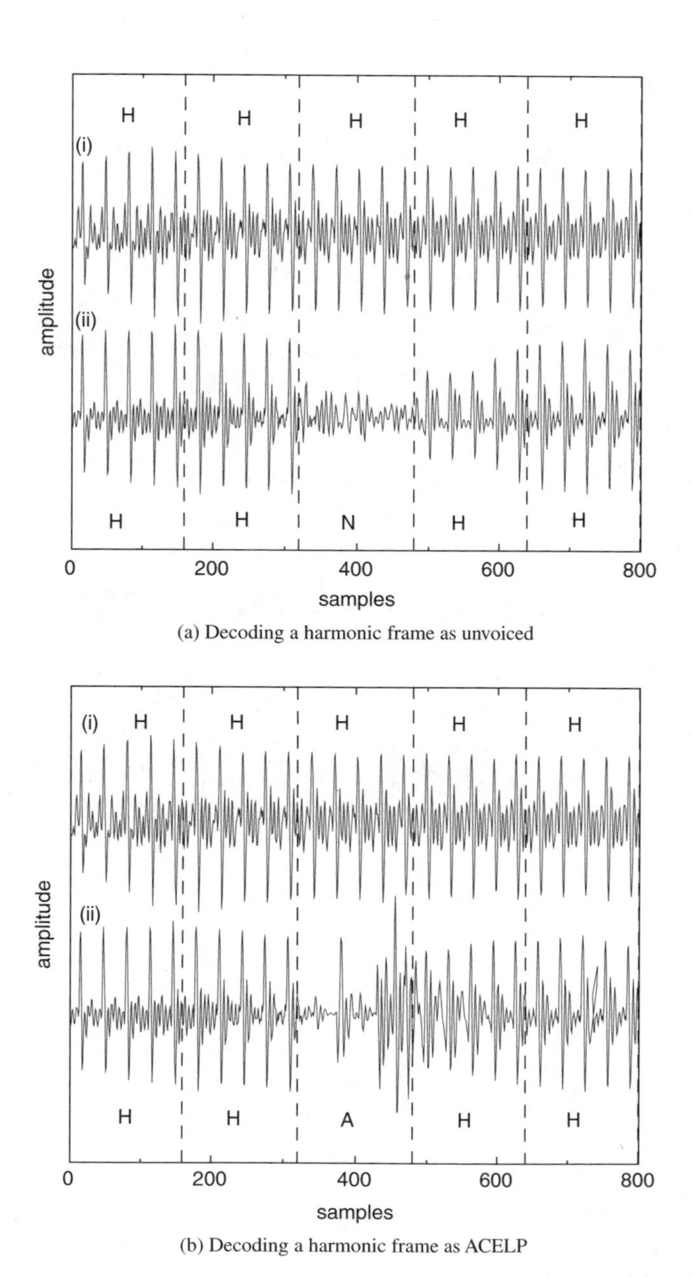

(a) Decoding a harmonic frame as unvoiced

(b) Decoding a harmonic frame as ACELP

Figure 9.39 Erroneous decoding of harmonic excitation frames, (i) Original speech, (ii) synthesized speech: A (ACELP), H (harmonic), and N (noise excitation)

excitation memories. However in Figure 9.40b, the next frame after decoding an ACELP frame as harmonic is also harmonic. Hence, the error propagates into the next frame, due to the harmonic interpolation process.

The LPC filter may propagate the errors, when the filter response is highly resonant. However the bandwidth expansion of the LPC coefficients ensures that the LPC impulse response dies away more quickly. Therefore all the mode errors are localized and the output does not become unstable in the presence of mode errors. This is mainly due to the independent memory initialization procedures of the coding algorithm when switching between the modes. The white-noise excitation mode always sets the previous gain equal to the present one when switched from a different mode. The harmonic excitation mostly depends on the received harmonic parameters when switched from a different mode; only the amplitude quantizer memories are initialized using the previous excitation vector. The LTP buffer is refreshed, regardless of the mode, with the latest excitation vector.

9.11.3 Performance Improvement Under Channel Errors

During the experiments described in the preceding sections, the robustness to mode-bit errors was improved by limiting the LTP gain to 1.2 and using the same set of bits to transmit the LSFs of all the modes. The encoder and the decoder cannot synchronize the random number generators at the presence of mode-bit errors. This affects the performance of the LTP when switched from white-noise excitation. However the exact content of the white-noise excitation has no significance and can be represented by any noise excitation vector. Therefore, the performance of the LTP was also improved by always reinitializing the LTP buffer to a fixed stored noise excitation vector when switching to ACELP from the white-noise excitation.

The robustness to mode-bit errors can be further improved by using error detection and correction techniques. If a mode error is only detected and not corrected, the concealment techniques based on waveform substitution can be used to reduce the resulting annoying artifacts [62]. The decoded parameters and the synthesized waveform may also be used to detect mode errors. As can be seen in Figures 9.38, 9.39, and 9.40, mode errors generally result in sudden changes in the waveform shape and the signal level, which are unusual for speech signals. Moreover certain mode patterns are more common than the others, e.g. for many speech utterances, ACELP to harmonic and back to ACELP occur, while the silence segments before and after are synthesized with the white-noise excitation. The transition from white noise to harmonic mode is extremely rare, since generally the onsets request ACELP. Consequently in order to assist in detecting mode errors, one can limit the possible switching combinations.

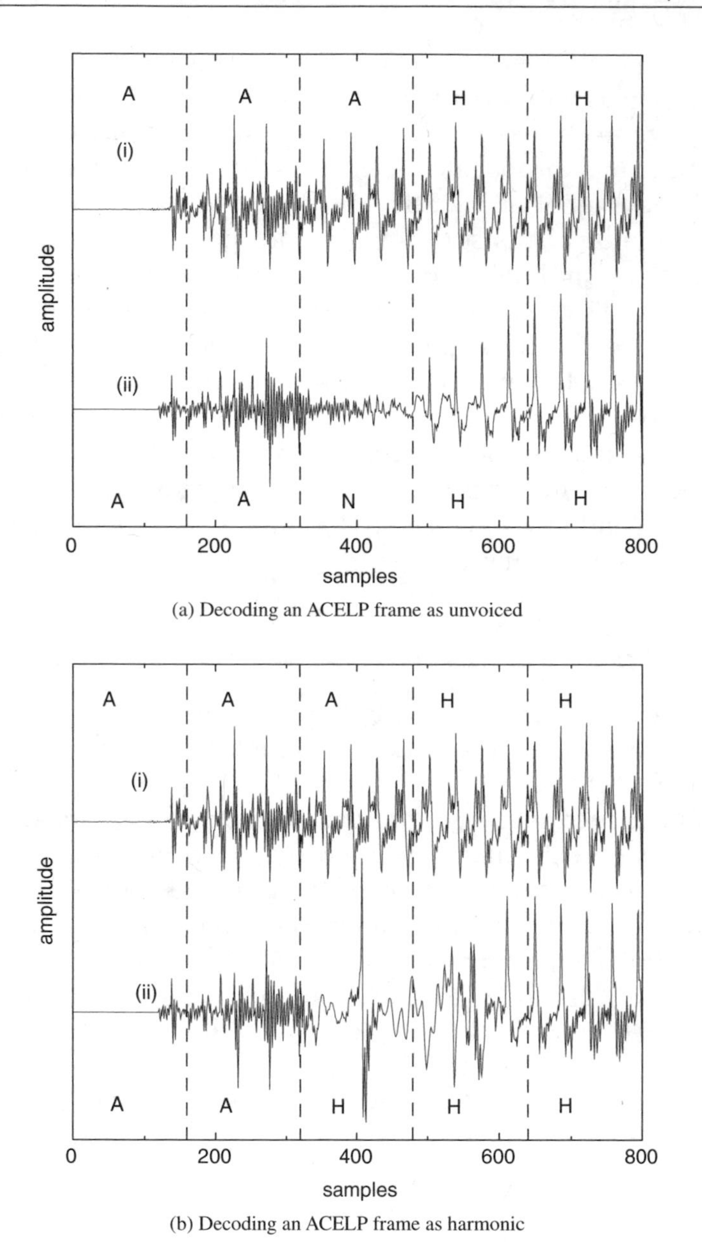

(a) Decoding an ACELP frame as unvoiced

(b) Decoding an ACELP frame as harmonic

Figure 9.40 Erroneous decoding of ACELP frames, (i) Original speech, (ii) synthesized speech: A (ACELP), H (harmonic), and N (noise excitation)

9.12 Summary

In this chapter the principle techniques behind an advanced hybrid coding algorithm, which integrates harmonic coding and waveform coding, have been presented. The two important design issues are speech classification and, when mode-switching, proper coder synchronization. Provided that these two processing stages are carried out successfully, the quality of speech produced by a hybrid coding method is of good to toll quality at around 3.5–5 kb/s (average). Simple informal subjective listening test results confirm that the hybrid model eliminates the limitations of the existing single-model-based coders.

The robustness of the hybrid coding algorithm under acoustic noise and channel error conditions is another important issue which requires significant research effort. The difficulties specific to hybrid coders are the speech classification under background noise, and the mode-bit errors due to random channel errors. Although the classification algorithm is capable of selecting the best mode under noisy background conditions, there is a significant bias towards ACELP in the presence of noise compared to clean speech conditions. This is due to the inability of the white-noise excitation or the harmonic excitation to encode the corrupted signals. The noisy speech synthesized using the harmonic mode sounds metallic, which can be improved by introducing a proper voicing mixture classification when harmonic mode is selected.

The robustness of the hybrid coder to mode errors has been tested by simulating all the possible mode errors. The coder is capable of isolating the mode errors and return to normal decoding almost immediately. This is mainly due to the independent memory reinitialization of the modes when switched from a different mode.

Finally it is important that each element or coding mode of the hybrid model is redesigned with the knowledge that the noise, ACELP and harmonic excitation models will be used during noise (or silence), transitions, and steady state voiced speech parts respectively. In this case the LPC parameters of ACELP and harmonic modes will have different vector quantizer tables which will be trained over transitional and steady state voiced speech only respectively, thus improving the quantization performance. In addition, using the LTP in ACELP mode at the onsets may not be necessary. Instead more pulses with phase spreading may be used to improve quality.

Bibliography

[1] R. J. McAulay and T. F. Quatieri (1995) 'Sinusoidal coding', in *Speech coding and synthesis* by W. B. Kleijn and K. K. Paliwal (Eds), pp. 121–74. Amsterdam: Elsevier Science

[2] R. J. McAulay and T. F. Quatieri (1986) 'Speech analysis/synthesis based on a sinusoidal representation', in *IEEE Trans. on Acoust., Speech and Signal Processing*, 34(4):744–54.

[3] D. Griffin and J. S. Lim (1988) 'Multiband excitation vocoder', in *IEEE Trans. on Acoust., Speech and Signal Processing*, 36(8):1223–35.

[4] I. Atkinson, S. Yeldener, and A. Kondoz (1997) 'High quality split-band LPC vocoder operating at low bit rates', in *Proc. of Int. Conf. on Acoust., Speech and Signal Processing*, pp. 1559–62. May 1997. Munich

[5] R. Salami, C. Laflamme, J.P. Adoul, A. Kataoka, S. Hayashi, T. Moriya, C. Lamblin, D, Massaloux, S. Proust, P. Kroon, and Y. Shoham (1998) 'Design and description of CS-ACELP: a toll quality 8 kbps speech coder', in *IEEE Trans. Speech and Audio Processing*, 6(2):116–30.

[6] C. Laflamme, J.-P. Adoul, H. Su, and S. Morissette (1990) 'On reducing computational complexity of codebook search through the use of algebraic codes', in *Int. Conf. on Acoust., Speech and Signal Processing*, pp. 177–80.

[7] W. B. Kleijn (1993) 'Encoding speech using prototype waveforms', in *IEEE Trans. Speech and Audio Processing*, 1:386–99.

[8] M. Schroeder and B. Atal (1985) 'Code excited linear prediction (CELP): high quality speech at very low bit rates', in *Proc. of Int. Conf. on Acoust., Speech and Signal Processing*, pp. 937–40. Tampa, FL

[9] D. K. Freeman, G. Cosier, C. B. Southcott, and I. Boyd (1989) 'The voice activity detector for the pan-European digital cellular mobile telephone service', in *Proc. of Int. Conf. on Acoust., Speech and Signal Processing*, pp. 369–72.

[10] S. Wang and A. Gersho (1992) 'Improved phonetically segmented vector excitation coding at 3.4 kbps', in *Proc. of Int. Conf. on Acoust., Speech and Signal Processing*, 1:349–352.

[11] T. E. Tremain (1982) 'The government standard linear predictive coding algorithm: LPC-10', in *Speech Technology*, 1:40–9.

[12] P. Kroon and B. Atal (1988) 'Strategies for improving CELP coders', in *Proc. of Int. Conf. on Acoust., Speech and Signal Processing*, 1:151–4.

[13] I. M. Trancoso, L. Almeida, and J. M. Tribolet (1986) 'A study on the relationships between stochastic and harmonic coding', in *Proc. of Int. Conf. on Acoust., Speech and Signal Processing*, pp. 1709–12.

[14] B. S. Atal and S. Singhal (1984) 'Improving performance of multipulse LPC coders at low bit rates', in *Proc. of Int. Conf. on Acoust., Speech and Signal Processing*, pp. 1.3.1–4.

[15] D. L. Thomson and D. P. Prezas (1986) 'Selective modelling of the LPC residual during unvoiced frames white noise or pulse excitation', in *Proc. of Int. Conf. on Acoust., Speech and Signal Processing*, pp. 3087–90.

[16] E. Shlomot, V. Cuperman, and A. Gersho (1998) 'Combined harmonic and waveform coding of speech at low bit rates', in *Proc. of Int. Conf. on Acoust., Speech and Signal Processing*.

[17] E. Shlomot, V. Cuperman, and A. Gersho (1997) 'Hybrid coding of speech at 4 kbps', in *Proc. IEEE Workshop on Speech Coding for Telecom*, pp. 37–8.

[18] J. Stachurski and A. McCree (2000) 'Combining parametric and waveform-matching coders for low bit-rate speech coding', in *X European Signal Processing Conf.*

[19] J. Stachurski and A. McCree (2000) 'A 4 kb/s hybrid MELP/CELP coder with alignment phase encoding and zero phase equalization', in *Proc. of Int. Conf. on Acoust., Speech and Signal Processing*, pp. 1379–82. May 2000. Istanbul

[20] A. V. McCree and T. P. Barnwell (1995) 'A mixed excitation LPC vocoder model for low bit rate speech coding', in *IEEE Trans. Speech and Audio Processing*, 3(4):242–50.

[21] J. Stachurski, A. V. McCree, and V. R. Viswanathan (1999) 'High quality MELP coding at bit-rates around 4 kbps', in *Proc. of Int. Conf. on Acoust., Speech and Signal Processing*.

[22] T. Moriya and M. Honda (1986) 'Speech coder using phase equalisation and vector quantisation', in *Proc. of Int. Conf. on Acoust., Speech and Signal Processing*, pp. 1701–4.

[23] J. Skoglund, W. B. Kleijn, and P Hedelin (1997) 'Audibility of pitch-synchronously modulated noise', in *Proc. IEEE Workshop on Speech Coding for Telecom*, pp. 51–2.

[24] H. Pobloth and W. B. Kleijn (1999) 'On phase perception in speech', in *Proc. of Int. Conf. on Acoust., Speech and Signal Processing*.

[25] Doh-Suk Kim (2000) 'Perceptual phase redundancy in speech', in *Proc. of Int. Conf. on Acoust., Speech and Signal Processing*.

[26] N. Katugampala, (2001) 'Multimode speech coding below 6 kb/s', Ph.D. thesis, CCSR, University of Surrey, UK.

[27] N. Katugampala and A. Kondoz (2002) 'Integration of harmonic and analysis by synthesis coders', in *IEE Proc. on Vision Image and Signal Processing*, pp. 321–6.

[28] N. Katugampala and A. Kondoz (2001) 'A hybrid coder based on a new phase model for synchronization between harmonic and waveform coded segments', in *Proc. of Int. Conf. on Acoust., Speech and Signal Processing*.

[29] TIA/EIA (1997) *Enhanced variable rate codec, speech service option 3 for wideband spread spectrum digital systems*, IS-127.

[30] W. Kleijn, P. Kroon, L. Cellario, and D. Sereno (1993) 'A 5.85 kbps CELP algorithm for cellular applications', in *Proc. of Int. Conf. on Acoust., Speech and Signal Processing*, 2:596–9.

[31] T. V. Ananthapadmanabha and B. Yegnanarayana (1979) 'Epoch extraction from linear prediction residual for identification of closed glottis interval', in *IEEE Trans. on Acoust., Speech and Signal Processing*, 27(4):309–19.

[32] Y. M. Cheng and D. O'Shaughnessy (1989) 'Automatic and reliable estimation of glottal closure instant and period', in *IEEE Trans. On Acoust., Speech and Signal Processing*, 37(12):1805–15.

[33] P. Satyanarayana Murthy and B. Yegnanarayana (1999) 'Robustness of group delay based method for extraction of significant instants of excitation from speech signals', in *IEEE Trans. Speech and Audio Processing*, 7(6):609–19.

[34] TIA/EIA (1997) 'Enhanced variable rate codec, speech service option 3 for wideband spread spectrum digital systems', IS-127.

[35] B. S. Atal and M. R. Schroeder (1974) 'Recent advances in predictive coding-applications to voiced speech synthesis', in *Speech Commun. Seminar*. Stockholm

[36] R. J. McAulay and T. F. Quatieri (1990) 'Pitch estimation and voicing decision based upon a sinusoidal speech model', in *Proc. of Int. Conf. on Acoust., Speech and Signal Processing*, 1:249–52.

[37] S. Villette, Y. D. Cho, and A. M. Kondoz (2000) 'Efficient parameter quantisation for 2.4/1.2 kbps split band LPC coding', in *Proc. IEEE Workshop on Speech Coding for Telecom*, pp. 32–4. September 2000. Wisconsin, USA

[38] M. Stefanovic, Y. D. Cho, S. Villette, and A. M. Kondoz (2000) 'A 2.4/1.2 kb/s speech coder with noise pre-processor', in *Proc. European Signal Processing Conference*. Tampere, Finland

[39] R. Hagen, E. Ekudden, B. Johansson, and W. Kleijn (1998) 'Removal of sparse excitation artifacts in CELP', in *Proc. of Int. Conf. on Acoust., Speech and Signal Processing*.

[40] C. Li and V. Cuperman (1998) 'Enhanced harmonic coding of speech with frequency domain transition modeling', in *Proc. of Int. Conf. on Acoust., Speech and Signal Processing*, pp. 581–4.

[41] N. S. Jayant and P. Noll (1984) *Digital Coding of Waveforms: Principles and applications to speech and video*. New Jersey: Prentice-Hall

[42] G. Kubin, B. S. Atal, and W. B. Kleijn (1993) 'Performance of noise excitation for unvoiced speech', in *Proc. IEEE Workshop on Speech Coding for Telecom*, pp. 35–6.

[43] I. Atkinson (1997) 'Advanced linear predictive speech compression at 3.0 kbit/s and below', Ph.D. thesis, CCSR, University of Surrey, UK.

[44] J. Sohn and W. Sung (1995) 'A voice activity detection employing soft decision based noise spectrum adaptation', in *Proc. of Int. Conf. on Acoust., Speech and Signal Processing*, pp. 365–8. Amsterdam

[45] A. Das, E. Paksoy, and A. Gersho (1995) 'Multimode and variable rate coding of speech', in *Speech coding and synthesis* by W. B. Kleijn and K. K. Paliwal (Eds), pp. 257–88. Amsterdam: Elsevier Science

[46] S. Wang, A. Sekey, and A. Gersho (1992) 'An objective measure for predicting subjective quality of speech coders', in *IEEE Journal on Selected Areas in Communications*, 10(5):819–829.

[47] S. V. Vaseghi (1990) 'Finite state CELP for variable rate speech coding', in *Proc. of Int. Conf. on Acoust., Speech and Signal Processing*, pp. 37–40.

[48] T. Eriksson and J. Sjoberg (1993) 'Evolution of variable rate speech coders', in *Proc. IEEE Workshop on Speech Coding for Telecom*, pp. 3–4.

[49] D. O'Shaughnessy (1987) *Speech communication: human and machine*. Addison Wesley

[50] NTT Group Available at http://www.ntt.co.jp/index_e.html.

[51] Y. D. Cho, S. Villette, and A. Kondoz (2001) 'Efficient spectral magnitude quantization for sinusoidal speech coders', in *Proc. of Vehicular Technology Conf.*

[52] B. H. Juang and A. H. Gray (1982) 'Multiple stage vector quantisation for speech coding', in *Proc. of Int. Conf. on Acoust., Speech and Signal Processing*, pp. 597–600. Paris

[53] P. Kroon and B. S. Atal (1991) 'On the use of pitch predictors with high temporal resolution', in *IEEE Trans. Signal Processing*, 39(3):733–5.

[54] Y. Ephraim and D. Malah (1985) 'Speech enhancement using a minimum mean square error log-spectral amplitude estimator', in *IEEE Trans. on Acoust., Speech and Signal Processing*, 33(2):443–5.

[55] K. Zeger and A. Gersho (1990) 'Pseudo-gray coding', in *IEEE Trans. on Communications*, 38(12):2147–58.

[56] K. A. Zeger and A. Gersho (1987) 'Zero redundancy channel coding in vector quantisation', in *IEE Electronics Letters*, 23(12):654–656.

[57] N. Farvardin (1990) 'A study of vector quantisation for noisy channels', in *IEEE Trans. Inform. Theory*, 36:799–809.

[58] N. Farvardin and V. Vaishampayan (1991) 'On the performance and complexity of channel optimised vector quantisers', in *IEEE Trans. Inform. Theory*, 37:155–60.

[59] T. Eriksson, J. Linden, and J. Skoglund (1999) 'Interframe LSF quantization for noisy channels', in *IEEE Trans. on Speech and Audio Processing*, 7(5):495–509.

[60] R. Cox, B. Kleijn, and P. Kroon (1989) 'Robust CELP coders for noisy backgrounds and noisy channels', in *Proc. of Int. Conf. on Acoust., Speech and Signal Processing*, pp. 739–42.

[61] S. Villette, M. Stefanovic, and A. Kondoz (1999) 'Split band LPC based adaptive multi rate GSM candidate', in *Proc. of Int. Conf. on Acoust., Speech and Signal Processing*.

[62] D. Goodman, G. Lockhart, O. Wasem, and W. Wong (1986) 'Waveform substitution techniques for recovering missing speech segments in packet voice communications', in *IEEE Trans. on Acoust., Speech and Signal Processing*, 34(6):1440–8.

10

Voice Activity Detection

10.1 Introduction

In voice communications, speech can be characterized as a discontinuous medium because of the pauses which are a unique feature compared to other multimedia signals, such as video, audio and data. The regions where voice information exists are classified as voice-active and the pauses between talk-spurts are called voice-inactive or silence regions. An example illustrating active and inactive voice regions for a speech signal is shown in Figure 10.1.

A voice activity detector (VAD) is an algorithm employed to detect the active and inactive regions of speech. When inactive regions are detected, transmission is generally stopped and only a general description of the background information is transmitted. At the decoder end, inactive frames are then reconstructed by means of comfort noise generation (CNG), which gives natural background sounds with smooth transitions from talk-spurts to pauses and vice versa. To enhance the naturalness of the generated background signal, regular updates of the average information on the background signal (especially necessary during noisy communication environments) is transmitted by the comfort noise insertion (CNI) module of the encoder. The overall structure of the silence compression scheme employing a VAD, CNG, and CNI is shown in Figure 10.2.

Speech communication systems which operate a VAD for compression of inactive speech regions provide various benefits especially useful for bandwidth-limited communication channels. These benefits can be summarized as given in the following list:

- Co-channel interference reduction in cellular communications: It is possible to suppress co-channel interference in cell-based wireless communication systems by decreasing transmission power during inactive regions (speech pauses).

Digital Speech. A. Kondoz
 ISBN 0-470-87007-9 (HB)

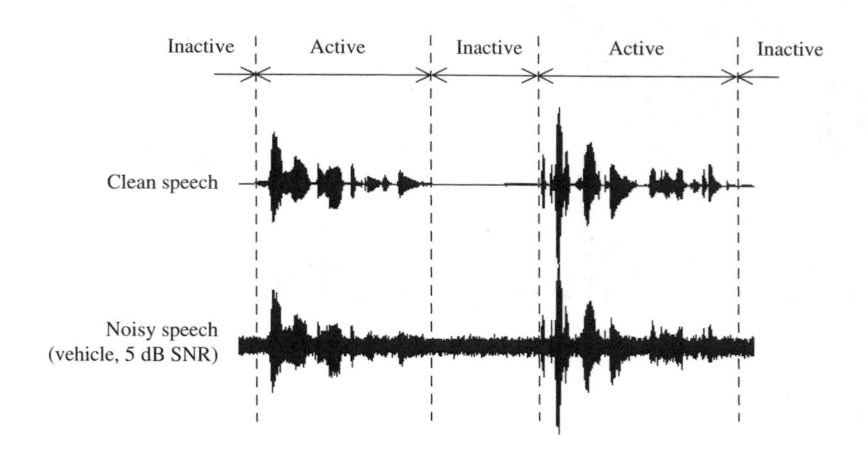

Figure 10.1 Voice active and inactive regions

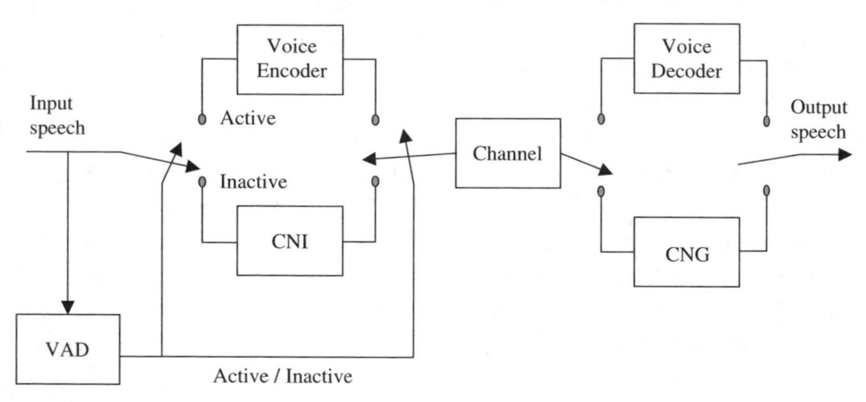

Figure 10.2 Overall structure of a speech coding system with silence compression

- Improvement of the soft channel capacity in the code division multiple access (CDMA) system: The theoretical capacity of a CDMA system is usually defined by the possible combinations of the spreading code. However, due to interference from other users, the CDMA capacity is limited to a value smaller than this theoretical limit, i.e. due to interference from other users, the error rates received by some users may be too high to enable accurate decoding. By reducing the transmission power during speech pauses, the interference on air can be reduced, which may automatically allow more users on the system, hence achieving an increase in the CDMA system's capacity.
- Power-saving for mobile terminals: Mobile terminals do not have to transmit radio signals during pauses. Thus, the battery life time of the terminals can be extended by conserving power during speech inactive periods.

- Increase in channel capacity by statistical multiplexing: A channel can be granted just during talk-spurts and released during pauses. Once granted, a user occupies a channel until the end of a talk-spurt and releases it immediately after the last active speech frame. To get the channel allocation again, the user makes a request at the start of the next talk-spurt. This way the channel resources can be utilized in a more efficient way by the statistical multiplexing scheme, which allows a number of users to communicate at the same time over limited channel resources. Note: in statistical multiplexing, there is a possibility that there are no free channel slots when a user makes a request. In this case, the new user may be rejected after a time-out, which may cause information loss resulting in some quality degradations.
- Reduction in packet losses when transmitting voice over packet-based networks: A packet-based system can be overloaded with more packets than it can handle. The congestion of packet-based systems can be reduced during voice communication by producing packets only during active speech regions and cutting out packets for the inactive speech regions.
- Bit-rate reduction: In addition to the bit-rate reduction achieved by speech compression techniques, the use of a VAD together with silence compression (cutting out the inactive speech regions) gives additional reduction in the bit-rate regardless of speech coders.

The VAD usually produces a binary decision for a given speech segment (usually 10–20 ms long) indicating either speech presence or absence, which is quite easy for clean background speech. For example, by checking the energy level of the input signal, it is possible to obtain a high speech/nonspeech detection performance. However, in real environments, the input signal may be mixed with noise characteristics which may be unknown and changing with time. In some cases where the background noise is significantly high, the speech may be obscured by this noise. Especially, the unvoiced sounds, which are important for speech intelligibility, may be misdetected in such noisy environments. Figure 10.1 shows an example for a noisy speech segment with vehicle noise of 5 dB signal to noise ratio (SNR). As can be seen from the figure, some low energy speech parts are fully submerged in noise, making it very difficult to discriminate these talk-spurts even by visual inspection. Incorrect classification of these talk-spurts can cause clipped sounds which may result in significantly degraded speech quality. On the other hand, the increase in false detection of silence loses the potential benefits of silence compression. There is a trade-off in VAD performance, maximizing the detection rate for active speech while minimizing the false detection rate of inactive speech regions.

10.2 Standard VAD Methods

In order to exploit the advantages of silence compression, a number of VAD algorithms have been proposed, some of which have been selected by standards organizations including ITU-T, ETSI, and TIA/EIA. ITU-T released G.729 Annex B (G.729B) [1] and G.723.1 Annex A (G.723.1A) [2] as extensions to the 8 kb/s G.729 [3] and 5.3/6.3 kb/s G.723.1 [4] speech coders for performing discontinuous transmission (DTX). ETSI recommended GSM-FR, -HR, and -EFR VAD methods for European digital cellular systems [5–7]. Recently, ETSI released two more VADs, adaptive multi-rate VAD option 1 (AMR1) and option 2 (AMR2) [8], with a view to using it in UMTS (the third generation mobile communications). The North American standards organisation, TIA/EIA, released two VADs one for IS-96 [9] and the other for IS-127 [10] and IS-733 [11] (the VADs suggested for IS-127 and IS-733 have the same structure). Table 10.1 shows standard VADs classified in terms of the input features mainly consisting of subband energies and the spectral shape. For example, the TIA/EIA VADs use a small number of subbands whereas the IS-96 VAD uses the overall signal energy. The IS-127 and IS-733 VAD, on the other hand, decomposes the input signal into two subbands only. Traditionally, ETSI VAD methods have been based on a more accurate spectral shape of the input signal. The reason behind this is that the energy of the predictive coding error increases when the spectral shapes between the background and input signal mismatch (i.e. when speech active). However, in the recent standard for AMR, ETSI adopted two kinds of VAD algorithms both of which are based on the spectral subband energies rather than the more accurate spectral shape. The ITU-T VAD standards, G.729B and G.723.1A, conduct the detection using four different features including both the spectral shape and subband energies.

Table 10.1 Classification of standard VAD methods depending on input features; the values in parentheses indicate the number of spectral subbands

Main features	VAD
Spectral shape	GSM-FR, GSM-HR, GSM-EFR
Sub-band energies	IS-96 (1), IS-127 (2), IS-733 (2)
	AMR1 (9), AMR2 (16)
Others	G.729B, G.723.1A

10.2.1 ITU-T G.729B/G.723.1A VAD

As an extension to the G.729 speech coder, ITU-T SG16 released G.729 Annex B in order to support DTX by means of VAD, CNI, and CNG. G.729B conducts a VAD decision every frame of 10 ms, using four different parameters:

- a full-band energy difference, $\Delta E_f = \overline{E_f} - E_f$
- a low-band energy difference, $\Delta E_l = \overline{E_l} - E_l$
- a spectral distortion, $\Delta LSF = \sum_{i=0}^{9}(\overline{LSF}_i - LSF_i)^2$
- a zero-crossing rate difference, $\Delta ZC = \overline{ZC} - ZC$

where E_f, E_l, LSF_i, and ZC are the full-band energy, low-band energy, i^{th} line spectral frequency, and zero-crossing rate of the input signal. $\overline{E_f}$, $\overline{E_l}$, $\overline{LSF_i}$, and $\overline{ZC}$ are the noise characterizing parameters updated using the background noise.

The block diagram of G.729B VAD is shown in Figure 10.3. The input parameters for the VAD can be obtained from the input signal or from the intermediate values of the speech encoder. Subsequently, the difference parameters, ΔE_f, ΔE_l, ΔLSF, and ΔZC, are computed from the input and noise parameters. A decision of voice activity is conducted over a four-dimensional hyper-space, based on a region classification technique, followed by a hangover scheme. The noise parameters are updated based on a first order autoregressive (AR) scheme, if the full-band energy difference is less than a certain fixed threshold. ITU-T G.723.1A VAD has a structure similar to G.729B VAD.

10.2.2 ETSI GSM-FR/HR/EFR VAD

The VAD algorithms of ETSI GSM-FR, -HR, and -EFR have a common structure, in which the predictive residual energy is compared with an adaptive

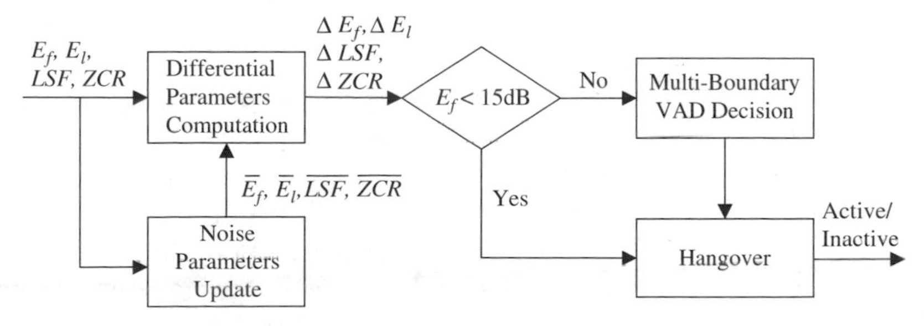

Figure 10.3 Block diagram of ITU-T G.729B VAD

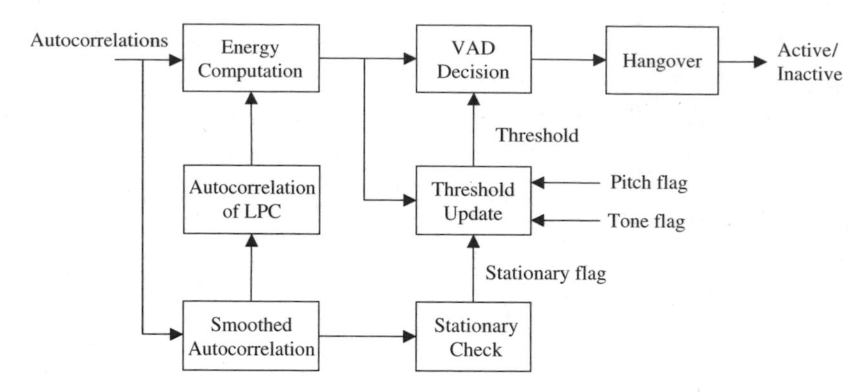

Figure 10.4 Block diagram of ETSI GSM-EFR VAD

threshold. The predictive residual energy is computed using the current and smoothed autocorrelation values which describe the spectral characteristics of the signal. The assumption is that if the signal is background noise only, which is fairly stationary, the average spectral shape will be similar to the current frame's shape and hence result in smaller residual signal energy. The threshold for VAD decision is updated during noise-only regions using the most recent noise signals in order to reflect up-to-date noise characteristics. A block diagram of the GSM-FR/HR/EFR VAD is shown in Figure 10.4.

10.2.3 ETSI AMR VAD

AMR1 decomposes the input signal into nine nonuniform subbands using filter banks where lower frequency bands have smaller bandwidths and higher frequency bands have larger bandwidths. Then it calculates each subband energy followed by its corresponding SNR estimate. The energy of the background noise used in calculating the SNR is computed by an adaptive method based on a first-order AR-model together with internal VAD logic. Finally, VAD decision is conducted by comparing the sum of the subband SNRs with an adaptive threshold, followed by a hangover. The block diagram of AMR1 is shown in Figure 10.5.

AMR2 has a structure similar to AMR1 in that VAD is performed using the subband energies together with the background noise energy. However, AMR2 transforms the input signal into the frequency domain using FFT, instead of the filter bank used in AMR1, and then calculates each subband energy in which the number of bands is 16 with a nonlinear scale in band grouping. Subsequently, SNRs for each subband are calculated using the input and the background noise spectra. The background noise energy for each band is adapted during noise frames using a first-order AR-based scheme. In order to prevent being over sensitive to nonstationary background noise conditions, AMR2 increases the threshold for final VAD decision for highly fluctuating signals, measured by the variance of their instantaneous frame-to-frame SNRs.

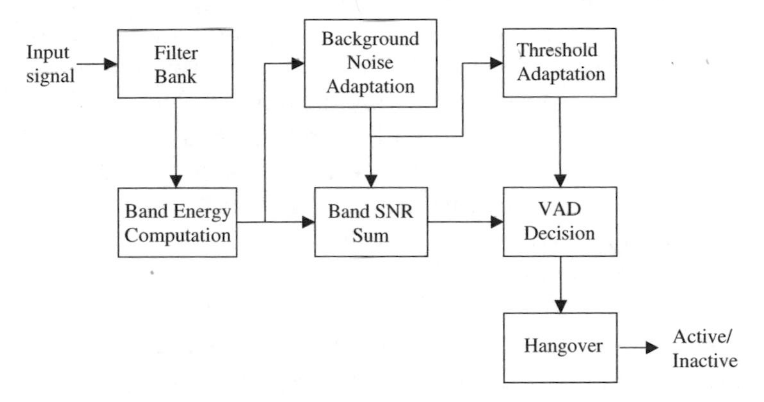

Figure 10.5 Block diagram of ETSI AMR VAD option 1

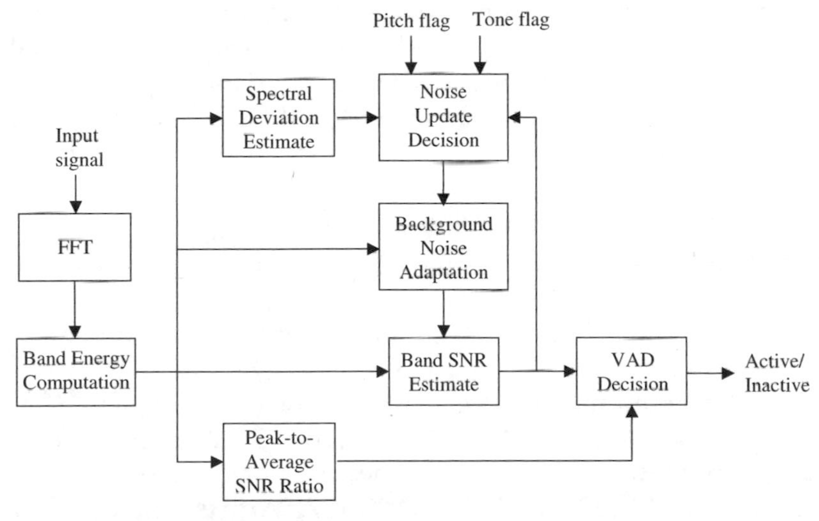

Figure 10.6 Block diagram of ETSI AMR VAD option 2

Furthermore, noise adaptation may not be accurately performed by measuring the spectral deviation when subband energies fluctuate rapidly. Thus, AMR2 changes the VAD threshold in an adaptive way together with the variation of burst and hangover counts. The hangover control is performed by measuring the peak-to-average SNR, in which the average SNR is calculated using AR-adaptation with the increased instantaneous SNR. In other words, for an increase of the peak-to-average SNR, it decreases the hangover and burst counts while increasing the VAD threshold. The block diagram of AMR2 is shown in Figure 10.6.

10.2.4 TIA/EIA IS-127/733 VAD

CDMA-based digital cellular systems have a natural structure for incorporating VAD, called a rate determination algorithm (RDA), which gives substantial improvement in channel capacity by controlling the radio transmission power to reduce co-channel interference. TIA/EIA released two kinds of RDA for IS-96 and IS-127, called 8 kb/s Qualcomm code-excited linear prediction (QCELP) and enhanced variable rate codec (EVRC), respectively. In the North American CDMA standard, IS-127 RDA supports three rates: 1, 1/2, and 1/8. Active speech is encoded in 1 or 1/2 rate and background noise is encoded in 1/8 rate. The RDA of IS-733, called 13 kb/s QCELP, is the same as IS-127.

As input parameters, IS-127 RDA uses two subband energies with the long-term prediction gain. Firstly, it calculates the smoothed subband energy using a first-order AR-model. Subsequently, the signal and noise energies for each subband are adapted depending on the long-term prediction gain. In other words, the signal energy is actively adapted to the current input if the prediction gain is relatively high. On the other hand, if the gain is relatively low, it increases the noise adaptation rate. Using the two subband energies of the signal and noise, each subband SNR is calculated. The final rate is determined by comparing the SNRs with adaptive thresholds depending on the level of background noise and the SNR of the previous frame, followed by a hangover. The block diagram of IS-127 RDA is shown in Figure 10.7.

10.2.5 Performance Comparison of VADs

The five standard VAD algorithms have been evaluated in terms of detection error rates for speech and silence. The test data was 96 seconds of speech, filtered by the modified IRS, and then mixed with vehicle and babble noises of 5, 10, 15, and 25 dB SNR. The active and inactive regions of the speech material were marked manually. The proportions of the inactive and active regions of the speech material were 0.43 and 0.57, respectively. The VAD decision is carried out every 10 ms in the cases of G.729B and AMR2, and every 20 ms in GSM-EFR, AMR1 and IS-127. With slight modification to the AMR2 source code, it is possible to obtain 10 ms results because AMR2 basically conducts the detection every 10 ms and then returns 20 ms results using a logical combination of the two 10 ms results. In handling the multiple

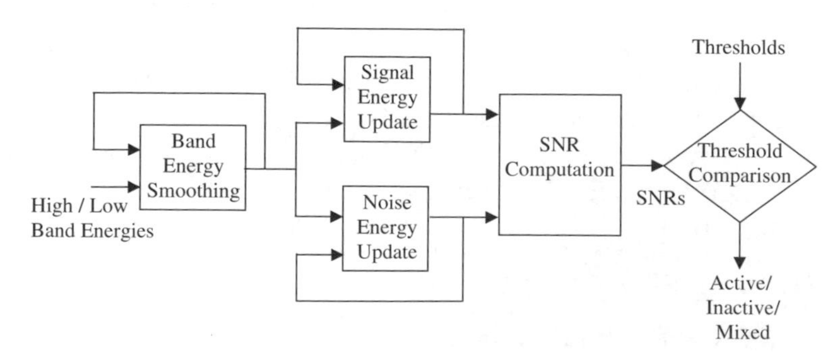

Figure 10.7 Block diagram of TIA/EIA IS-127 RDA

rates of IS-127, the upper two rates, 1 and 1/2, have been assumed to be voice active and the lowest rate, 1/8, is treated as voice inactive.

Performance in a vehicle noise environment are shown in Figures 10.8 and 10.9, and performance for babble noise are shown in Figures 10.10

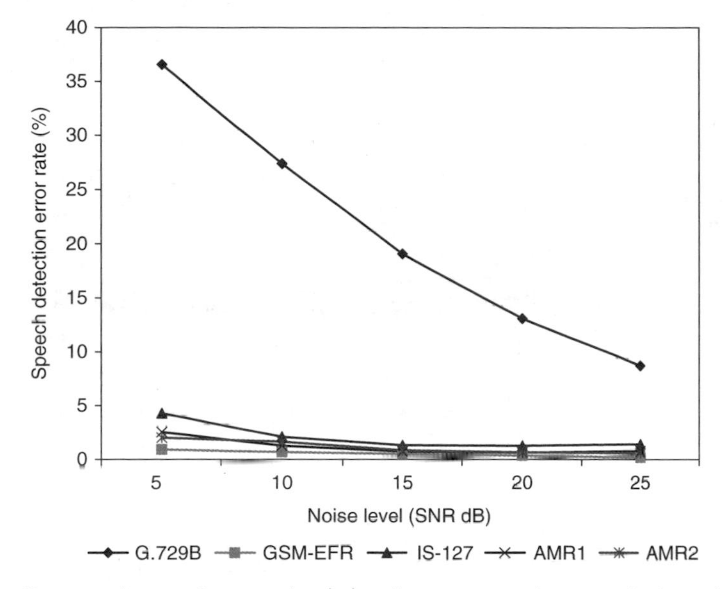

Figure 10.8 Comparison of speech detection error rates against various vehicle noise levels

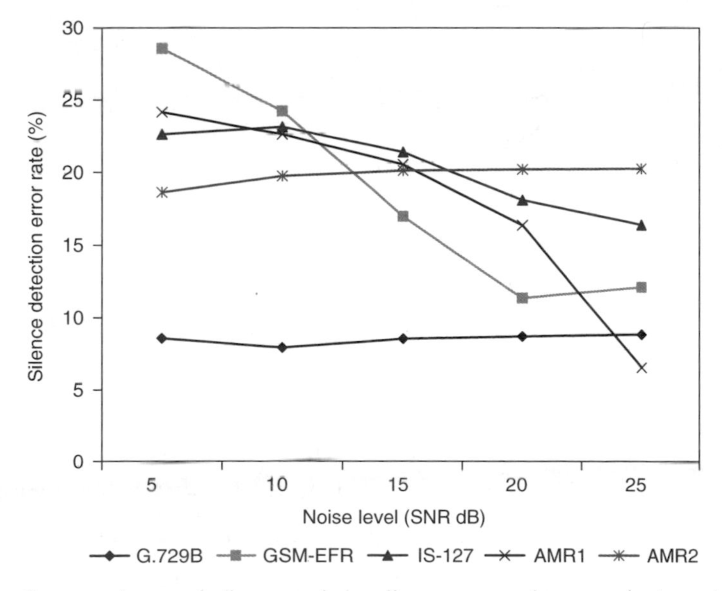

Figure 10.9 Comparison of silence detection error rates against various vehicle noise levels

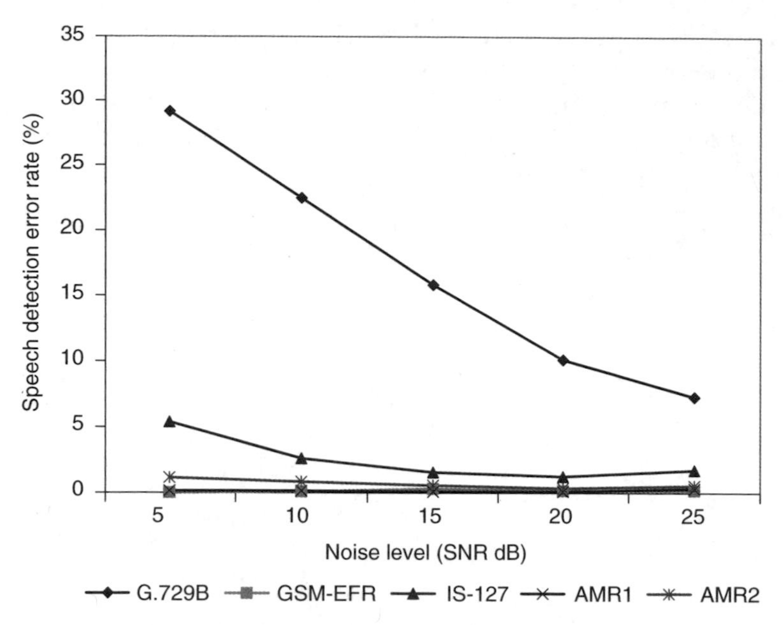

Figure 10.10 Comparison of speech detection error rates against various babble noise levels

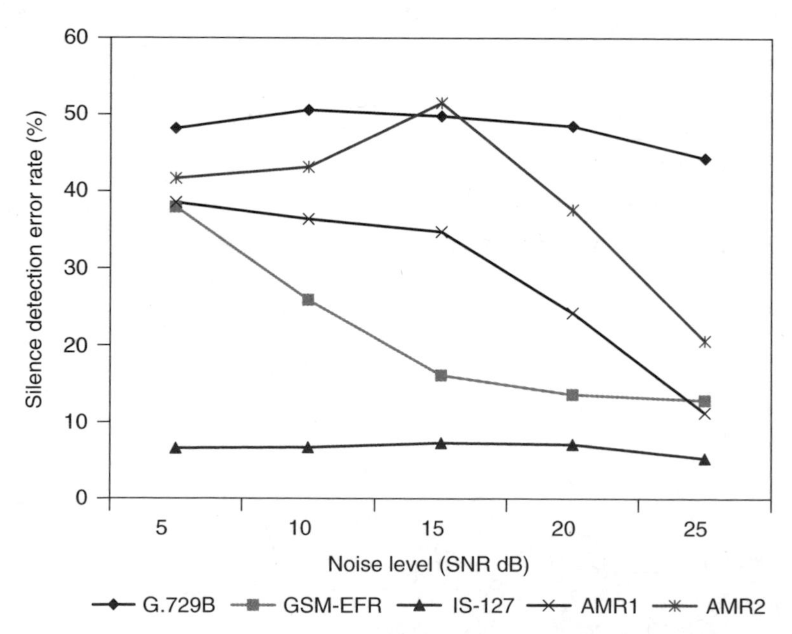

Figure 10.11 Comparison of silence detection error rates against various babble noise levels

and 10.11. G.729B exhibits the worst performance compared with other methods, especially for low SNRs. G.729B produces high speech detection errors, which can cause severe clipping of speech. IS-127 exhibits relatively high error rates for speech detection compared with those of ETSI VADs. However, it produces quite reasonable performances in silence detection for babble noisy speech. ETSI VAD methods, i.e. GSM-EFR, AMR1, and AMR2, exhibit similar performances in speech detection, while giving quite variable performances in silence detection. GSM-EFR produces the most desirable performances for relatively high SNRs, i.e. greater than 15 dB. However, the error rates of silence detection increase substantially for decreasing SNR. AMR2 produces

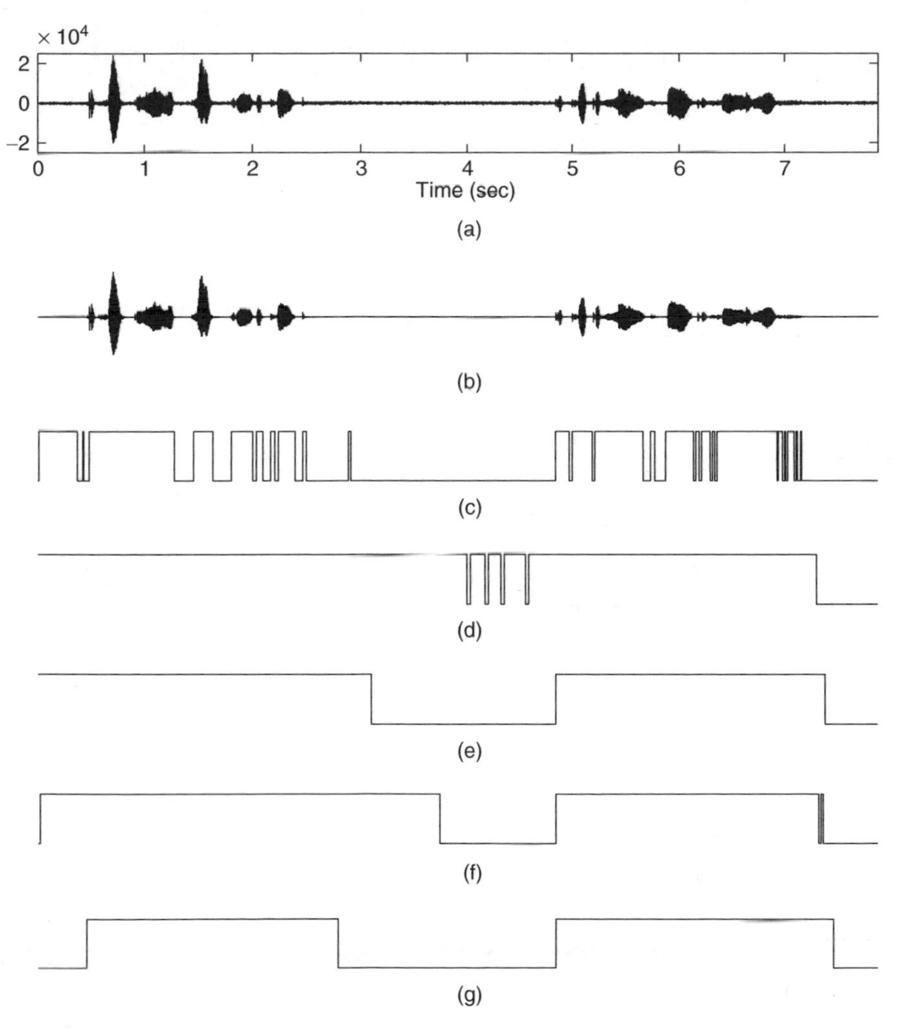

Figure 10.12 Comparison of VAD results over vehicle noise of 15 dB SNR: (a) noisy input speech, (b) clean speech, (c) G.729B, (d) IS-127, (e) GSM-EFR, (f) AMR1, and (g) AMR2

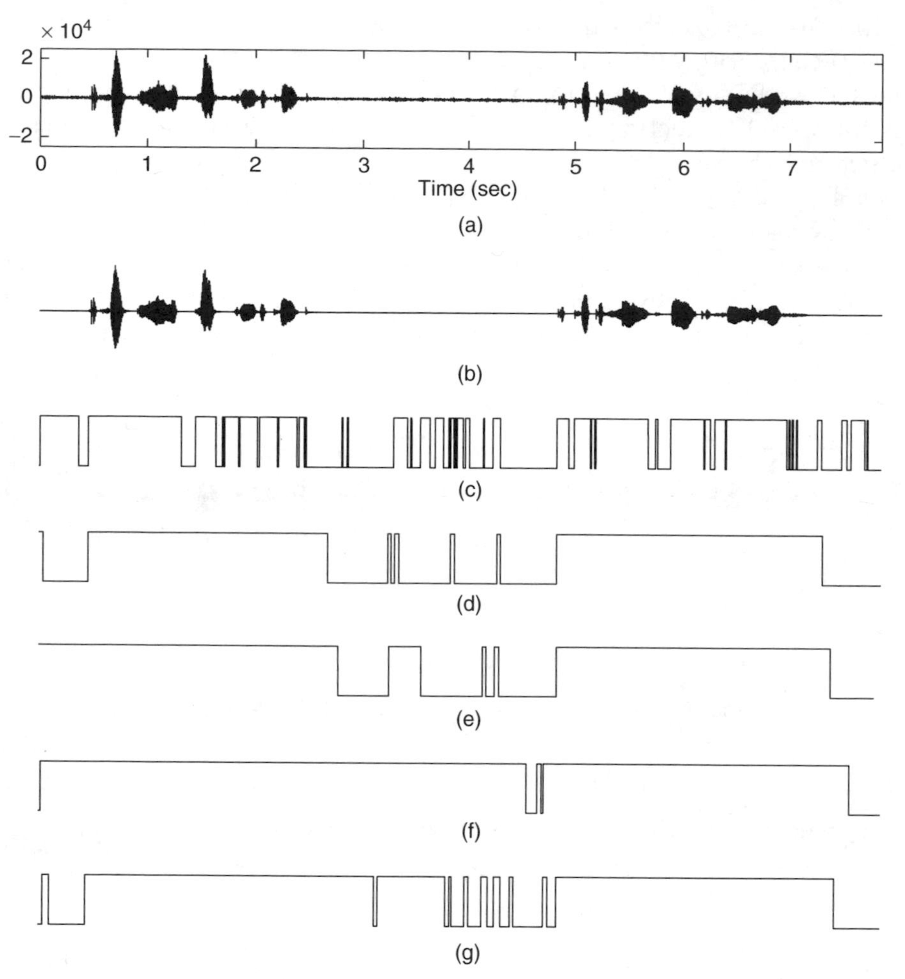

Figure 10.13 Comparison of VAD results over babble noise of 15 dB SNR: (a) noisy input speech, (b) clean speech, (c) G.729B, (d) IS-127, (e) GSM-EFR, (f) AMR1, and (g) AMR2

relatively consistent results regardless of the noise levels in silence detection for vehicle noisy speech. The performance of AMR1 is between GSM-EFR and AMR2. The characteristics of frame-wise voice activity decisions for various noise sources and levels are shown in Figures 10.12 and 10.13.

10.3 Likelihood-Ratio-Based VAD

Sohn *et al.* have proposed a novel method which, unlike traditional VAD methods, is based on a statistical model. They report that it can produce a high detection accuracy [12]. The reason for the high performance is attributed to the adoption of Ephraim and Malah's noise suppression rules [13] for the voice activity decision rules.

A voice activity decision can be considered as a test of hypotheses: H_0 and H_1, which indicate speech absence and presence, respectively. Assuming that each spectral component of speech and noise has complex Gaussian distribution [13], in which the noise is additive and uncorrelated with speech, the conditional probability density functions (PDF) of a noisy spectral component Y_k, given $H_{0,k}$ and $H_{1,k}$, are:

$$p(Y_k|H_{0,k}) = \frac{1}{\pi\lambda_{N,k}} \exp\left\{-\frac{|Y_k|^2}{\lambda_{N,k}}\right\} \tag{10.1}$$

$$p(Y_k|H_{1,k}) = \frac{1}{\pi(\lambda_{N,k} + \lambda_{X,k})} \exp\left\{-\frac{|Y_k|^2}{\lambda_{N,k} + \lambda_{X,k}}\right\} \tag{10.2}$$

where k indicates the spectral bin index, and $\lambda_{N,k}$ and $\lambda_{X,k}$ denote the variances of the noise and speech spectra, respectively.

The likelihood ratio (LR) of the kth spectral bin, Λ_k, is defined from the above two PDFs as [12]:

$$\Lambda_k = \frac{p(Y_k|H_{1,k})}{p(Y_k|H_{0,k})} = \frac{1}{1+\xi_k} \exp\left\{\frac{(1+\gamma_k)\xi_k}{1+\xi_k}\right\} \tag{10.3}$$

where γ_k and ξ_k are the *a posteriori* and *a priori* SNRs defined as, $\gamma_k = |Y_k|^2/\lambda_{N,k} - 1$ and $\xi_k = \lambda_{X,k}/\lambda_{N,k}$. Note that the definition of the *a posteriori* SNR is slightly different from the original one, $\gamma_k = |Y_k|^2/\lambda_{N,k}$ [13]. The noise variance is assumed to be known through noise adaptation (see Section 10.3.2). However, the variance of the speech is unknown, thus the *a priori* SNR of the nth frame, $\xi_k^{(n)}$, is estimated using the decision-directed (DD) method [13] as:

$$\hat{\xi}_k^{(n)} = \alpha \frac{\left|\hat{X}_k^{(n-1)}\right|^2}{\lambda_{N,k}^{(n-1)}} + (1-\alpha)MAX\{\gamma_k^{(n)}, 0\} \tag{10.4}$$

where α is a weighting term, e.g. 0.98, and the clean speech spectral amplitude, $|\hat{X}_k|$, is estimated using the minimum mean square error of the log spectral amplitude estimator [14]. The decision about the voice activity is performed by the geometric mean of the Λ_k over all spectral bins as:

$$\Lambda = \exp\left\{\frac{1}{K}\sum_{k=1}^{K} \log \Lambda_k\right\} \tag{10.5}$$

where K denotes the number of spectral bins.

The *a posteriori* SNR γ_k fluctuates highly from frame to frame because of the high fluctuation of the short-time spectral amplitude $|Y_k|$. On the other hand, the *a priori* SNR $\hat{\xi}_k$ changes slowly due to the smoothing effect. As the value of α increases, $\hat{\xi}_k$ becomes smoother. The variations of γ_k and $\hat{\xi}_k$ balance each other in the calculation of Λ_k and, consequently, result in enhanced performance for the VAD. The DD estimator for the *a priori* SNR is therefore useful not only for avoiding the musical noise phenomenon in speech enhancement [15], but also for reducing the error rate in voice activity detection.

10.3.1 Analysis and Improvement of the Likelihood Ratio Method

The behaviour of the LR in equation (10.3) with respect to the *a priori* and *a posteriori* SNRs, is shown in Figure 10.14. The ML estimator [12] results in lower performance in comparison with the DD estimator because of the inherent high-fluctuation of the *a posteriori* SNR. The LR employing the DD estimator has the following properties:

- If the *a posteriori* SNR is very high, i.e. $\gamma_k \gg 1$, and the range of the *a priori* SNR is limited, the LR becomes very high, i.e. $\Lambda_k \gg 1$.
- If the *a posteriori* SNR is low, i.e. $\gamma_k < 1$, the *a priori* SNR becomes a key parameter in the calculation of the LR.

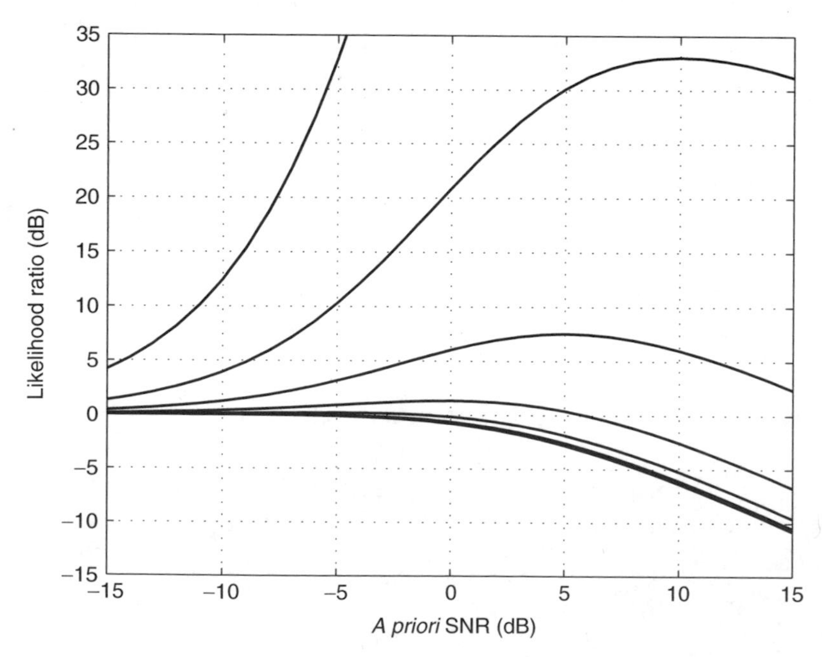

Figure 10.14 Likelihood ratio vs *a priori* SNR vs *a posteriori* SNR (the solid lines from the top represent *a posteriori* SNRs of 15, 10, 5, 0, −5, −10, and −15 dB, respectively)

In practice, the threshold of the LR is set between 0.2 dB and 0.8 dB, and both the *a posteriori* and the *a priori* SNRs are bounded between −15 dB and 15 dB.

Assuming that the noise characteristics change slowly, delay in estimation of the noise variance $\lambda_{N,k}^{(n-1)}$ in equation (10.4) does not seriously affect the *a priori* SNR $\hat{\xi}_k^{(n)}$. However, the spectral amplitude of the speech signal may change abruptly, particularly in onset and offset regions, in which the power of the spectral bins can increase and decrease rapidly, respectively. At the offset region, γ_k can be low but $\hat{\xi}_k$ can be much higher than γ_k due to the delay in $|\hat{X}_k^{(n-1)}|^2$ as given in equation (10.4). Thus Λ_k becomes too low, according to the second property above, and, consequently, Λ may become lower than the threshold of VAD. On the other hand, the delay rarely causes a problem at the onset regions, according to the first property above, as $\gamma_k^{(n)}$ in equation (10.3) is usually large enough.

It is possible to consider an adaptive weighting factor in the estimation of the *a priori* SNR in equation (10.4). In other words, a lower α can be assigned for the active region, and a higher α for the inactive region. When a low α is assigned at the offset region, it reduces the effect of the delay in equation (10.4), producing a lower $\hat{\xi}_k$, and therefore may prevent the abrupt decay of Λ_k. However, it is not easy to design a generalized adaptive rule that will result in good performance over various kinds of speech and noise signals. Instead, Cho [16, 17] has suggested a smoothed likelihood ratio (SLR) $\Psi_k^{(n)}$ which is defined as

$$\Psi_k^{(n)} = \exp\left\{\kappa \log \Psi_k^{(n-1)} + (1-\kappa)\log \Lambda_k^{(n)}\right\} \qquad (10.6)$$

where κ is a smoothing factor and $\Lambda_k^{(n)}$ is defined in equation (10.3) for the nth frame. The decision of the voice activity is finally carried out by computing,

$$\Psi^{(n)} = \exp\left\{\frac{1}{K}\sum_{k=1}^{K} \log \Psi_k^{(n)}\right\} \qquad (10.7)$$

and comparing it against a threshold. An nth input frame is classified as voice-active if $\Psi^{(n)}$ is greater than a threshold and voice-inactive otherwise.

Examples of the LR and the SLR over a segment of speech are shown in Figure 10.15. The SLR seems to overcome the problem outlined for the LR. As shown in Figure 10.15b, the SLR is relatively higher than the LR at the offset regions. The comparison over inactive frames is also shown in Figure 10.15c, which indicates that the SLR fluctuates less than the LR.

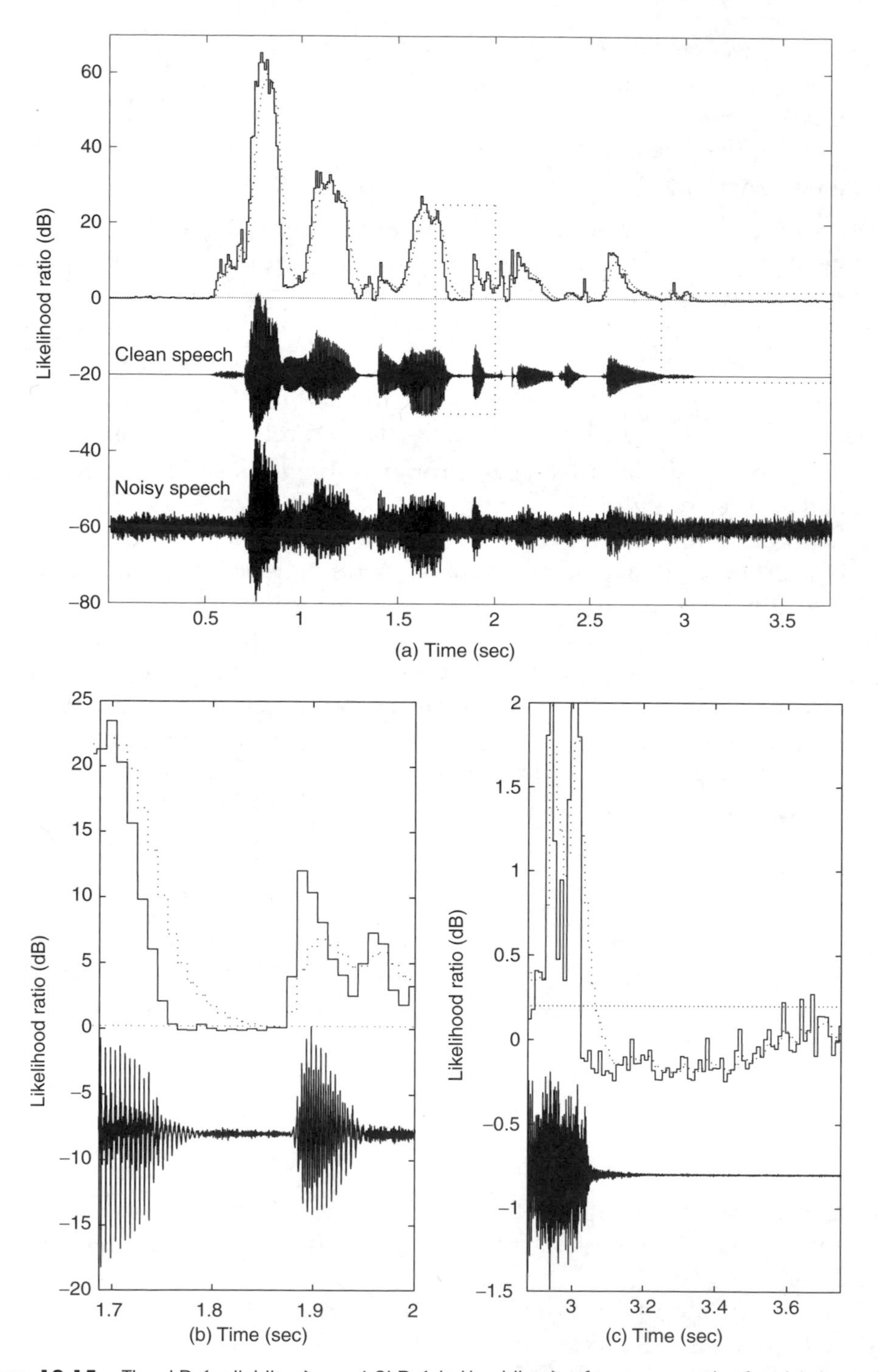

Figure 10.15 The LR (solid line) and SLR (dotted line) of a segment of vehicle noise signals of 5 dB SNR; the dotted horizontal-line indicates the VAD threshold and the boxed regions in (a) are enlarged in figures (b) and (c)

10.3.2 Noise Estimation Based on SLR

Depending on the characteristics of the noise source, the short-time spectral amplitudes of the noise signal can fluctuate strongly from frame to frame. In order to cope with time-varying noise signals, the variance of the noise spectrum is adapted to the current input signal by a soft decision-based method. The speech absence probability (SAP) of the kth spectral bin, $p(H_{0,k}|Y_k)$, can be calculated by Bayes' rule as:

$$p(H_{0,k}|Y_k) = \frac{p(H_{0,k})p(Y_k|H_{0,k})}{p(H_{0,k})p(Y_k|H_{0,k}) + p(H_{1,k})p(Y_k|H_{1,k})} = \frac{1}{1 + \frac{p(H_{1,k})}{p(H_{0,k})}\Psi_k} \quad (10.8)$$

where $p(H_{1,k}) = 1 - p(H_{0,k})$, and the unknown *a priori* speech absence probability (PSAP), $p(H_{0,k})$, is estimated in an adaptive manner given by:

$$\hat{p}(H_{0,k}^{(n)}) = MIN\{MAX\{\beta\hat{p}(H_{0,k}^{(n-1)}) + (1-\beta)p(H_{0,k}^{(n)}|Y_k^{(n)}), H_0^{(L)}\}, H_0^{(U)}\} \quad (10.9)$$

where β is a smoothing factor, e.g. 0.65. The lower and upper limits, $H_0^{(L)}$ and $H_0^{(U)}$, of the PSAP are determined through experiments, e.g. 0.2 and 0.8. Note that, for SLR, Ψ_k is applied to the calculation of the SAP instead of LR, Λ_k.

The variance of the noise spectrum of the kth spectral component in the nth frame, $\lambda_{N,k}^{(n)}$, is updated in a recursive way as:

$$\lambda_{N,k}^{(n)} = \eta\lambda_{N,k}^{(n-1)} + (1-\eta)E(|N_k^{(n)}|^2|Y_k^{(n)}) \quad (10.10)$$

where η is a smoothing factor, e.g. 0.95. The expected noise power-spectrum $E(|N_k^{(n)}|^2|Y_k^{(n)})$ is estimated by means of a soft-decision technique [18] as:

$$\begin{aligned} E(|N_k^{(n)}|^2|Y_k^{(n)}) &= E(|N_k^{(n)}|^2|H_{0,k})p(H_{0,k}|Y_k^{(n)}) + E(|N_k^{(n)}|^2|H_{1,k})p(H_{1,k}|Y_k^{(n)}) \\ &= |Y_k^{(n)}|^2 p(H_{0,k}|Y_k^{(n)}) + \lambda_{N,k}^{(n-1)} p(H_{1,k}|Y_k^{(n)}) \end{aligned} \quad (10.11)$$

where $p(H_{1,k}|Y_k^{(n)}) = 1 - p(H_{0,k}|Y_k^{(n)})$. During some tests, it is observed that SLR-based adaptation is useful for the estimation of the noise spectra with high variations, such as a babble noise source.

10.3.3 Comparison

The effect of the smoothing factor κ in equation (10.6) is shown in Figure 10.16. Note that the case of $\kappa = 0$ reduces equation (10.6) to the LR-based method. It is obvious from the results that the detection accuracy increases with increase in κ, at the offset regions without serious degradations in the performance

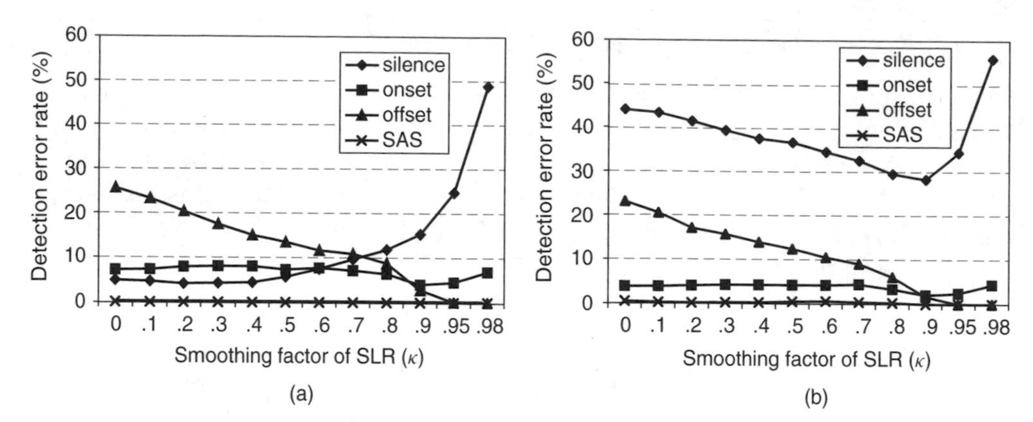

Figure 10.16 Analysis of the smoothing factor κ of the SLR with respect to detection error rates; the noise level is 10 dB SNR and the noise sources are (a) vehicle and (b) babble; SAS indicates speech active sections

at the onset regions for both vehicle and babble noisy signals. In the case of vehicle noisy signals, as κ increases, the false alarm rate in the inactive frames increases gradually for $\kappa < 0.9$, and then substantially for $\kappa > 0.9$. However, in the case of babble noisy signals, it can be seen that the error rate decreases gradually as κ increases for $\kappa < 0.9$, and then increases like the case of the vehicle noisy signal, for $\kappa > 0.9$. Therefore, if κ is selected properly, SLR-based method can give significantly improved performances over the LR-based method.

Under various noise levels and sources, the performance of VAD methods such as SLR-based VAD [16, 17], ITU-T G.729 annex B VAD (G.729B) [1], ETSI AMR VAD option 2 (AMR2) [8], and LR-based VAD with and without the hangover scheme [12] have been compared as shown in Table 10.2. Original AMR2 produces the detection result every 20 ms by the *logical OR* operation of two 10 ms detection results, thus the 10 ms result can be obtained easily by slight modification of the original code. Taking into account the results in Figure 10.16, $\kappa = 0.9$ is selected for SLR-based VAD. G.729B generates considerably high error rates at the active regions in comparison with other methods. It is important to note that frequent detection errors of speech frames lead to serious degradation in speech quality, thus the error rate of speech frame detection should be as low as possible. LR-based VAD gives consistently superior performance to G.729B, but VAD without the hangover scheme produces relatively high detection error rates in the active regions. The hangover scheme can considerably alleviate this problem, but the speech detection error rate is still somewhat high in comparison with the results of both SLR-based VAD and AMR2. The performance of SLR-based VAD and AMR2 seems to be comparable.

Table 10.2 Comparison of speech and silence detection error rates of SLR-based, LR-based, AMR2, and G.729B VADs

SNR (dB)	VAD	Detection error rate (%)							
		Vehicle noise				Babble noise			
		Inactive	Onset	Offset	SAS	Inactive	Onset	Offset	SAS
5	SLR	13.87	6.42	7.51	0.00	29.40	2.43	4.93	0.52
	LR	4.49	12.88	30.92	0.00	46.25	6.90	27.77	2.49
	LR + HO*	5.33	12.05	12.86	0.00	46.50	4.52	11.28	1.48
	AMR2	18.64	9.13	0.00	0.00	41.66	4.75	0.26	0.00
	G.729B	8.58	70.23	60.21	5.14	48.17	56.79	45.88	5.12
15	SLR	17.12	3.48	0.73	0.00	29.20	2.05	0.00	0.00
	LR	5.07	5.34	19.85	0.00	41.76	3.70	16.83	0.08
	LR + HO	7.52	4.75	6.80	0.00	42.67	3.32	4.18	0.00
	AMR2	20.15	3.78	0.00	0.00	51.53	2.19	0.26	0.00
	G.729B	8.57	31.19	39.41	0.00	49.79	25.90	32.73	0.00
25	SLR	23.01	2.82	0.00	0.00	30.77	1.54	0.00	0.00
	LR	6.64	3.29	11.79	0.00	34.38	1.54	8.75	0.00
	LR + HO	10.94	1.56	2.75	0.00	36.45	0.89	1.59	0.00
	AMR2	20.28	2.68	0.00	0.00	20.61	2.31	0.12	0.00
	G.729B	8.85	12.75	19.06	0.00	44.30	11.34	15.49	0.00

LR + HO means LR-based VAD with the hangover scheme

10.4 Summary

In this chapter, standard VAD techniques as well as LR- and SLR-based VAD have been reviewed. Through performance evaluation of the standard VAD methods, including G.729B, GSM-EFR, AMR1, AMR2, and IS-127, it has been shown that both AMR1 and AMR2 produce relatively high and consistent performances over various noise sources and levels. On the other hand, statistical-model-based LR VAD, performs well but may have a problem at the offset regions of speech signals which may be solved with a hangover in the decision making. The SLR method newly-proposed by Cho [16, 17] has overcome this problem without the need for a hangover. SLR VAD has comparable performance to AMR2.

Bibliography

[1] ITU-T (1996) *A silence compression scheme for G.729 optimised for terminals conforming to ITU-T V.70*, ITU-T Rec. G.729 Annex B.

[2] ITU-T (1996) *Dual rate speech coder for multimedia communications transmitting at 5.3 and 6.3 kbit/s. Annex A: Silence compression scheme*, ITU-T Rec. G.723.1 Annex A.

[3] ITU-T (1996) *Coding of speech at 8 kbit/s using conjugate-structure algebraic-code-excited linear prediction (CS-ACELP)*, ITU-T Rec. G.729.

[4] ITU-T (1996) *Dual rate speech coder for multimedia communications transmitting at 5.3 and 6.3 kbit/s*, ITU-T Rec. G.723.1.

[5] ETSI (1998) *Digital cellular telecommunications system (phase 2+); Voice activity detector (VAD) for full rate speech traffic channels*, GSM 06.32 (ETSI EN 300 965 v7.0.1).

[6] ETSI (1999) *Digital cellular telecommunications system (phase 2+); Voice activity detector (VAD) for full rate speech traffic channels*, GSM 06.42 (draft ETSI EN 300 973 v8.0.0).

[7] ETSI (1997) *Digital cellular telecommunications system; Voice activity detector (VAD) for enhanced full rate (EFR) speech traffic channels*, GSM 06.82 (ETS 300 730), March.

[8] ETSI (1998) *Digital cellular telecommunications system (phase 2+); Voice activity detector (VAD) for adaptive multi-rate (AMR) speech traffic channels*, GSM 06.94 v7.1.1 (ETSI EN 301 708).

[9] P. DeJaco, W. Gardner, and C. Lee (1993) 'QCELP: The North American CDMA digital cellular variable rate speech coding standard', in *IEEE Workshop on Speech Coding for Telecom*, pp. 5–6.

[10] TIA/EIA (1997) *Enhanced variable rate codec, speech service option 3 for wideband spread spectrum digital systems*, IS-127.

[11] TIA/EIA (1998) *High rate speech service option 17 for wideband spread spectrum communication systems*, IS-733.

[12] J. Sohn, N. S. Kim, and W. Sung (1999) 'A statistical model-based voice activity detection', in *IEEE Signal Processing Letters*, 6(1):1–3.

[13] Y. Ephraim and D. Malah (1984) 'Speech enhancement using a minimum mean square error short-time spectral amplitude estimator', in *IEEE Trans. on Acoust., Speech and Signal Processing*, 32(6):1109–20.

[14] Y. Ephraim and D. Malah (1985) 'Speech enhancement using a minimum mean square error log-spectral amplitude estimator', in *IEEE Trans. on Acoust., Speech and Signal Processing*, 33(2):443–5.

[15] O. Cappé (1994) 'Elimination of musical noise phenomenon with the Ephraim and Malah noise suppression', in *IEEE Trans. Speech and Audio Processing*, 2(2):345–9.

[16] Y. D. Cho (2001) 'Speech detection enhancement and compression for voice communications', Ph.D. thesis, CCSR, University of Surrey, UK.

[17] Y. Cho and A. M. Kondoz (2001) 'Analysis and improvement of a statistical model-based voice activity detector', in *IEEE Signal Processing Letters*, 8(10):276–8.

[18] J. Sohn and W. Sung (1995) 'A voice activity detection employing soft decision based noise spectrum adaptation', in *Proc. of Int. Conf. on Acoust., Speech and Signal Processing*, pp. 365–8. Amsterdam

11

Speech Enhancement

11.1 Introduction

In voice communications, speech signals can be contaminated by environmental noise and, as a result, the communication quality can be affected making the speech less intelligible. Furthermore, compression of the noisy speech with a low bit-rate vocoder may result in considerable quality degradation due to frequent estimation errors of speech production model parameters required by the vocoder. This problem can be reduced significantly by speech enhancement (or noise cancellation), which may enable more pleasant voice communication by suppressing the noise components in input signals.

Generally, it is assumed that the noisy speech signal is formed additively by speech and noise signals in which the noise is generated by environmental sources such as vehicles, street noise, babble, etc. Therefore, in real environments, complete noise cancellation is not feasible as it is not possible to completely track varying noise types and characteristics that change with time. However, by assuming that the noise characteristics change slowly in comparison with speech, it is possible to achieve significant reduction in the background noise levels producing more pleasant and intelligible speech quality. Speech enhancement techniques can help the speech model parameter extraction process used in low bit-rate vocoders and hence they are becoming an integral part of low bit-rate speech coding systems.

Speech enhancement techniques can be classified, depending on the number of available microphones, into single and multiple channels. In the case of a single channel, the reference noise is not available explicitly. The noise statistics are typically characterized during voice-inactive regions between talk spurts using a voice activity detector. On the other hand, when dual channels are available, one microphone senses the noisy speech, but the other can be used mainly to catch the noise. By eliminating the noise factor collected by the second microphone from the first, it would be possible to

Digital Speech. A. Kondoz
 ISBN 0-470-87007-9 (HB)

cancel the noise more efficiently. However, in real environments, the multiple microphone scheme can be limited in its use. In the following, we consider the single microphone case only.

Over the last three decades, many kinds of speech enhancement techniques have been proposed [1–4], mostly based on transform domain techniques, adaptive filtering, and model-based methods. The transform-based technique transforms the time-domain signal into other domains, suppresses noise components, and then applies the corresponding inverse transform to reconstruct enhanced speech signals. Discrete Fourier transform (DFT), discrete cosine transform (DCT), Karhunen–Loève transform (KLT), and wavelet transform (WT) are widely-known transform methods. DFT-based techniques have been intensively investigated based on short-time spectral amplitudes (STSA). KLT-based techniques, called signal subspace-based methods [5], decomposes the space into signal (or speech) and noise subspaces by means of eigen decomposition, and then suppresses the noise component in the eigenvalues. DCT-based techniques [6, 7] are of lower computational complexity and higher frequency resolution than DFT-based methods. It is also possible to consider WT-based methods in order to simultaneously exploit the time and frequency characteristics of noisy speech signals. Adaptive filtering, on the other hand, cancels the noise using adaptive filters such as the Kalman filter. A Kalman filter models noisy speech signals in terms of state space and observation equations, which represent the speech production process and the noise addition model together with channel distortion, respectively [4]. Kalman filters normally assume a white Gaussian noise distribution; however, Gibson *et al.* proposed a generalization of Kalman-filtering over coloured noise signals [8, 9]. Finally, model-based techniques classify the noisy signal using an *a priori* speech model, such as hidden Markov and voiced/unvoiced models, and then conducts the enhancement depending on classified speech models [2]. This method can be useful for improving noise reduction performance for various kinds of speech signals. However, it requires extra training to build the model with intensive computation. In addition, it may exhibit model selection errors which cause significant speech quality degradation. Fundamentally, it is not easy to handle complicated speech signals with a finite number of speech models.

Amongst the speech enhancement techniques, DFT (or STSA)-based methods have been well investigated in the forms of spectral subtraction, Wiener filtering, maximum likelihood-STSA estimation, and minimum mean square error STSA estimation. The reason for the popularity of STSA-based speech enhancement is due not only to its computational simplicity but also to recent technical advances in this technique producing significant speech quality improvement. In the following section, details of some of the most used STSA-based speech enhancement techniques are reviewed.

11.2 Review of STSA-based Speech Enhancement

Assuming that the noise $d(n)$ is additive to the speech signal $x(n)$, the noisy speech $y(n)$ can be written as,

$$y(n) = x(n) + d(n), \quad \text{for } 0 \leq n \leq K-1 \tag{11.1}$$

where n is the time index. The objective of speech enhancement is to find the enhanced speech $\hat{x}(n)$ given $y(n)$, with the assumption that $d(n)$ is uncorrelated with $x(n)$. The time-domain signals can be transformed to the frequency domain as,

$$Y_k = X_k + D_k, \quad \text{for } 0 \leq k \leq K-1 \tag{11.2}$$

where Y_k, X_k, and D_k denote the short-time DFT of $y(n)$, $x(n)$, and $d(n)$, respectively. The STSA-based speech enhancement filters out the noise by modifying the spectral amplitudes of Y_k in equation (11.2). Therefore, the enhanced spectrum $\hat{X}_k$ can be written in terms of the modification factor (gain) G_k and the noisy spectrum Y_k as,

$$\hat{X}_k = G_k Y_k, \quad \text{for } 0 \leq G_k \leq 1 \tag{11.3}$$

The gain G_k is a function of *a posteriori* SNR,

$$\gamma_k \equiv \frac{|Y_k|^2}{E(|D_k|^2)} \tag{11.4}$$

and *a priori* SNR,

$$\xi_k \equiv \frac{E(|X_k|^2)}{E(|D_k|^2)} \tag{11.5}$$

where $E(|D_k|^2)$ and $E(|X_k|^2)$ are the statistical variances of the kth spectral components of the noise and the speech, respectively. The function definition of the gain G_k depends on specific enhancement methods. The *a posteriori* SNR γ_k in equation (11.4) can be obtained easily as Y_k is the input noisy spectrum and $E(|D_k|^2)$ can be obtained through a noise adaptation procedure discussed in Section 11.3. However, the speech variance $E(|X_k|^2)$ for the estimation of ξ_k in equation (11.5) is not available. As a solution, Ephraim and Malah [10] proposed the decision-directed (DD) method given by,

$$\hat{\xi}_k^{(t)} = \alpha \frac{|\hat{X}^{(t-1)}|^2}{E(|D_k^{(t)}|^2)} + (1-\alpha) MAX(\gamma_k^{(t)} - 1, 0) \tag{11.6}$$

where $0 \leq \alpha < 1$ and t is the frame index.

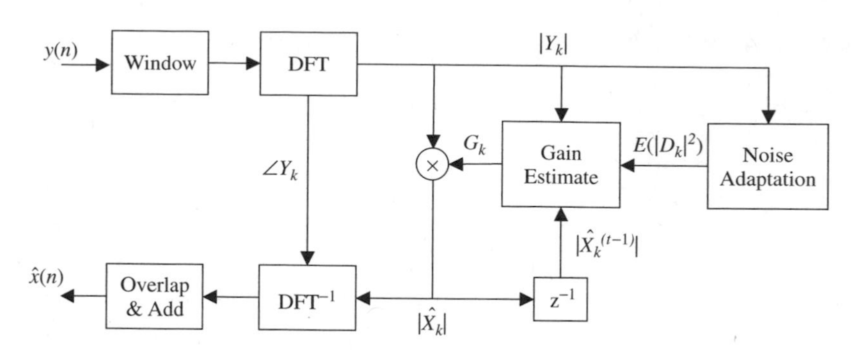

Figure 11.1 Block diagram of general STSA-based speech enhancement method

The main aim of speech enhancement can be stated as an optimization problem, where the residual noise is minimized while maintaining the speech quality. The optimization process therefore requires a trade-off between noise reduction and speech quality. For example, over-estimation of the noise statistics may degrade the speech quality or intelligibility. On the other hand, estimation of noise may not be accurate, leading to considerable residual noise. The most typical residual noise in speech enhancement is musical noise, also called tonal noise, which is composed of narrowband signals appearing and disappearing with time-varying amplitudes and frequencies.

The overall block diagram of a generalized STSA-based speech enhancement method is shown in Figure 11.1. The noisy speech, $y(n)$, is first converted into STSA, $|Y_k|$, by a DFT with windowing. The enhanced spectral amplitude, $|\hat{X}_k|$, is estimated by multiplying the noisy signal spectral components, Y_k, with their corresponding estimated gains, G_k. Enhanced speech, $\hat{x}(n)$, is then reconstructed by applying the inverse DFT to the enhanced STSA, $|\hat{X}_k|$, with the noisy speech phase, $\angle Y_k$, followed by an appropriate overlap-and-add procedure to compensate for the window effect and to alleviate abrupt signal changes between two consecutive frames. The most critical part of this process is the accurate estimation of the gains, G_k, which is discussed next.

11.2.1 Spectral Subtraction

The noisy spectrum Y_k in equation (11.2) can be converted to the power spectrum as,

$$|Y_k|^2 = |X_k|^2 + |D_k|^2 + X_k^* D_k + X_k D_k^* \tag{11.7}$$

where X_k^* and D_k^* denote the complex conjugates of X_k and D_k, respectively. In order to estimate $|X_k|^2$ in equation (11.7), the statistically expected values are applied since $|D_k|^2$, $X_k^* D_k$, and $X_k D_k^*$ are not available. We therefore get,

$$|Y_k|^2 = |\hat{X}_k|^2 + E(|D_k|^2) + E(X_k^* D_k) + E(X_k D_k^*) \tag{11.8}$$

where $E(\cdot)$ is the ensemble average and $|\hat{X}_k|^2$ is the enhanced power spectrum. The expected noise $E(|D_k|^2)$ can be estimated by a noise adaptation procedure as shown in Section 11.3. Due to the assumption that $x(n)$ is uncorrelated with $d(n)$, $E(X_k^* D_k) = 0$ and $E(X_k D_k^*) = 0$. Thus, equation (11.8) can be rewritten as,

$$|Y_k|^2 = |\hat{X}_k|^2 + E(|D_k|^2) \tag{11.9}$$

The enhanced power spectrum $|\hat{X}_k|^2$ can be estimated by subtracting $E(|D_k|^2)$ from $|Y_k|^2$, which is called power spectral subtraction.

The spectral power subtraction can be generalized with an arbitrary spectral order, called generalized spectral subtraction (GSS), as,

$$|Y_k|^\nu = |\hat{X}_k|^\nu + E(|D_k|^\nu) \tag{11.10}$$

where ν is the spectral order. In the cases of $\nu = 1$ and $\nu = 2$, GSS in equation (11.10) can be reduced to the magnitude and power spectral subtractions, respectively.

In practice, GSS-based speech enhancement may exhibit severe musical noise due to the high fluctuation of the STSA of noisy signals. In some cases the estimated noise magnitude can be larger than the input spectral magnitude, where the enhanced spectral magnitudes are clamped to zero in order to prevent the spectral magnitude from being negative. The clamping which happens irregularly with frequency and time leads to producing the sound of musical tones.

Berouti *et al.* [11] proposed a method for alleviating the musical noise phenomenon, where $|\tilde{X}_k|^\nu = |Y_k|^\nu - \alpha E(|D_k|^\nu)$ and Berouti's GSS (GBSS) is given by,

$$|\hat{X}_k|^\nu = \begin{cases} |\tilde{X}_k|^\nu & ; \ \text{if } |\tilde{X}_k|^\nu > \beta E(|D_k|^\nu) \\ \beta E(|D_k|^\nu) & ; \ \text{otherwise} \end{cases} \tag{11.11}$$

where α and β are the spectral over-subtraction and floor factors, respectively, with $\alpha \geq 1$ and $0 \leq \beta \leq 1$. Note that GBSS reduces to GSS when $\alpha = 1$ and $\beta = 0$, and to Power Spectral Subtraction (PSS) if $\nu = 2$, $\alpha = 1$, and $\beta = 0$. GBSS is capable of reducing the overall residual noise level as well as typical musical noises by appropriately adjusting α and β. The GBSS gain $G_k^{(GBSS)}$ becomes,

$$G_k^{(GBSS)} = \begin{cases} \left[1 - \alpha\left(\frac{1}{\gamma_k}\right)^{\frac{\nu}{2}}\right]^{\frac{1}{\nu}} & ; \ \text{if } \gamma_k^{\frac{\nu}{2}} > \alpha + \beta \\ \beta^{\frac{1}{\nu}} \frac{1}{\sqrt{\gamma_k}} & ; \ \text{otherwise} \end{cases} \tag{11.12}$$

The noise floor factor β contributes to the reduction of musical noise sounds. It has the effect of converting the narrowband musical noise into a wider band noise. Although higher β values give less musical noise, if β is set too high it may result in an increase of the level of other artifacts of residual noise. The over-subtraction factor, α, is useful for reducing the overall level of residual noise. Higher α values give lower levels of residual noise. However, too high α values may cause distortion in perceived speech quality. Through experiments, it is found that GBSS with $\nu = 2$, $\alpha = 4 \sim 8$, and $\beta = 0.1$ give a moderate level of musical noise reduction while maintaining the perceived speech quality.

In GBSS, both spectral over-subtraction and floor factors are fixed to constant values. However, each set of parameters exhibits different noise reduction performances depending on the selection of these two factors. There are approaches to obtain the optimal factors based on the psycho-acoustic model and a parametric formulation. In the psycho-acoustic approach, both α and β change each frame depending on the psychoacoustic masking threshold for each spectral component [12]. In the parametric formulation, α is derived using the MMSE-based metric [13].

11.2.2 Maximum-likelihood Spectral Amplitude Estimation

In DFT-based speech enhancement, given $Y_k = X_k + D_k$, the optimum estimate of the speech magnitude $|X_k|$ is obtained from the noisy spectrum Y_k, in which $X_k = |X_k| \exp(j\theta_k)$ where θ_k is the phase of X_k. Assuming that the noise D_k has complex Gaussian distribution, the probability density function (PDF) of Y_k conditioned over $|X_k|$ and θ_k is,

$$p(Y_k|\,|X_k|, \theta_k) = \frac{1}{\pi E(|D_k|^2)} \exp\left\{-\frac{|Y_k|^2 - 2|X_k|\mathrm{Re}(e^{-j\theta_k} Y_k) + |X_k|^2}{E(|D_k|^2)}\right\} \quad (11.13)$$

McAulay [14] has shown that the maximum likelihood (ML) estimate of $|X_k|$ can be obtained from the derivative of PDF with respect to $|X_k|$, where the ML estimate $|\hat{X}_k|$ is given by,

$$|\hat{X}_k| = \frac{1}{2}\left(|Y_k| + \sqrt{|Y_k|^2 - E(|D_k|^2)}\right) \quad (11.14)$$

which can be written in terms of the gain as,

$$G_k^{(ML)} = \frac{1}{2} + \frac{1}{2}\sqrt{1 - \frac{1}{\gamma_k}} \quad (11.15)$$

11.2.3 Wiener Filtering

The Wiener filter (WF) is a minimum mean square error (MMSE) estimate of a desired signal in the time domain [1, 4]. Given a noisy signal $y(n)$, for $0 \leq n \leq N-1$, the Wiener filter produces the MMSE estimate $\hat{x}(n)$ of speech $x(n)$ as,

$$\underbrace{\begin{pmatrix} \hat{x}(0) \\ \hat{x}(1) \\ \vdots \\ \hat{x}(N-1) \end{pmatrix}}_{\hat{\mathbf{x}}} = \underbrace{\begin{pmatrix} y(0) & y(-1) & \cdots & y(1-P) \\ y(1) & y(0) & \cdots & y(2-P) \\ \cdots & \cdots & \cdots & \cdots \\ y(N-1) & y(N-2) & \cdots & y(N-P) \end{pmatrix}}_{\mathbf{Y}} \underbrace{\begin{pmatrix} w_0 \\ w_1 \\ \vdots \\ w_{P-1} \end{pmatrix}}_{\mathbf{w}} \tag{11.16}$$

where w_k are the filter coefficients for $0 \leq k \leq P-1$ with the filter order P. Equation (11.16) can be rewritten in the algebraic form as,

$$\hat{\mathbf{x}} = \mathbf{Yw} \tag{11.17}$$

The Wiener filter error signal $\mathbf{e}$ is the difference between the desired and estimated speech signals given by,

$$\mathbf{e} = \mathbf{x} - \hat{\mathbf{x}} \tag{11.18}$$

The error metric ε is defined as,

$$\begin{aligned} \varepsilon &= \mathbf{e}^{\mathrm{T}}\mathbf{e} \\ &= (\mathbf{x} - \mathbf{Yw})^{\mathrm{T}}(\mathbf{x} - \mathbf{Yw}) \\ &= \mathbf{x}^{\mathrm{T}}\mathbf{x} - \mathbf{w}^{\mathrm{T}}\mathbf{Y}^{\mathrm{T}}\mathbf{x} - \mathbf{x}^{\mathrm{T}}\mathbf{Yw} - \mathbf{w}^{\mathrm{T}}\mathbf{Y}^{\mathrm{T}}\mathbf{Yw} \end{aligned} \tag{11.19}$$

The filter coefficients $\mathbf{w}$ are derived by setting the derivative of ε to zero with respect to $\mathbf{w}$,

$$\frac{\partial \varepsilon}{\partial \mathbf{w}} = -2(\mathbf{x}^{\mathbf{T}}\mathbf{Y} - \mathbf{w}^{\mathbf{T}}\mathbf{y}^{\mathbf{T}}\mathbf{Y}) = 0 \tag{11.20}$$

Then, the optimal $\mathbf{w}$ is given by,

$$\mathbf{w} = (\mathbf{Y}^T\mathbf{Y})^{-1}\mathbf{Y}^T\mathbf{x} \tag{11.21}$$

in which $\mathbf{Y}^T\mathbf{Y}$ and $\mathbf{Y}^T\mathbf{x}$ are the autocorrelation matrix R_{yy} of $y(n)$ and the cross-correlation vector r_{yx} between $y(n)$ and $x(n)$, respectively. Thus, equation (11.21) can be written as,

$$\mathbf{w} = R_{yy}^{-1} r_{yx} \tag{11.22}$$

Note that because of the assumption that the speech is uncorrelated with noise, $R_{yy} = R_{xx} + R_{dd}$ and $r_{yx} = r_{xx}$. Thus, equation (11.22) becomes,

$$\mathbf{w} = (R_{xx} + R_{dd})^{-1} r_{xx} \tag{11.23}$$

Equation (11.23) can be interpreted in the frequency domain as,

$$\begin{aligned} G_k^{(\mathrm{WF})} &= \frac{E(|X_k|^2)}{E(|X_k|^2) + E(|D_k|^2)} \\ &= \frac{\xi_k}{1 + \xi_k} \end{aligned} \tag{11.24}$$

Note that the Wiener filter gain $G_k^{(\mathrm{WF})}$ in equation (11.24) is defined in terms of the *a priori* SNR ξ_k only.

11.2.4 MMSE Spectral Amplitude Estimation

The Wiener filter is a time-domain MMSE estimation while McAulay's method is a frequency-domain ML estimation technique. Thus, it is possible to consider the MMSE estimate of the spectral amplitude [10], which minimizes

$$\varepsilon = (|X_k| - |\hat{X}_k|)^2 \tag{11.25}$$

The MMSE-STSA estimate, $|\hat{X}_k|$ given Y_k, is,

$$\begin{aligned} |\hat{X}_k| &= E(|X_k|\,|Y_k) \\ &= \frac{\displaystyle\int_0^{\infty}\int_0^{2\pi} \alpha_k p(Y_k|\alpha_k, \theta_k) p(\alpha_k, \theta_k) d\theta_k d\alpha_k}{\displaystyle\int_0^{\infty}\int_0^{2\pi} p(Y_k|\alpha_k, \theta_k) p(\alpha_k, \theta_k) d\theta_k d\alpha_k} \end{aligned} \tag{11.26}$$

where,

$$p(Y_k|\alpha_k, \theta_k) = \frac{1}{\pi E(|D_k|^2)} \exp\left\{ -\frac{|Y_k - \alpha_k e^{j\theta_k}|^2}{E(|D_k|^2)} \right\} \tag{11.27}$$

and,

$$p(\alpha_k, \theta_k) = \frac{\alpha_k}{\pi E(|X_k|^2)} \exp\left\{ -\frac{\alpha_k^2}{E(|X_k|^2)} \right\} \tag{11.28}$$

in which α_k and θ_k are dummy variables for the spectral amplitude and phase, respectively, of X_k. The amplitude has the Rayleigh distribution given by,

$$p(\alpha_k) = \frac{2\alpha_k}{E(|X_k|^2)} \exp\left\{-\frac{\alpha_k^2}{E(|X_k|^2)}\right\} \tag{11.29}$$

and the phase has the uniform distribution given by,

$$p(\theta_k) = \frac{1}{2\pi} \tag{11.30}$$

Through derivation given in [10], equation (11.26) can be rewritten as,

$$|\hat{X}_k| = \Gamma(1.5)\frac{\sqrt{v_k}}{\gamma_k} \exp\left(-\frac{v_k}{2}\right)\left\{(1+v_k)I_0\left(\frac{v_k}{2}\right) + v_k I_1\left(\frac{v_k}{2}\right)\right\}|Y_k| \tag{11.31}$$

where $\Gamma(\cdot)$ is the gamma function with $\Gamma(1.5) = \sqrt{\pi}/2$, $I_0(\cdot)$ and $I_1(\cdot)$ denote the modified Bessel functions of zero and first order, respectively, and $v_k \equiv \frac{\xi_k}{1+\xi_k}\gamma_k$.

As a variant, Ephraim and Malah [15] proposed an MMSE log spectral amplitude (MMSE-LSA) estimator, based on the well-known fact that a distortion measure with the log spectral amplitudes is more suitable for speech processing. The MMSE-LSA estimator minimizes the following distortion measure,

$$\varepsilon = \left\{\log|X_k| - \log|\hat{X}_k|\right\}^2 \tag{11.32}$$

with

$$|\hat{X}_k| = \exp\left[E\{\log(|X_k|)\,|Y_k\}\right] \tag{11.33}$$

From [15], the final estimate becomes,

$$|\hat{X}_k| = \frac{\xi_k}{1+\xi_k}\exp\left\{\frac{1}{2}\int_{v_k}^{\infty}\frac{e^{-t}}{t}dt\right\}|Y_k| \tag{11.34}$$

11.2.5 Spectral Estimation Based on the Uncertainty of Speech Presence

The conventional speech enhancement methods can be extended by incorporating the uncertainty of speech presence [14, 15]. The absence and presence of speech, H_0 and H_1, respectively, can be defined as,

$$H_0 : Y_k = D_k \tag{11.35}$$

$$H_1 : Y_k = X_k + D_k \tag{11.36}$$

Assuming that each spectral component of speech and noise has complex Gaussian distribution, and that the noise is additive to and uncorrelated with the speech signal, the conditional probability density functions observing a noisy spectral component Y_k, given H_0 and H_1, are

$$p(Y_k|H_0) = \frac{1}{\pi E(|D_k|^2)} \exp\left\{-\frac{|Y_k|^2}{E(|D_k|^2)}\right\} \tag{11.37}$$

$$p(Y_k|H_1) = \frac{1}{\pi (E(|D_k|^2) + E(|X_k|^2))} \exp\left\{-\frac{|Y_k|^2}{E(|D_k|^2) + E(|X_k|^2)}\right\} \tag{11.38}$$

where k is the spectral bin index, $0 \leq k \leq K/2$, and $E(|D_k|^2)$ and $E(|X_k|^2)$ denote the variances of the kth spectral components of noise and speech, respectively.

The probability of speech presence can be given by Bayes' rule,

$$\begin{aligned} p(H_1|Y_k) &= \frac{p(Y_k|H_1)p(H_1)}{p(Y_k|H_0)p(H_0) + p(Y_k|H_1)p(H_1)} \\ &= \frac{\mu\Lambda}{1 + \mu\Lambda_k} \end{aligned} \tag{11.39}$$

where,

$$\mu = \frac{p(H_1)}{p(H_0)} \tag{11.40}$$

in which $p(H_1)$ and $p(H_0)$ denote the *a priori* probability of speech presence and absence, respectively. The likelihood ratio of the kth spectral bin Λ_k can be defined from the above two likelihood ratios,

$$\begin{aligned} \Lambda_k &= \frac{p(Y_k|H_1)}{p(Y_k|H_0)} \\ &= \frac{1}{1 + \xi_k} \exp\left\{\frac{(1 + \gamma_k)\xi_k}{1 + \xi_k}\right\} \end{aligned} \tag{11.41}$$

The enhanced spectrum based on the probability of speech presence is written as,

$$\hat{X}_k = E(X_k|Y_k, H_0)p(H_0|Y_k) + E(X_k|Y_k, H_1)p(H_1|Y_k) \tag{11.42}$$

where $p(H_0|Y_k)$ denotes the probability of speech absence given Y_k. Since the expected speech spectrum under speech absence is zero, i.e. $E(X_k|Y_k, H_0) = 0$, equation (11.42) can be simplified to,

$$\hat{X}_k = E(X_k|Y_k, H_1)p(H_1|Y_k) \tag{11.43}$$

$E(X_k|Y_k, H_1)$ and $p(H_1|Y_k)$ can be computed by a conventional spectral estimator and equation (11.39), respectively.

11.2.6 Comparisons

Objective speech qualities for voice-active regions are evaluated in terms of both segmental SNR (SEGSNR) improvement and Itakura–Saito distortion (ISD). The SEGSNR improvement indicates the difference between the SEGSNRs of the enhanced speech and the noisy input signals, in which the SEGSNR is defined by,

$$\text{SEGSNR(dB)} = \frac{10}{M}\left[\sum_{m=0}^{M-1} \log_{10}\left\{\sum_{n=mN}^{(m+1)N-1} \frac{x^2(n)}{(x(n) - \hat{x}(n))^2}\right\}\right] \tag{11.44}$$

where N and M are the frame size and the total number of frames, respectively. The ISD is defined as,

$$\text{ISD(dB)} = 10\log_{10}\left\{\frac{\mathrm{a}_x^T \mathrm{R}_{\hat{x}} \mathrm{a}_x}{\mathrm{a}_{\hat{x}}^T \mathrm{R}_{\hat{x}} \mathrm{a}_{\hat{x}}}\right\} \tag{11.45}$$

where a_x and $\mathrm{a}_{\hat{x}}$ are the LPC coefficients of the desired and estimated speech signals, respectively, and $\mathrm{R}_{\hat{x}}$ is the autocorrelation matrix of the estimated signal.

For comparison, speech material of 64 seconds, mixed with vehicle and helicopter noises of 0, 5 and 10 dB SNR were used. Enhancement processing was applied every 10 ms in the frequency domain by the five types of spectral estimator: PSS, GBSS, ML, WF, and MMSE-LSA. The MMSE-LSA is further classified, depending on the adoption of the speech presence uncertainty, into MMSE-LSA-HD and MMSE-LSA-SD in which HD and SD denote the hard and soft decision methods, respectively. The reference (the best possible processed signal) is obtained using the original spectral amplitudes with the phases of the noisy signal, because the ideal speech enhancement is achieved with the original speech spectral amplitudes and the phases of the noisy input speech.

The SEGSNR improvement and ISD for the vehicle and the helicopter noisy signals are shown in Figures 11.2, 11.3, 11.4, and 11.5. From the analysis, it is

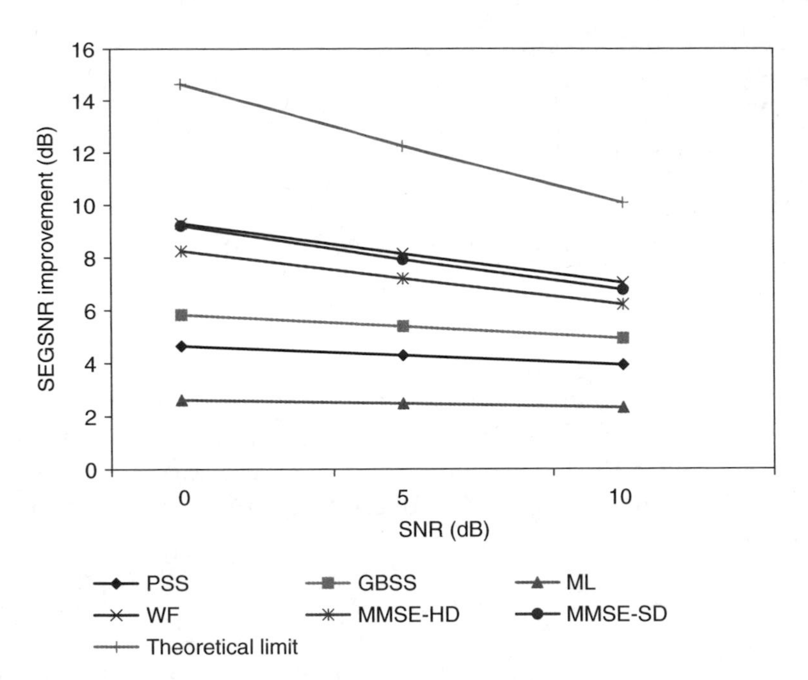

Figure 11.2 SEGSNR improvements from STSA-based speech enhancement methods in vehicle noise environments

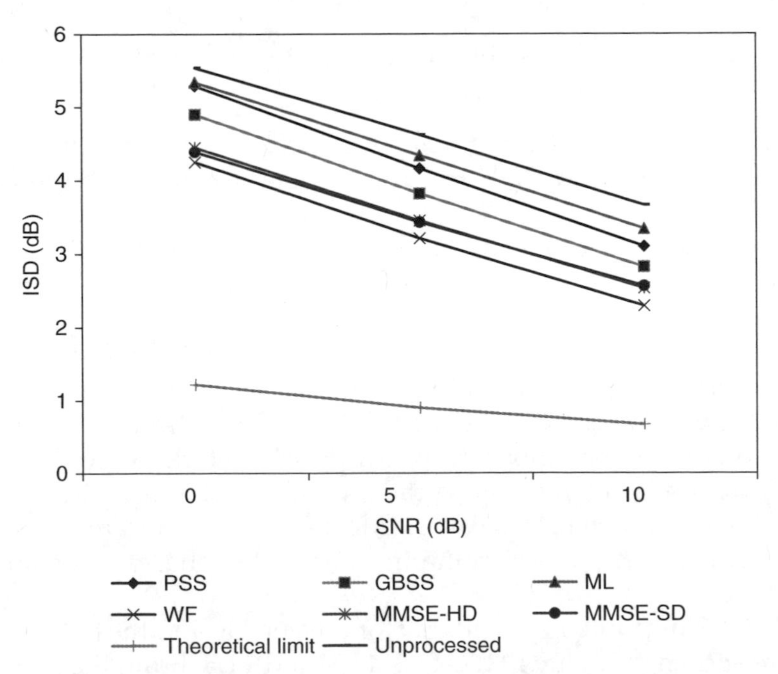

Figure 11.3 ISDs of STSA-based speech enhancement methods in vehicle noise environments

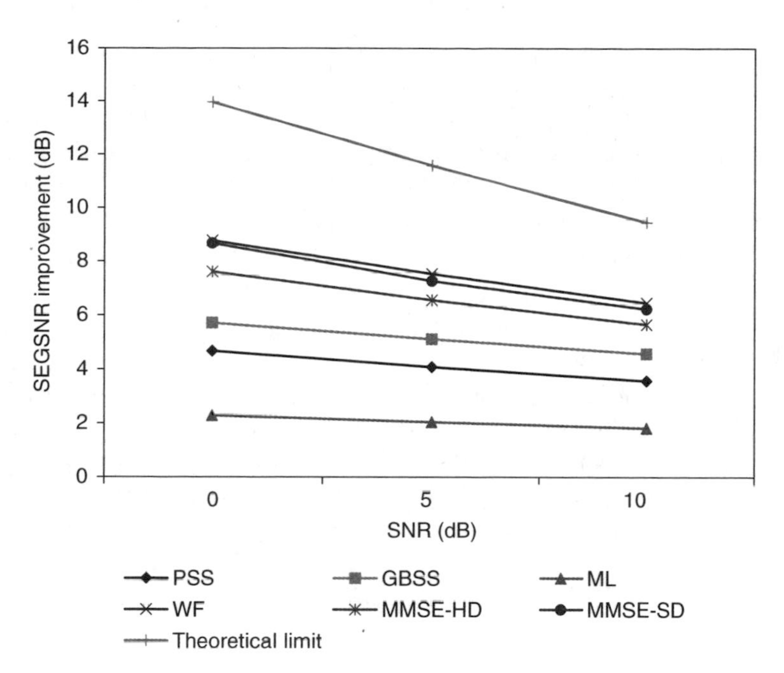

Figure 11.4 SEGSNR improvements from STSA-based speech enhancement methods in helicopter noise environments

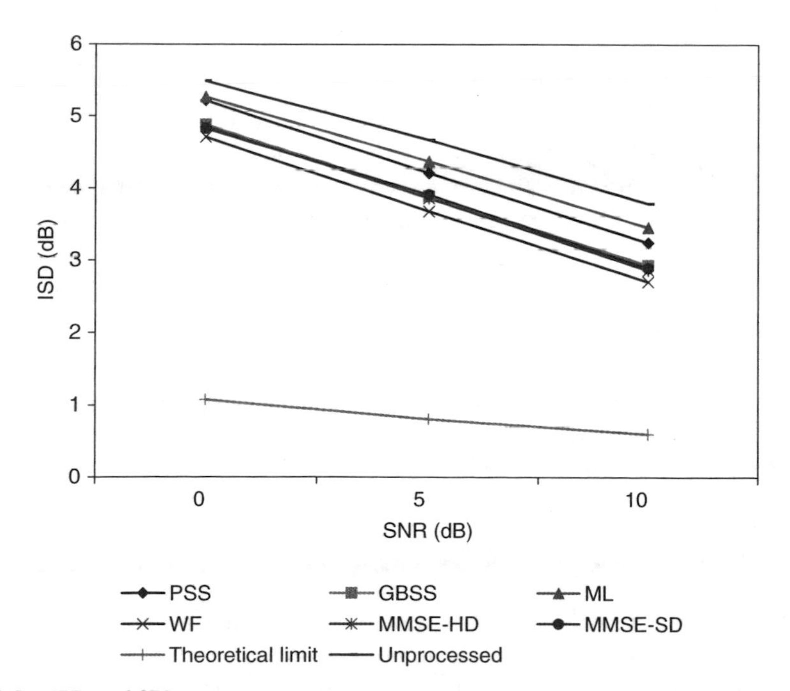

Figure 11.5 ISDs of STSA-based speech enhancement methods in helicopter noise environments

found that WF- and MMSE-based methods give better results than the other methods.

For the noisy input signal in Figure 11.6, the spectrograms of different enhancement methods, showing the characteristics of the residual noise, are shown in Figures 11.9, 11.10, 11.11, 11.12, 11.13, and 11.14. The spectrograms of the noise-free and reference signals are shown in Figures 11.7 and 11.8, respectively. For the PSS and ML-based methods, severe musical noise gives irregular spots in the spectrograms in Figures 11.9 and 11.11, respectively. The GBSS method with $\nu = 2, \alpha = 4$, and $\beta = 0.1$ reduces the musical tones to a moderate level (see Figure 11.10), compared with the PSS and ML methods. The WF-based method gives a further reduction in the level of the residual noise as shown in Figure 11.12. Using the MMSE-STSA-based method, it is possible to further eliminate the musical noises (see Figure 11.13). Even though the level of the overall residual noise of the MMSE-STSA is slightly higher than that of the WF method, the sound quality of MMSE-STSA is perceptually more comfortable than that of the WF method. The higher speech quality is due to further reduction in tonal signals. Combining the soft-decision technique with the MMSE-based method, it is possible to reduce the overall level of the residual noise as shown in Figure 11.14.

11.2.7 Discussion

Ephraim and Malah's speech enhancement method gives higher performance mainly due to the DD-based *a priori* SNR estimation. Cappe [16] has shown its usefulness for eliminating musical noise phenomena through behavioural analysis. From interpretation of equation (11.6), it is not difficult to see that $\hat{\xi}_k$ is a smoothed version of γ_k. The *a posteriori* SNR γ_k shows high fluctuation from frame to frame, while $\hat{\xi}_k$ changes slowly. By exploiting the characteristics of the two SNRs, γ_k and $\hat{\xi}_k$, improved performance in speech quality is achieved.

The WF produces better performance than either GBSS- or ML-based methods. The reason behind this better performance is also due to the DD-based *a priori* SNR estimation used in the gain function of the WF. The usefulness of the DD-based *a priori* SNR can also be applied to *a posteriori* SNR-based speech enhancement methods, such as GBSS- and ML-based spectral estimators, by replacing the *a posteriori* SNR with the *a priori* SNR [17] as,

$$\gamma_k = \hat{\xi}_k + 1 \tag{11.46}$$

Although substantial reduction of musical noise is achieved by the WF-based method, it is observed that the musical noise is not completely removed (see Figure 11.12). It is also possible to show that the musical noise phenomenon exists in the *a priori* SNR-based speech enhancement using equation (11.46).

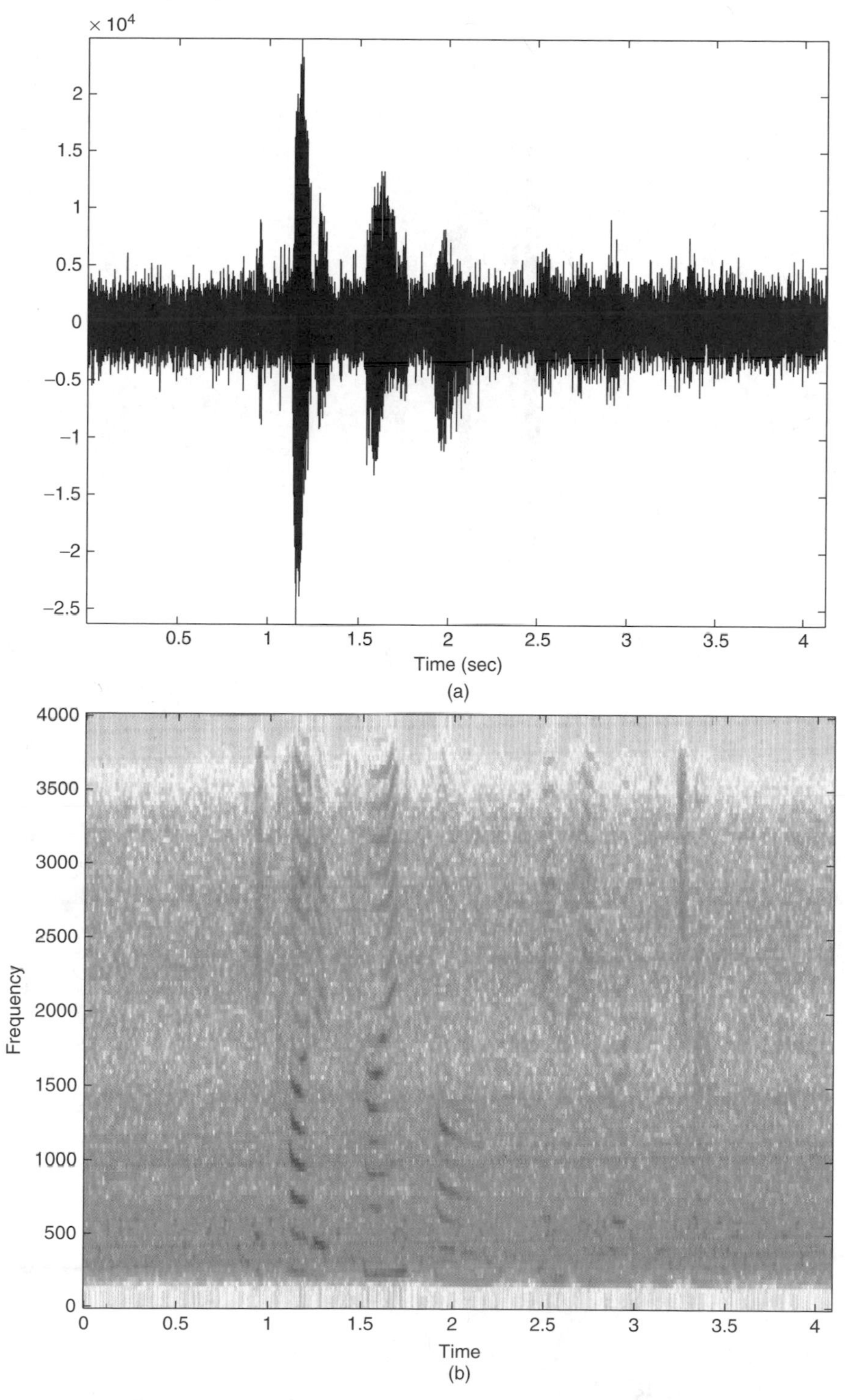

Figure 11.6 Noisy speech: (a) time waveform and (b) spectrogram at 5 dB SNR vehicle noise

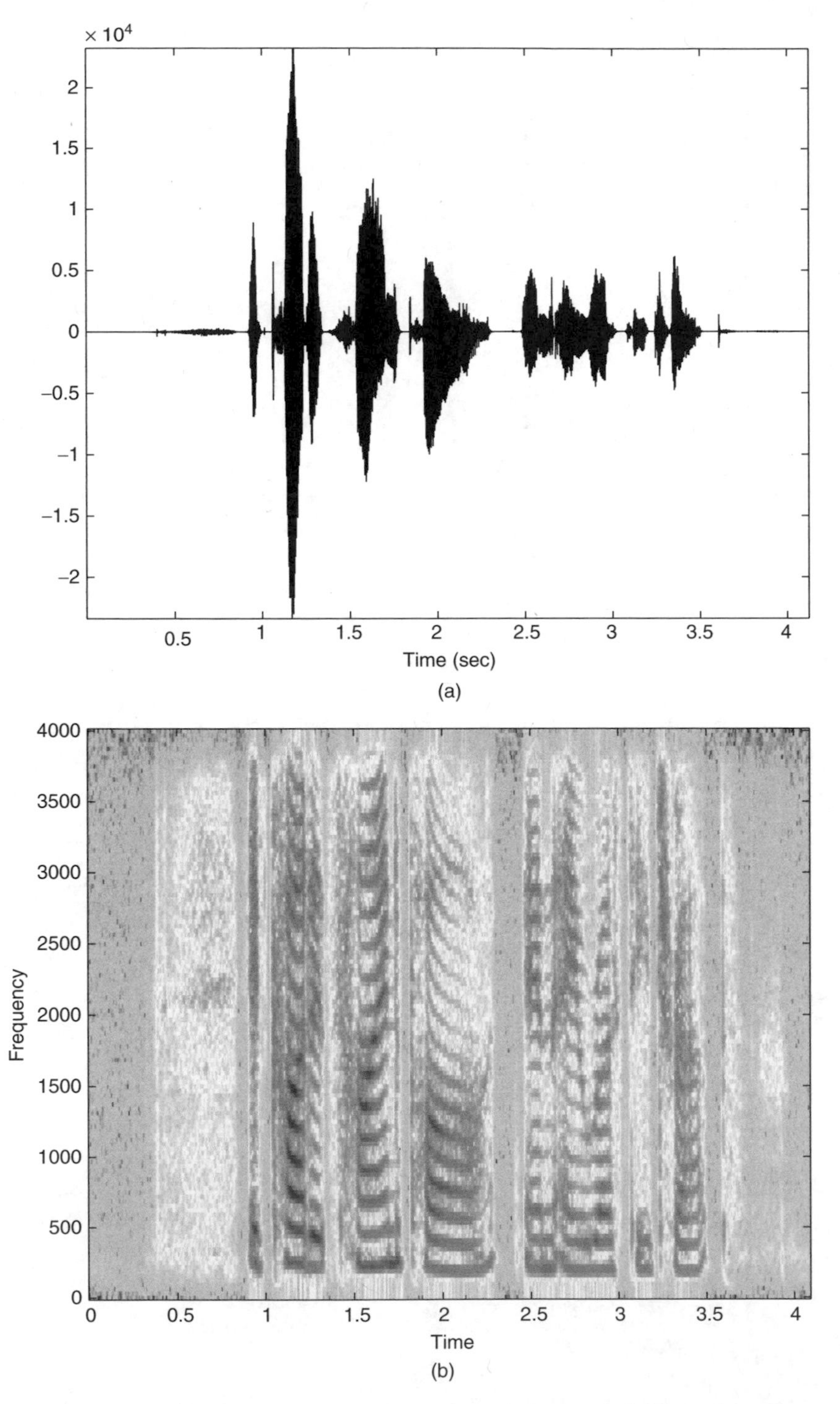

Figure 11.7 Noise-free speech: (a) time waveform and (b) spectrogram

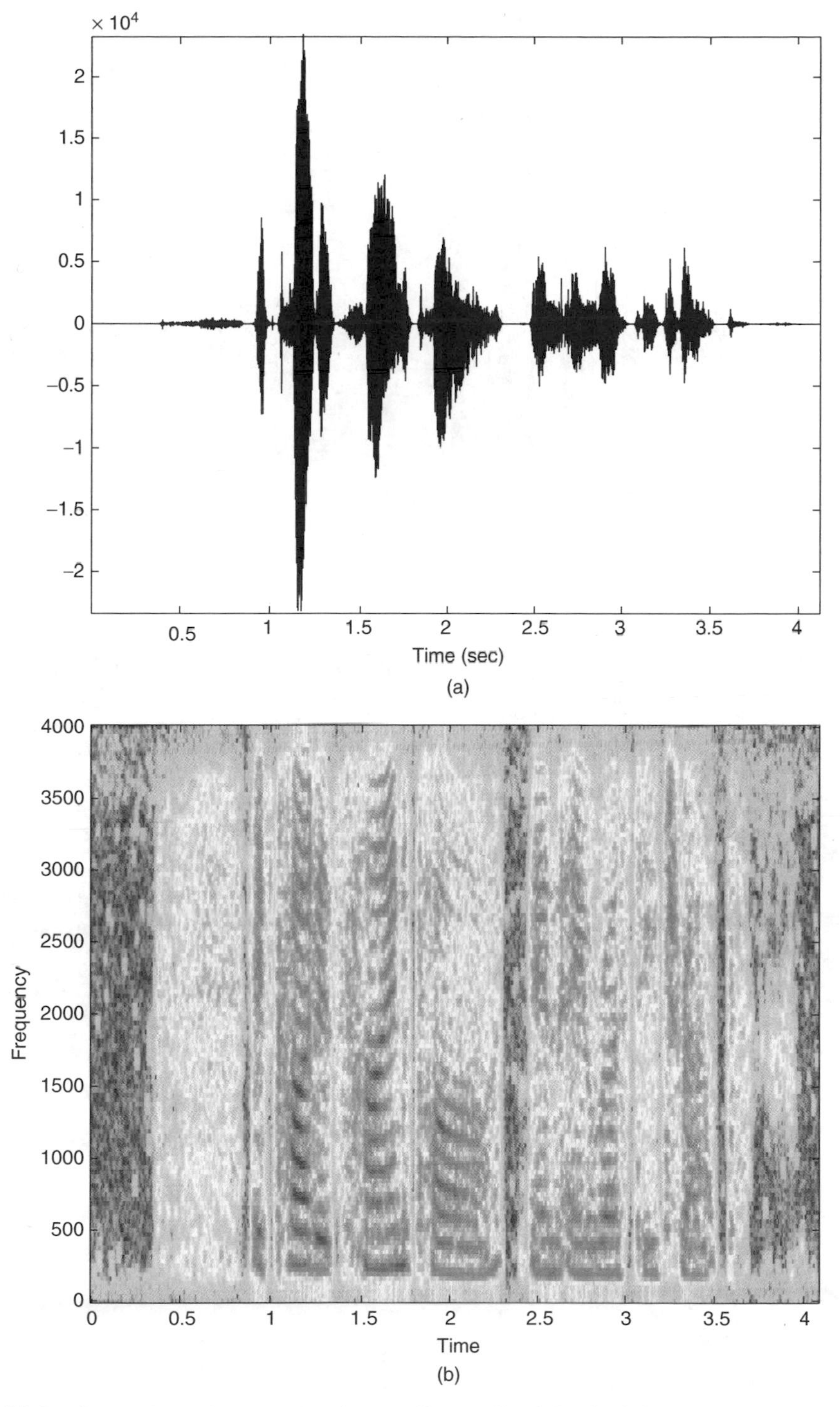

Figure 11.8 Speech enhanced using a theoretical limit: (a) time waveform and (b) spectrogram

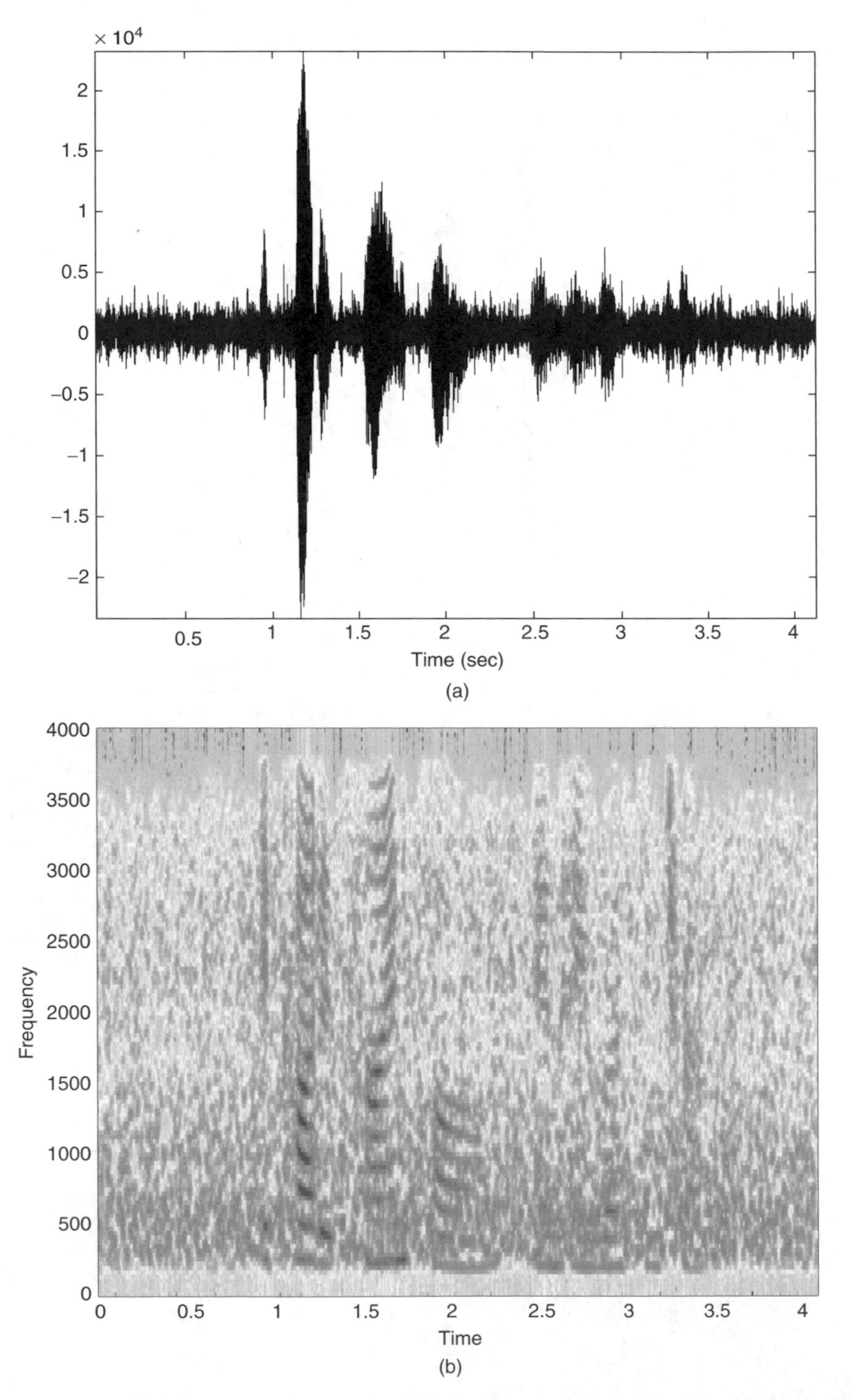

Figure 11.9 Speech enhanced by PSS: (a) time waveform and (b) spectrogram

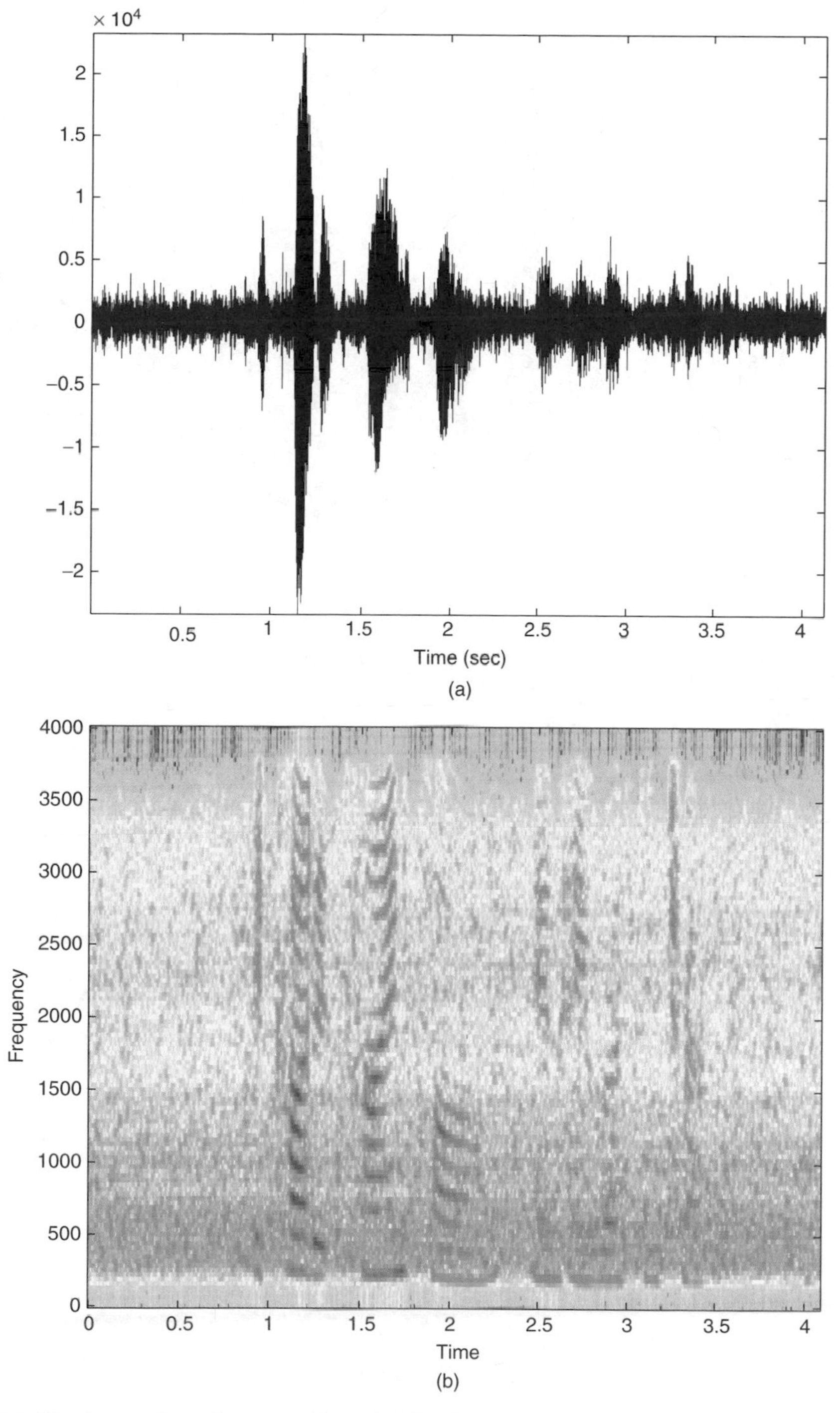

Figure 11.10 Speech enhanced by GBSS with $\nu = 2$, $\alpha = 4$, and $\beta = 0.1$: (a) time waveform and (b) spectrogram

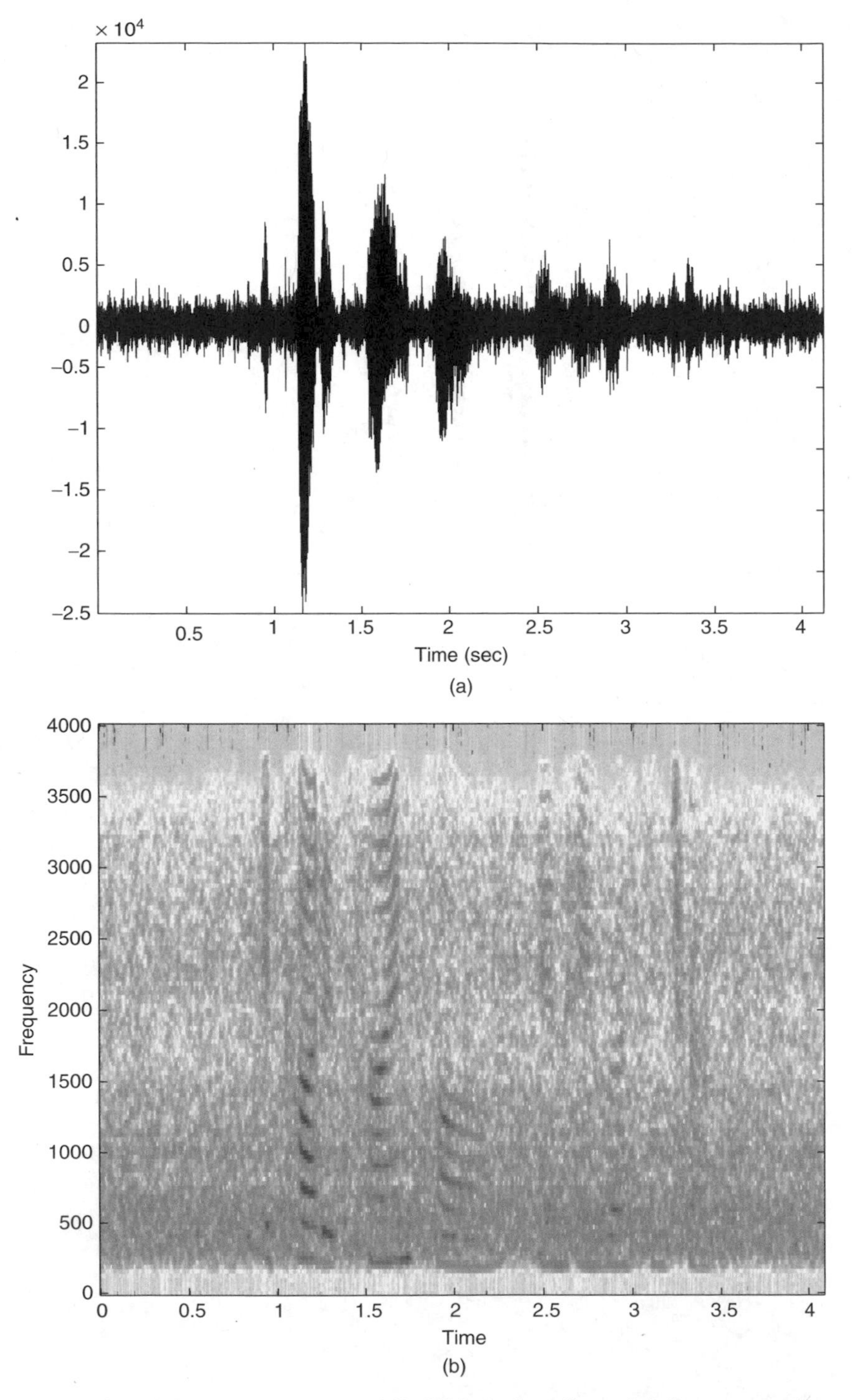

Figure 11.11 Speech enhanced by ML-STSA estimation: (a) time waveform and (b) spectrogram

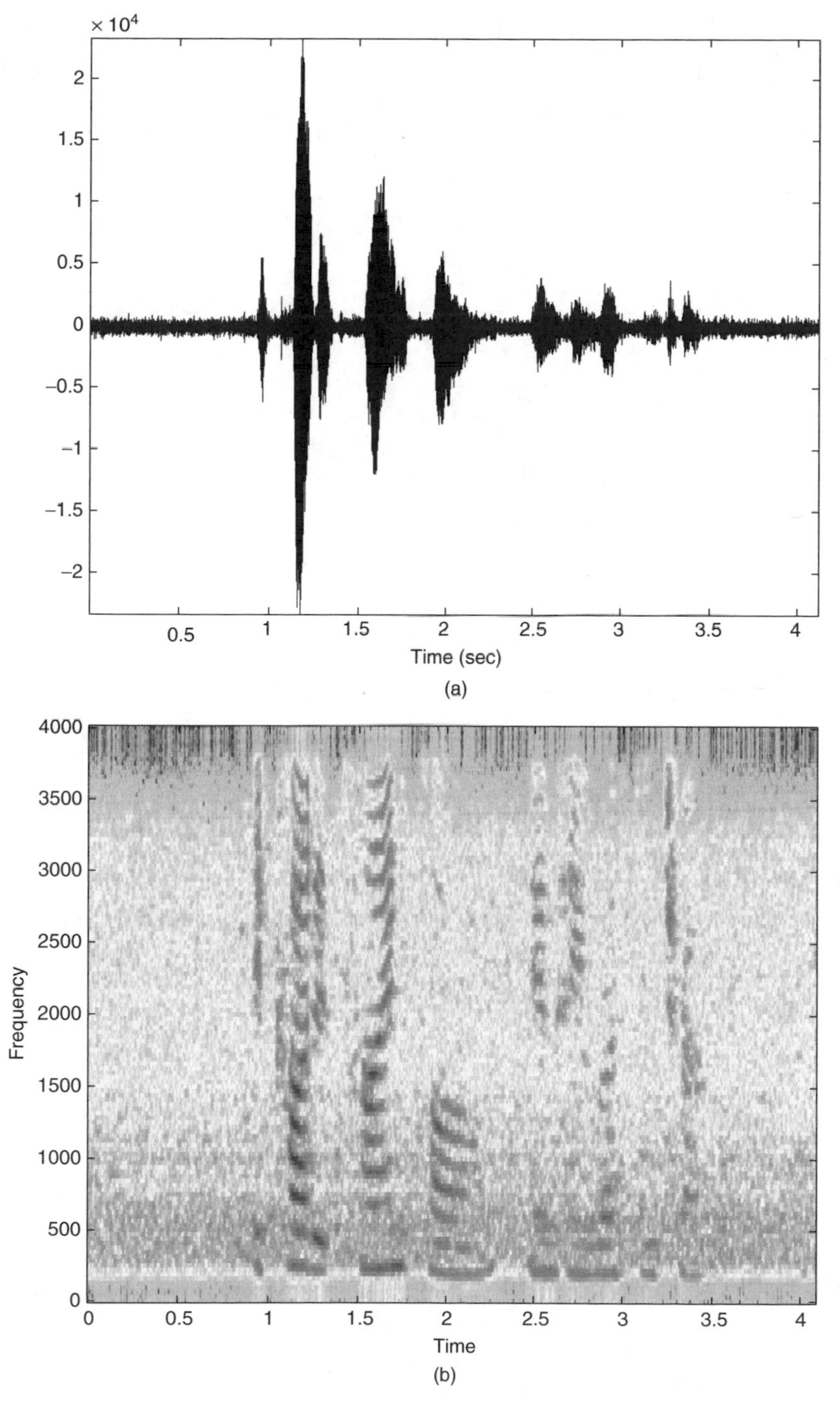

Figure 11.12 Speech enhanced by WF: (a) time waveform and (b) spectrogram

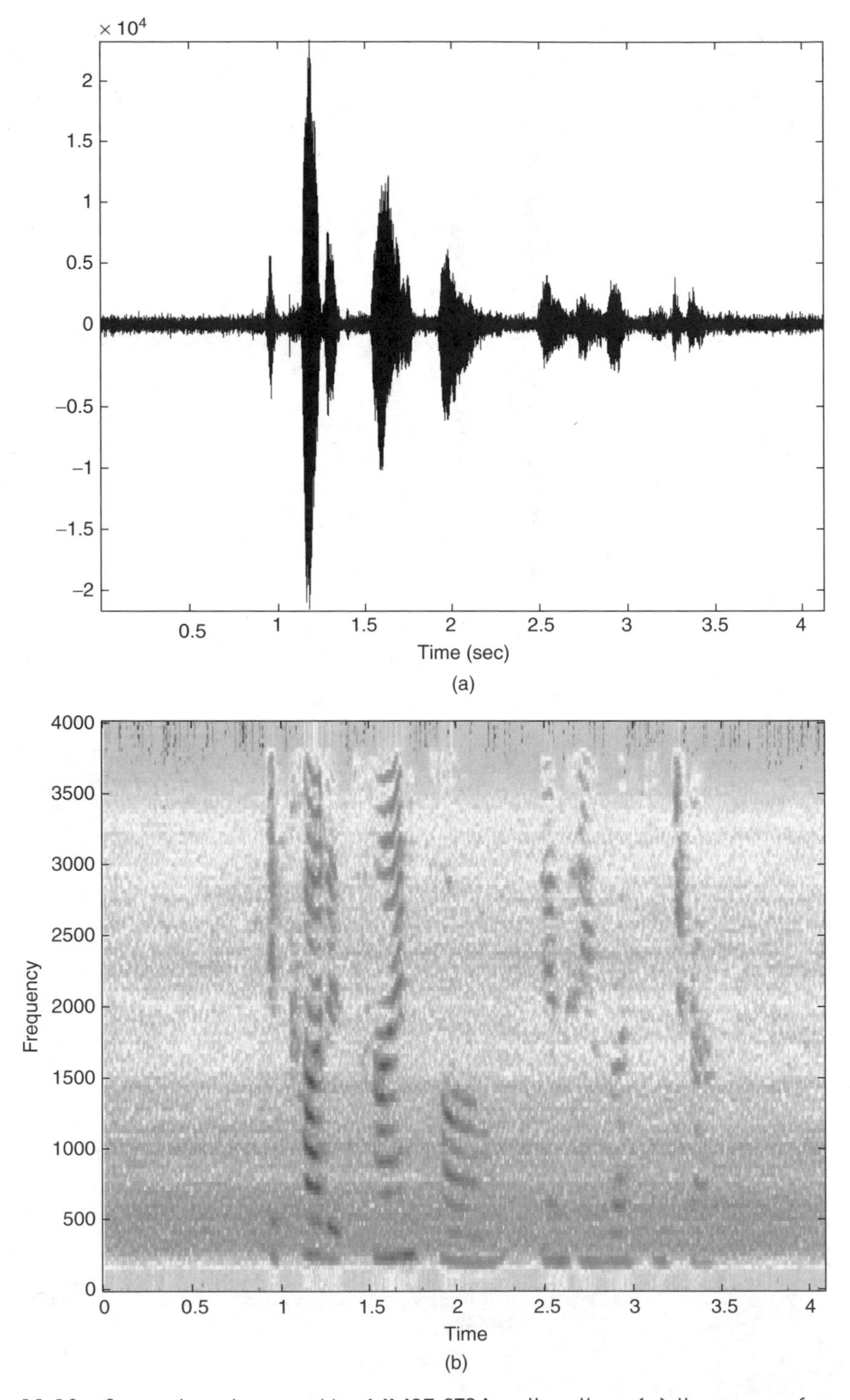

Figure 11.13 Speech enhanced by MMSE-STSA estimation: (a) time waveform and (b) spectrogram

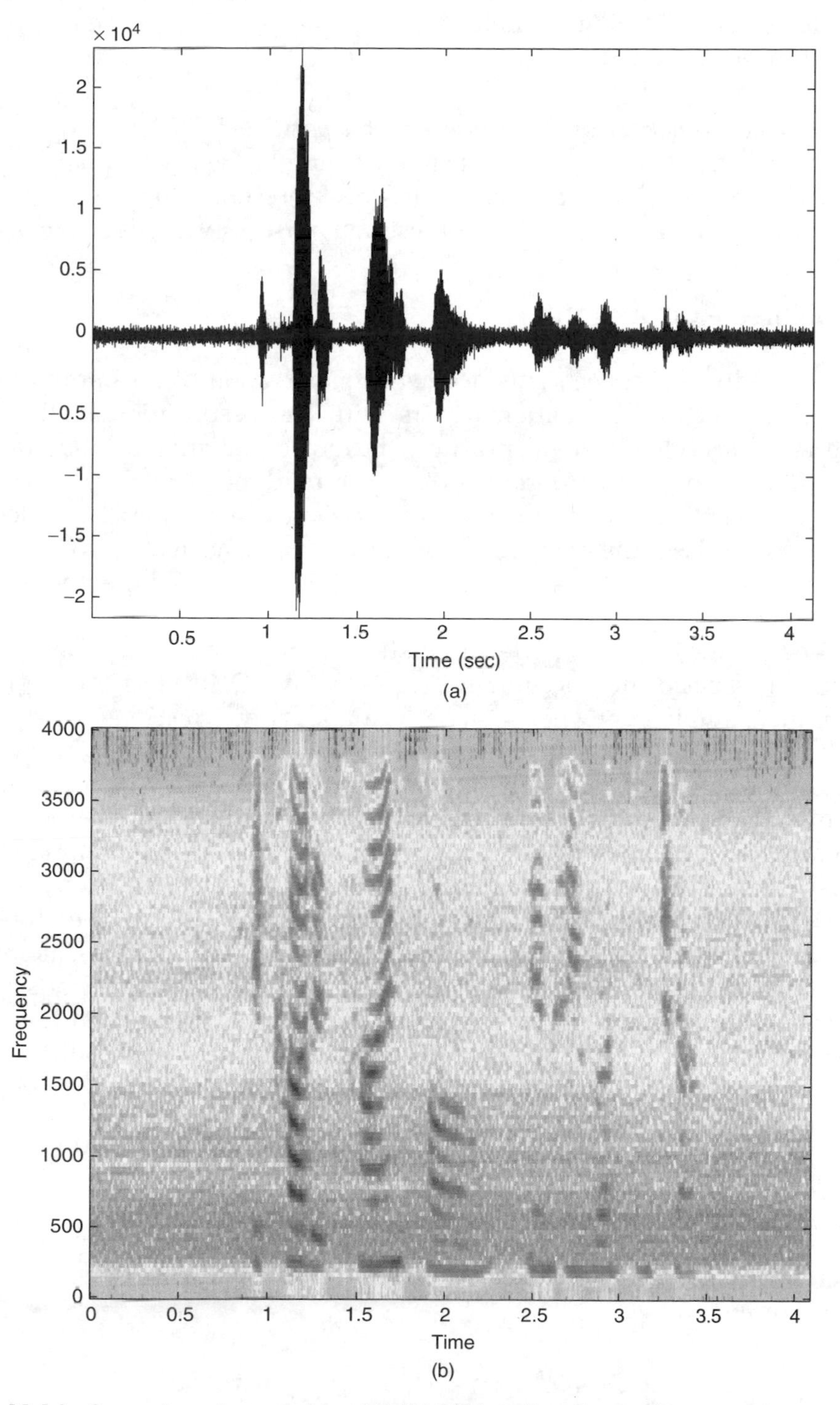

Figure 11.14 Speech enhanced by MMSE-STSA estimation with speech presence uncertainty: (a) time waveform and (b) spectrogram

Therefore, the following guidelines for designing a speech enhancement algorithm can be stated:

- Proper combination of the *a priori* and *a posteriori* SNRs is important to eliminate the musical noise while maintaining high speech quality.
- The soft-decision technique based on speech presence uncertainty is useful for further suppressing the level of residual noise for voice-inactive regions.

11.3 Noise Adaptation

Frequency-domain speech enhancement focuses mainly on improved estimation of spectral attenuation factors with the assumption of the given noise statistics. However, in practice, the noise statistics exhibit frame to frame fluctuations which require robust estimation for good performance. Noise estimation methods can be classified into two types: hard decision (HD), which adapts the noise variance during voice-inactive regions by voice activity detection (VAD), and soft decision (SD), which adapts the noise all the time.

HD-based methods are quite successful when voice activity classification of speech is performed accurately. However, VAD itself is a complicated technique to implement when high performances under various noise sources and levels are required. Thus, speech detection errors due to VAD may cause over-estimation or under-estimation of the noise statistics, which may lead to degradation of speech quality. The performance of the HD-based method is therefore heavily dependent on the performance of the VAD method used.

SD-based methods adapt the noise statistics based on the uncertainty of speech absence, instead of the hard-limited function used in the HD-based methods [18, 19]. SD-based methods do not rely on VAD decisions and update the noise statistics even in the presence of speech. SD-based methods rely on the accurate estimation of the mixture ratio between speech and noise. The inaccurate measurement of speech absence (or presence), especially in voice-active regions, can seriously distort the enhanced speech. Cho [20] proposed a mixed-decision-based noise adaptation, combining the characteristics of the HD- and SD-based methods.

11.3.1 Hard Decision-based Noise Adaptation

The HD-based method conducts noise adaptation during speech absence regions only,

$$E(|D_k^{(t)}|^2) = \begin{cases} \eta E(|D_k^{(t-1)}|^2) + (1-\eta)|Y_k^{(t)}|^2 & \text{if } Y^{(t)} \in H_0 \\ E(|D_k^{(t-1)}|^2) & \text{otherwise} \end{cases} \tag{11.47}$$

where the superscript t indicates the frame index, η is the smooth adaptation factor, e.g. 0.95, and Y is the noisy spectrum. In the case of speech presence, usually indicated by a VAD, it does not update the noise variance. HD-based noise adaptation has been widely used in speech enhancement.

11.3.2 Soft Decision-based Noise Adaptation

The SD-based noise estimation, the estimated noise given by Y_k, is formulated as,

$$\begin{aligned} E(D_k|Y_k) &= E(D_k|Y_k, H_0)p(H_0|Y_k) + E(D_k|Y_k, H_1)p(H_1|Y_k) \\ &= \{p(H_0|Y_k) + p(H_1|Y_k)G_{D,k}\}Y_k \end{aligned} \quad (11.48)$$

where $E(D_k|Y_k, H_0) = Y_k$, $E(D_k|Y_k, H_1) = G_{D,k}Y_k$. The probability of speech presence $p(H_1|Y_k)$ is defined in equation (11.39) and $p(H_0|Y_k) = 1 - p(H_1|Y_k)$. The optimal noise gain $G_{D,k}$ can be derived from the Wiener estimator W in the time domain. It can be shown that $W = R_{dd}(R_{xx} + R_{dd})^{-1}$, in which R_{dd} and R_{xx} denote the covariance matrices of the noise and speech signals resulting in the filter frequency response given by,

$$\begin{aligned} G_{D,k} &= \frac{E(|D_k|^2)}{E(|X_k|^2) + E(|D_k|^2)} \\ &= \frac{1}{1 + \xi_k} \end{aligned} \quad (11.49)$$

where ξ_k is the *a priori* SNR which can be estimated using the decision-directed method defined in equation (11.6). Here, the estimation of noise gain G_D is an independent task within the noise estimation process which may be used in other kinds of enhanced spectral estimation techniques, such as MMSE, MMSE-LSA, etc. The noise variance of the SD-based method may be estimated in a recursive manner as given below,

$$E(|D_k^{(t)}|^2) = \eta E(|D_k^{(t-1)}|^2) + (1 - \eta)|E(D_k^{(t)}|Y_k^{(t)})|^2 \quad (11.50)$$

11.3.3 Mixed Decision-based Noise Adaptation

In order to alleviate the problems in the HD- and SD-based methods, the MD-based method is proposed [20] for noise adaptation as

$$E(|D_k^{(t)}|^2) = \begin{cases} \eta E(|D_k^{(t-1)}|^2) + (1 - \eta)|Y_k^{(t)})|^2 & ; \text{if } Y^{(t)} \in H_0 \text{ and } \Lambda^{(t)} \leq \theta \\ \eta E(|D_k^{(t-1)}|^2) + (1 - \eta)|E(D_k^{(t)}|Y_k^{(t)})|^2 & ; \text{if } Y^{(t)} \in H_0 \text{ and } \Lambda^{(t)} > \theta \\ E(|D_k^{(t-1)}|^2) & ; \text{ otherwise} \end{cases} \quad (11.51)$$

where $\Lambda^{(t)} = \{\prod_{k=1}^{K} \Lambda_k^{(t)}\}^{1/K}$, as defined in equation (11.41). The threshold θ is set to a sufficiently small value, i.e. $\theta < 1$, that rarely classifies the speech as silence.

11.3.4 Comparisons

In order to show the robustness of the noise adaptation techniques, speech quality is compared in terms of both SEGSNR improvement and ISD with respect to the speech-detection error-rate of VAD (E_d). Various E_d are calibrated by a voice activity detector [21], and then frame-by-frame VAD results are given to each noise adaptation method. For the experiment, speech material of 64 seconds was mixed with vehicle noise at 5 dB SNR, and then processed every 10 ms in the frequency domain by the MMSE estimator [10] employing noise adaptation methods. Finally, the enhanced speech signal is obtained by the inverse DFT of the enhanced spectrum, followed by the overlap-and-add procedure.

SEGSNR improvement and ISD between the clean and enhanced speech signals for vehicle noisy speech signals of 0, 5, and 10 dB SNR are shown in Figures 11.15, 11.16, and 11.17, respectively. The experiments confirm that

- The SD-based method results in worse performance compared with both the MD- and the HD-based methods, for low E_d.
- The HD-based method exhibits significant degradation in performance with increases in E_d.
- The MD-based method produces, regardless of the VAD performance, robust and superior performance in comparison with the HD- and SD-based methods.

Note that for very low E_d, i.e. $0.0 \leq E_d < 0.1$, the performances of the MD and HD are slightly worse than in the case of $E_d = 0.2$. This is caused by less frequent adaptation of the noise frames because of the increased false alarm rate of the VAD. n other words, VAD produces the low E_d at the expense of an increased false alarm rate during pauses.

Results for helicopter noisy speech with levels of 0, 5, and 10 dB SNR are shown in Figures 11.18, 11.19, and 11.20, respectively. They exhibit performance patterns similar to the vehicle noisy signals despite differences in the absolute values being measured.

In conclusion we can say that the STSA-based spectral enhancement techniques including GSS, GBSS, ML, WF, and MMSE-based algorithms together with the estimate of speech presence uncertainty have various advantages and disadvantages. The MMSE-based STSA method combined with speech presence uncertainty is perhaps the best currently available method for

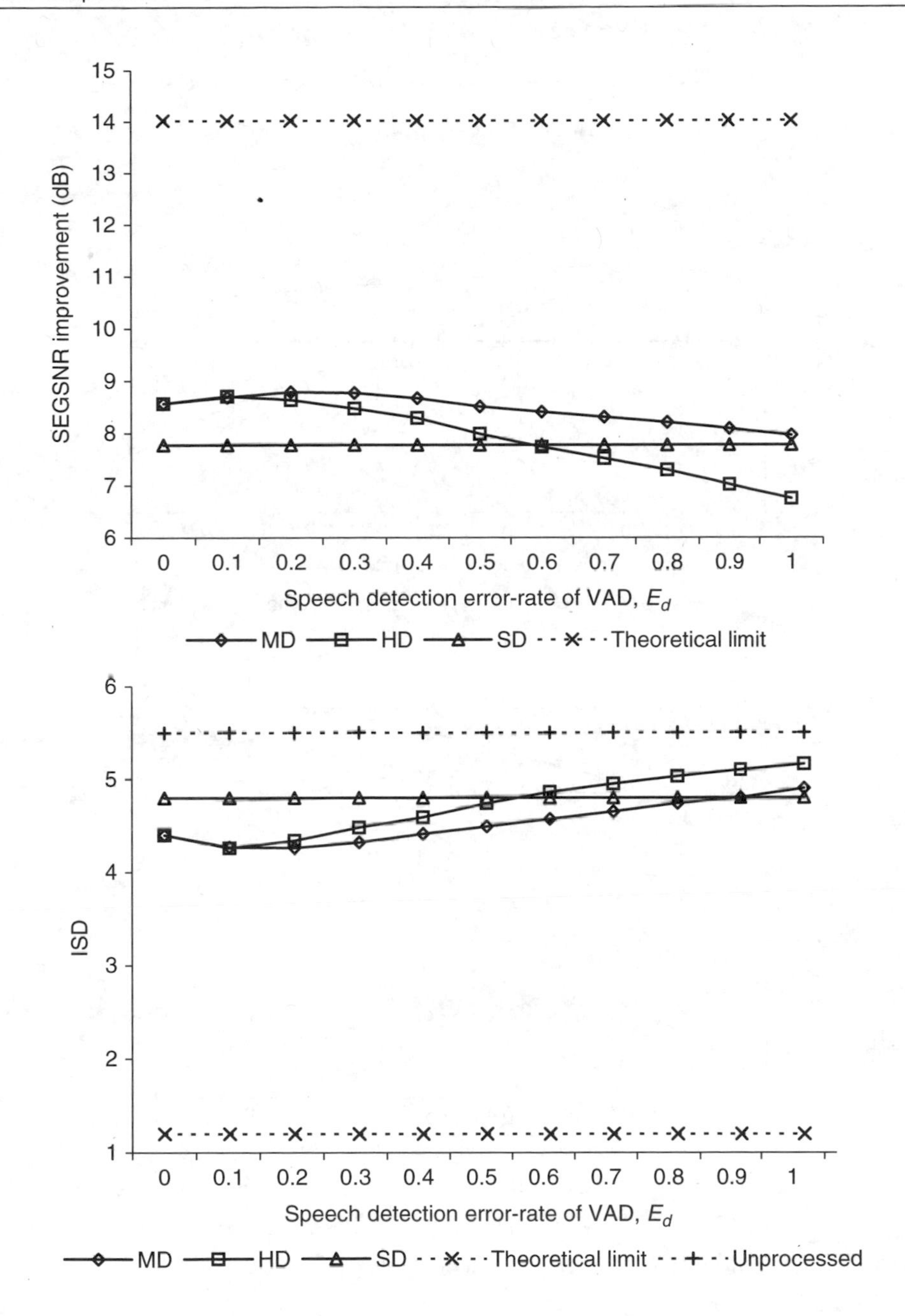

Figure 11.15 Comparison of SEGSNR improvement and ISD against the speech-detection error-rate of VAD for vehicle noisy speech of 0 dB SNR

noise reduction. In speech enhancement systems, accurate noise estimation/adaptation is necessary to keep track of the noise characteristics. Noise estimation and adaptation is the most important area that requires further research for better speech enhancement techniques.

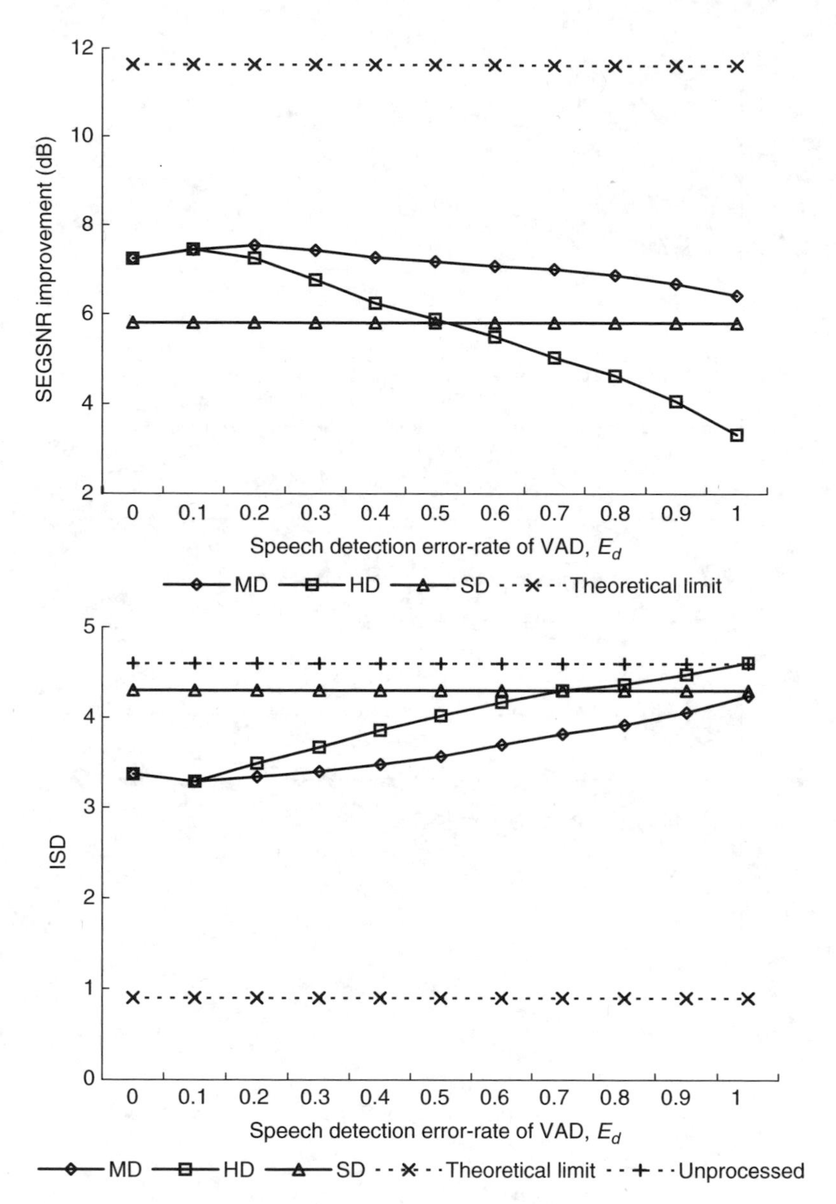

Figure 11.16 Comparison of SEGSNR improvement and ISD against the speech-detection error-rate of VAD for vehicle noisy speech of 5 dB SNR

11.4 Echo Cancellation

Echo in a telecommunications system is the delayed and distorted sound which is reflected back to the source. In telecommunications, there are two types of echo: acoustic echo, which results from the reflection of sound waves and acoustic coupling between the microphone and loudspeaker,

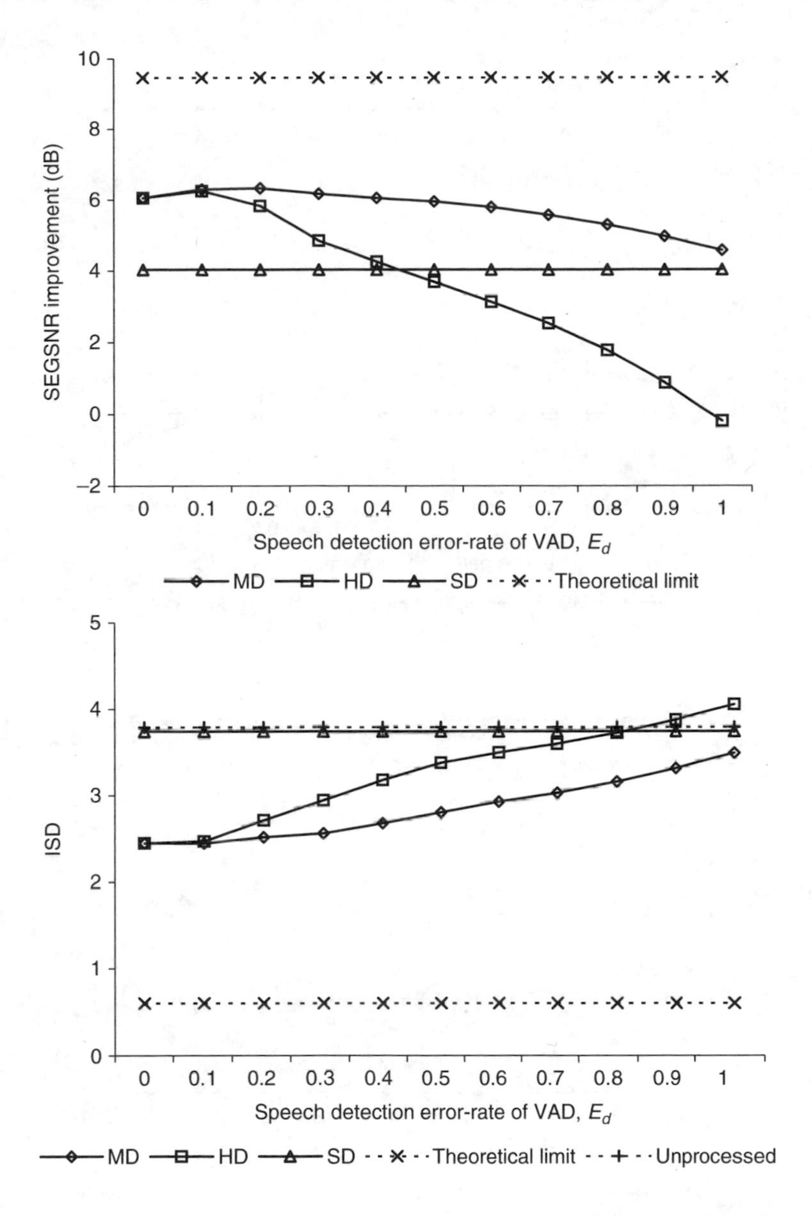

Figure 11.17 Comparison of SEGSNR improvement and ISD against the speech-detection error-rate of VAD for vehicle noisy speech of 10 dB SNR

and electrical echo, generated at the two-to-four wire conversion hybrid transformer due to imperfect impedance matching. Here, we will develop cancellation for the electrical echo which will be equally applicable for acoustic echo cancellation.

The source of electrical echo can be understood by considering a simplified block diagram of a connection between a pair of subscribers, S1 and S2,

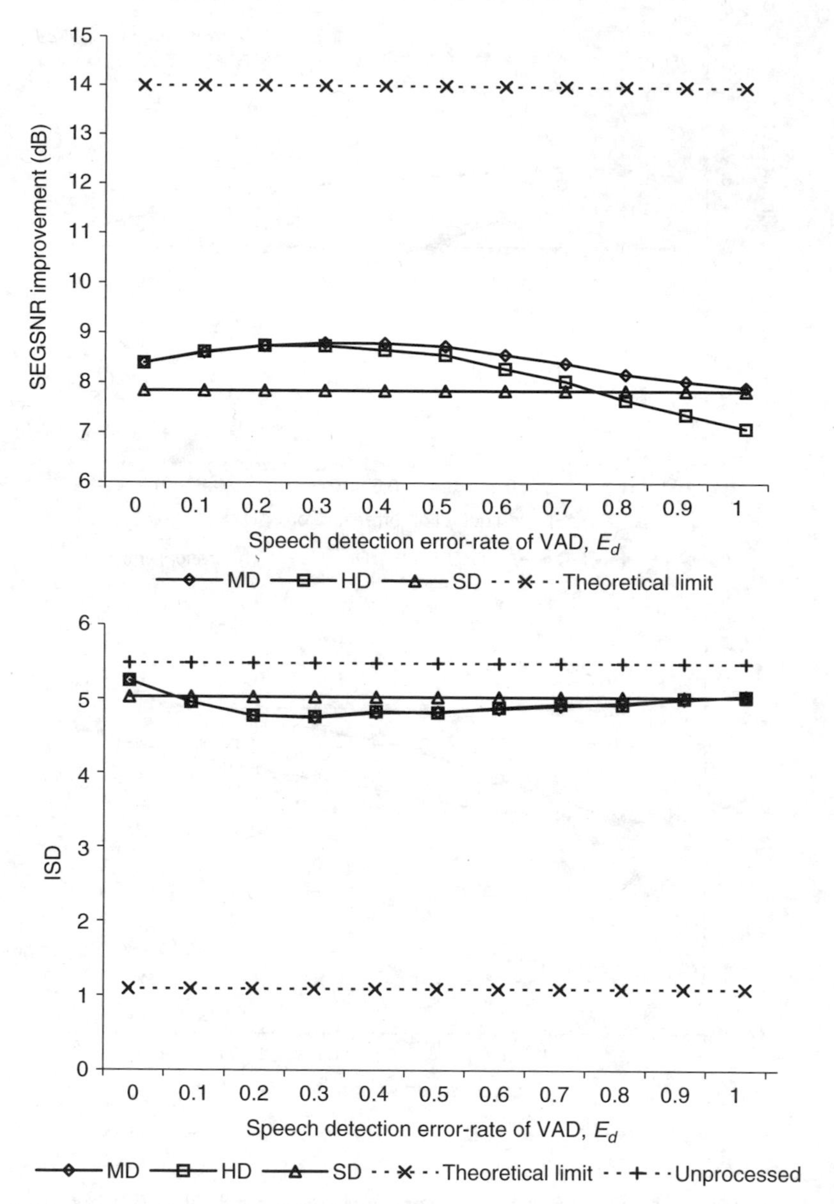

Figure 11.18 Comparison of SEGSNR improvement and ISD against the speech-detection error-rate of VAD for helicopter noisy speech of 0 dB SNR

as shown in Figure 11.21. It can be seen from this block diagram that each subscriber has a two-wire loop over which both the received signal and transmitted signals travel. On the four-wire part of the line, the two directions of transmission are separated. The speech from S1 travels on the upper path and the speech from S2 travels on the lower path, as indicated by the arrows. The converter device between the two- and four-wire sections

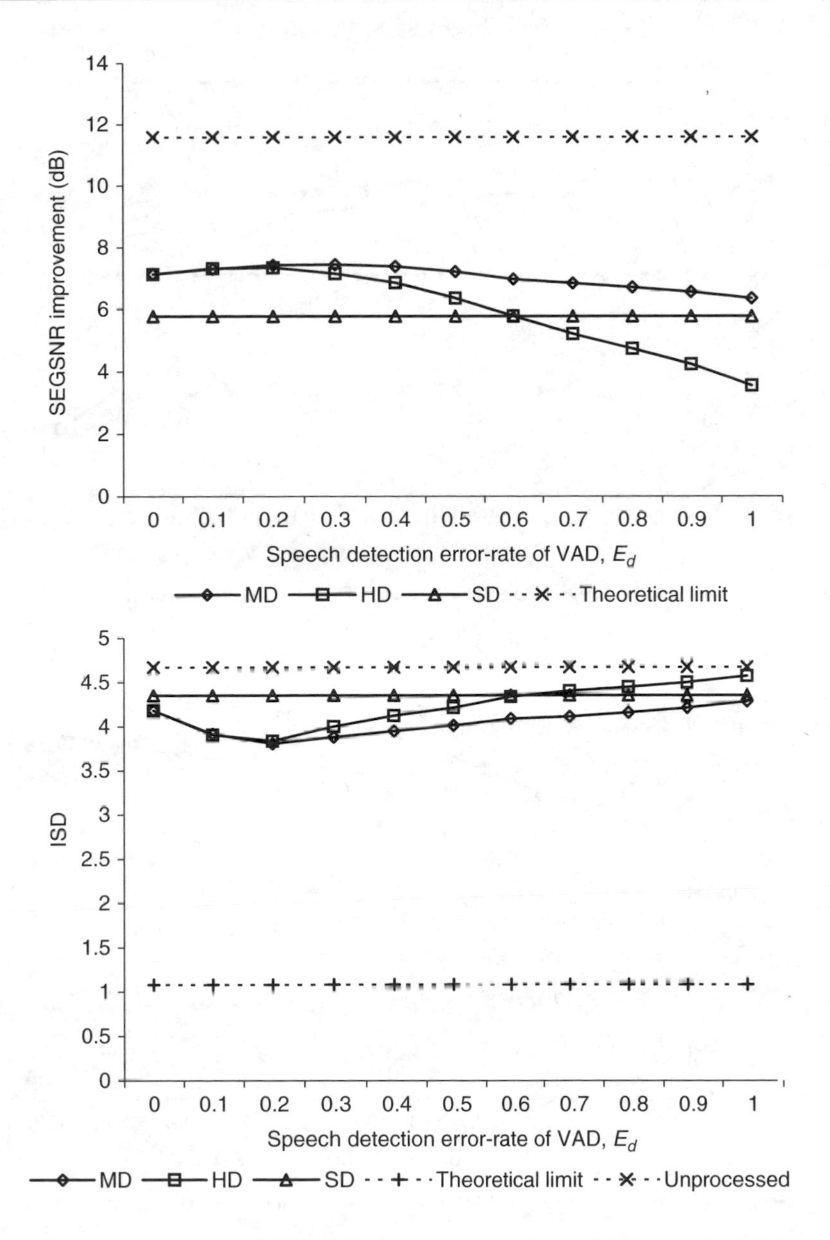

Figure 11.19 Comparison of SEGSNR improvement and ISD against the speech-detection error-rate of VAD for helicopter noisy speech of 5 dB SNR

is called the hybrid. The role of the hybrid is to direct the signal energy arriving from S1 or S2 to the upper or lower path of the four-wire circuit, without allowing any leakage back to the source over the opposite direction line. Because of impedance mismatching, however, some of the transmitted signal returns to the original source, which hears a delayed version of its own

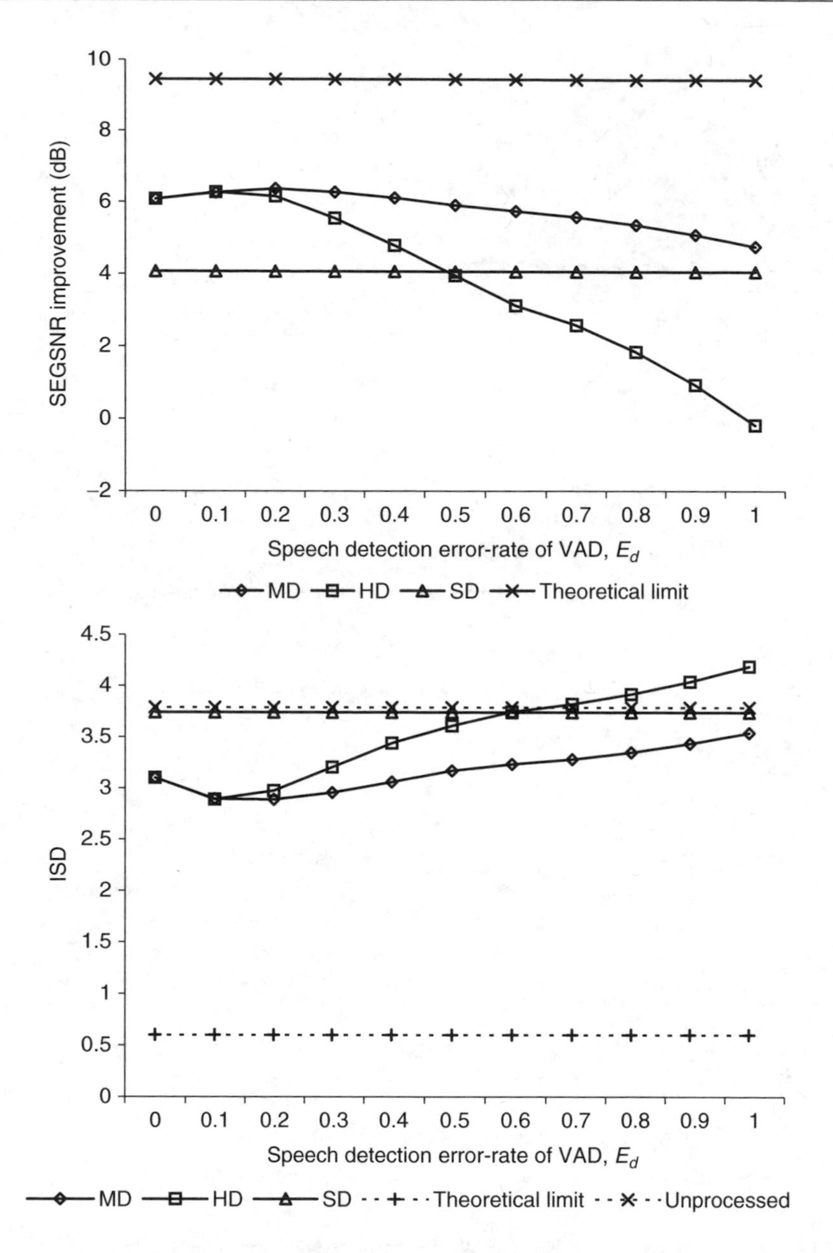

Figure 11.20 Comparison of SEGSNR improvement and ISD against the speech-detection error-rate of VAD for helicopter noisy speech of 10 dB SNR

speech. This is called the talker echo and its subjective effect depends on the round trip delay around the loop. For short delays and reasonable attenuation (6 dB or more) the talker echo cannot be distinguished from the normal side tone of the telephone and hence does not cause problems. In applications such as satellite communications however, as a consequence of high altitude,

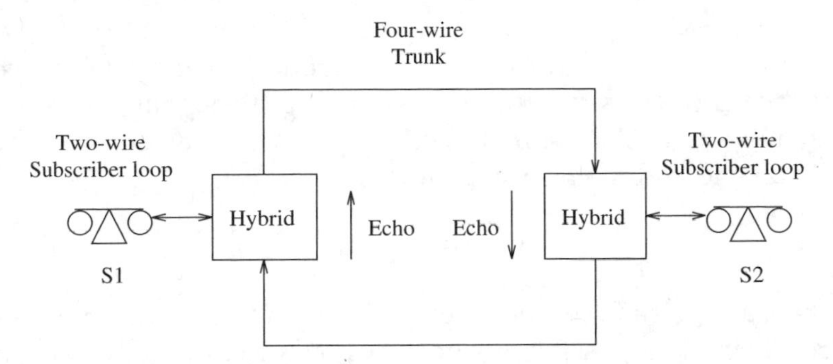

Figure 11.21 Block diagram of duplex connection between two subscribers

a round trip delay of approximately 540 ms (270 ms each way) is possible; this makes the echo very disturbing and may in fact make it impossible to carry out a conversation. In such cases, it is essential to control or remove the echo. Since the subjective disruption of echo is proportional to the round trip delay as well as the echo energy level, the echo control techniques usually depend on the circuit length.

Some international connections use a switch (called an echo suppressor) operated by the voice activity, which attempts to impose an open circuit on the return path from listener to talker when the listener is silent. However, an echo suppressor cannot operate during double-talk and hence produces choppy echo. For this reason, echo suppressors are now being replaced by echo cancellers that are based on adaptive filtering techniques.

11.4.1 Digital Echo Canceller Set-up

A block diagram of an echo canceller for one direction of transmission is shown in Figure 11.22, where the far-end talker signal is denoted by $y(i)$, the unwanted echo signal by $r(i)$, and the near-end talker signal by $x(i)$. The

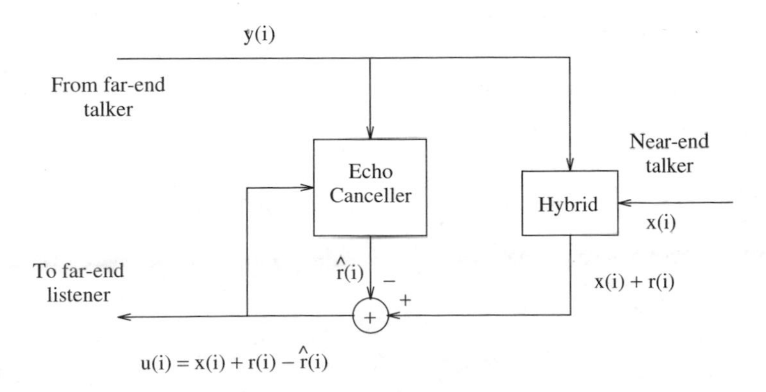

Figure 11.22 Block diagram of an echo canceller

near-end signal and the echo are added together at the output of the hybrid. Since the far-end signal is available as a reference for the echo canceller, the replica of the echo $\hat{r}(i)$ is estimated by matching the signals on both paths of the four-wire section. The estimated echo is then subtracted from the total of the returned echo and the near-end signal,

$$u(i) = x(i) + r(i) - \hat{r}(i) \tag{11.52}$$

The difference between $r(i)$, the returned echo, and $\hat{r}(i)$, the estimated echo, should be as small as possible for good echo cancellation performance. The echo canceller estimates the echo by using the far-end reference signal in a transversal filter such as the one shown in Figure 11.23. This filter basically acts as a tapped delay line. If the impulse response of the filter is same as the echo path response, then the estimated echo and the returned echo become identical, resulting in perfect echo cancellation. Since the echo path response is not known in advance and may vary slowly with time, the coefficients of the transversal filter are adapted. In order to produce no distortion on the near-end talker signal, the filter coefficients are only updated when there is no near-end activity.

The number of filter coefficients, which may be very significant from a complexity point of view, is usually determined by the length of the echo path impulse response, which typically lasts 2 to 4 ms, requiring 32 taps ($4 \times 10^{-3}/125 \times 10^{-6}$) approximately. However, the impulse response of the echo path may be delayed by some time depending on the distance between the position of the echo canceller and the hybrid in the system. Moreover more than a few taps may be needed to accurately model the response of the hybrid. Therefore, the use of 64 or 128-tap filters are typical.

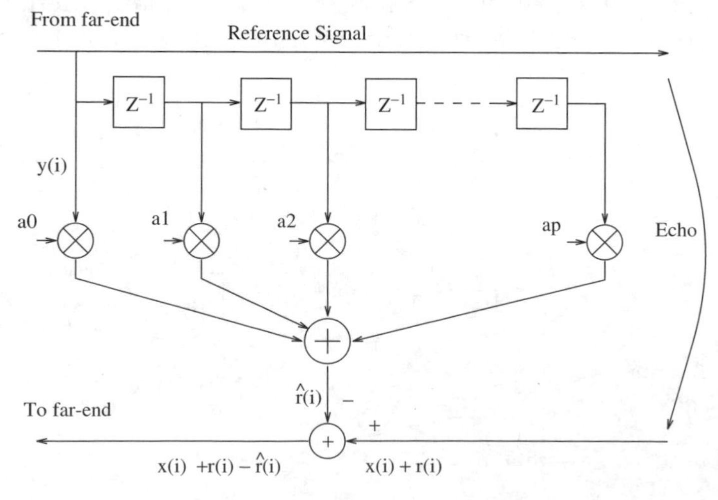

Figure 11.23 Block diagram of a transversal filter used in echo cancellation

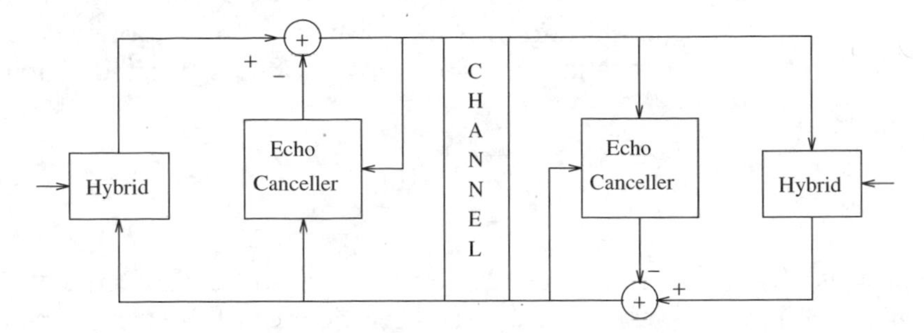

Figure 11.24 Block diagram showing echo cancellation applied to both ends

In practice, echo cancellers are applied on both ends to cancel the echoes in each direction as shown in Figure 11.24. An echo canceller should, in general, satisfy the following fundamental requirements:

- Rapid convergence of the filter coefficients when turned on.
- Very low echo when there is no near-end speech.
- Slow divergence when there is no far- or near-end speech.
- Little divergence when both near- and far-end signals are present.

The ITU-T G.165 recommendations [22], which summarize the above requirements, are as follows:

- After convergence with no near-end speech, with input noise level between −10 dBm0 and −30 dBm0, final echo return loss (ERL) should be −40 dB.
- After 500 ms of first start up, the parameters should converge to give at least 27 dB echo reduction with no near-end speech.
- Degradation of residual echo after 2 minutes from the time all signals are removed from the fully-converged canceller should not be more than 10 dB.
- The returned echo level, 500 ms after interruption of the echo path, should reach −40 dBm0.

11.4.2 Echo Cancellation Formulation

An echo canceller can be split into the following parts: adaptive transversal filter, near-end speech detection, and residual error suppression.

Adaptive Transversal Filter

In a digital echo canceller both the reference and echo signals are available in digital form. Therefore the echo path impulse response can be represented in digital form by a_k,

$$r(i) = \sum_{k=0}^{N-1} a_k y(i-k) \tag{11.53}$$

Assuming the system is linear and the echo path impulse response is of finite length N, then the echo canceller forms the replica of the returned echo using,

$$\hat{r}(i) = \sum_{k=0}^{N-1} a_k y(i-k) \tag{11.54}$$

When $a_k = h_k$, for $k = 0, 1, \ldots, N-1$ the returned and estimated echoes are identical resulting in no residual echo. The coefficients of the transversal filter are updated to match the slowly time-varying echo path impulse response by minimizing the mean squared residual error given by:

$$e^2(i) = [r(i) - \hat{r}(i)]^2 \tag{11.55}$$

When there is no near-end speech ($x(i) = 0$), the filter coefficients are updated in such a way that the residual error tends to a minimum. The update of the coefficients at each iteration is controlled by a step size β,

$$h_k(i+1) = h_k(i) + 2\beta e(i) y(i-k) \tag{11.56}$$

The convergence of the algorithm is determined by the step size β and the power of the far-end signal $y(i)$. In general, making β large speeds up the convergence, while a smaller β reduces the asymptotic cancellation error. It has been shown that the convergence time constant is inversely proportional to the power of $y(i)$ and that the algorithm will converge very slowly for low-signal levels [23]. To overcome this situation, the loop gain is usually normalized by an estimate of the far-end signal power,

$$2\beta = 2\beta(i) = \frac{\beta_1}{P_y(i)} \tag{11.57}$$

where β_1 is a compromise value of the step size constant and $P_y(i)$ is an estimate of the average power in $y(i)$ at time i. The far-end signal power can be estimated by

$$P_y(i) = [L_y(i)]^2 \tag{11.58}$$

where,

$$L_y(i+1) = (1-\rho) L_y(i) + \rho |y(i)| \tag{11.59}$$

and a typical value of $\rho = 2^{-7}$. The above equation is only an estimate of the average signal level, which is updated for every sample using the approximation for ease of implementation in real-time.

Near-End Speech Detection

The quality of the echo canceller can be affected significantly if the near-end speech is not detected accurately. This is because the filter coefficients will be adjusted wrongly and hence will distort the near-end speech. Therefore, the coefficients are updated only when there is no near-end speech; they are kept fixed during near-end activity to prevent divergence. The power estimate $\hat{s}(i)$ of the near-end composite signal $s(i) = x(i) + r(i)$ is usually compared with the power estimate $\hat{y}(i)$ of the far-end signal $y(i)$ to decide if there is near-end activity. The power estimate is computed as

$$\hat{s}(i+1) = (1-\alpha)\hat{s}(i) + \alpha|s(i)| \tag{11.60}$$

and,

$$\hat{y}(i+1) = (1-\alpha)\hat{y}(i) + \alpha|y(i)| \tag{11.61}$$

where a typical value for α is $1/32$. Near-end speech is declared when

$$\hat{s}(i) \geq MAX[\hat{y}(i), \hat{y}(i-1), \ldots, \hat{y}(i-N)] \tag{11.62}$$

In order to avoid continuous switching, every time near-end speech is detected, it is assumed to last for some time (typically 600 samples).

Residual Echo Suppressor

Due to nonlinearities in the echo path, the convergence of the filter coefficients and hence the accuracy of the echo path modelling is limited to around 30 to 35 dB. In order to further enhance the performance of the echo canceller, a residual echo suppressor can be used. This can be done simply by comparing the returned signal power with a threshold relative to the far-end signal, and completely eliminating it if it falls below the threshold. Again the returned signal power is estimated using

$$L_u(i+1) = (1-\rho)L_u(i) + \rho|u(i)| \tag{11.63}$$

Whenever $L_u(i)/L_y(i) < 2^{-4}$, the residual echo suppressor is activated. In some applications however, it may be perceptually more acceptable to leave a very low level of random signal to indicate that the line is not dead.

11.4.3 Improved Performance Echo Cancellation

Echo cancellation based on the NLMS algorithm (or other variants of the general LMS algorithm) performs well with both acoustic and electrical echoes, provided that near-end speech is not present. The performance, however, degrades when near-end speech is present (and is even worse if near-end speech cannot be detected correctly). Significant performance degradation is also expected when echo is contaminated with background noise.

Echo cancellers generally stop filter coefficient adaptation when near-end speech is present. An accurate near-end speech detector is therefore necessary to avoid divergence of the filter coefficients, which may have two drawbacks. First, the cancellation performance strongly depends on the accuracy of the near-end speech detector. The second drawback is related to the length of the near-end speech presence. In cases where a near-end speech segment is long, the echo characteristics may change considerably and if the filter coefficients are not continually adapted during those segments, then the filter will lose synchronization with the echo path changes, leading to a large change when filter coefficient adaptation is resumed. This may result in temporary filter divergence causing performance reduction.

An adaptive normalized least mean squared (ANLMS) algorithm has been suggested by Al-Naimi [24] to overcome these problems. It is based on the NLMS algorithm (with a 128-tap transversal adaptive filter [25]). The NLMS of [25] differs from the general NLMS in that filter coefficients are updated less frequently with a thinning factor, M, resulting in

$$h_k(i+1) = h_k(i) + \beta \frac{\sum_{m=0}^{M-1} e(i+M-m)y(i+M-m-k)}{\sigma(i)^2} \tag{11.64}$$

The ANLMS includes a number of enhancements to the system in [25] which are: increased robustness to noise contamination, continuous filter coefficient adaptation, and elimination of the need for a near-end speech detector. The ANLMS is given by,

$$h_k(i+1) = h_k(i) + w_k(i)\beta \frac{\sum_{m=0}^{M-1} e(i+M-m)y(i+M-m-k)}{\psi(i)^2 \rho(i)^2} \tag{11.65}$$

where $\psi(i)$ and $\rho(i)$ are given by,

$$\psi(i) = \alpha_e \psi(i-1) + (1-\alpha_e) \left|y(i)\right| \tag{11.66}$$

and,

$$\rho(i) = \alpha_e \rho(i-1) + (1-\alpha_e) \left|z(i)\right| \tag{11.67}$$

respectively. The weighting function $w_k(i)$ is

$$w_k(i) = \exp\left\{-\left(\frac{\hat{h}_k(i) - \overline{h}_k(i)}{\gamma\beta}\right)^2\right\} \tag{11.68}$$

where $\hat{h}_k(i)$ is the unweighted estimate and $\overline{h}_k(i)$ is the average track of filter coefficient k at time i, given by,

$$\hat{h}_k(i) = h_k(i) + \beta \frac{\sum_{m=0}^{M-1} e(i+M-m)y(i+M-m-k)}{\psi^2(i)\rho^2(i)} \tag{11.69}$$

$$\overline{h}_k(i) = \alpha_h \overline{h}_k(i-1) + (1-\alpha_h)h_k(i). \tag{11.70}$$

Note that $0 \leq \alpha_e \leq 1, 0 \leq \alpha_h \leq 1$ and $\gamma > 0$ are tuning parameters which need to be optimized for a given application.

The performance improvement with the ANLMS method stems from the soft-decision weighting function, $w_k(i)$. This weighting function removes the need for a near-end speech detector and its associated problems. It also provides a soft-decision means of continuous filter coefficient adaptation so as not to lose synchronization with echo path changes. In addition, it results in increased robustness to background noise contamination.

At time i, the weighting function $w_k(i)$ depends, for its calculation, on the average track of filter coefficient k (as given in equation (11.70)) and on the unweighted estimate of filter coefficient $h_k(i)$ at time i (as given in equation (11.69)). If the difference between the estimated and the related average filter coefficient track is large (which mostly occurs due to the presence of noise, near-end speech or both), then the weighting will be small. On the other hand, when the difference is small the $w_k(i)$ will be close to one. The weighting $w_k(i)$ and the step size β determine the adaptive step size. The adaptive step size is close to β for changes that follow smoothly the evolution of each filter coefficient track, whilst being much less than β for changes that are generally not related to the echo path change over time. The variance of the weighting function $w_k(i)$, i.e. how fast it will decay from the unity value, is controlled by the value $\gamma\beta$.

Note that the performance of this method depends on correct estimation of the average value of each filter coefficient track and therefore requires an initialization period that is dependent on the application. This initial period is essential in getting a reliable average filter coefficient track and for the overall system convergence.

The ITU-T recommendation G.165 defines the echo canceller performance requirements using band-limited white noise (300–3400 Hz) test signals for

Table 11.1 ITU-T recommendations and ANLMS system performance results

Tests	ITU-T recommendation G.165		ANLMS
	Input levels	Recommendation	Results
Steady state residual echo level	−30 dBm0 −20 dBm0 −10 dBm0	−48 dBm0 −42 dBm0 −36 dBm0	−83 dBm0 −72 dBm0 −60 dBm0
Convergence	−30 dBm0 −20 dBm0 −10 dBm0	attenuation $\geq$ 27 dB attenuation $\geq$ 27 dB attenuation $\geq$ 27 dB	30 dB 30 dB 30 dB
Leak rate (i.e. slow divergence when no signal)	−30 dBm0 −20 dBm0 −10 dBm0	(For all input levels, residual echo level should not increase more than 10 dB)	(Echo level increase of 6 dB was evident for all input levels)
Infinite return loss convergence (i.e. rapid return to convergence after an interrupt to echo path)	−30 dBm0 −20 dBm0 −10 dBm0	$\leq$ −37 dBm0 $\leq$ −37 dBm0 $\leq$ −37 dBm0	−78 dBm0 −68 dBm0 −57 dBm0

far-end and near-end ports. A test is devised (see [22]) for each of the requirements in Table 11.1, listed with the results obtained for the various tests.

The requirements for echo canceller performance for double-talk situations is subdivided into two tests. The first is related to the double-talk detection part of the echo canceller. As there is no such double-talk detector used in the ANLMS system this test is not performed. The second part of the test is aimed at ensuring that, in double-talk situations, the divergence is low. The requirement for this part is that only a 10 dB increase in residual echo level of the results listed in the steady state test (Test No. 1 of [22]) are permitted. The ANLMS is well within this requirement.

Note that this does not mean that systems based on either LMS or NLMS do not satisfy the ITU-T requirements. On the contrary, they do satisfy them, but the advantages of ANLMS are the continued filter coefficient adaptation even during cross-talk scenarios and that there is no need for switching or VADs, which results in more consistency.

In order to improve the overall system performance, a noise suppressor and an echo canceller can be used jointly. The noise suppressor may be integrated either prior to the echo canceller or after it. Integrating prior to the echo canceller in order to remove the noise from the near-end signal

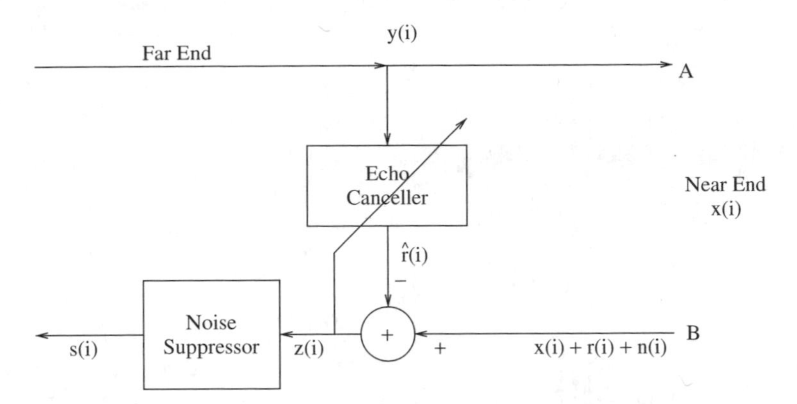

Figure 11.25 Block diagram of cascaded echo cancellation and noise suppression

usually distorts the echo signal in a nonlinear manner, which may make echo cancellation more difficult. By placing the noise suppressor after the echo canceller, to remove the residual echo error as well as noise, may therefore be more appropriate as shown in Figure 11.25.

The performance of this set-up has been tested both subjectively and objectively. Subjective testing was carried out through informal listening tests, while objective testing was conducted through various filter coefficient convergence behaviours. Two different echoes were used for this purpose. The first was a simple echo resulting from a single delay and attenuation of the far-end speech signal and the second was the sum of three different delayed and attenuated versions of the far-end speech. Each echo was mixed with the near-end speech signal along with vehicle noise contamination resulting in SNRs of 0, 5, 10, 15 and 20 dB.

Results obtained using the simple echo case are shown in Figures 11.26–11.31. The echo was generated by delaying the far end speech by 40 samples and attenuated through a factor of 0.48. Part (a) of Figures 11.26–11.31 shows the input to the cascaded system and the corresponding output signals and part (b) shows the convergence track of filter coefficients h_{40} and h_0. The robustness of the system under noisy conditions and the convergence of the filter coefficients (h_{40} and h_0), even in the presence of near-end speech, are quite evident in Figures 11.26–11.31. Note that, as also highlighted above, neither a near-end speech detector nor a switch for filter coefficient adaptation is needed. All that is needed is an initial training period for which the $w_k(i)$ are set to one. In this setup, the initial period is 1 second for which the near-end speech is assumed to be absent. The weighting function is switched on after that and is responsible for convergence of the filter coefficients during near-end speech presence and silences in the near-end signal. Based on the average track of each filter coefficient (i.e. $\overline{h}_k(i)$) and the selection of the $\gamma\beta$ value in the $w_k(i)$ definition, only the step changes that follow the average

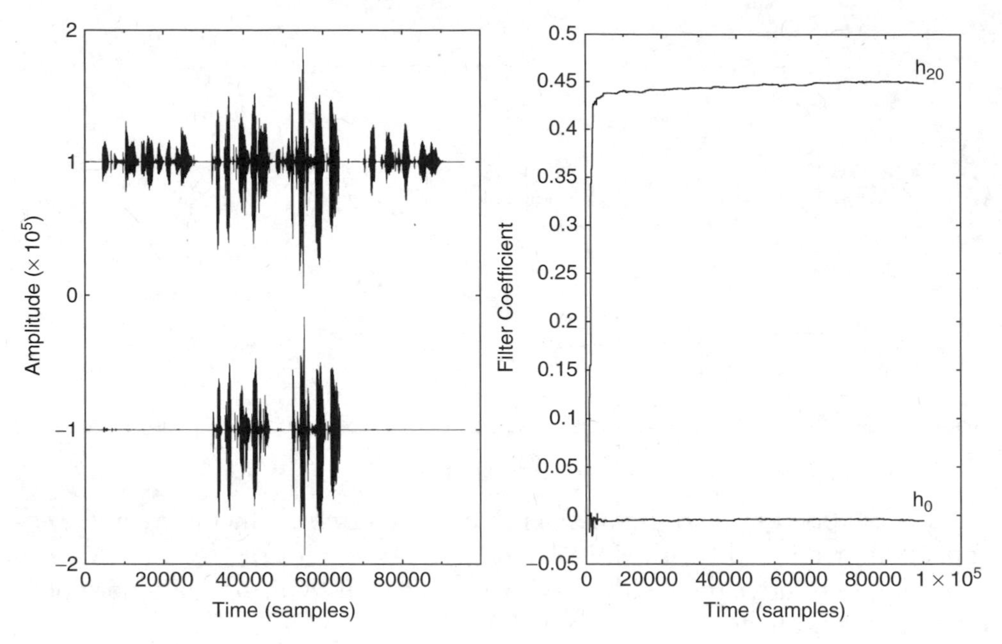

Figure 11.26 Performance of the noise-echo canceller for clean speech

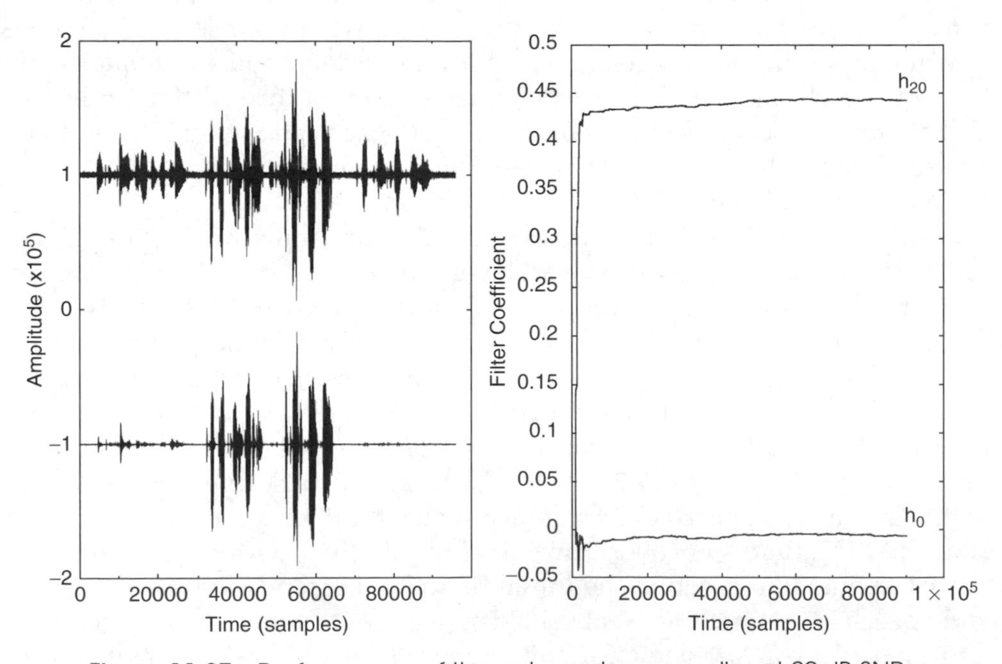

Figure 11.27 Performance of the noise-echo canceller at 20 dB SNR

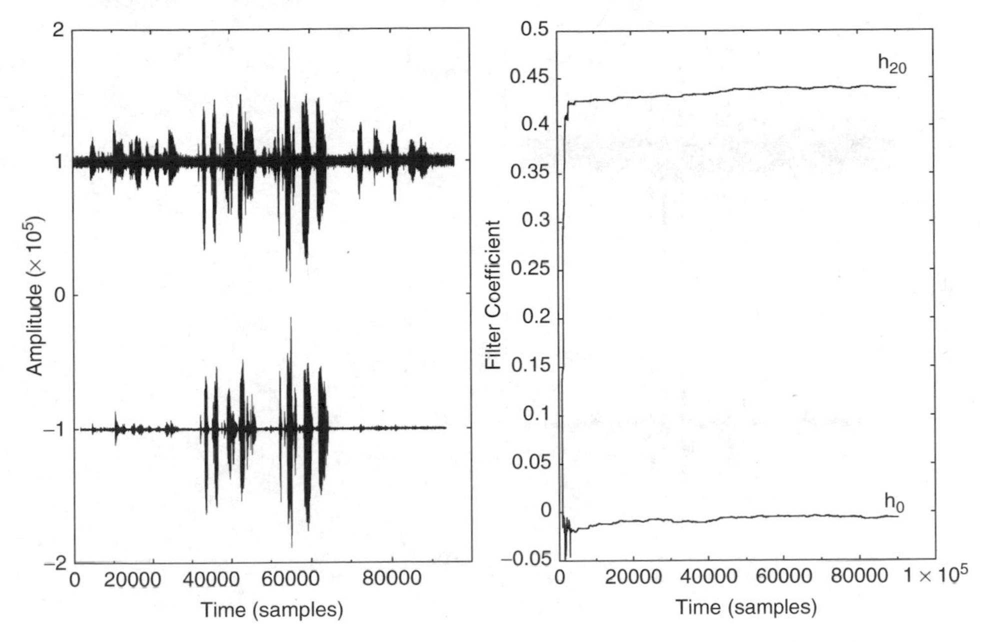

Figure 11.28 Performance of the noise-echo canceller at 15 dB SNR

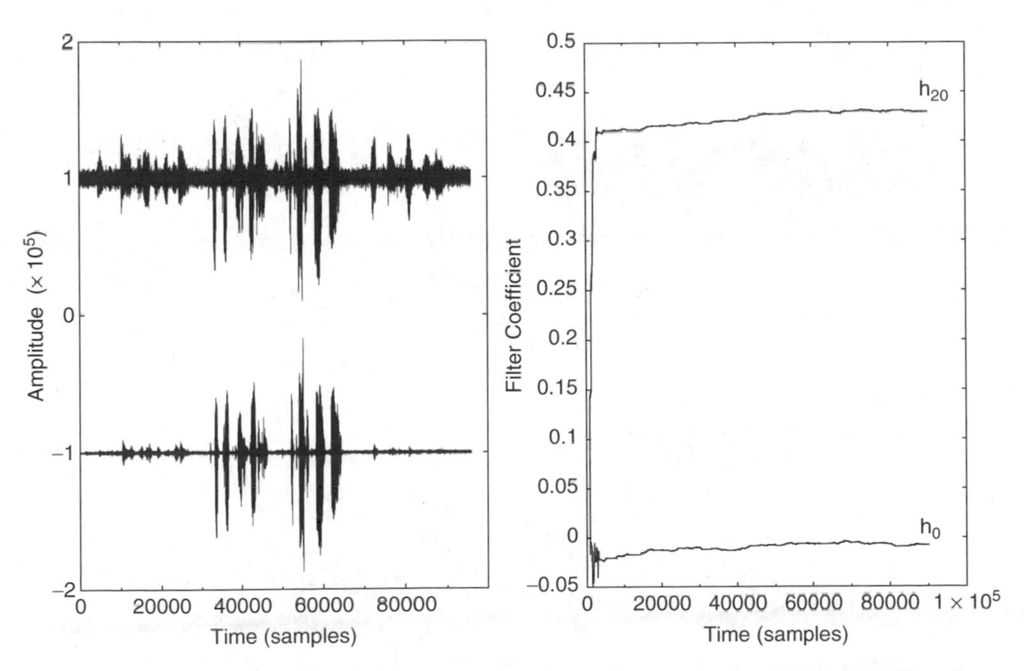

Figure 11.29 Performance of the noise-echo canceller at 10 dB SNR

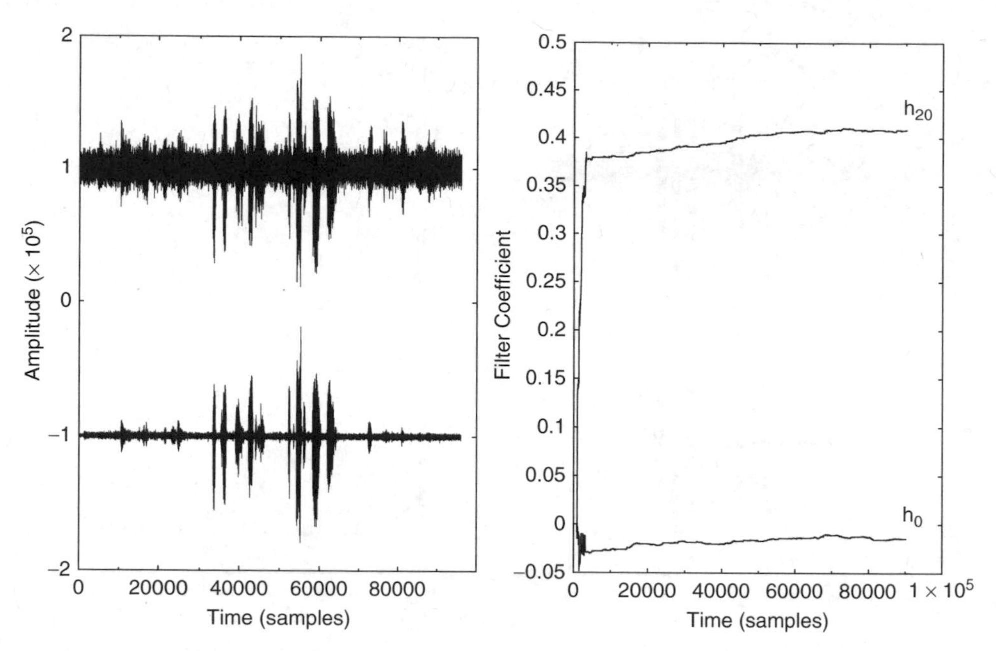

Figure 11.30 Performance of the noise-echo canceller at 5 dB SNR

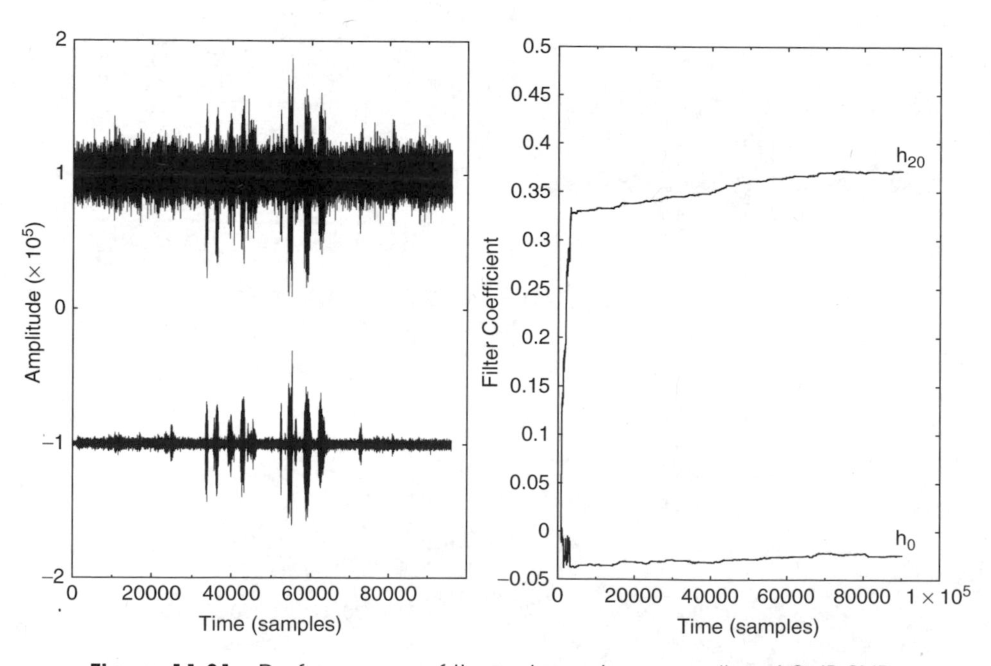

Figure 11.31 Performance of the noise-echo canceller at 0 dB SNR

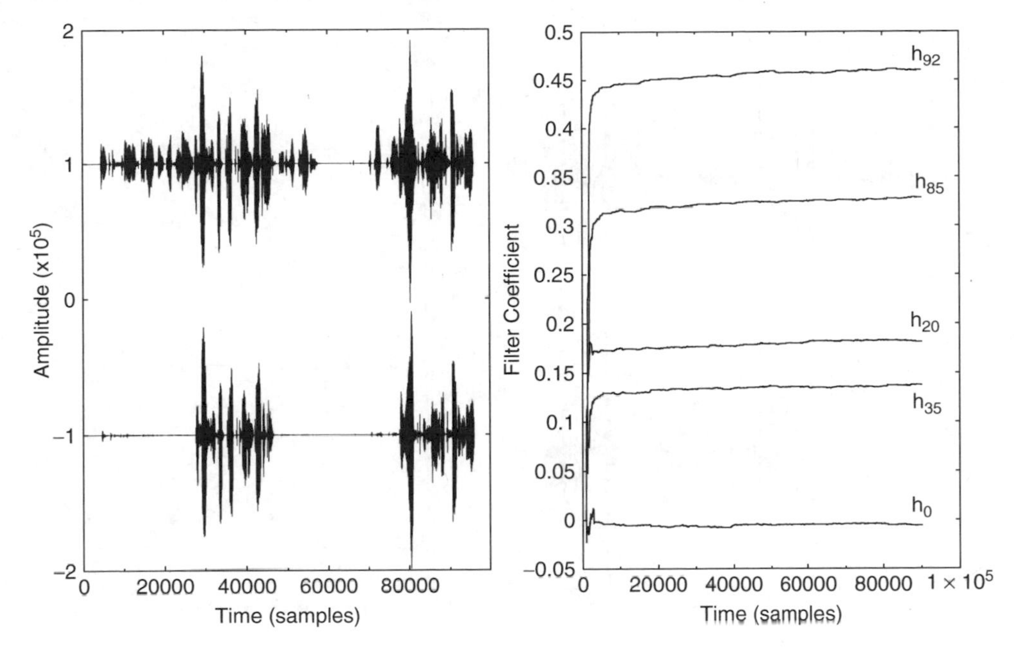

Figure 11.32 Performance of noise-echo canceller for clean speech

track have a considerable effect on the filter coefficient adaptation. Otherwise the overall step size due to $w_k(i)$ (i.e. $w_k(i)\beta$) will be small, therefore not changing the previous filter coefficient value by much and thus reducing the likelihood of divergence.

A similar result was obtained in the second experiment when a more complex echo was used. The echo used for this setup was generated through the sum of three different delays: 20, 40 and 60 samples with corresponding attenuation factors of 0.2, 0.48 and 0.35 respectively. Figures 11.32–11.37 show the results obtained for the second setup which proves the effectiveness of the new adaptation algorithm proposed by Al-Naimi [24].

11.5 Summary

With advanced signal processing algorithms and techniques it is possible to improve the quality of speech communications significantly. Both echo and noise cancellation/suppression algorithms have been reasonably well developed to tackle high levels of echo and noise present in communication systems. It is, of course, important to adapt the existing algorithms to specific communication systems to maximize their performances. When both acoustic noise and echo are present it is important to tune the overall enhancement algorithms (noise suppressor and echo canceller) jointly to maximize performance. Another important issue is the convergence time of

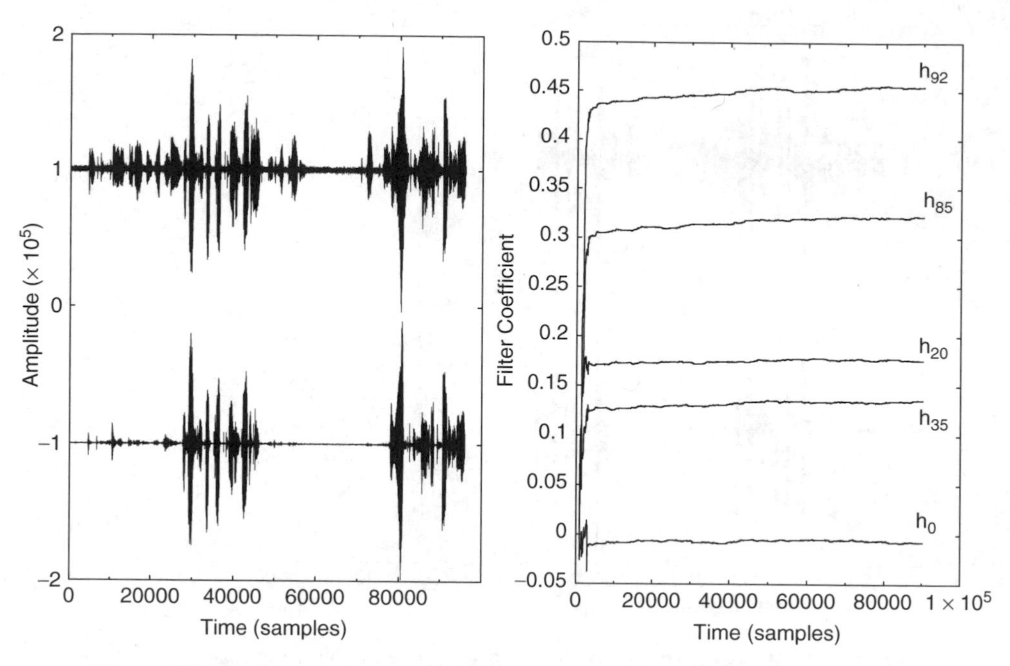

Figure 11.33 Performance of the noise-echo canceller at 20 dB SNR

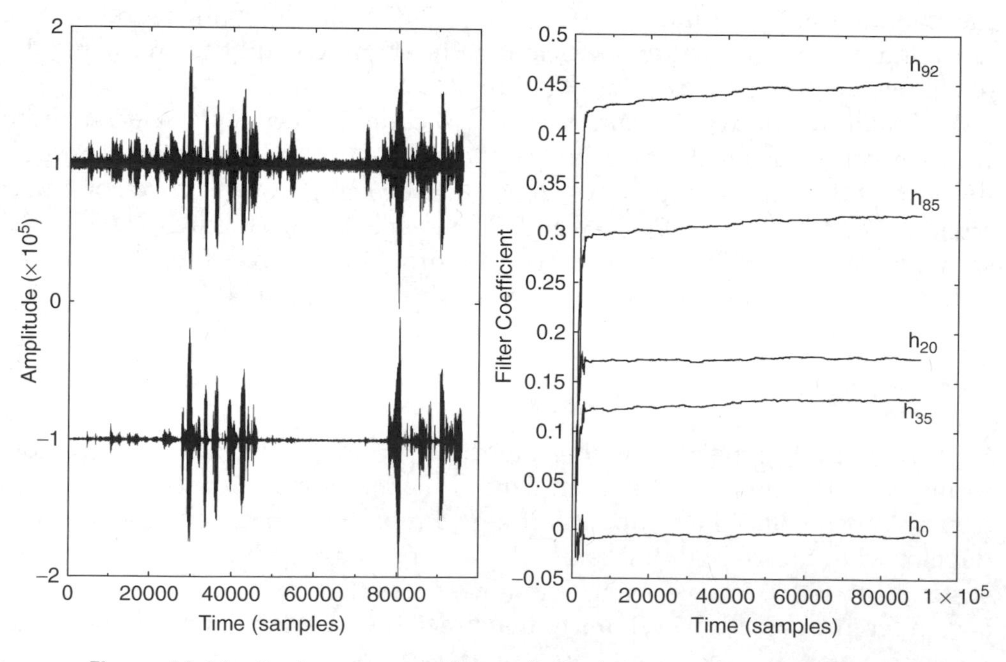

Figure 11.34 Performance of the noise-echo canceller at 15 dB SNR

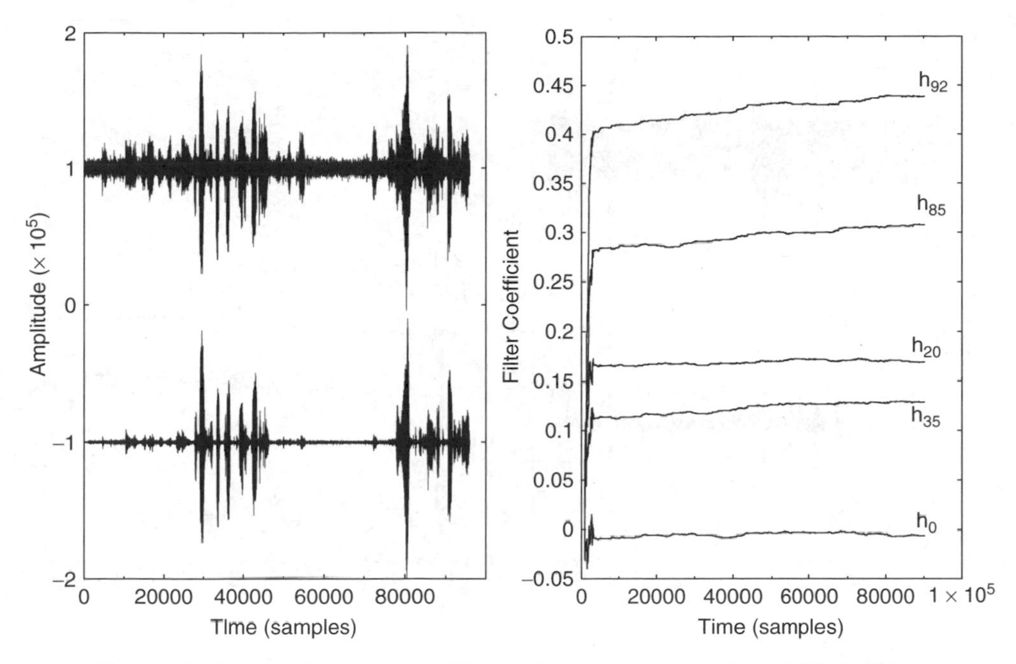

Figure 11.35 Performance of the noise-echo canceller at 10 dB SNR

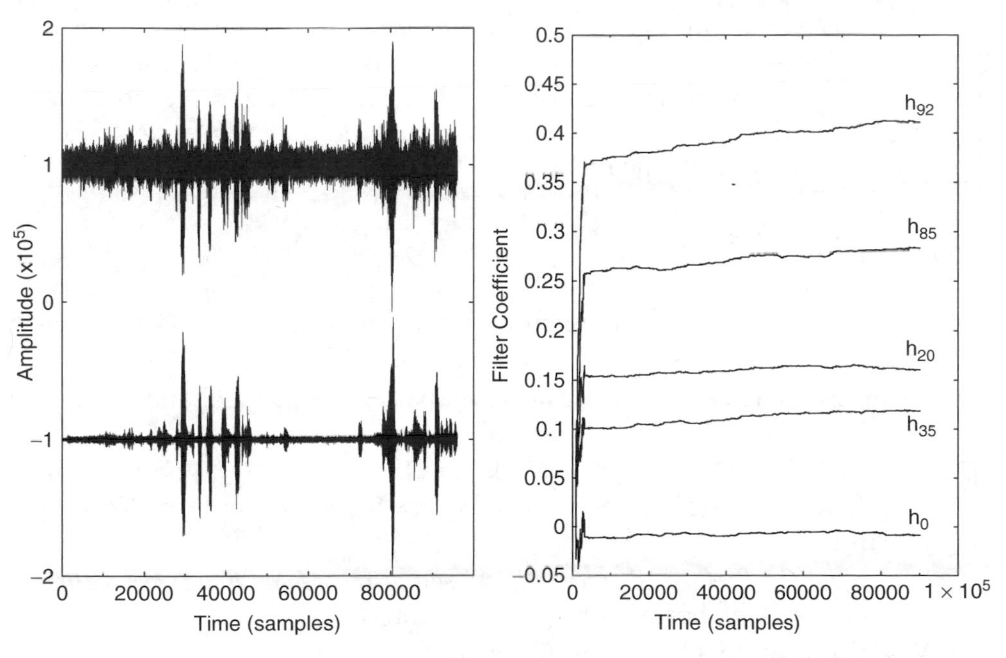

Figure 11.36 Performance of the noise-echo canceller at 5 dB SNR

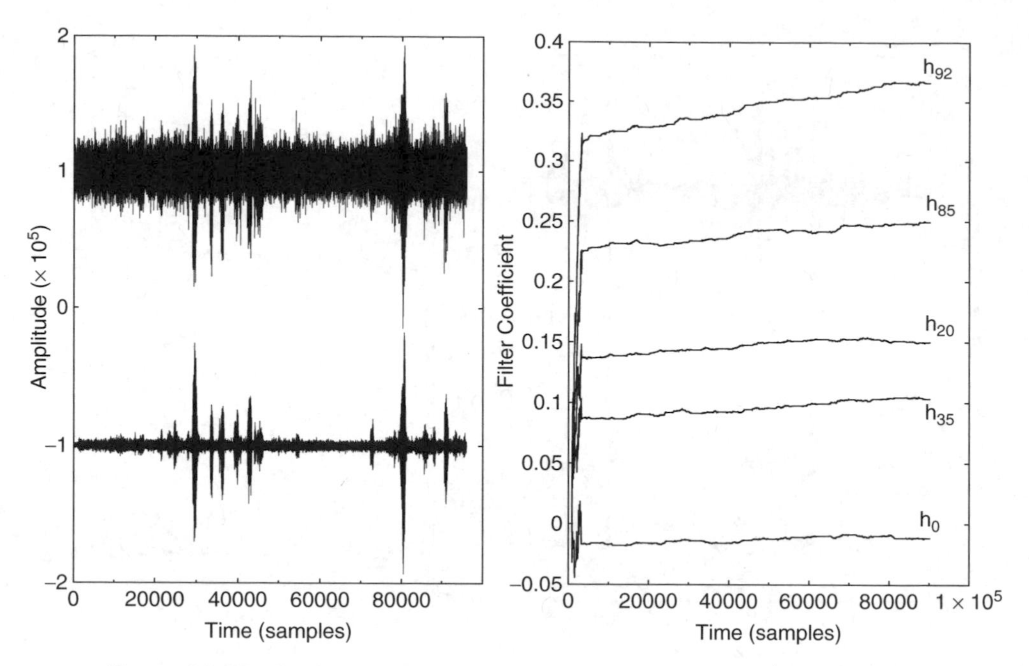

Figure 11.37 Performance of the noise-echo canceller at 0 dB SNR

the adaptive filtering used in the enhancement algorithms. It is crucial that the filters converge rapidly and do not diverge under any circumstances (any level of acoustic noise and echo). Normalized LMS algorithms usually provide adequate performance. The newly-proposed adaptive normalized LMS algorithm [24], discussed in this chapter, has shown robust performance under significant levels of acoustic noise and echo.

Bibliography

[1] J. S. Lim and A. V. Oppenheim (1979) 'Enhancement and bandwidth compression of noisy speech', in *Proc. IEEE*, 67(12):1568–1604.

[2] Y. Ephraim (1992) 'Statistical-model-based speech enhancement systems', in *Proc. IEEE*, 80(10):1526–55.

[3] J. R. Deller, H. G. Proakis, and J. H. L. Hansen, (1993) 'Speech enhancement', in *Discrete-Time Processing of Speech Signals*, Chapter 8. New York: Macmillan

[4] S. V. Vaseghi (2000) *Advanced digital signal processing and noise reduction*, 2nd edition. Chichester: John Wiley & Sons Ltd

[5] Y. Ephraim and H. L. Van Trees (1995) 'A signal subspace approach for speech enhancement', in *IEEE Trans. Speech and Audio Processing*, 3(4):251–66.

[6] I. Y. Soon, S. N. Koh, and C. K. Yeo (1998) 'Noisy speech enhancement using discrete cosine transform', in *Speech Communications*, 24(3):249–57.

[7] I. Y. Soon and S. N. Koh (2000) 'Low distortion speech enhancement', in *IEE Proc. on Vision, Image and Signal Processing*, 147(3):247–53, June.

[8] J. D. Gibson, B. Koo, and S. D. Gray (1991) 'Filtering of colored noise for speech enhancement and coding', in *IEEE Trans. Signal Processing*, 39:1732–42.

[9] Z. Goh, K. C. Tan, and B. T. G. Tan (1999) 'Kalman-filtering speech enhancement method based on a voicedunvoiced speech model', in *IEEE Trans. Speech and Audio Processing*, 7(5):510–24.

[10] Y. Ephraim and D. Malah (1984) 'Speech enhancement using a minimum mean square error short-time spectral amplitude estimator', in *IEEE Trans. on Acoust., Speech and Signal Processing*, 32(6):1109–20.

[11] M. Berouti, R. Schwartz, and J. Makhoul (1979) 'Enhancement of speech corrupted by acoustic noise', in *Proc. of Int. Conf. on Acoust., Speech and Signal Processing*, pp. 208–11.

[12] N. Virag (1999) 'Single channel speech enhancement based on masking properties of the human auditory systems', in *IEEE Trans. Speech and Audio Processing*, 7(2):126–37.

[13] B. L. Sim, Y. C. Tong, J. S. Chang, and C. T. Tan (1998) 'A parametric formulation of the generalised spectral subtraction method', in *IEEE Trans. Speech and Audio Processing*, 6(4):328–37.

[14] R. J. McAulay and M. L. Malpass (1980) 'Speech enhancement using a soft-decision suppression filter', in *IEEE Trans. on Acoust., Speech and Signal Processing*, 28(2):137–45.

[15] Y. Ephraim and D. Malah (1985) 'Speech enhancement using a minimum mean square error log-spectral amplitude estimator', in *IEEE Trans. on Acoust., Speech and Signal Processing*, 33(2):443–5.

[16] O. Cappé (1994) 'Elimination of musical noise phenomenon with the Ephraim and Malah noise suppression', in *IEEE Trans. Speech and Audio Processing*, 2(2):345–9.

[17] P. Scalart and J. V. Filho (1996) 'Speech enhancement based on a priori signal to noise estimation', in *Proc. of Int. Conf. on Acoust., Speech and Signal Processing*, pp. 629–31. Atlanta, GA, USA

[18] J. Sohn and W. Sung (May 1998) 'A voice activity detection employing soft decision based noise spectrum adaptation', in *icassp*, Seattle, WA, USA, 365–8.

[19] N. S. Kim and J. H. Chang (2000) 'Spectral enhancement based on global soft decision', in *IEEE Signal Processing Letters*, 7(5):108–110.

[20] Y. D. Cho (2001) 'Speech detection enhancement and compression for voice communications', Ph.D. thesis, CCSR, University of Surrey, UK.

[21] Y. D. Cho, K. Al-Naimi, and A. Kondoz (2001) 'Improved voice activity detection based on a smoothed statistical likelihood ratio', in *Proc. of Int. Conf. on Acoust., Speech and Signal Processing*. Salt Lake City, UT

[22] ITU-T (1993) *Echo cancellers*, ITU-T Rec. G.165.

[23] D. G Messerschmitt (1984) 'Echo cancellation in speech and data transmission', in *IEEE Journal on Selected Areas in Communications*, 2(2):283–303.

[24] K. T. Al-Naimi (2002) 'Advanced speech processing and coding techniques', Ph.D. thesis, CCSR, University of Surrey, UK.

[25] D. Messerschmitt, D. Hedberg, C. Cole, A. Haoui, and P. Winship (1989) 'Digital voice echo canceller with a TMS32020', Application Report: SPRA129, in *Digital Signal Processing Solutions*, p. 32. Texas Instruments

Index

Digital Speech. A. Kondoz
© 2004 John Wiley & Sons, Ltd ISBN 0-470-87007-9 (HB)

D

V

W

Z